T0323937

Bioassays

Bioassays

Advanced Methods and Applications

Edited by

Donat-P. Häder
Friedrich-Alexander University, Erlangen–Nürnberg, Germany

Gilmar S. Erzinger
University of Joinville Region–UNIVILLE, Joinville, Brazil

Elsevier
Radarweg 29, PO Box 211, 1000 AE Amsterdam, Netherlands
The Boulevard, Langford Lane, Kidlington, Oxford OX5 1GB, United Kingdom
50 Hampshire Street, 5th Floor, Cambridge, MA 02139, United States

British Library Cataloguing-in-Publication Data
A catalogue record for this book is available from the British Library

Library of Congress Cataloging-in-Publication Data
A catalog record for this book is available from the Library of Congress

ISBN: 978-0-12-811861-0

For Information on all Elsevier publications
visit our website at https://www.elsevier.com/books-and-journals

Publisher: Candice Janco
Acquisition Editor: Laura Kelleher
Editorial Project Manager: Karen R. Miller
Production Project Manager: Omer Mukthar
Cover Designer: Greg Harris

Typeset by MPS Limited, Chennai, India

Contents

List of Contributors

Roberto Abdala-Díaz Malaga University, Malaga, Spain

F. Gabriel Acién Almeria University, Almeria, Spain

Jawad Ali University of Swabi, Anbar, Pakistan

Félix Álvarez-Gómez Malaga University, Malaga, Spain

Luiz Américo University of Joinville Region—UNIVILLE, Joinville, SC, Brazil

Azizullah Azizullah Kohat University of Science and Technology (KUST), Kohat, Pakistan

Murray T. Brown Plymouth University, Plymouth, United Kingdom

Soyeon Choi Incheon National University, Incheon, South Korea

Stephen Depuydt Ghent University Global Campus, Incheon, South Korea

Tino Dornbusch LemnaTec GmbH, Aachen, Germany

Nils G.A. Ekelund Malmö University, Malmö, Sweden

Gilmar S. Erzinger University of Joinville Region—UNIVILLE, Joinville, SC, Brazil

Félix L. Figueroa Malaga University, Malaga, Spain

Juan Luis Gómez-Pinchetti University of Las Palmas de GC, Las Palmas, Spain

Elena G. Govorunova University of Texas Health Science Center at Houston, Houston, TX, United States

Donat-P. Häder Friedrich-Alexander University, Erlangen-Nürnberg, Germany

Taejun Han Incheon National University, Incheon, South Korea; Ghent University Global Campus, Incheon, South Korea

Dieter Hanelt University of Hamburg, Hamburg, Germany

Peter-Diedrich Hansen Technical University of Berlin, Berlin, Germany

Jens Hauslage Institute of Aerospace Medicine German Aerospace Center (DLR), Cologne, Germany

Ruth Hemmersbach Institute of Aerospace Medicine German Aerospace Center (DLR), Cologne, Germany

Gerda Horneck German Aerospace Center (DLR), Cologne, Germany

Marcus Jansen LemnaTec GmbH, Aachen, Germany

Sarzamin Khan University of Swabi, Anbar, Pakistan

Nathalie Korbee Malaga University, Malaga, Spain

Hojun Lee Incheon National University, Incheon, South Korea

Kevin Nagel LemnaTec GmbH, Aachen, Germany

Therezinha M. Novais Oliveira University of Joinville Region–UNIVILLE, Joinville, SC, Brazil

Jihae Park Ghent University Global Campus, Incheon, South Korea

Stefan Paulus LemnaTec GmbH, Aachen, Germany

Luiz H. Pinto University of Joinville Region–UNIVILLE, Joinville, SC, Brazil

Peter Richter Friedrich-Alexander University, Erlangen-Nürnberg, Germany

Francine Schmoeller University of Joinville Region–UNIVILLE, Joinville, SC, Brazil

Oleg A. Sineshchekov University of Texas Health Science Center at Houston, Houston, TX, United States

Cleiton Vaz Universidade do Estado de Santa Catarina–UDESC, Pinhalzinho, SP, Brazil

Preface

While most of the water on our planet is either too salty for human consumption or is locked up in snow and ice, only a small fraction is potable. Increasing demands from a fast growing human population, and by industry and agriculture, are an additional burden on the dwindling resources. At the same time disposal of toxic substances from municipal and industrial sources as well as from agriculture result in pollution of rivers and lakes as well as ground water reservoirs. Quality assessment and monitoring of freshwater resources are of high priority in order to avoid damage to human health and ecosystem integrity. Chemical analyses are time consuming, expensive, and usually limited to a few classes of substances, which is in contrast to the growing number of potentially toxic chemicals that count in the tens of thousands.

Furthermore, toxins combined with other substances or other environmental stress factors may have synergistic effects that escape routine chemical analyses. Upper limits for toxins vary between countries and may change over time and, what is more important, they may not reflect the real threat for human health and the biota.

As an alternative to chemical analyses, the presence of toxic substances and pollutants can be monitored by using bioassays, which utilize organisms as bioindicators. One of the classical examples was the use of fish which were placed in potentially polluted water; when they showed abnormal swimming behavior or died this was an indication of the presence of lethal or sublethal concentrations of pollutants in the water. Today many organisms and biological materials are employed in bioassays including biomolecules, cell lines, bacteria, microorganisms, lower and higher plants, as well as invertebrates and vertebrates. Different endpoints can be assayed as indicators for toxicity including mortality, motility, behavior, growth, and reproduction, as well as physiological responses such as photosynthesis, protein biosynthesis, and genetic alteration.

Bioassays do not provide information on the chemical nature of the pollutant, but they indicate the presence of a toxin that may pose a threat to human health or ecosystem function and integrity.

This volume describes the principles and functioning of many bioassays for water, sediment, air, and soil, monitoring the effects of pollutants and other hazardous environmental stress factors such as solar UV radiation. Sensitivity, cost efficiency, speed of analysis, and ease of use of commercially available bioassays are compared and legal regulations are discussed for a number of developed and developing countries.

Donat-P. Häder and
Gilmar S. Erzinger

Introduction

1

Donat-P. Häder[1] *and Gilmar S. Erzinger*[2]
[1]Friedrich-Alexander University, Erlangen-Nürnberg, Germany,
[2]University of Joinville Region—UNIVILLE, Joinville, SC, Brazil

The occurrence of humans on this planet and its—in evolutionary terms—rapid expansion and explosive population growth has shaped the Earth and the environment in most cases in a negative way. The conquest of all continents and the (mis-) usage of the oceans have led to alterations of almost all ecosystems with only a few regions left in their original status [1]. This unprecedented expansion into all fields of the biosphere takes its toll on the quality of the atmosphere, the terrestrial and aquatic ecosystems, and even the vast glaciers and snow-covered areas on the poles and in high alpine regions. It has also resulted in mass destruction of native populations and started a rapidly enhancing extinction of species in all taxa [2]. Typical examples are the extinction of large vertebrates such as the mammoth, the mastodon and the saber-tooth tiger during the last millennia [3] and the loss of the passenger pigeon, of which hundreds of millions of these once most abundant birds on this planet were killed [4]. In addition, we are losing many microbial, plant, and animal species every day often without even knowing them. This loss is increased by the effects of global climate change: Extinction risks for some sample areas covering some 20% of the Earth's terrestrial surface have been estimated as 15%−37% over the next three to four decades [5].

1.1 Freshwater ecosystems

Most of the Earth's water is located in the oceans where it is too salty for human consumption. Large quantities are bound in the form of glaciers and snow covering the poles and high mountains. Thus only a small fraction of less than 1% of the global water is available for human usage [6,7]. Simultaneously the need for potable and uncontaminated freshwater for households, industry, and agriculture has multiplied during the past few centuries. Even with a stabilization of the human population further needs for freshwater are predicted [8].

In contrast to the growing need for freshwater, the limited resources are diminished by pollution from domestic, agricultural, and industrial wastes [9]. Industrial wastes include persistent organic pollutants (POPs) such as chlorinated organic chemicals and microplastics [10,11] as well as heavy metals such as Hg, Pb, Cu, Cr, and As which accumulate in lakes, rivers, and coastal waters [12]. POPs have been linked with type 2 diabetes [13]. Contamination by heavy metal pollutants may cause cardiovascular problems, damaged or reduced mental and central nervous functions, lower energy levels, and damage to blood composition, and may affect

Bioassays. DOI: http://dx.doi.org/10.1016/B978-0-12-811861-0.00001-2

lungs, kidneys, liver, and other vital organs [14,15]. Especially in developing countries these effluents are often dispatched into rivers, lakes, or the groundwater untreated or only filtrated to remove particulate substances (cf. Chapter 18: Ecotoxicological monitoring of wastewater and Chapter 21: Environmental monitoring using bioassays, this volume).

Arsenic pollution has become a major problem in many countries. In Asia alone at least 140 million people drink arsenic-polluted water [16]. More than 18 million small wells have been dug into the soil in India over the past 30 years in order to avoid surface water which is often contaminated by bacteria or other pollutants. Rapid pumping of water from these wells has changed the courses of previously clean underground streams so that they now flow through arsenic-containing sediments. While developed countries with arsenic-polluted groundwater such as the Southwest US have the means to filter out the toxicant from the water, developing countries lack that option because of the high costs of, e.g., the conventional aluminum-based drinking-water treatment [17]. The upper limit of arsenic in drinking water has been set to $10\,\mu g\,L^{-1}$ by the World Health Organization (WHO) but the Indian government allows $50\,\mu g\,L^{-1}$; however, even this value is often far exceeded in many wells [18]. Pyrite minerals containing high concentrations of arsenic are eroded from the Himalayan Mountains and carried into India, Bangladesh, China, Pakistan, and Nepal. After reaction with oxygen and heavy metals such as iron it forms granules which are concentrated in the sediments from which it leaches out into the water which is tapped by the newly dug wells.

Arsenic taken up with the drinking water or ingested with vegetables which have been irrigated with polluted water causes a number of serious chronic diseases in animals and humans [19]. One of the first symptoms is scarring of the skin [20]. When it accumulates over time in the body it causes brain damage, heart disease, and cancer [21,22]. Heavy metals accumulate in the aquatic food web. They are taken up by phyto- and zooplankton, which in turn are ingested by secondary consumers such as crustaceans, fish, birds, and mammals—which are finally consumed by humans. This bioaccumulation may pose a major threat for human health [23,24]. The degree of bioaccumulation can be determined by calculating a biomagnification (or bioconcentration) factor [25]. E.g., B, Ba, Cd, Co, Cr, Cu, and Ni have been calculated to accumulate in muscle and fat tissue of fish such as carp and tilapia in the Yamuna river, Delhi [26]. Similar concentrations of heavy metals were found in rivers in Pakistan and India [27,28].

Organic pollutants as well as inorganic toxic substances accumulate in sediments and pose considerable long-term risks for human health and the biota [29]. Chlorophenol compounds produced by degradation of pesticides and chlorinated hydrocarbons [30] are among the most toxic pollutants in aquatic ecosystems because of their chemical stability and low degradability [31,32].

Even in developed countries industrial wastes are often not completely removed from the effluents and cause pollution of groundwater, drinking-water reservoirs, and natural ecosystems. In addition, raw oil and refined petrol components pose a major threat for the dwindling freshwater resources [33]. Climate change, water acidification, and exposure to solar UV radiation transform petrol components which have reached the water by oil spills [34]. These derivatives can be even more toxic than the original substances.

The excessive employment of fertilizers in agriculture causes accumulation of nitrogen and phosphorous compounds in surface and groundwater due to runoff from fields and gardens [35]. Nitrate has become a major problem in many countries. The upper limit of 50 mg L^{-1} in groundwater (European Union, 44 mg L^{-1} in the US) [36] can often only be reached by dilution with clean water from mountain streams before usage as drinking water. In Germany about 30% of the country distributes drinking water which is close to or exceeds this limit concentration. Nitrate itself is not toxic but can be converted to nitrite via the nitrate-nitrite-nitric oxide pathway [37]. Even at low concentration in the water nitrate accumulates in the blood and muscles of e.g., insects, mollusks, crustaceans, fish, and mammals including humans, causing acute and chronic toxicity [38–40].

Uncontrolled use of pesticides such as chemicals against insects, nematodes, mollusks, and fungi increase the level of toxicants in the water of artificial reservoirs and natural ecosystems. For a long time mosquitoes have been attacked by spraying oil products on the surface of infested water reservoirs [41]. These residues as well as the mosquitocidal essential oils nowadays being used [42] have also been found to be toxic to aquatic organisms [43]. Alternatively, dichlordiphenyltrichlorethan (DDT) has been employed as an organochlorine insecticide mainly against malaria-transmitting mosquitoes as contact or food poison since the 1940s. The production is fairly simple and inexpensive and the chemical was used under many trade names for several decades [44]. More than 1.8 million tonnes have been produced globally. It shows low toxicity in mammals but later was found to accumulate in adipose tissues in humans and animals as endpoints in the food chain [45]. In addition to hormone-like effects and suspicion of being cancerogenic [46], in predatory birds it resulted in eggshell thinning, massively decreasing populations of eagles, condors, falcons, and other birds of prey [47]. Therefore it was banned in the 1970s in most industrial nations. In 2004 the Stockholm Convention on POPs formally banned the use of DDT with the controversial exemption of application against parasite-carrying insects such as mosquitoes transmitting malaria and visceral leishmaniasis [48] and more recently against dengue transmitted by the tiger mosquito *Aedes aegypti* which also transmits yellow fever, chikungunya, and Zika fever [49]. However, many populations of *A. aegypti* have been found to have developed resistance against DDT. Today India is the only producer of DDT and is also its largest consumer [50].

Cleaning products such as detergents, soaps, and disinfectants are widely used in households, institutions, and industry. After usage they are discharged into the wastewater and thus reach aquatic environments polluting domestic and municipal wastewater [51,52]. Detergents are usually mixtures of several components such as surfactants, bleaching agents, enzymes, and fillers [53], and their concentrations have been increasing worldwide over the last few decades [54]. In municipal effluents detergents have been found at concentrations ranging from 0.008 to 6.2 mg L^{-1}. After discharge into rivers detergent concentrations were found in the range of 0.084 to 5.592 mg L^{-1} [55–58]. However there are large differences in detergent concentrations between rivers in various countries [56]. Many components and breakdown products of detergents have been found to be toxic to the aquatic biota. Surfactants, bleaching chemical, fillers and builders, fabric brighteners, enzymes, and coloring agents are discharged into the wastewater and after reaching rivers and ecosystems they affect a wide range of aquatic organisms [59,60]. The toxicity of liquid detergents in the water has been analyzed using the green flagellate *Euglena gracilis* in the Ecotox bioassay system [58,61] (cf. Chapter 10: Ecotox, this volume).

The increasing use of personal care products and pharmaceuticals results in the discharge of these chemicals and their derivatives into household wastewaters [7,62]. The widespread use of contraceptive drugs such as estrogens causes the accumulation in terrestrial and aquatic habitats [63,64]. Because in heavily populated areas the river water is recycled several times, the concentration of these hormones increases uncontrolled since they are not removed from the water by filtration; bioremediation and bioassays for these substances are only just emerging [65]. Substances such as estrogens have been found to have effects on sexual development and feminization of animals such as fish [62,65,66] and to alter the sex ratio in frogs [67]. Furthermore, chemicals such as antibiotics have been found to accumulate in crop plants when irrigated with wastewaters containing those substances [68].

An estimated 780 million people—mostly in developing countries—do not have access to clean freshwater and about 2.2 billion lack safe sanitation [69]. Due to these circumstances about 5 million people die each year from diseases induced by polluted drinking water, such as diarrhoea and infection by water-borne parasites (http://www.who.int/topics/mortality/en/). Application of simple and inexpensive measures could prevent many hundreds of thousands premature deaths by improving freshwater quality for human consumption [70].

1.2 Marine waters

Environmental pollution of marine ecosystems affects growth and productivity in all prokaryotic and eukaryotic organisms. Phytoplankton is the major producer in the oceans and its productivity rivals that of all terrestrial ecosystems taken together [71]. Pollution is more severe in coastal ecosystems than in the open oceans [72,73], but the latter are also stricken by the accumulation of plastic material which has been calculated to amount to 250,000 t distributed over the oceans [74]. Coastal ecosystems are affected by terrestrial run-off which includes municipal and industrial effluents, and fertilizers and pesticides from agriculture. The weed killer atrazine inhibits the photosynthetic electron transport chain and has been found to impair productivity in phytoplankton. This effect can be quantified monitoring chlorophyll *a* fluorescence e.g., by pulse amplitude modulated fluorescence (PAM) [75] (cf. Chapter 9: Photosynthesis assessed by chlorophyll fluorescence, this volume). However, natural phytoplankton populations have been shown to develop an induced community tolerance to atrazine. The molecular mechanism of this resistance based on a genetic adaptation of the phytoplankton organisms has been clarified by Chamovitz et al. with the herbicide norflurazon [76,77].

Natural phytoplankton communities are affected by accidental crude oil spills especially in shallow waters such as the Arctic Ocean. A water-soluble fraction of crude oil is formed by pyrenes which accumulate in the sediments. They are very toxic to phytoplankton which can be shown using a microwell bioassay [78,79]. Since oil production will increase especially in coastal ecosystems, more pollution and damage to phytoplankton is expected [80]. This effect is augmented by increasing solar UV radiation and climate change induced higher temperatures [81].

Polychlorinated biphenyls (PCB) are major pollutants in marine ecosystems. These lipophilic chemicals can easily cross cell membranes of phytoplankton and therefore accumulate in the cells, as demonstrated in four species from the Baltic Sea [82]. PCBs are also found in remote marine ecosystems where they are introduced into the water by air-water exchange [83]. POPs were detected during a cruise in the Greenland Current and Arctic Ocean to accumulate in phytoplankton [84]. Depending on the temperature, cell size, and hydrophobicity, POPs may enter cells, but are broken down by bacteria and phytoplankton. Other toxic substances found in marine waters are polycyclic aromatic hydrocarbons (PAHs), polychlorinated dioxins, furans (PCDD/Fs), and polybrominated diphenyl ethers (PBDEs), where they impair phytoplankton [85]. Results from the Mediterranean Sea, the Atlantic, Arctic, and Southern Ocean indicate that solar UV increases the toxicity of PAHs from combustion engines and other pollutants [86]. Antifouling paints on ship hulls such as tributyltin are further toxic agents for phytoplankton communities. As a consequence, recently new chemicals such as 4,5-dichloro-2-n-octyl-isothiazoline-3-one (DCOI) have been developed which rapidly degrade when released from ship hulls [87].

1.3 Terrestrial ecosystems

Human activities have a major impact on terrestrial ecosystems and even alter major biogeochemical cycles such as the global nitrogen cycle [88]. Soils in many developing and developed countries are being contaminated by heavy metals such as Cu, Pb, Hg, Cr, and Ni inadvertently released from mining and industry [89]. E.g., Cr has been found to affect seed germination, root elongation, root-tip mitosis, and micronucleus induction in several crop plants including cabbage (*Brassica oleracea*), cucumber (*Cucumis sativus*), lettuce (*Lactuca sativa*), wheat (*Triticum aestivum*), and corn (*Zea mays*) using a soil plate bioassay [90].

These hazards require a close monitoring of soils and the biota [91]. Plants can be used as bioindicators for toxicity in the soil [92] (cf. Chapter 8: Pigments and Chapter 13: Image processing for bioassays, this volume). E.g., the presence of high Cu concentrations results in a discoloration of leaves due to changes in the pigment content [93]. Inside the cell heavy metals bind to proteins and peptides

[94]. Land snails have been employed as bioassay for monitoring heavy metal concentrations in the soil [95]. E.g. *Helix aspersa* is a major herbivore that tolerates high concentrations of lead which is transferred through a polluted ecosystem. Many of the toxicants in the soil are brought in by atmospheric pollution and can be carried by wind over long distances. Sulfur dioxide, carbon monoxide, and nitrogen dioxide from exhaust pipes of coal-burning power plants are transported over hundreds of kilometers [96] and can be traced from satellites [97].

Ibuprofen and perfluorooctanoic acid (PFOA) are toxic pollutants in terrestrial ecosystems. PFOA was more toxic than Ibuprofen to the monocotyledonous *Sorgum bicolor* [98]. At low levels the two toxins induced a synergism as shown by the Combination Index method and the ecological risk was assessed by calculating the Hazard Quotient as the ratio between the concentration measured in the soil and the no observed effect concentration (NOEC) predicted from EC_{50} curves. This result stresses the notion that in evaluating toxicity of chemicals in the soil synergistic effects of multiple toxicants in mixtures need to be taken into account [99]. Imatinib mesylate is currently the most widely used cytostatic drug in Europe. It accumulates in soils and the genotoxicity and acute toxic effect have been determined in two widely-used plant bioassays: micronucleus (MN) assays with meiotic tetrad cells of *Tradescantia* and in mitotic root tip cells of *Allium cepa*. Additionally, acute toxic effects (inhibition of cell division and growth of roots) were monitored in onions [100].

Given the large number of toxic chemicals from diverse classes both monitoring and phytoremediation of soils is necessary in agriculturally used areas as well as natural habitats. Bioassays with *Lupinus luteus* and associated endophytic bacteria showed no toxicity when exposed to heavy metals and benzopyren, but were impaired when grown in landfill soil containing these materials in addition to PCB and Diesel oil [101].

1.4 Air

Air is being polluted by hazardous emission of toxic substances from traffic, industry, households, and natural sources such as volcano eruptions. These pollutants can be gaseous or particulate. Nanoparticles in particular have become focus of attention since they may cause respiratory risks even at low concentrations [102,103]. Epidemiological studies have revealed a close relationship between high concentrations of air pollution particles such as organic and metal compounds and human morbidity and mortality [104]. Particulate substances in air may either result in the production of reactive oxygen species (ROS) or by inducing ROS production by the host response.

Urban air contains a large number of mutagenic pollutants. This was demonstrated by short-term mutagenicity tests using bacteria, human cells and plants [104]. Especially in the megacities in China the air pollution has reached a dangerous potential, and the International Agency for Research on Cancer (IARC) focuses on the evaluation of the carcinogenicity of outdoor air pollution in China [105].

As in water and soil, the toxicities in air are based on multipollution exposure with a number of components, resulting in synergistic effects [106]. Since chemical analysis of these components is often difficult or impossible, bioassays have been developed to indicate the health hazard of air. The commercially available bioassays are often marketed in the form of ready-to-use *toxkits*. They have the advantage of being easy to use and relatively inexpensive in comparison with conventional analytical methods. The bioindicator organisms include unicellular systems such as cell lines, fungi, and bacteria, as well as multicellular organisms, such as invertebrate and vertebrate animals, and plants [107]. One of the bioassays uses the luminescent bacterium *Aliivibrio fischeri* [108] (cf. Chapter 12: Bioluminescence systems in environmental biosensors, this volume). Genetic damage and bioaccumulation of trace elements in flower buds of *Tradescantia pallida* has been used as bioassay to determine genotoxicity of polluted air in and around São Paulo, Brazil [109].

Another stress factor is the increasing solar UV-B radiation (280–315 nm) due to stratospheric ozone depletion through anthropogenic release of fluorochlorocarbons (CFCs) and other gaseous pollutants [110]. This effect started last century in the late 1970s, culminating during the first decade of the current century. Solar UV-B levels are predicted to slowly return to pre-1970s levels by 2065 because of the long lifetimes of the ozone-depleting substances in the stratosphere [111]. Solar UV-B levels have increased over the poles (ozone holes) and the mid latitudes, while they were stable in the tropics. UV-B impairs aquatic and terrestrial ecosystems, biogeochemical cycles, and human health [111–113]. In addition to radiometers that measure solar UV irradiances, bioassays have been developed to monitor the radiation exposure level of human beings (cf. Chapter 16: Bioassays for solar UV radiation, this volume).

1.5 Need for bioassays

Monitoring of the environment, including marine and freshwater resources, soil, and air, has a high priority because of the increasing demand and urgent drive to improve the quality of the environment, which is important for our health. This can be done by chemical analyses for water, soil, and air (cf. Chapter 2: Chemical analysis of air and water, this volume). However, the vast number of organic and inorganic molecules, both of natural origin and those produced by industry, and numbering in the hundreds of thousands [114]—as well as the large areas and numerous habitats—prevent large-scale analyses of the environmental health and biodiversity. While many of these substances are toxic to microbes, plants, animals, and humans on acute or chronic exposure, only a very few are being monitored in the environment. A recent census lists 223 organic chemicals with are monitored in freshwater ecosystems on a continental scale in 4000 European sites [115]. These substances include pesticides, tributyltin, PAHs, and brominated flame retardants, which are major contributors to the risk of causing acute lethal and chronic effects on algae, invertebrates, and fish which have been found in 14% and 42% of the sites, respectively.

In the past, the toxicity of certain chemicals was underestimated or the substances escaped routine chemical analyses or obtained toxicity only when in contact with other substances or operated synergistically [99]. The catastrophic effects of the Seveso poisoning by the inadvertent release of high levels of 2,3,7,8-tetrachlorodibenzo-p-dioxin (2,3,7,8-TCDD) went undetected for some time because the substance was not on the list of monitored chemicals [116].

Another problem is that permissive limits for toxic substances in the water, soil, and air differ substantially between countries and are subject to changes over time (cf. Chapter 3: Historical development of bioassays and Chapter 4: Regulations, political and societal aspects, toxicity limits, this volume) [117]. The examples for arsenic and nitrate have been mentioned above. Upper limits for many other pollutants such as heavy metals, POPs, and pharmaceuticals have been defined; however, there is no consistency from country to country [118]. But for most of the several 100,000 chemicals no permissive limits exist. A fairly new source of toxic substances is the recycling of electronic wastes [119,120]. Even though the Basel Convention on the Control of Transboundary Movement of Hazardous Wastes and their Disposal was adopted in 1989 and enforced in 1992, many industrialized countries, including the United States which did not ratify the convention, export their e-wastes into China and African countries. One of the recommendations to decrease the uncontrolled release of toxic substances is to ban uses of deca-BDE (bromodiphenyl ether) in addition to penta- and octa-BDEs. polyvinyl chloride (PVC) in electronic products should be replaced with nonchlorinated polymers.

A summary of air polluting substances has been defined in 1964 [121]. The IARC and WHO have defined the limit concentrations of fine and ultrafine atmospheric particles such as sulfate and black carbon, which enter the lungs by respiration and further penetrate into the blood, where they cause cancer, DNA mutations, heart attacks, and premature death [122]. But the limits for fine and ultrafine particles in the air are often exceeded mostly in large cities even in industrialized countries, especially during dry, calm periods in the summer [123]. Organic solvents of glues and lacquers pose another problem of pollutants in air [124].

As a consequence it is very important to define upper limits for pollutants in water, soil, and air. These limits should be defined on a global basis and adjusted on a regular basis when new results indicate potential hazards. In addition, the production and emission of these substances should be limited and replaced by less toxic materials.

Since it is obviously not possible to fully monitor water, soil, and air in our environment using chemical analyses, one option is to use bioassays for this purpose. The basic idea is that a biosystem, which can be a biological molecule, a prokaryotic or eukaryotic organism, responds to the stress inflicted by toxic substances when in contact. This response is recorded by a suitable instrument. By definition, a bioassay does not identify the nature of the pollutant or toxic substance [125]. But also chemical analyses of potentially polluted samples often reveal only the class of chemicals involved. It rather indicates the presence of a stress factor indicating a potential challenge for the ecosystem or health hazard. However, when several endpoints are being used in the analysis the qualified response could give some hints on the nature of the toxicant.

There are a number of prerequisites for an effective modern bioassay system:

- The endpoint(s) of the system should be very sensitive for the toxicant being analyzed
- The response time should be fast
- The bioassay should be usable for acute (short-term) or long-term measurements
- The instrument should be easy to use and not require a lengthy training of personnel
- The price should not be excessive especially when it is intended to be used in developing countries
- The biological material should be easily accessible and the running costs should be low

An early example for a bioassay was the deployment of fish in potentially polluted water. When they died or showed abnormal behavior this was an indication for the presence of toxic substances in the water. Today many biological substances are being used in bioassays including DNA, enzymes, proteins, and pigments [126–129] as well as different organisms ranging from viruses, bacteria, protists, lower and higher plants, to invertebrates and vertebrates [108,130–134].

Potential endpoints used in different bioassays include mortality [135], motility and behavior [136–138], growth [139] and reproduction [99] as well as physiological parameters such as photosynthesis [140], protein biosynthesis [141] and genetic alteration of organisms [142]. Using a bioassay, effects of different concentrations of toxic substances are determined and EC_{50} curves constructed. These contain important information on the toxicity of a substance (or mixture of pollutants) such as NOEC (no observed effect concentration, LD (lethal dose) [143,144], and the EC_{50} value (concentration at which a 50% inhibition is found [145] (Fig. 1.1).

In this volume a number of established commercially available bioassays are described. The Microtox test monitors the decrease of bioluminescence produced by genetically modified bacteria under the effect of toxic substances [146]. Lemnatox is a bioassay which analyzes the growth and pigmentation of the aquatic angiosperm *Lemna* affected by toxic substances [147]. This plant was recently also used to monitor the effects of four herbicides (atrazine, diuron, paraquat, and simazine) on three *Lemna* species (*L. gibba, L. minor and L. paucicostata*) [148]. The end points were increase in frond area, root length, photosynthetic quantum yield and maximal electron transport rate after 72 h exposure to the herbicides. Diuron and paraquat were the most toxic substances with EC_{50} values of 6.0–12.3 $\mu g\ L^{-1}$ for root length. Also the other end points were sensitive enough to detect concentrations above the allowed thresholds determined by international standards.

Ulvatox has recently been developed to monitor the effects of municipal wastewater and industrial effluents. It detects the onset of zoospore release from marginal thallus disk cut from the marine green alga *Ulva pertusa* (cf. Chapter 7: Toxicity testing using the marine macroalga *Ulva pertusa*: method development and application, this volume). The zoospore release follows an exact timing under the experimental conditions during a 96 h period [149,150]. This release is delayed under the effect of toxic substances in the samples from municipal or industrial wastewater.

The bioassay Ecotox monitors several motility, orientation, and cell form parameters of the photosynthetic unicellular flagellate *E. gracilis* using a fully automatic computer-controlled image analysis system for cell tracking [143,151,152]

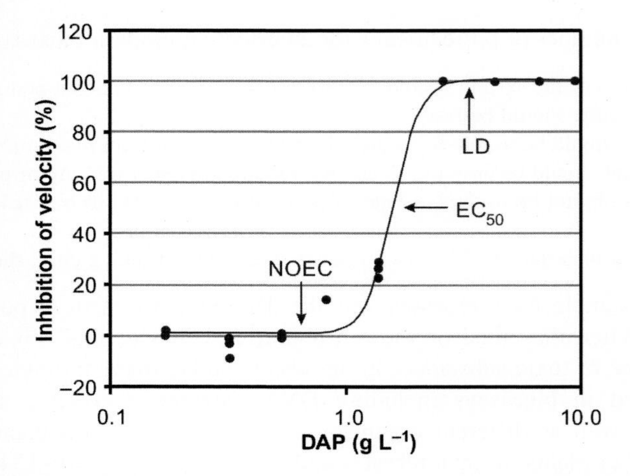

Figure 1.1 Inhibition of the velocity of *Eglena gracilis* by the fertilizer di-ammonium phosphate (DAP) after 1 h of exposure. The figure shows the experimental data (*circles*) and the fitted curve (*solid line*). The ordinate indicates the percentage inhibition of velocity in dependence of the concentration of DAP. The NOEC (no observed effect concentration) is 0.63 g L^{-1}, The EC_{50} (concentration which causes a 50% inhibition) is 2.11 g L^{-1} and the LD (lethal dose) is 3.2 g L^{-1}.
Source: Redrawn after Azizullah A. Ecotoxicological assessment of anthropogenically produced common pollutants of aquatic environments [PhD thesis]. Erlangen, Germany: Friedrich-Alexander University; 2011.

(cf. Chapter 10: Ecotox, this volume). This system has been used to monitor toxicity of municipal industrial wastewater and natural ecosystems, and to validate the efficiency of water treatment plants, and monitor pollution by heavy metals [28,153–161]. The organism used in this bioassay has also been employed to determine the toxicity of phenolic substances [132]. Also the crustacean *Daphnia* is being utilized in bioassays based on mortality, motility, orientation, and form factor parameters using computerized image analysis [162,163].

Some of the disadvantages of some commercially available bioassays are high investment costs and/or high prices for consumables. Some do not have a high sensitivity towards specific toxic substances and some tests are devaluated by long analysis times, which may be on the order of several days as in the case of bioassays based on the growth of organisms. The current volume discusses the advantages and disadvantages of bioassays, their characteristics, as well as their hardware and software.

References

[1] Hughes TP, Carpenter S, Rockström J, Scheffer M, Walker B. Multiscale regime shifts and planetary boundaries. Trends Ecol Evolut 2013;28(7):389–95.

[2] Ceballos G, Ehrlich PR, Barnosky AD, García A, Pringle RM, Palmer TM. Accelerated modern human–induced species losses: entering the sixth mass extinction. Sci Adv 2015;1(5):e1400253.
[3] Stuart AJ. Late Quaternary megafaunal extinctions on the continents: a short review. Geol J 2015;50(3):338–63.
[4] Hung C-M, Shaner P-JL, Zink RM, Liu W-C, Chu T-C, Huang W-S, et al. Drastic population fluctuations explain the rapid extinction of the passenger pigeon. Proc Natl Acad Sci USA 2014;111(29):10636–41.
[5] Thomas CD, Cameron A, Green RE, Bakkenes M, Beaumont LJ, Collingham YC, et al. Extinction risk from climate change. Nature 2004;427(6970):145–8.
[6] Kumar HD, Häder D-P. Global aquatic and atmospheric environment. Berlin, Heidelberg, New York: Springer-Verlag; 1999.
[7] Gleick PH. The world's water volume 8: The biennial report on freshwater resources. Washington, DC: Island Press; 2014.
[8] Kasei RA, Barnabas A, Ampadu B. The nexus of changing climate and impacts on rainfed water supply and fresh water availability for the inhabitants of Densu Basin and parts of Accra-Ghana, West Africa. J Environ Earth Sci 2014;4 (20):84–96.
[9] Jones JAA. Global hydrology: processes, resources and environmental management. London and New York: Routledge; 2014.
[10] Lee D-H, Porta M, Jacobs Jr DR, Vandenberg LN. Chlorinated persistent organic pollutants, obesity, and type 2 diabetes. Endocr Rev 2014;35(4):557–601.
[11] Bakir A, Rowland SJ, Thompson RC. Enhanced desorption of persistent organic pollutants from microplastics under simulated physiological conditions. Environ Pollut 2014;185:16–23.
[12] Förstner U, Wittmann GT. Metal pollution in the aquatic environment. Berlin: Springer Science & Business Media; 2012.
[13] Taylor KW, Novak RF, Anderson HA, Birnbaum LS, Blystone C, DeVito M, et al. Evaluation of the association between persistent organic pollutants (POPs) and diabetes in epidemiological studies: a national toxicology program workshop review. Environ Health Perspect 2013;121(7):774.
[14] Solenkova NV, Newman JD, Berger JS, Thurston G, Hochman JS, Lamas GA. Metal pollutants and cardiovascular disease: mechanisms and consequences of exposure. Am Heart J 2014;168(6):812–22.
[15] Verma R, Dwivedi P. Heavy metal water pollution-A case study. Recent Res Sci Technol 2013;5(5):98–9.
[16] Daigle K. Death in the water. Sci Am 2016;314(1):42–51.
[17] Gregor J. Arsenic removal during conventional aluminium-based drinking-water treatment. Water Res 2001;35(7):1659–64.
[18] Berg M, Winkel LHE, Amini M, Rodriguez-Lado L, Hug SJ, Podgorski J, et al. (June 2016) Regional to sub-continental prediction modeling of groundwater arsenic contamination. In: Bhattacharya P, Vahter M, Jarsjö J, Kumiene J, Ahmad A, et al., editors. Arsenic research and global sustainability: proceedings of the sixth international congress on arsenic in the environment (As 2016). Stockholm, Sweden: CRC Press; 2016. p. 21. June 19–23, 2016.
[19] Bamberger M, Oswald RE. Impacts of gas drilling on human and animal health. New Solut 2012;22(1):51–77.
[20] Guadanhim LR, Gonçalves RG, Bagatin E. Observational retrospective study evaluating the effects of oral isotretinoin in keloids and hypertrophic scars. Int J Dermatol 2016;55(11):1255–8.

[21] Ibeto LK, Love PB. Squamous cell carcinoma. Clinical cases in skin of color. Berlin Heidelberg: Springer; 2016. p. 89–97.
[22] Naujokas MF, Anderson B, Ahsan H, Aposhian HV, Graziano JH, Thompson C, et al. The broad scope of health effects from chronic arsenic exposure: update on a worldwide public health problem. Environ Health Perspect 2013;121(3):295.
[23] Alava JJ, Ross PS, Gobas FA. Food web bioaccumulation model for resident killer whales from the Northeastern Pacific Ocean as a tool for the derivation of PBDE-sediment quality guidelines. Arch Env Contamin Toxicol 2016;70:155–68.
[24] Mailman M, Bodaly R, Paterson MJ, Thompson S, Flett RJ. Low-level experimental selenite additions decrease mercury in aquatic food chains and fish muscle but increase selenium in fish gonads. Arch Env Contamin Toxicol 2014;66(1):32–40.
[25] Korsman JC, Schipper AM, de Vos MG, van den Heuvel-Greve MJ, Vethaak AD, de Voogt P, et al. Modeling bioaccumulation and biomagnification of nonylphenol and its ethoxylates in estuarine–marine food chains. Chemosphere 2015;138:33–9.
[26] Singh A, Srivastava S, Verma P, Ansari A, Verma A. Hazard assessment of metals in invasive fish species of the Yamuna River, India in relation to bioaccumulation factor and exposure concentration for human health implications. Environ Monit Assess 2014;186(6):3823–36 English
[27] Muhammad S, Shah MT, Khan S. Health risk assessment of heavy metals and their source apportionment in drinking water of Kohistan region, northern Pakistan. Microchem J 2011;98(2):334–43.
[28] Azizullah A, Khattak MNK, Richter P, Häder D-P. Water pollution in Pakistan and its impact on public health—a review. Environ Int 2011;37(2):479–97.
[29] Kukučka P, Audy O, Kohoutek J, Holt E, Kalábová T, Holoubek I, et al. Source identification, spatio-temporal distribution and ecological risk of persistent organic pollutants in sediments from the upper Danube catchment. Chemosphere 2015;138:777–83.
[30] Karci A. Degradation of chlorophenols and alkylphenol ethoxylates, two representative textile chemicals, in water by advanced oxidation processes: the state of the art on transformation products and toxicity. Chemosphere 2014;99:1–18.
[31] Wei D, Wang Y, Wang X, Li M, Han F, Ju L, et al. Toxicity assessment of 4-chlorophenol to aerobic granular sludge and its interaction with extracellular polymeric substances. J Hazard Mater 2015;289:101–7.
[32] Lindholm-Lehto PC, Knuutinen JS, Ahkola HS, Herve SH. Refractory organic pollutants and toxicity in pulp and paper mill wastewaters. Environ Sci Pollut Res 2015;22 (9):1–27.
[33] Liu Y, Kujawinski EB. Chemical composition and potential environmental impacts of water-soluble polar crude oil components inferred from ESI FT-ICR MS. PLoS One 2015;10(9):e0136376.
[34] Nikinmaa M. Climate change and ocean acidification—Interactions with aquatic toxicology. Aquat Toxicol 2013;126:365–72.
[35] Wu L, Ma Y, Fu B, Zhang W, Zhang X, Lu Q, et al. Reviews on agricultural nonpoint source pollution monitoring techniques and methods. Agric Sci Technol 2014;15 (12):2214.
[36] Espejo-Herrera N, Kogevinas M, Castaño-Vinyals G, Aragonés N, Boldo E, Ardanaz E, et al. Nitrate and trace elements in municipal and bottled water in Spain. Gac Sanit 2013;27(2):156–60.
[37] Lidder S, Webb AJ. Vascular effects of dietary nitrate (as found in green leafy vegetables and beetroot) via the nitrate-nitrite-nitric oxide pathway. Br J Clin Pharmacol 2013;75(3):677–96.

[38] Wuertz S, Schulze S, Eberhardt U, Schulz C, Schroeder J. Acute and chronic nitrite toxicity in juvenile pike-perch (*Sander lucioperca*) and its compensation by chloride. Comp Biochem Physiol C Toxicol Pharmacol 2013;157(4):352–60.

[39] Barbieri E, Bondioli ACV, Melo CB, Henriques MB. Nitrite toxicity to *Litopenaeus schmitti* (Burkenroad, 1936, Crustacea) at different salinity levels. Aquacult Res 2016;47(4):1260–8.

[40] Özen H, Kamber U, Karaman M, Gül S, Atakişi E, Özcan K, et al. Histopathologic, biochemical and genotoxic investigations on chronic sodium nitrite toxicity in mice. Exp Toxicol Pathol 2014;66(8):367–75.

[41] Bonah C. 'Health crusades': environmental approaches as public health strategies against infections in sanitary propaganda films, 1930–60. Environment, health and history. Heidelberg, Berlin: Springer; 2012. p. 154–77.

[42] Conti B, Flamini G, Cioni PL, Ceccarini L, Macchia M, Benelli G. Mosquitocidal essential oils: are they safe against non-target aquatic organisms? Parasitol Res 2014;113(1):251–9.

[43] Lewis M, Pryor R. Toxicities of oils, dispersants and dispersed oils to algae and aquatic plants: review and database value to resource sustainability. Environ Pollut 2013;180:345–67.

[44] Makowa HB. The relationship between the insecticide dichloro-diphenyl-trichloroethane and chloroquine in *Plasmodium falciparum* resistance [PhD thesis]. Stellenbosch: Stellenbosch University; 2012.

[45] Ciliberti A, Martin S, Ferrandez E, Belluco S, Rannou B, Dussart C, et al. Experimental exposure of juvenile savannah monitors (*Varanus exanthematicus*) to an environmentally relevant mixture of three contaminants: effects and accumulation in tissues. Environ Sci Pollut Res 2013;20(5):3107–14.

[46] Cohn BA, La Merrill M, Krigbaum NY, Yeh G, Park J-S, Zimmermann L, et al. DDT exposure in utero and breast cancer. J Clin Endocrinol Metab 2015;100 (8):2865–72.

[47] Burnett LJ, Sorenson KJ, Brandt J, Sandhaus EA, Ciani D, Clark M, et al. Eggshell thinning and depressed hatching success of California Condors reintroduced to central California. The Condor 2013;115(3):477–91.

[48] Hung H, Katsoyiannis AA, Guardans R. Ten years of global monitoring under the Stockholm convention on persistent organic pollutants (POPs): trends, sources and transport modelling. Environ Pollut 2016;217:1–3.

[49] Vontas J, Kioulos E, Pavlidi N, Morou E, Della Torre A, Ranson H. Insecticide resistance in the major dengue vectors *Aedes albopictus* and *Aedes aegypti*. Pestic Biochem Physiol 2012;104(2):126–31.

[50] Yadav IC, Devi NL, Syed JH, Cheng Z, Li J, Zhang G, et al. Current status of persistent organic pesticides residues in air, water, and soil, and their possible effect on neighboring countries: a comprehensive review of India. Sci Total Environ 2015;511:123–37.

[51] Cirelli AF, Ojeda C, Castro MJ, Salgot M. Surfactants in sludge-amended agricultural soils: a review. Environ Chem Lett 2008;6(3):135–48.

[52] Liwarska-Bizukojc E, Miksch K, Malachowska-Jutsz A, Kalka J. Acute toxicity and genotoxicity of five selected anionic and nonionic surfactants. Chemosphere 2005;58 (9):1249–53.

[53] Hellmuth H, Dreja M. Understanding interactions of surfactants and enzymes: Impact of individual surfactants on stability and wash performance of protease enzyme in detergents. Tenside Surfact Det 2016;53(5):502–8.

[54] Rosen MJ, Kunjappu JT. Surfactants and interfacial phenomena. Hoboken, NJ: John Wiley & Sons; 2012.

[55] Dorn PB, Salanitro JP, Evans SH, Kravetz L. Assessing the aquatic hazard of some branched and linear nonionic surfactants by biodegradation and toxicity. Environ Toxicol Chem 1993;12(10):1751–62.

[56] Minareci O, Öztürk M, Egemen Ö, Minareci E. Detergent and phosphate pollution in Gediz River, Turkey. Afr J Biotechnol 2009;8(15):3568.

[57] Azizullah A, Richter P, Jamil M, Häder D-P. Chronic toxicity of a laundry detergent to the freshwater flagellate *Euglena gracilis*. Ecotoxicology 2012;21(7):1957–64.

[58] Azizullah A, Richter P, Ullah W, Ali I, Häder D-P. Ecotoxicity evaluation of a liquid detergent using the automatic biotest ECOTOX. Ecotoxicology 2013;22(6):1043–52.

[59] Aizdaicher N, Markina ZV. Toxic effects of detergents on the alga *Plagioselmis prolonga* (Cryptophyta). Russ J Mar Biol 2006;32(1):45–9.

[60] Sánchez-Fortún S, Marvá F, D'ors A, Costas E. Inhibition of growth and photosynthesis of selected green microalgae as tools to evaluate toxicity of dodecylethyldimethylammonium bromide. Ecotoxicology 2008;17(4):229–34.

[61] Azizullah A, Richter P, Häder D-P. Photosynthesis and photosynthetic pigments in the flagellate *Euglena gracilis*—As sensitive endpoints for toxicity evaluation of liquid detergents. J Photochem Photobiol B Biol 2014;133:18–26.

[62] Erzinger GS, Brasilino FF, Pinto L, Häder D-P. Environmental toxicity caused by derivatives of estrogen and chemical alternatives for removal. Pharm Anal Acta 2014;5(8):1–2.

[63] Matthiessen P, Sumpter JP. Effects of estrogenic substances in the aquatic. Fish Ecotoxicol 2013;86:319.

[64] Zhang F-S, Xie Y-F, Li X-W, Wang D-Y, Yang L-S, Nie Z-Q. Accumulation of steroid hormones in soil and its adjacent aquatic environment from a typical intensive vegetable cultivation of North China. Sci Total Environ 2015;538:423–30.

[65] Bhandari RK, Deem SL, Holliday DK, Jandegian CM, Kassotis CD, Nagel SC, et al. Effects of the environmental estrogenic contaminants bisphenol A and 17α-ethinyl estradiol on sexual development and adult behaviors in aquatic wildlife species. Gen Comp Endocrinol 2015;214:195–219.

[66] Valdés ME, Amé MV, de los Angeles Bistoni M, Wunderlin DA. Occurrence and bioaccumulation of pharmaceuticals in a fish species inhabiting the Suquía River basin (Córdoba, Argentina). Sci Total Environ 2014;472:389–96.

[67] Hogan NS, Duarte P, Wade MG, Lean DR, Trudeau VL. Estrogenic exposure affects metamorphosis and alters sex ratios in the northern leopard frog (*Rana pipiens*): identifying critically vulnerable periods of development. Gen Comp Endocrinol 2008;156 (3):515–23.

[68] Pan M, Wong CK, Chu L. Distribution of antibiotics in wastewater-irrigated soils and their accumulation in vegetable crops in the Pearl River delta, Southern China. J Agric Food Chem 2014;62(46):11062–9.

[69] J. Eliasson, Deputy Secretary-General of the United Nations. WHO. Progress on drinking water and sanitation—2014 update. 2014.

[70] Apte JS, Marshall JD, Cohen AJ, Brauer M. Addressing global mortality from ambient PM2. 5. Environ Sci Technol 2015;49(13):8057–66.

[71] Häder D-P, Williamson CE, Wangberg S-A, Rautio M, Rose KC, Gao K, et al. Effects of UV radiation on aquatic ecosystems and interactions with other environmental factors. Photochem Photobiol Sci 2015;14:108–26.

[72] Gómez N, O'Farrell I. Phytoplankton from urban and suburban polluted rivers. Adv Limnol 2014;127–42.

[73] Munir S, Zaib-un-nisa Burhan TN, Morton SL, Siddiqui PJA. Morphometric forms, biovolume and cellular carbon content of dinoflagellates from polluted waters on the Karachi coast, Pakistan. Indian. J Geo-Mar Sci 2015;44:1.
[74] Eriksen M, Lebreton LC, Carson HS, Thiel M, Moore CJ, Borerro JC, et al. Plastic pollution in the world's oceans: more than 5 trillion plastic pieces weighing over 250,000 tons afloat at sea. PLoS One 2014;9(12):e111913.
[75] Hancke K, Dalsgaard T, Sejr MK, Markager S, Glud RN. Phytoplankton productivity quantified from chlorophyll fluorescence. 18 Danske Havforskermøde; Copenhagen, Denmark 2015.
[76] Chamovitz D, Pecker I, Hirschberg J. The molecular basis of resistance to the herbicide norflurazon. Plant Mol Biol 1991;16(6):967–74.
[77] López-Rodas V, Flores-Moya A, Costas E. Genetic Adaptation of Phytoplankters to Herbicides. INTECH Open Access Publisher; 2011.
[78] Ozhan K, Bargu S. Can crude oil toxicity on phytoplankton be predicted based on toxicity data on benzo (a) pyrene and naphthalene? Bull Environ Contam Toxicol 2014;92 (2):225–30.
[79] Petersen K, Heiaas HH, Tollefsen KE. Combined effects of pharmaceuticals, personal care products, biocides and organic contaminants on the growth of *Skeletonema pseudocostatum*. Aquat Toxicol 2014;150:45–54.
[80] Hylander S, Grenvald JC, Kiørboe T. Fitness costs and benefits of ultraviolet radiation exposure in marine pelagic copepods. Funct Ecol 2014;28:149–58.
[81] Grenvald J, Nielsen T, Hjorth M. Effects of pyrene exposure and temperature on early development of two co-existing Arctic copepods. Ecotoxicology 2013;22(1):184–98. English
[82] Gerofke A, Kömp P, McLachlan MS. Bioconcentration of persistent organic pollutants in four species of marine phytoplankton. Environ Toxicol Chem 2005;24 (11):2908–17.
[83] Dachs J, Eisenreich SJ, Hoff RM. Influence of eutrophication on air-water exchange, vertical fluxes, and phytoplankton concentrations of persistent organic pollutants. Environ Sci Technol 2000;34(6):1095–102.
[84] Galbán-Malagón CJ, Cabrerizo A, Berrojálbiz N, Ojeda MJ, Dachs J. Air-water exchange and phytoplankton accumulation of persistent organic pollutants in the Greenland current and Arctic Ocean 2012. Available from: http://132.246.11.198/2012-ipy/pdf-all/ipy2012arAbstract00801.pdf.
[85] Del Vento S, Dachs J. Prediction of uptake dynamics of persistent organic pollutants by bacteria and phytoplankton. Environ Toxicol Chem 2002;21(10):2099–107.
[86] Echeveste P, Agustí S, Dachs J. Cell size dependence of additive versus synergetic effects of UV radiation and PAHs on oceanic phytoplankton. Environ Pollut 2011;159 (5):1307–16.
[87] Bérard A, Benninghoff C. Pollution-induced community tolerance (PICT) and seasonal variations in the sensitivity of phytoplankton to atrazine in nanocosms. Chemosphere 2001;45(4):427–37.
[88] Erisman JW, Galloway JN, Seitzinger S, Bleeker A, Dise NB, Petrescu AR, et al. Consequences of human modification of the global nitrogen cycle. Phil Trans R Soc B 2013;368(1621):20130116.
[89] Stankovic S, Kalaba P, Stankovic AR. Biota as toxic metal indicators. Environ Chem Lett 2014;12(1):63–84.
[90] Hou J, Liu GN, Xue W, Fu WJ, Liang BC, Liu XH. Seed germination, root elongation, root-tip mitosis, and micronucleus induction of five crop plants exposed to chromium in fluvo-aquic soil. Environ Toxicol Chem 2014;33(3):671–6.

[91] Luepke N-P. Monitoring environmental materials and specimen banking. In: Proceedings of the International Workshop, Berlin (West), 23–28 October 1978: Springer Science & Business Media; Berlin, 2012.
[92] Prasad MNV. Heavy metal stress in plants: from biomolecules to ecosystems. Berlin: Springer Science & Business Media; 2013.
[93] Zhao J-S, Zhang Q, Liu J, Zhang P-D, Li W-T. Effects of copper enrichment on survival, growth and photosynthetic pigment of seedlings and young plants of the eelgrass *Zostera marina*. Mar Biol Res 2016;12(7):695–705.
[94] Reddy GN, Prasad MNV. Heavy metal-binding proteins/peptides: occurrence, structure, synthesis and functions. A review. Environ Exp Bot 1990;30:251–64.
[95] Nica DV, Bordean D-M, Borozan AB, Gergen I, Bura M, Banatean-Dunea I. Use of land snails (Pulmonata) for monitoring copper pollution in terrestrial ecosystems. Reviews of Environmental Contamination and toxicology, 225. Heidelberg: Springer; 2013. p. 95–137.
[96] Vallero D. Fundamentals of air pollution. Waltham, MA: Academic Press; 2014.
[97] Schmidt A, Leadbetter S, Theys N, Carboni E, Witham CS, Stevenson JA, et al. Satellite detection, long-range transport, and air quality impacts of volcanic sulfur dioxide from the 2014–2015 flood lava eruption at Bárðarbunga (Iceland). J Geophys Res 2015;120(18):9739–57.
[98] González-Naranjo V, Boltes K. Toxicity of ibuprofen and perfluorooctanoic acid for risk assessment of mixtures in aquatic and terrestrial environments. Int J Environ Sci Technol 2014;11(6):1743–50.
[99] Chen C, Wang Y, Qian Y, Zhao X, Wang Q. The synergistic toxicity of the multiple chemical mixtures: Implications for risk assessment in the terrestrial environment. Environ Int 2015;77:95–105.
[100] Pichler C, Filipič M, Kundi M, Rainer B, Knasmueller S, Mišík M. Assessment of genotoxicity and acute toxic effect of the imatinib mesylate in plant bioassays. Chemosphere 2014;115:54–8.
[101] Gutiérrez-Ginés M, Hernández A, Pérez-Leblic M, Pastor J, Vangronsveld J. Phytoremediation of soils co-contaminated by organic compounds and heavy metals: bioassays with *Lupinus luteus* L. and associated endophytic bacteria. J Environ Manage 2014;143:197–207.
[102] Gordon SB, Bruce NG, Grigg J, Hibberd PL, Kurmi OP, Lam K-bH, et al. Respiratory risks from household air pollution in low and middle income countries. Lancet Respir Med 2014;2(10):823–60.
[103] Santos APM, Segura-Muñoz SI, Nadal M, Schuhmacher M, Domingo JL, Martinez CA, et al. Traffic-related air pollution biomonitoring with *Tradescantia pallida* (Rose) Hunt. cv. *purpurea* Boom in Brazil. Environ Monit Assess 2015;187 (2):1–10.
[104] Ceretti E, Zani C, Zerbini I, Viola G, Moretti M, Villarini M, et al. Monitoring of volatile and non-volatile urban air genotoxins using bacteria, human cells and plants. Chemosphere 2015;120:221–9.
[105] Loomis D, Huang W, Chen G. The International Agency for Research on Cancer (IARC) evaluation of the carcinogenicity of outdoor air pollution: focus on China. Chin J Cancer 2014;33:189–96.
[106] Oakes M, Baxter L, Long TC. Evaluating the application of multipollutant exposure metrics in air pollution health studies. Environ Int 2014;69:90–9.
[107] Wieczerzak M, Namieśnik J, Kudłak B. Bioassays as one of the Green Chemistry tools for assessing environmental quality: A review. Environ Int 2016;94:341–61.

[108] Ma XY, Wang XC, Ngo HH, Guo W, Wu MN, Wang N. Bioassay based luminescent bacteria: interferences, improvements, and applications. Sci Total Environ 2014;468:1–11.

[109] da Costa GM, Petry CT, Droste A. Active versus passive biomonitoring of air quality: genetic damage and bioaccumulation of trace elements in flower buds of *Tradescantia pallida*. Water Air Soil Pollut 2016;227(7):1–12.

[110] Bernhard G, Manney G, Fioletov V, Grooß J-U, Heikkilä A, Johnsen B, et al. Ozone and UV radiation. In: State of the Climate in 2011. Bull Am Meteor Soc 2012;93(7): S129-S132.

[111] UNEP. Environmental effects of ozone depletion and its interactions with climate change: progress report, 2008. Photochem Photobiol Sci 2009;8:13–22. Pubmed Central PMCID: 060131.

[112] Ballare CL, Caldwell MM, Flint SD, Robinson A, Bornman JF. Effects of solar ultraviolet radiation on terrestrial ecosystems. Patterns, mechanisms, and interactions with climate change. Photochem Photobiol Sci 2011;10(2):226–41. PubMed PMID: WOS:000286835400003.

[113] Häder D-P, Helbling EW, Williamson CE, Worrest RC. Effects on aquatic ecosystems and interactions with climate change. Photochem Photobiol Sci 2011;8:242–60.

[114] Ginebreda A, Kuzmanovic M, Guasch H, de Alda ML, López-Doval JC, Muñoz I, et al. Assessment of multi-chemical pollution in aquatic ecosystems using toxic units: Compound prioritization, mixture characterization and relationships with biological descriptors. Sci Total Environ 2014;468:715–23.

[115] Malaj E, Peter C, Grote M, Kühne R, Mondy CP, Usseglio-Polatera P, et al. Organic chemicals jeopardize the health of freshwater ecosystems on the continental scale. Proc Natl Acad Sci U S A 2014;111(26):9549–54.

[116] Mitoma C, Mine Y, Utani A, Imafuku S, Muto M, Akimoto T, et al. Current skin symptoms of Yusho patients exposed to high levels of 2, 3, 4, 7, 8-pentachlorinated dibenzofuran and polychlorinated biphenyls in 1968. Chemosphere 2015;137:45–51.

[117] Stokinger HE. Modus operandi of threshold limits committee of ACGIH. Am Ind Hyg Assoc J 1964;25(6):589–94.

[118] Häder D-P. A grand challenge for environmental toxicity: What are the permissive limits of toxic pollutants in the environment? Front Environ Sci 2013;1:1–2. English.

[119] Kolias K, Hahladakis JN, Gidarakos E. Assessment of toxic metals in waste personal computers. Waste Manage 2014;34(8):1480–7.

[120] Naidu SL, Sullivan JH, Teramura AH, DeLucia EH. The effects of ultraviolet-B radiation on photosynthesis of different aged needles in field-grown loblolly pine. Tree Physiol 1993;12:151–62.

[121] Stern AC. Summary of existing air pollution standards. J Air Pollut Control Assoc 1964;14(1):5–15.

[122] Samoli E, Andersen ZJ, Katsouyanni K, Hennig F, Kuhlbusch TA, Bellander T, et al. Exposure to ultrafine particles and respiratory hospitalisations in five European cities. Eur Respirat J 2016;48(3):674–82.

[123] Samara C, Kantiranis N, Kollias P, Planou S, Kouras A, Besis A, et al. Spatial and seasonal variations of the chemical, mineralogical and morphological features of quasi-ultrafine particles (PM 0.49) at urban sites. Sci Total Environ 2016;553:392–403.

[124] Barman ML, Sigel NB, Beedle DB, Larson RK. Acute and chronic effects of glue sniffing. Calif Med 1964;100(1):19.

[125] Häder D-P, Erzinger GS. Ecotox – Monitoring of pollution and toxic substances in aquatic ecosystems. J Ecol Environ Sci 2015;3(2):22–7.

[126] Lanzone V, Compagnone D, Tofalo R, Fasoli G, Corrado F. DNA-based bioassay for the detection of Benzo [a] pyrene oxidation products. Sensors. Heidelberg: Springer; 2014. p. 153–7.
[127] Zhang L, Cao X, Wang L, Zhao X, Zhang S, Wang P. Printed microwells with highly stable thin-film enzyme coatings for point-of-care multiplex bioassay of blood samples. Analyst 2015;140(12):4105–13.
[128] Hu H-y, Zhang L, Wang Y. Crop development based assessment framework for guiding the conjunctive use of fresh water and sewage water for cropping practice—A case study. Agric Water Manag 2016;169:98–105.
[129] Wang Y, Suzek T, Zhang J, Wang J, He S, Cheng T, et al. PubChem bioassay: 2014 update. Nucleic Acids Res 2013:gkt978, http://dx.doi.org/10.1093/nar/gkt978.
[130] de Castro-Català N, Kuzmanovic M, Roig N, Sierra J, Ginebreda A, Barceló D, et al. Ecotoxicity of sediments in rivers: invertebrate community, toxicity bioassays and the toxic unit approach as complementary assessment tools. Sci Total Environ 2015;540:2097–396.
[131] Hafner C, Gartiser S, Garcia-Käufer M, Schiwy S, Hercher C, Meyer W, et al. Investigations on sediment toxicity of German rivers applying a standardized bioassay battery. Environ Sci Pollut Res 2015;22(21):1–13.
[132] Kottuparambil S, Kim Y-J, Choi H, Kim M-S, Park A, Park J, et al. A rapid phenol toxicity test based on photosynthesis and movement of the freshwater flagellate, *Euglena agilis* Carter. Aquat Toxicol 2014;155:9–14.
[133] Nabeela F, Azizullah A, Bibi R, Uzma S, Murad W, Shakir SK, et al. Microbial contamination of drinking water in Pakistan—a review. Environ Sci Pollut Res 2014;21 (24):13929–42.
[134] Feigin L, Svergun D, Dembo A. Symmetry and structure of bacteriophage T7. Crystal symmetries: Shubnikov Centennial papers. 1988;17:617.
[135] Oxborough RM, N'Guessan R, Jones R, Kitau J, Ngufor C, Malone D, et al. The activity of the pyrrole insecticide chlorfenapyr in mosquito bioassay: towards a more rational testing and screening of non-neurotoxic insecticides for malaria vector control. Malar J 2015;14(1):1.
[136] Azizullah A, Richter P, Häder D-P. Comparative toxicity of the pesticides carbofuran and malathion to the freshwater flagellate *Euglena gracilis*. Ecotoxicology 2011;20 (6):1442–54.
[137] Azizullah A, Richter P, Häder D-P. Sensitivity of various parameters in *Euglena gracilis* to short-term exposure to industrial wastewaters. J Appl Phycol 2012;24 (2):187–200.
[138] Häder D-P, Jamil M, Richter P, Azizullah A. Fast bioassessment of wastewater and surface water quality using freshwater flagellate *Euglena gracilis*—a case study from Pakistan. J Appl Phycol 2014;26(1):421–31.
[139] Killberg-Thoreson L, Sipler RE, Bronk DA. Anthropogenic nutrient sources supplied to a Chesapeake Bay tributary support algal growth: a bioassay and high-resolution mass spectrometry approach. Estuar Coasts 2013;36(5):966–80.
[140] Sjollema SB, van Beusekom SA, van der Geest HG, Booij P, de Zwart D, Vethaak AD, et al. Laboratory algal bioassays using PAM fluorometry: Effects of test conditions on the determination of herbicide and field sample toxicity. Environ Toxicol Chem 2014;33(5):1017–22.
[141] Gong Y, Tian H, Wang L, Yu S, Ru S. An integrated approach combining chemical analysis and an in vivo bioassay to assess the estrogenic potency of a municipal solid waste landfill leachate in Qingdao. PLoS One 2014;9(4):e95597.

[142] Davies JC, Mazurek J. Pollution control in United States: evaluating the system. London: Routledge; 2014.
[143] Azizullah A, Richter P, Häder D-P. Ecotoxicological evaluation of wastewater samples from Gadoon Amazai Industrial Estate (GAIE), Swabi, Pakistan. Int J Environ Sci 2011;1(5):959–76.
[144] Azizullah A, Richter P, Häder D-P. Effects of long-term exposure to industrial wastewater on photosynthetic performance of *Euglena gracilis* measured through chlorophyll fluorescence. J Appl Phycol 2014;27:303–10.
[145] Sebaugh J. Guidelines for accurate EC50/IC50 estimation. Pharm Stat 2011;10(2):128–34.
[146] Fernández-Piñas F, Rodea-Palomares I, Leganés F, González-Pleiter M, Muñoz-Martín MA. Evaluation of the ecotoxicity of pollutants with bioluminescent microorganisms. Bioluminescence: Fundamentals and Applications in Biotechnology-Volume 2. Heidelberg: Springer; 2014. p. 65–135.
[147] Häder D-P, Erzinger GS, Aziz A. Advanced methods for biomonitoring aquatic ecosystems. In: Govil JN, editor. Environmental Science & Engineering. Studium Press LLC; in press.
[148] Park J, Brown MT, Depuydt S, Kim JK, Won D-S, Han T. Comparing the acute sensitivity of growth and photosynthetic endpoints in three *Lemna* species exposed to four herbicides. Environ Pollut 2017;220(Pt B):818–27.
[149] Han T, Choi G-W. A novel marine algal toxicity bioassay based on sporulation inhibition in the green macroalga *Ulva pertusa* (Chlorophyta). Aquat Toxicol 2005;75(3):202–12.
[150] Kim Y-J, Han Y-S, Kim E, Jung J, Kim S-H, Yoo S-J, et al. Application of the *Ulva pertusa* bioassay for a toxicity identification evaluation and reduction of effluent from a wastewater treatment plant. Front Env Sci 2015; http://dx.doi.org/10.3389/fenvs.2015.00002.
[151] Tahedl H, Häder D-P. Automated biomonitoring using real time movement analysis of *Euglena gracilis*. Ecotoxicol Environ Saf 2001;48(2):161–9.
[152] Azizullah A, Murad W, Adnan M, Ullah W, Häder D-P. Gravitactic orientation of *Euglena gracilis* - a sensitive endpoint for ecotoxicological assessment of water pollutants. Front Env Sci 2013;1:4.
[153] Azizullah A. Ecotoxicological assessment of anthropogenically produced common pollutants of aquatic environments [PhD thesis]. Erlangen, Germany: Friedrich-Alexander University; 2011.
[154] Azizullah A, Jamil M, Richter P, Häder D-P. Fast bioassessment of wastewater and surface water quality using freshwater flagellate *Euglena gracilis*—a case study from Pakistan. J Appl Phycol 2014;26(1):421–31.
[155] Azizullah A, Nasir A, Richter P, Lebert M, Häder D-P. Evaluation of the adverse effects of two commonly used fertilizers, DAP and urea, on motility and orientation of the green flagellate *Euglena gracilis*. Environ Exp Bot 2011;74:140–50.
[156] Azizullah A, Richter P, Häder D. Ecotoxicological evaluation of wastewater samples from Gadoon Amazai Industrial Estate (GAIE), Swabi, Pakistan. Int J Environ Sci Technol 2011;1:1–17.
[157] Ahmed H. Biomonitoring of aquatic ecosystems [PhD thesis]. Erlangen-Nürnberg: Friedrich-Alexander-Universität; 2010.
[158] Ahmed H, Häder D-P. Rapid ecotoxicological bioassay of nickel and cadmium using motility and photosynthetic parameters of *Euglena gracilis*. Environ Exp Bot 2010;69:68–75.

[159] Ahmed H, Häder D-P. Monitoring of waste water samples using the ECOTOX biosystem and the flagellate alga *Euglena gracilis*. J Water Air Soil Pollut 2011; 216(1-4):547–60.
[160] Ahmed H, Häder D-P. A fast algal bioassay for assessment of copper toxicity in water using *Euglena gracilis*. J Appl Phycol 2010;22(6):785–92.
[161] Ahmed H, Häder D-P. Short-term bioassay of chlorophenol compounds using *Euglena gracilis*. SRX Ecol 2009;2010.
[162] Chen L, Fu X, Zhang G, Zeng Y, Ren Z. Influences of temperature, pH and turbidity on the behavioral responses of *Daphnia magna* and Japanese Medaka (*Oryzias latipes*) in the biomonitor. Procedia Environ Sci 2012;13:80–6.
[163] Struewing KA, Lazorchak JM, Weaver PC, Johnson BR, Funk DH, Buchwalter DB. Part 2: Sensitivity comparisons of the mayfly *Centroptilum triangulifer* to *Ceriodaphnia dubia* and *Daphnia magna* using standard reference toxicants; NaCl, KCl and $CuSO_4$. Chemosphere 2015;139:597–603.

Chemical analysis of air and water

2

Sarzamin Khan and Jawad Ali
University of Swabi, Anbar, Pakistan

Part I Water analysis

2.1 Overview

About 70% of the Earth's surface consists of water (about 97% of the total water by volume), but most of it is in the oceans and thus not available for direct consumption due to the high salinity and, in many cases, high concentration of pollutants. Of the remaining 3% (freshwater) of the total water, 0.003% is in the ice caps, glaciers, and deep under the Earth surface, which is very difficult to utilize. Only 2.997% of the water is accessible and is present as soil moisture, groundwater, water vapor, and in lakes and rivers [1]. The total estimated freshwater is about 28.8 million km^3, which is used by the whole population of the world, which amounts nowadays to more than 7 billion individuals.

According to its origin, water is classified into various types such as groundwater, rain water, and wastewater. It can also be classified according to its use such as potable water (water used in most industries), process water, or pure water (used in households, food, and pharmaceutical industries) and water used for special purposes called ultra-pure water. Water plays fundamental roles in different fields and in different ways and after use it becomes polluted by different impurities to different degrees.

The continuous increasing population along with their socio-economic ascendance demands freshwater. In contrast, degradation in quality of this fundamental resource has become more evident because of the increase of economic activity (which requires roads, dams, electric power, etc.) as well as the lack of proper sanitation, particularly in very densely populated areas of the world. In developing countries an important source of water contamination is the discharge of untreated or partially treated industrial and domestic wastewater. Wastewaters are often discharged to nearby rivers and streams without any proper treatment, traveling long distances until they contaminate large bodies of freshwater downstream, even affecting underground water [2]. Economic growth, in most developing countries, is not accompanied by proper allocation of resources to improve sanitation and water treatment facilities.

Water pollution creates serious health issues for humans and other organisms due to transmission of bacteria, viruses and water-borne pathogens, and noxious and toxic pollutants. In developing countries, it is estimated that about 1.2 billion people are suffering from different diseases due to the use of poor quality water. According to a survey, about 30% of natural deaths can be attributed to polluted

Bioassays. DOI: http://dx.doi.org/10.1016/B978-0-12-811861-0.00002-4

water. Besides, 80% of all food production in developing countries use polluted water for irrigation [3,4]. Innumerous socio-economic benefits are linked to water, such as crop irrigation, industrial activities, and transport, in addition to the fact that water plays a main role in the ecosystems. The rapid economic and industrial growth has seriously affected large bodies of water such as lakes, rivers, and other water reservoirs. These water bodies are the ultimate recipients of pollutants that impose, due to degradation resistance, risks to the aquatic system due to bioaccumulation and bio-amplification [5].

The origin of the water effluents decides the chemical constituents and pollutants present in polluted water bodies. Industrial effluents have a complex nature of pollutants and include organic and inorganic substances like pharmaceutical residues, dyes, and metals, requiring treatment before discharge in wastewater streams. To evaluate the quality of water, qualitative and quantitative information about the constituents (pollutants) must be known. Extensive farming generates organic pollutants such as pesticides and other agrochemicals that are carried to water bodies and groundwater.

Pollutants present in wastewater are dissolved or undissolved (Fig. 2.1). The undissolved pollutants include precipitated components (e.g., soil residues, solid

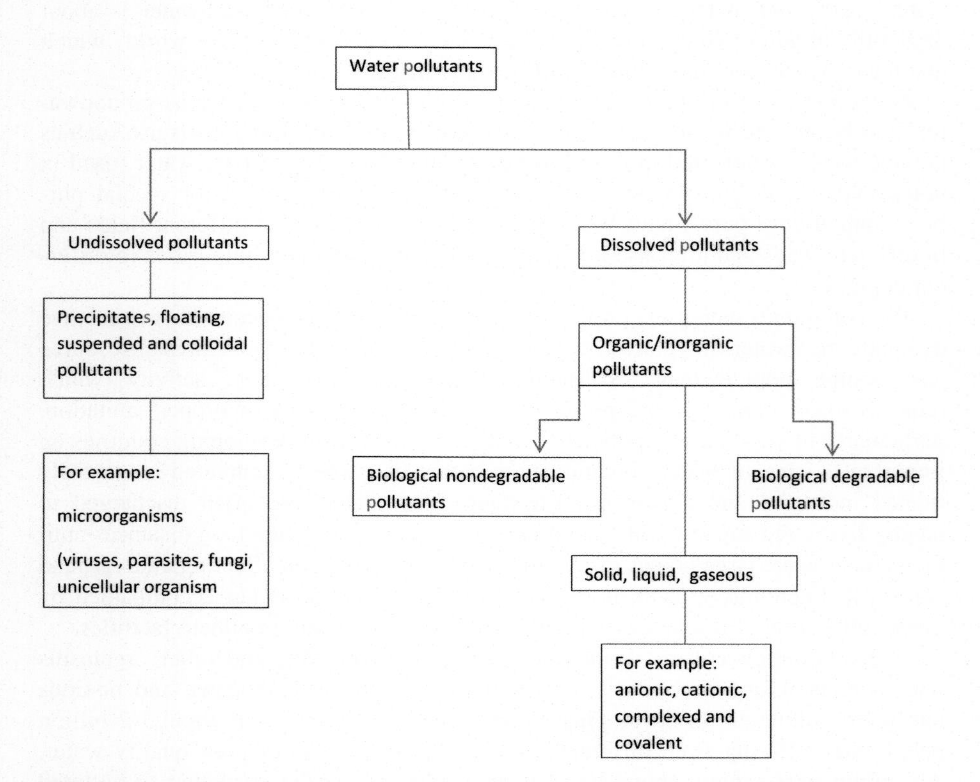

Figure 2.1 Classification of water pollutants as dissolved and undissolved substances.

food wastes, human waste products). Besides floating materials such as plants and animal residues, oil, grease, foams, suspended solids, colloidal materials, and clay minerals are all included as undissolved pollutants. Microorganisms such as phytoplankton, zooplankton, and fungi along with viruses, bacteria, and parasites are also undissolved forms of pollutants in wastewater and in natural water.

Dissolved pollutants can be divided in two categories: organic and inorganic matter. Such pollutants may or may not be biodegradable. The substances that are dissociable and converted into anions and cations (complexes included) are all categorized in dissolved pollutants. Some gases easily react with water to form pollutants, comprising a part in the formation of water pollution (carbon dioxide, nitrogen dioxide, sulfur dioxide, hydrogen sulfide, and ammonia) affecting water pH conditions.

The organic matter is divided into two types (dissolved and particulate organic matter) based on their origin and solubility (Fig. 2.2). These may come from natural sources or from anthropogenic sources. Naturally organic matter comes into contact with water bodies from plants and microbial residues or produced (in situ) in water through biological processes. It should be mentioned that the contribution of natural and anthropogenic organic matter are different for other pollutants. For example, chlorinated hydrocarbons are continuously added to water bodies not only through industrial activity but also due to natural phenomena. Another case is that most of nitrate ions (NO_3^-) are added to water from natural resources, despite that anthropogenic sources such as farming are also important. Cationic, anionic, and neutral species contributions come from different natural sources, but human activities are important. In particular, human and veterinary drug residues may be present in lakes down to 250 m depth, which create toxicological and ecological risks.

2.2 Physical and chemical characteristics

Depending on its destination, it is necessary to test the quality of water by monitoring different physicochemical parameters. Physical tests include the measurement of pH and turbidity and evaluation of color and odor. Chemical tests include biological oxygen demand (BOD), chemical oxygen demand (COD), dissolved oxygen, alkalinity, and hardness. Other characteristics may also be determined for specific quality control in order to protect ecosystems and other biota. For instance, water should be tested for trace quantities of metal ions, anions, and organic substances like pesticides, surfactants, etc.

2.2.1 Odor, color

The color of water should be transparent and there should be no suspended organic or inorganic particles; moreover, for best quality the water should be odorless.

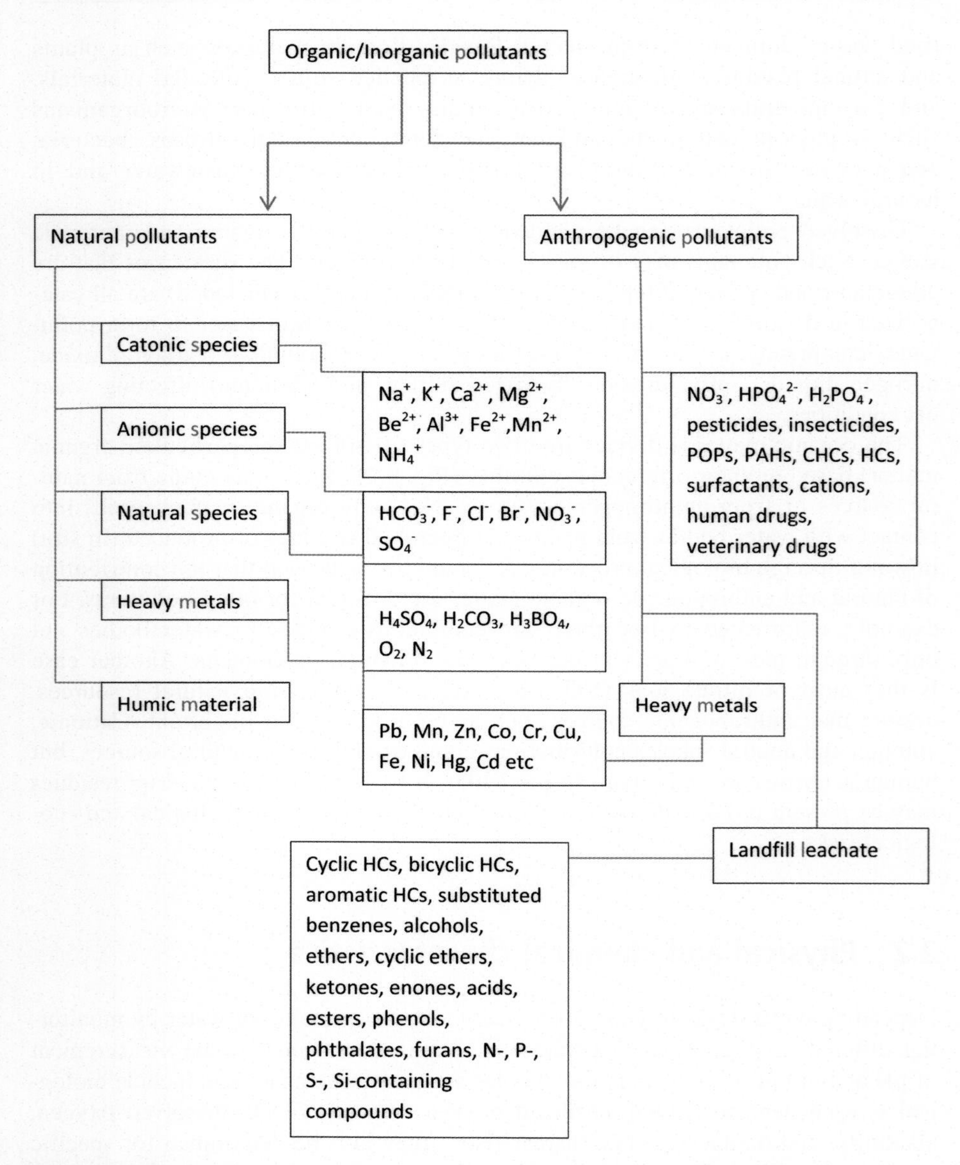

Figure 2.2 Classification of water pollutants as organic and inorganic substances.

2.2.2 *pH*

The pH value plays an important role: decreasing pH values cause higher corrosion. The pH has also been related with electrical conductance and total alkalinity. The pH value may change due to various phenomena; for example higher pH values suggest that the carbon dioxide, carbonate-bicarbonate equilibrium is affected due to a change in the physicochemical conditions [6].

2.2.3 Turbidity

Turbidity is another parameter that can affect the quality of water. It can be determined from a comparison of the intensity of scattered light in the sample to that of a standard reference suspension under the same conditions.

2.2.4 Biochemical oxygen demand (BOD)

BOD is a measure of organic material contamination in water. The procedure of BOD is based on determining the quantity of dissolved oxygen before and after the incubation of samples for a specific interval of time at a controlled temperature (20°C). The net change in quantity of oxygen in expressed in the mass of oxygen per unit volume (mg L^{-}) [7].

2.2.5 Chemical oxygen demand (COD)

The COD is the estimate of oxygen required for the portion of organic matter in wastewater that is subjected to oxidation and also the amount of oxygen consumed by organic matter from boiling acid potassium dichromate solution. COD is a water quality measure used not only to determine the amount of biologically active substances such as bacteria but also biologically inactive organic matter in water [8]. It is an important and rapidly measured variable for characterizing water bodies, sewage, industrial wastes, and treatment plant effluents. For COD analysis the water is subjected to reflux using silver sulfate and potassium dichromate as catalysts. Organic matter will partially reduce dichromate and the remainder can be measured after titrating with ferrous ammonium sulfate. COD can be determined for both filtered and unfiltered samples while the difference between the results of both gives the COD of the particulate matter [7].

2.3 Metals in water bodies

Some cations are essential to human health and to the physiological functions of other organisms. However, exposure to high concentrations has adverse effects on the health of organisms, as well as causing damage to ecosystems. Ions such as Hg^{2+}, Hg_2^{2+}, Cr^{6+}, Pb^{2+}, or Cd^{2+} are toxic as they are not required for biological functions and they accumulate in the organism. These and other ions are sometimes classified as heavy metals, although no accurate definition exists. Household and industrial effluents are common sources of those pollutants, which directly affect the structure and productivity of ecosystems. The importance of controlling the level of toxic metals in natural and potable water has generated the development of novel sensitive analytical methods that also include those able to perform speciation analysis [9].

For the determination of toxic metals at low levels, atomic spectrometry is commonly employed, which includes atomic absorption spectrometry (AAS) and the techniques based on inductively coupled plasma (ICP) as ion reservoir/source, such as optical emission spectrometry (ICP OES) and mass spectrometry (ICP-MS). These techniques can be coupled with powerful separation techniques such as high-performance liquid chromatography (HPLC) and capillary electrophoresis (CE). Other approaches such as potentiometry using ion-selective electrodes, ion chromatography (IC), and redox titrations can be used, depending on the analyzed substances and their levels.

2.3.1 Spectrometric methods

AAS and ICP OES are based on the electronic transitions that occur in free atom clouds in a reservoir at high temperatures. In AAS the measured absorbance (which is a log function of transmittance) is a result of the absorption by atoms and photons from an electromagnetic radiation source. In optical emission, ground state atoms are transformed to excited state ions due to thermal and collisional processes that occur in the Ar^+ rich ICP. The return of ions to the ground state emits characteristic photons that can be related to the nature and quantity of the target species. Both techniques allow determinations at the $\mu g\ L^{-1}$ (parts per billion) levels and present established protocols to determine more than 70 metals and metalloids in different types of water samples using approaches that vary upon the employed atomization reservoirs (atomizer) and ways to introduce the analyte into atomizers. ICP is the standard source for OES, and for AAS different atomizers are available and used according to the nature of the analytes, their levels in samples, and expected interferences expected from sample matrices. In Flame Atomic Absorption Spectrophotometry (FAAS) the aqueous sample is aspirated into a flame where the process of atomization takes place. The air/acetylene flame is usually employed to determine elements such as Al, Co, Fe, Zn, and others, while hot nitrous oxide/acetylene is utilized for the determination of Cu, Ag, Au, Ga, In, Ti, Si, As, Sb, Se, V, and others. Analyte preconcentration on solid phases, such as silica gel chemically modified with N-(1-carboxy-6-hydroxy) benzylidenepropylamine, enables the determination even at sub $\mu g\ L^{-1}$ levels for analytes such as Cr, Mn, Fe, Co, Ni, Cu, Zn, Cd, and Pb in natural water samples [10].

Electrothermal atomizers, with the graphite furnace (GF) being the most important, are used when lower detection limits are required because of the high atom density cloud and stabilized platform temperature furnace conditions that allow best performance towards potential interferences. Small sample volumes (usually between 10–50 μL) are introduced into a graphite tube (in many cases onto a graphite platform) where a heating temperature program is applied. GFAAS allows the use of chemical modifiers as well as an optimized heating program to perform pyrolysis of matrix components, both improving selectivity in determinations.

Some speciation has been attempted for chromium (Cr(VI) and Cr(III)) in water samples by using a selective adsorption in a solid phase prior to the sampling into the furnace along with selecting proper chemical modifiers and the heating program. Hydride generation is a means to transfer hydride forming elements (As, Se, Sb, and a myriad of others) into the atomizer to perform AAS and ICP-OES, in this way avoiding matrix interferences. In applying hydride generation in AAS, elements like As, Se, and Sb are reacted with sodium borohydride to form their respective hydrides (AsH_3, H_2Se, SbH_3) that are transported to an atomization quartz cell, heated to about 900°C, where they are decomposed to liberate the original atoms to be probed by the incident excitation radiation. This method may decrease detection limits by 10–100 times. Chemical speciation using HG-AAS has been accomplished in a fast way for As(III) and As(V) in a number of contaminated groundwater samples [7]. Mercury, the only volatile metal even at room temperature, is determined by the cold vapor AAS technique that relies on the transfer of the mercury vapor—using inert gas—to a quartz cell, where it is probed by the incident radiation. Mercury in the sample should be converted into Hg^{2+} prior to the reduction to elemental mercury by a reduction reagent such as $SnCl_2$. Commercial dedicated instruments for the quantification of total mercury and mercury species are available. For total mercury, detection even at pg L^{-1} has been achieved by using compact spectrometers with a multipath cell that increases the optical length by several times. A dedicated system that comprises an atomic fluorescence detector in tandem with a gas chromatographic system is available for speciation analysis after chemical derivatization of the mercury species (Hg_2^+, CH_3Hg^+, and $CH_3CH_2Hg^+$) with, for instance, sodium tetrapropyl borate. After preconcentration and percolation through the column, the separated derivatizated mercury species are decomposed into elemental mercury through heating prior to detection [11].

ICP-MS has become the most powerful analytical tool for elemental determination in water samples. It covers almost all elements of the periodic table and enables isotopic analysis. Modern analytical systems allow a very efficient collection of the ions formed in the ICP. Reaction and collisional cells have been employed to minimize isobaric interferences that impair the detection of the most abundant isotopes of many elements. The spectral (mass-to-charge) resolution has been improved over the years resulting in the quadrupole analyzer being the one with the best cost-benefit for routine analysis. Sector field mass spectrometers allow very high resolution to enable the minimization of isobaric interferences. Nowadays, the coupling of ICP-MS with either HPLC or CE has become routine in speciation analysis allowing even preconcentration before introduction into the ICP. The possibility to apply isotopic dilution for quantification, the extremely low detection limits, and the system robustness acquired over time has made ICP-MS essential for any environmental control laboratory. Applying both techniques, and after optimization of experimental conditions, Li, Be, B, Al, V, Cr, Mn, Fe, Co, Ni, Cu, Zn, As, Se, Rb, Sr, Ag, Cd, Sb, Cs, Ba, Hg, Tl, Pb, and Bi have been determined in natural water samples in the μg L^{-1} range [12].

2.3.2 Ion chromatography

Ion chromatography (IC) is applied for separation and analysis of both anions and cations in environmental samples. The separation of analytes on a column leads to the identical analytical results as in other separation techniques. The differences in ion-exchange affinities of different species of ions result in their separation, and the most common detection is through conductivity detectors. The most widely used approach is the ion (cation) exchange with reversible complexation. However, for complex samples anion exchange with irreversible complexation is also employed. But nowadays the most effective one is the bifunctional ion exchange column that is mostly utilized to separate heavy and transition metals. Application of the chelation IC, and the online sample pretreatment quantification of lead, copper, cadmium, and other transition metals, have been carried out in water samples. The analysis of drinking water resulted in quantities of transition metals and heavy metals in the ng mL^{-1} range such as Pb (2.0), Cu (0.2), Cd (0.6), Co (0.2), and Ni (0.2) [13].

2.3.3 Anions in water

The anion content in water can cause environmental problems. Anions such as nitrates, sulfates, phosphates, chloride, fluoride, boron, and cyanide may be present at various concentrations in potable and wastewater and can pose toxic risks to aquatic systems. For the anions of most environmental concern limited quantification methods are available for their determination in comparison to metallic pollutants. In most cases, for ion determination like, NO_3^-, NO_2^-, PO_4^{3-}, SO_4^{2-}, SO_4^{2-}, F^-, Cl^-, Br^-, and I^-, ion chromatography is applied for sequential determination in environmental matrices. However, for anions NO_3^-, NO_2^-, and PO_4^3, the cheaper method is determination with the help of a UV-visible spectrophotometer, as described in the literature [14].

2.3.4 Chloride

The presence of chloride in municipal wastes and sewage effluent ultimately increases the chloride content in fresh and wastewater. It comes from activities carried out in agricultural areas, as well as from industrial activities and chloride stores. The high content of chloride is mostly because of human activities. The permissible limit of chloride in drinking water is between 250 and 1000 mg L^{-1} [15]. The widely employed approaches for determination of chloride include ion-selective electrodes, IC (multiple ion determination), gravimetric, titrimetric, and colorimetric methods. For example in the ferricyanide-based colorimetric method a reaction between the chloride ion and mercuric thiocyanate takes place that gives a mercuric chloride and a thiocyanate ion (SCN^-). The thiocyanate ion then reacts with Fe^{3+}; the product of this reaction is a highly colored ferric thiocyanate. The chloride ion can be analyzed from the concentration of ferric thiocyanate, since the concentration of chloride in the sample is exactly equal to the concentration of $Fe(SCN)_3$ [9].

2.3.5 Fluoride

In nature fluoride exists in the form of fluorspar (fluorite), rock phosphate, triphite, phosphorite crystals, etc. The climate of the area and the chemical composition of rock minerals where water circulates are responsible for the concentration of fluoride. According to the WHO the presence of fluoride in water at lower concentrations can prevent dental carries but higher levels of fluoride may have adverse effects. The permissible limit of fluoride in drinking water is 1.5 mg L^{-1}. Besides IC, various spectrophotometric approaches have been reported for quantitative determination of fluoride in different water samples. For example, colorimetric SPADNS applies zirconyl-dye lake (zirconyl chloride octahydrate ($ZrOCl_2 \cdot 8H_2O$) and sodium 2-(parasulfophenylazo)-1,8-dihydroxy-3,6-naphthalene disulfunate (SPADNS) in the acidic medium for fluoride quantification. Fluoride displaces the Zr^{2+} ion from the dye lake and forms a colorless complex anion $(ZrF_6)^{2-}$ along with the dye. The concentration of the fluoride ion in the water sample can be determined as the color intensity of the solution lightens as a function of the increasing fluoride ion concentration [16].

2.3.6 Cyanide

Due to the toxic effects of cyanide, which inhibits metabolic enzymes like cytochrome oxidase (which transports electrons to molecular oxygen), sensitive and selective quantification of cyanide has gained great interest among researchers. Human activities are mostly responsible for the release of cyanide in water bodies. It is important to mention that the main sources are the pesticides industries. The allowable concentration of cyanide for drinking water is 0.02 mg L^{-1} [16]. The major methods for determination of cyanide are titrimetric, colorimetric and ion-selective electrode methods as well as IC. For example in the colorimetric method cyanide is allowed to react with chloramine-T at pH 8 to produce cyanogen chloride (CNCl); this is treated with pyridine-barbituric acid or pyridine-pyrazolone. From the color of the final product the concentration of the cyanide in the sample can be calculated at ppm levels [17,18].

2.3.7 Sulfate

The high solubility of sulfates provides a basis for their presence in natural and freshwater. The oxidation of various ores results in the production of sulfate ions, due to which these ions are mostly present in the industrial wastes. Besides IC, UV spectrophotometric and turbidimetric methods are widely employed for sulfate ions determination in different environmental samples. The limit for sulfate is 200 and 400 mg L^{-1} [19]. For example, a turbidimetric flow-injection method has been developed for the quantification of sulfate in natural and residual waters using a 7.0% (w/v) barium chloride solution prepared in 0.1% (w/v) polyvinyl alcohol as precipitating agent. The detection limit was below 5 mg L^{-1} [20].

2.3.8 Nitrate

Nitrates are present in raw water mostly as nitrogenous compounds and their presence at low or moderate concentrations is harmful to human health while high concentrations may cause methemoglobinemia. These ions enter the water bodies from chemical and fertilizer industries, animal excretions, decaying vegetables, and domestic and industrial discharge. Nitrate ions are analyzed quantitatively by IC, nitrate selective electrodes, and brucine and UV spectrophotometry. The maximal limit for nitrate is 45 mg L^{-1} [19]. The cadmium reduction approach makes use of cadmium as a catalyst to reduce nitrate (NO_3^-) to nitrite (NO_2^-) followed by diazotization with sulfanilamide. After this the product is treated with *N*-(1-naphthyl)-ethylenediamine which gives a highly colored azo dye as a complex product. This colored product is measured at a wavelength of 540 nm. The advantage of the method is the differentiation between nitrate and nitrite [21].

2.3.9 Phosphate

Phosphorus is an essential nutrient for plant growth; it is also responsible for the growth of plants in freshwater and in other water bodies. Normally groundwater contains only a minimal phosphorus level because of the low solubility of native phosphate minerals and the ability of soils to retain phosphate. This element is present in water samples in three different forms: orthophosphate, condensed phosphate (meta-, pyro-, and poly-phosphate) and lastly, organically bound phosphorus. The permissible limit of this ion is 0.05 mg L^{-1} [19]. The different forms of phosphates can be measured accurately by colorimetric analysis and IC. In the colorimetric method phosphate determination is made from the complex formation of phosphomolybdate after the reaction between molybdate and phosphate followed by its reduction with thiourea in acidic medium. Under the optimized experimental conditions the color intensity was found to be proportional to the concentration of phosphate in different water samples ranging from 1.5 to 2.4 μg L^{-1} [22].

2.4 Organic pollutants in water

The monitoring of organic pollutants in water bodies is very difficult as these are released into the environment by numerous industrial applications and other human activities. They have long been a matter of concern among the environmental community, regulatory agencies, and the general public. Among the organic pollutants, those related to oil and grease along with pesticides, nitrosamines, and some of the so-called emerging pollutants are the ones that require most attention.

2.4.1 Pesticides

Modern farming requires pesticides to control insects and other parasites in order to guarantee the needs of a growing population. Although important advances in biological control and use of natural substances in pest control have been achieved, the use of diverse classes of synthetic pesticides is and for long time is going to be the principal means to guarantee high crop productivity, resulting in an impact on the environment especially in contamination of groundwater, rivers, and lakes.

2.4.2 Nitrogen-containing compounds (carbamate, triazine and urea based pesticides)

Nitrogen-containing pesticides are characterized by the presence of carbamate, urea, or a triazine ring in their structures which comprise the three most common classes of nitrogen-containing pesticides. Many instrumental techniques, like gas chromatography (GC) with different detectors—specifically with mass detectors (GC/MS), and HPLC with positive ion electrospray mass spectrometry, have been applied generally to analyze these substances. Three water samples of lake water, well water, and tap water were investigated for carbamate (Aldicarb-ADC); the detected quantities were 4.53, 4.69, and 3.57 ng L^{-1} respectively, while sulfoxide (ASX) was 5, 2.45 and 1.58 ng L^{-1} and Aldicarb sulfoxide (ASX) contamination was 5.43, 4.17, and 2.77 ng L^{-1}. All these detections were carried out with the ion electrospray mass spectrometry technique.

Besides, using a capillary column of appropriate polarity and a nitrogen-phosphorus detector (NPD) in the N-specific mode, many types of nitrogen-containing pesticides can be analyzed. Moreover, carbamate pesticides are selectively analyzed by HPLC using post column chemical derivatization. In most cases carbamates are separated on a C-18 analytical column and then hydrolyzed with 0.05 N sodium hydroxide. Hydrolysis converts the carbamates to their methylamines that after chemical reaction with derivative reagent *o*-phthalaldehyde and 2 mercaptoethanol form highly fluorescent compounds. Urea is also a class of pesticides that are structurally similar to carbamates. Aqueous samples can be analyzed by the U. S. EPA method 553 using a reverse phase HPLC column in combination with a mass spectrometer with a particle beam interface. In this method the selected aqueous samples are extracted by liquid–liquid extraction (LLE) with methylene chloride. The extract is concentrated by evaporating methylene chloride and exchanging the solvent to methanol. Alternatively, the sample may be extracted by solid-phase extraction (SPE), using a sorbent cartridge packed with C-18 impregnated on silica or a neutral polystyrene/divinyl-benzene polymer. The sample is mixed with ammonium acetate (0.8 g L^{-1}) prior to extraction. The cartridge is flushed with methanol and then the sample is passed through the cartridge. The analytes are eluted from the cartridge with methanol and concentrated by evaporation. The sample extract is injected into a HPLC containing a reverse phase HPLC column. The analytes are identified by comparing their retention times with that of a standard. The target urea compounds, after subjecting to SPE, were detected in a concentration range of 1 to 50 μg L^{-1} in environmental water samples like drinking water and surface waters [23]. For weed control the triazine herbicides are used but they are also a ubiquitous environmental problem in soil and water samples due to their mobility and solubility in water. Due to their hazardous effects the EPA and European Union (EU) legislation have established a maximum concentration of a few parts per billion. Generally for extraction of triazines from water samples, LLE, supercritical fluid extraction, and SPE are used. The widely employed analytical methods for their determination are HPLC and GC. Different triazines such as prometryne propazine and simazine were detected by using GC with glass columns containing 500 mg of 50–100 pm cross-bonded porous silica. The detected values for prometryne, propazine, and simazin in tap and lake water were 97.3, 95.2, and 71.4 ng L^{-1}, respectively [24].

Organochlorine pesticides include all chlorine-containing organics employed for pest control. The majority of chlorinated pesticides that were used in the past are no longer in use because of their harmful toxic effects on human health and the contamination of the environment. However, due to their persistent nature such pesticides and their residues are still found in the environment in trace quantities in groundwaters, soils, sediments, and wastewaters. These substances are stable, bioaccumulative, and toxic, and some are also carcinogenetic. Many of these common chlorinated pesticides are listed as priority pollutants by U.S. EPA. Chlorinated pesticides in aqueous and nonaqueous matrices may be determined by the U.S. EPA methods 608, 625, 505, 508, 8080, and 8270 (U.S. EPA 1984–94). Analysis of these pesticides requires the extraction of the aqueous or nonaqueous samples by a suitable organic solvent, concentration and cleanup of the extracts, and the determination of the analytes in the extracts [25].

Another important class of pesticides are organophosphorus compounds. These compounds can cause moderate to severe acute poisoning. These pesticides inhibit the function of the enzym acetylcholinesterase by phosphorylating or binding to the enzyme at its esteratic site. The symptoms of acute toxicity include the tightening of the chest, increased salivation and lacrimation, nausea, abdominal cramps, diarrhea, pallor, the elevation of blood pressure, headache, insomnia, and tremor. The ingestion of large quantities may cause convulsions, coma, and death. The toxicity, however, varies from substance to substance. Phorate, demeton, and disulfoton are among the most toxic organophosphorus insecticides, while malathion, ronnel, and tokuthion are much less toxic. The most commonly employed techniques to analyze organophosphorus pesticides also include GC and GC/MS. The detector required for GC analysis is either a NPD operated in the phosphorus specific mode, or a flame photometric detector (FPD) operated in the phosphorus specific mode, or an alkali flame ionization detector. The analysis of organophosphorus pesticides by GC/MS should be the method of choice wherever possible. Various organophosphorus pesticides such as cumaphos dimethoate and diazinon were determined by GC. Cumaphos was detected in 82.4–98.4 ng L^{-1}, dimethoate, 9.0–10.2 ng L^{-1} and diazinon 90.4–89.1 ng L^{-1} [24].

2.4.3 Nitrosamines

Nitrosamines (or nitrosoamines) are nitroso derivatives of amines in which a nitroso (NO) group is attached to the nitrogen atom of the amine. Nitrosamines are toxic compounds as well as potent animal and human carcinogens. The U.S. EPA classified some of these compounds as priority pollutants in industrial wastewaters, potable waters, and hazardous wastes. Such pollutants occurring in environmental samples can be determined by the U.S. EPA's analytical procedures. The analysis involves the extraction of these compounds into a suitable solvent followed by their detection on the GC by using either a NPD, or a reductive Hall electrolytic conductivity detector (HECD), or a thermal energy analyzer (TEA). These compounds can also be detected by an flame ionization detector (FID), but at a lower sensitivity. For example, it has been reported that the detector response to the lowest amount of N-nitrosodi-*n*-propylamine on FID is in the range of 200 ng, in comparison to 5 ng on the NPD. The existence of nitrosamines in samples such as raw water, prechlorinated water, sand filtered water, groundwater wells, and ozonated water were detected as 7.9, 2.6, 17.7, 0, 11.5, and 7.3 ng L^{-1}, using GC/MS under high-resolution MS conditions [26]. It is important to note that the nitrosodiphenylamine breaks down to diphenylamine at a temperature above 200°C in the GC inlet and is measured as diphenylamine; therefore the latter compound must be removed or separated before analysis [26].

2.4.4 Oil and grease

Oil and grease refer to groups of substances that a solvent can extract from a sample and that do not volatilize during evaporation of the solvent at 100°C. These substances are soluble in *n*-hexane and consist of fatty acids, animal fats, soaps, greases, vegetable oils, waxes, mineral oils, etc. Many industrial effluents have an oily nature, such as in the industries of oil prospecting, petrochemicals, edible oils, dairy products, wastes from slaughterhouses, and refrigerators etc. Sanitary sewers also present concentrations of oils and greases generally in the range of 50–100 mg L^{-1}. There are also oils discharged into natural waters in specific situations, spills from maritime and river accidents. In the natural waters, oils and greases accumulate on the surfaces and can cause serious ecological problems by making the gas exchanges that occur between the liquid and the atmosphere difficult, especially for oxygen. Oil and grease in aqueous samples are determined by the partition-gravimetric method, partition-infrared method, soxhlet extraction method, and solid phase partition-gravimetric method. The selection of the method depends on several factors including the nature of the sample, volatile substances present in the sample, the availability of materials and equipment for analysis, and the time and cost of analysis. The most commonly utilized method is solvent extraction, known as the soxhlet method. In this the sample is initially acidified to promote emulsion breaking and facilitate separation of the oil. For example, Travis et al. detected oil and grease in gray water (untreated kitchen and bathroom effluents) of up to 200 mg L^{-1} using the gravimetric analysis method [27].

2.4.5 Emerging pollutants

Emerging pollutants include any synthetic or naturally occurring substances that are not commonly subjected to analysis but may have the potential to cause adverse effects on ecological and human health. Identification and quantification of these chemicals and their transformation products in different environmental compartments is becoming essential in gaining knowledge about their presence and ultimate fate. The identification and quantification of such pollutants at trace concentrations have therefore assumed significant importance similar to other classes. For example, the presence of many steroids, hormones, and their metabolites has been detected in the environment in waters. The U.S. EPA developed an analytical procedure for the determination of steroids and hormones in aqueous, solid, and bio--solid samples [28]. The biological compounds from the class of steroids and hormones like estrone, estriol, and estradiol were quantitatively determined in different water samples including both drinking and surface water from diverse sources employing liquid chromatography-mass spectrometry (LC-MS). The separation column was LiChrospher 100 Rp-18 (250 × 4 mm, 5 μm) and the MS system was HP1100 and the sample pretreatment was done using SPE with C18. The detected quantities of these emerging pollutants varied from 2-500 ng L^{-1} depending on the source and nature of samples.

Many pharmaceutical and personal care products are also of growing concern in the environmental matrices and have been included as a class of emerging pollutants; the methods of choice have always been chromatography-mass spectrometry techniques. For example analgesics (phenazone), β-blockers (propranolol), antineoplastics (ifosfamide), and lipid lowering agents (simvastatin) were detected in tap and surface water by using LC-MS. The detected level of these substances varied from 8 to 44 ng L^{-1}. The same procedure was carried out for antibiotics such as sulfadiazine and amoxicillin, where the detected levels for sulfadiazine were in the range from 3.7 to 11 ng L^{-1}, while amoxicillin ranged in between 15 to 21 ng L^{-1}. It is important to note that for pharmaceutical residues the U.S. EPA method 1694 [28] also describes an analytical method to measure these substances in multimedia environmental samples by HPLC combined with tandem mass spectrometry (HPLC/MS) [9].

Part II Air analysis

2.5 Review

The atmosphere is the gaseous matrix from the Earth up to space that consists of a mixture of various gases such as nitrogen, oxygen, argon, carbon dioxide, neon, helium, methane, sulfur dioxide, etc., termed as air. Besides other gaseous molecules, water (in solid and liquid form), salt crystals and dust particles are also present in the air. Negatively and positively charged ions are also present in the upper layers of atmosphere. The introduction of pollutants from natural and anthropogenic sources and propagation of polluting components affects the chemical composition and characteristics of the atmosphere. The pollutants in the atmosphere may be present in the form of the particles or gaseous molecules. When pollutants are present in air that may adversely affect the environment it is termed as polluted air or air pollution. Due to human activities or natural phenomena, direct or indirect introduction of substances into the atmosphere takes place, which can cause damage to human health, as well as other organisms and ecosystems, and disturb the proper use of the natural and human environment. The two main reasons for air pollution may be the chemical processes that take place in the atmosphere or the different substances that are directly emitted to the air. The chemical species (pollutants) that are responsible for air pollution are categorized into two main types: the substances or pollutants which are directly released to the air from human activities or natural sources are termed primary pollutants, while those which are formed from primary pollutants in the atmosphere itself (chemical reactions that take place) are called secondary pollutants. These reactions occur as a result of mutual interactions between elements and compounds or by the action of different types of the energy. The pollutants in the atmosphere, especially those that present a threat to human health, have gained great attention and all of these types of substances are considered to be harmful substances. These substances include carbon monoxide, sulfur, dust, ammonia, nitrogen oxides, chlorine, formaldehyde, carbon disulfide, fluorine and their compounds, phenol, hydrogen sulfide, sulfuric acid, arsenic compounds, and lead, all of which are toxic if present in large concentrations or if they exceed their permissible limits. The presence of some of the substances in the air above the permissible limit increases the overall temperature of the Earth's surface, causing global warming, for example CH_4, C_2H_6, CF_2Cl_2, CCl_4, CH_3Cl, N_2O, CO_2, and SO_2. Most air pollutants are in gaseous form and most of these (about 90%) came from natural sources, and 10% are due to human activities [29]. However, the contribution from anthropogenic sources is regularly increasing, due to rapid increase in industries and urbanization, which enhances the natural process, polluting the air at a much higher rate.

2.6 Air sampling

The sampling is the most critical task for the quantification of pollutants in air samples. Moreover, the preservation and delivery of the collected samples to the laboratories is also an important factor for accurate results. Portable equipments to analyze the sample on site enhance the reliability of the analytical results. The air samples are collected by "air sample collectors" which are composed of particulate matter with different particle size: e.g., PM10 ($<10\ \mu m$)—where PM represents particulate matter—that are used to collect the largest particles, PM2.5 ($<2.5\ \mu m$) to collect medium-size particles and PM1.0 ($<1\ \mu m$) to collect smaller particles. The collection of each type of sample such as particulates, aerosols, and gases is used for different analytical purposes. Various types of air sample collectors and particulate matter are used for different air sampling, for example the sampling of particulate matter and gases are different from each other. For reliable sampling and quantitative analysis the sampling area or the site information is important. A general informative description should be obtained of the region. After that, data about samples, e.g., the nature of the samples, such as dust, indoor, outdoor, gaseous, particulate, radioactive, etc. must be known. Finally, information is ascertained about the analysis and what kind of analysis is required for samples like heavy metals, anion, and other organic compounds [7].

2.7 Analysis of air pollutants

The pollutants in the air that adversely affect the air quality comprise sulfur, nitrogen, carbon and halogen compounds, radioactive substances, and particles. To maintain air quality of both in indoor and outdoor environments the quantitative information of pollutants is essential.

2.7.1 Sulfur compounds

Various substances are present in the atmosphere that contain sulfur and act as pollutants, like sulfur dioxide, sulfur trioxide, sulfuric acid, hydrogen sulfide, and carbon disulfide. Each sulfur compound has specific characteristics to be considered as a pollutant. E.g., the permissible limit of sulfur dioxide in air is 125.0 $\mu g\ m^{-3}$ on a daily basis and below 50.0 $\mu g\ m^{-3}$ on an annual basis. The techniques for determination of SO_2 in air samples used previously were spectrophotometry, chemiluminescence, IC, spectrofluorometric, potentiometric, and amperometric methods. 15.15 $\mu g\ m^{-3}$ of SO_2 were detected in the air of Al-Ain city (UAE) over a year through IC coupled with a potentiometric detector by a passive sampling method [30].

2.7.2 Nitrogen compounds

Different nitrogenous compounds emitted to the air that cause pollution include: N_2O, NO, NO_2, NH_3, NH_4^+, and NO_3^-. Natural sources contribute more as compared to anthropogenic sources. For example ammonia is a nitrogenous compound which can cause pollution when it exceeds 7 $\mu g\ m^{-3}$. For quantitative NO_2 determination in air samples different analytical techniques have been developed such as the use of a spectrophotometer coupled with azo dyes, chemiluminescence, laser-induced fluorescence, optical sensors, electrochemical sensors, and applying different sampling approaches like a passive sampling method, etc. According to Salem et al. monitoring of NO_2 was carried out in the air of Al-Ain city (UAE) over a year through IC attached to a potentiometric detector, and the sampling was done through passive sampling. The level detected was more than 59.26 $\mu g\ m^{-3}$ [30].

2.7.3 Carbon compounds

The presence of carbon-containing compounds like carbon dioxide, carbon monoxide, hydrocarbons, etc. above the permissible level also contributes to air pollution. As a result of different biological activities in nature and also due to forest fires, more carbon compounds enter the environment as a natural source than from anthropogenic sources. It is important to note that the established permissible limits of aromatic hydrocarbons like benzene and toluene are 40 and 47 $\mu g\ m^{-3}$ respectively in air [31]. The presence of different aromatic hydrocarbons is determined using a differential optical absorption spectrometer (DOAS) and an automatic gas chromatograph (VOC analyzer), and the detection limits were found at 128 and 138 $\mu g\ m^{-3}$ [32].

2.7.4 Halogens and their compounds

The already reported toxicity data available for hydrochloric acid, hydrofluoric acid, and silicon tetrafluride revealed that these compounds are the important halogen compounds, which cause air pollution at different concentrations. Both natural and human activities are the cause of the release of such compounds. Using the techniques of electron capture GC, CCl_3F and $CC1_2F_2$ in the range of modern tropospheric air were found to be 220 ppt and 400 ng L^{-1}, respectively [33].

2.7.5 Radioactive substances

During different research processes and industrial methods such as physical, chemical, biological and other processes, radioactive substances are produced, accumulated, and continuously released into the air. The most attractive site for the researcher is the study of gases emitted from nuclear power plants. Various gases and aerosols are responsible for atmospheric radioactivity. Radioactive substances may form solid or liquid particles either directly or by sorption on the surface of nonradioactive particles. Inductively coupled air plasma–atomic emission spectrometry (air-plasma ICPAES) is used to determine the presence of radioactive substances in air samples. Three different types of radioactive substances (Th, U, and V) were measured by using the above mentioned technique; the concentration of Th was 0.0023 mg m^{-3}, that of U was 0.2 mg m^{-3} and that of V 0.02 mg m^{-3} [34].

2.7.6 Particles

Large contributions of particle contamination in the atmosphere are due to natural sources, which are reported to be up to 89%; only 11% is due to human activities. Particles may be solid or may be in liquid phase. From these particles, so-called secondary particles are formed directly in the atmosphere during a change in the state of matter or perhaps due to the interaction of these substances with other compounds forming liquid or solid products. Different particles such as microorganisms, pollen, spores, and salt spray from the oceans, terpenes, metals, dust particles, and ashes of forest fires are the pollutants due to natural sources, while humans are responsible for the secondary particles formed from gaseous pollutants, sulfates, and nitrates, as well as metals etc. In air samples various metals such as Cd, Co, Cr, Hg, and Pb were detected using analytical techniques like neutron activation analysis, X-ray fluorescence, and ICP emission spectroscopy. The detection limits for Cd were 4, 6, and 0.4 ng m^{-3} respectively; 0.02, 0.4, and 1 ng m^{-3} were the detection limits for Co. Cr has 0.2, 1 and 1 ng m^{-3} detection limits. Hg and Pb were only detected by X-ray fluorescence and inductively coupled plasma emission spectroscopy: Results show that 1.26 and 1.10 ng m^{-3} were detected, respectively [35].

References

[1] Perry G, Duffy P, Miller N. An extended data set of river discharges for validation of general circulation models. J Geophys Res 1996;101(D16):21339–49.

[2] Davis RJ. The mitogen-activated protein kinase signal transduction pathway. J Biol Chem 1993;268:14553.

[3] Elhance AP. Hydropolitics in the third world: conflict and cooperation in international river basins. Washington, DC: US Institute of Peace Press; 1999.

[4] United Nations Commission for Sustainable Development. Comprehensive assessment of the fresh water resources of the world. Geneva, Switzerland: World Meteorological Organization; 1997.

[5] Rhind S. Anthropogenic pollutants: a threat to ecosystem sustainability? Philos Transact Royal Soc London B 2009;364(1534):3391–401.

[6] Gupta PK. Methods in environmental analysis; water, soil and air. Jodhpur, RJ: Updesh Purohit for Agrobios; 2004.

[7] Mahmood MB. Environmental applications of instrumental chemical analysis. Boca Raton, FL: CRC Press Taylor & Francis Group; 2015.

[8] Khuhawari M, Mirza MA, Leghari S, Arain R. Limnological study of Baghsar Lake district Bhimber, Azad Kashmir. Pak J Bot 2009;41(4):1903–15.

[9] Patnaik P. Handbook of environmental analysis: chemical pollutants in air, water, soil, and solid wastes. Boca Raton, FL: CRC Press; 2010.

[10] Mahmoud ME, Kenawy I, Hafez MA, Lashein R. Removal, preconcentration and determination of trace heavy metal ions in water samples by AAS via chemically modified silica gel N-(1-carboxy-6-hydroxy) benzylidenepropylamine ion exchanger. Desalination 2010;250(1):62–70.

[11] Leopold K, Foulkes M, Worsfold P. Methods for the determination and speciation of mercury in natural waters—a review. Anal Chim Acta 2010;663(2):127–38.

[12] Sneddon J, Vincent MD. ICP-OES and ICP-MS for the determination of metals: application to oysters. Anal Lett 2008;41(8):1291–303.

[13] Lu H, Mou S, Yan Y, Tong S, Riviello J. On-line pretreatment and determination of Pb, Cu and Cd at the μg l^{-1} level in drinking water by chelation ion chromatography. J Chromatogr A 1998;800(2):247–55.

[14] Muhammad S, Shah MT, Khan S. Health risk assessment of heavy metals and their source apportionment in drinking water of Kohistan region, northern Pakistan. Microchem J 2011;98(2):334–43.

[15] Raviprakash SL, Krishna RG. The chemistry of ground water Paravada area with regard to their suitability for domestic and irrigation purpose. India J Geochem 1989;4 (1):39–54.

[16] Goulden PD, Afghan BK, Brooksbank P. Determination of nanogram quantities of simple and complex cyanides in water. Anal Chem 1972;44(11):1845–9.

[17] Agrawal O, Sunita G, Gupta V. A sensitive colorimetric reagent for the determination of cyanide and hydrogen cyanide in various environmental samples. J Chin Chem Soc 2005;52(1):51–7.

[18] López F, Giménez E, Hernández F. Analytical study on the determination of boron in environmental water samples. Fresenius' J Anal Chem 1993;346(10):984–7.
[19] Baker J, Campbell K, Johnson H, Hanway J. Nitrate, phosphorus, and sulfate in subsurface drainage water. J Environ Qual 1975;4(3):406–12.
[20] Morais I, Rangel AO, Souto MRS. Determination of sulfate in natural and residual waters by turbidimetric flow-injection analysis. J AOAC Int 2001;84(1):59–64.
[21] Margeson JH, Suggs JC, Midgett MR. Reduction of nitrate to nitrite with cadmium. Anal Chem 1980;52(12):1955–7.
[22] Shyla B, Nagendrappa G. A simple spectrophotometric method for the determination of phosphate in soil, detergents, water, bone and food samples through the formation of phosphomolybdate complex followed by its reduction with thiourea. Spectrochim Acta Mol Biomol Spectrosc 2011;78(1):497–502.
[23] Wang N, Budde WL. Determination of carbamate, urea, and thiourea pesticides and herbicides in water. Anal Chem 2001;73(5):997–1006.
[24] Molto J, Pico Y, Font G, Manes J. Determination of triazines and organophosphorus pesticides in water samples using solid-phase extraction. J Chromatogr A 1991;555 (1–2):137–45.
[25] Júnior JLR, Re-Poppi N. Determination of organochlorine pesticides in ground water samples using solid-phase microextraction by gas chromatography-electron capture detection. Talanta 2007;72(5):1833–41.
[26] Zhao Y-Y, Boyd J, Hrudey SE, Li X-F. Characterization of new nitrosamines in drinking water using liquid chromatography tandem mass spectrometry. Environ Sci Technol 2006;40(24):7636–41.
[27] Travis MJ, Weisbrod N, Gross A. Accumulation of oil and grease in soils irrigated with greywater and their potential role in soil water repellency. Sci Total Environ 2008;394 (1):68–74.
[28] Blair BD, Crago JP, Hedman CJ, Klaper RD. Pharmaceuticals and personal care products found in the Great Lakes above concentrations of environmental concern. Chemosphere 2013;93(9):2116–23.
[29] Tölgyessy J. Chemistry and biology of water, air and soil: environmental aspects. Amsterdam: Elsevier; 1993.
[30] Salem AA, Soliman AA, El-Haty IA. Determination of nitrogen dioxide, sulfur dioxide, ozone, and ammonia in ambient air using the passive sampling method associated with ion chromatographic and potentiometric analyses. Air Qual Atmos Health 2009;2 (3):133–45.
[31] Duarte-Davidson R, Courage C, Rushton L, Levy L. Benzene in the environment: an assessment of the potential risks to the health of the population. Occup. Environ. Med 2001;58(1):2–13.
[32] Brocco D, Fratarcangeli R, Lepore L, Petricca M, Ventrone I. Determination of aromatic hydrocarbons in urban air of Rome. Atmos Environ 1997;31(4):557–66.
[33] Bullister J, Weiss R. Determination of CCl_3F and CCl_2F_2 in seawater and air. Deep Sea Res Part 1 Oceanogr Res Pap 1988;35(5):839–53.
[34] Baldwin DP, Zamzow DS, D'Silva AP. Detection limits for hazardous and radioactive elements in airborne aerosols using inductively coupled air plasma–atomic emission spectrometry. J Air Waste Manag Assoc 1995;45(10):789–91.
[35] Ondov JM, Dodd JA, Tuncel G. Nuclear analysis of trace elements in size-classified submicrometer aerosol particles from a rural airshed. Aerosol Sci Technol 1990;13 (2):249–63.

Historical development of bioassays

3

Donat-P. Häder
Friedrich-Alexander University, Erlangen-Nürnberg, Germany

3.1 Introduction

Long before computers were available to record data and automate measurements, bioassays have been developed to monitor toxicity in the environment and changes in human physiology [1,2]. Two bioassay methods have been described to determine the toxicity of the seed oil and glycosides as insecticides consisting of residual film application and leaf dipping [3]. The seed oil was analyzed by gas chromatography and mass spectroscopy. Munch and Grantham have determined the toxicity of a tincture and extract of the poisonous aconite (*Aconitum napellus*) using guinea pigs and rats as bioindicators [4]. The Scottish bacteriologist Fleming found the toxic action of the mold *Penicillium* on the growth of bacteria by accident [5]. He had been working with *Staphylococcus* cultures growing on culture disks which he inspected from time to time. During these examinations the plates became infected with air-borne fungal spores (*Asperigillus, Sporotrichum, Cladosporium*, and others), which started growing on the plates. Obviously the molds had excreted a substance which diffused rapidly in the agar and had stopped the growth of the bacteria such as *Staphylococcus*. This method of using agar plates on which molds block the bacterial growth is still being used to determine the effectivity of antibiotics against various bacterial strains [6]. Sir Alexander Fleming received the Nobel Prize for medicine in 1945 for his discovery of the antibiotic substance benzylpenicillin (Penicillin G) [7].

3.2 Bioassays for environmental toxicity

Some relatively simple bioassays have been developed to determine the toxicity and biodegradability by subjecting the material to anaerobic treatment [8]. The assays determine the cumulative methane production of the material in a chemically defined medium and have been used to monitor processed samples of peat. Brine shrimp (*Artemia salina* Leach) is being used as a simple bioassay to determine the toxicity of extracts from plants [9]. Seed extracts from more than 40 plants of the Euphorbiaceae family were analyzed using this rapid, inexpensive, and reliable bioassay. The brine shrimp lethality test has also been used in parallel to the inhibition of crown gall tumors induced by *Argrobacterium tumefaciens* on potato disk as an antitumor bioassay as well as frond growth in duckweed (*Lemna minor*), which is a

Bioassays. DOI: http://dx.doi.org/10.1016/B978-0-12-811861-0.00003-6

bioassay for herbicides and plant growth stimulants [10]. These tests were applied to indicate the presence of piceatannol, which is an antileukaemic agent derived from *Euphorbia lagascae*, saponins from *Chenopodium quinoa*, and annonaceous acetogenins extracted from *Asimina triloba* plants and *Annona muricata* seeds, which constitute a new group of botanically derived pesticides. In addition, very effective antitumor acetogenins from *Annona bullata* and usnic acid from lichens as potent biodegradable herbicides have been tested with these bench-top bioassays.

Zinc is a notorious pollutant in freshwater [11]. Zebrafish have been used as bioassay to detect zinc sulfate at concentrations from 5 to 40 ppm [12]. However, it is interesting to note that the toxicity depends on the age of the fish: newly laid eggs survived the longest time while 4–13 days-old fish survived the shortest time and many adult fish showed no mortality.

Bioassays have also been used to indicate nutrient limitation in tropical rain forest soil [13]. For this purpose six shrub species and a large herb were used in a Costa Rican lowland rain forest soil. The growth response of the herb *Phytolacca rivinoides* indicated high concentrations of N but limiting concentrations of P, K, Mg, and Ca which was confirmed by soil measurements. In contrast, the shrubs including *Miconia* and *Piper* spp. did not show a strong growth response. The lack of a response to P fertilization can be explained by assuming that the plants receive this mineral from mycorrhizal associations. In addition, these woody plants grow at slower rates.

A review compared a total of 275 toxicity tests for pollutant toxicity to fish developed to determine acute toxicity [14]. Most of the tests showed that acute lethality ends within 3 days. In order to compare results it is recommended to determine the LC_{50} (lethal concentration for 50% of the population) after long-term exposure. In order to shorten the required time the LC_{50} value after 4 days can be used, which is often similar to the long-term LC_{50} value. Technically, the values (e.g., mean survival times) should be drawn against increasing concentrations on a logarithmic plot. The LC_{50} value is then estimated by choosing an exposure time from the asymptotic part of the toxicity curve. For this exposure time mortality is plotted against the concentration and the LC_{50} value is determined from the fitted curve. In addition, confidence limits of the LC_{50} value should be determined. The statistics of bioassays have been reviewed describing the processes, reaction, principles, and approaches in biological assays [15]. One important conclusion is that the response should be more or less linear when plotted against the log dose. Thus a quantitative knowledge of the dose-response relation is an important prerequisite for the design of any bioassay.

The company MicroBioTests Inc. (Mariakerke-Gent, Belgium) follows a different philosophy. They provide Toxkit microbiotests for applications in aquatic and terrestrial ecotoxicology which do not rely on culturing and maintenance of live organisms [16]. These tests were used in an international comparison which involved several laboratories in eight European countries monitoring toxicity in wastes. They have also been used in southern Italy for toxicity screening of surface waters collected in late spring and late autumn at 19 sampling stations [17]. The Algaltoxkit follows the ISO guidelines. In essence it is a 72 h algal growth inhibition test based on *Pseudokichneriella subcapitata* which is provided in the form of

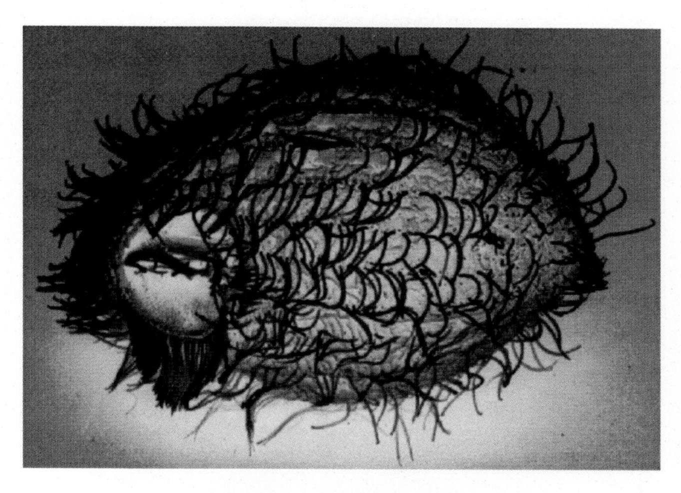

Figure 3.1 Anostracan crustacean *Thamnocephalus platyurus*.
Source: Courtesy MicroBioTests Inc., with permission.

algal beads [18]. The Daphtoxkit contains dormant eggs (ephippia) of *Daphnia* [19]. The Thamnotoxkit determines the mortality of larval Anostracan crustacean *Thamnocephalus platyurus* (Fig. 3.1) hatched from cysts after 24 h exposure to a potentially toxic pollutant [20]. This test has also been used to detect cyanobacterial toxins, and compared the results with the rat hepatocyte test and mouse test [21]. The rotifer *Brachionus calyciflorus* is also hatched from cysts and is used in the Rotoxkit which measures the inhibition of reproduction after 48 h exposure following the French standard AFNOR NF 90-377, 2000 [22,23]. The Protoxkit is a 24 h growth inhibition test using the ciliate *Tetrahymena thermophila* which is supplied in a stock culture vial (Fig. 3.2). Another microbiotest in this group is the Phytotoxkit which determines germination and early plant development using image analysis of shoot and root growth in a monocotyledon (*Sorghum saccharatum*) and two dicotyledon (*Lepidium sativum* and *Sinapis alba*) during three days [24]. This test can also be performed with *Avena sativa* and *Brassica napus*, but the exposure time needs to be extended to four days. This test has been evaluated to determine the toxicity of sediments which contained oil derivatives, polycyclic aromatic hydrocarbons, and several heavy metals such as Cd, Cr, Cu, Hg, and Ni [25]. The results of these tests have been evaluated and compared with standard tests based on the determination of the geometrical means of the EC_{50} values together with standard deviation and coefficients of variance using the warning limit approach [26] (Table 3.1).

In order to achieve comparability between different bioassays and to classify toxicity of wastewaters, different systems have been developed in different countries. However, comparing several bioassays requires maintaining and culturing several live test organisms, which can be carried out only by large laboratories specialized in toxicity determination. Several research groups from ten central and

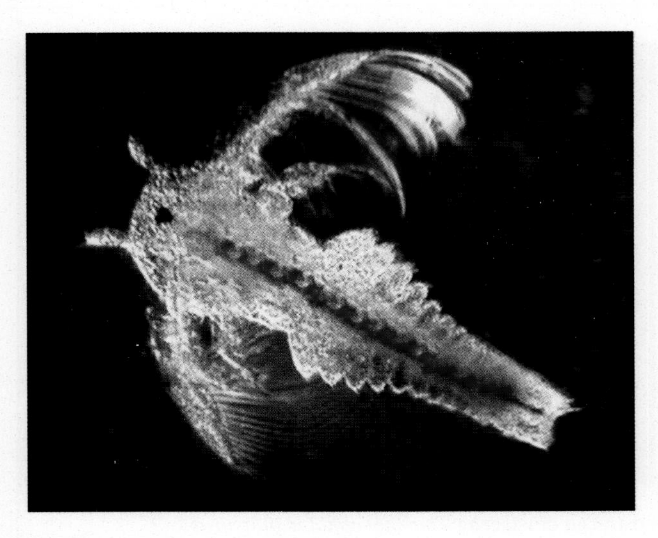

Figure 3.2 Ciliate *Tetrahymena thermophila*.
Source: Courtesy MicroBioTests Inc., with permission.

Table 3.1 Summary of the number of standard tests and Toxkit microbiotests performed using algae, *Daphnia* and rotifers and the number of acceptable tests (in parenthesis)

Substrate	Standard algal test	Algaltoxkit	Standard *Daphnia* test	Daphtoxkit	Standard rotifer test	Rotoxkit
INC	20 (17)	7 (5)	43 (39)	11 (8)	3 (3)	4 (4)
SOI	21 (14)	7 (7)	43 (40)	11 (11)	3 (3)	3 (3)
WOO	19 (16)	7 (6)	42 (40)	11 (11)	3 (3)	3 (3)
Sum	60 (47)	21 (18)	128 (119)	33 (30)	9 (9)	10 (10)
Percentage acceptable	78	86	93	90	100	100

eastern European countries have been coordinated by Ghent University (Belgium) to develop an alternative toxicity classification system using a battery of microbiotests which do not require the maintenance of live test organisms [16]. For this purpose the wastewater samples are used undiluted except when the effect is higher than 50%. This scoring system classifies wastewaters in five categories of increasing toxicity and calculates a weight factor. It is applicable to river waters, groundwaters, drinking waters, mine waters, sediment pore waters, industrial effluents, soil leachates, and waste dump leachates.

A number of bioassays including the brine shrimp assay, marine fungi, and algae, as well as some fish, have been employed to detect the effects of biologically active substances isolated from 40 marine sponge species collected from intertidal and subtidal habitats near San Diego, California [27]. Twenty-eight sponges contained substances with antimicrobial activity which have been isolated and identified. The actions identified were suppression of growth in marine fungi and algae, modification of adult invertebrates, toxicity in goldfish, inhibition of sexual reproduction in a brown alga and interference with the early development and late larvae metamorphosis in invertebrates. Another study also indicated antibacterial activity in marine Demosphongiae [28]. These findings may explain why marine sponges are rarely overgrown by algae or marine sessile animals.

3.3 Bioassays for human physiology and toxicity

A number of bioassays have been developed to monitor complex diseases in humans. A primary bioassay of human tumor stem cells has been developed to support stem cell colony growth in agar [29]. Tumor stem cell colonies show specific growth characteristics and morphology depending on the type of cancer they have been isolated from. This bioassay not only allows for the type of tumor cell to be determined but also for the study of the effects of irradiation and anticancer drugs. One such bioassay uses microcantilevers to detect a prostate-specific antigen (PDS) [30]. This is based on the mechanism that when a specific biomolecule binds to the surface of a microcantilever beam, intermolecular nanomechanics bend the cantilever, which can be detected optically, which allows for its use as a clinically relevant diagnostic tool for prostate cancer.

Many chemicals have been shown or are presumed to be carcinogenic to humans [31]. Others are expected to be noncarcinogenic. Twenty-nine chemicals for which no carcinogenicity has been demonstrated have been tested using rodents in bioassays for carcinogenicity [32]. Nineteen of these chemicals were positive and only one was negative, which demonstrated a low specificity of animal bioassays. On the other hand, these tests are very sensitive: all known carcinogenic substances (for humans) induced cancers in the animals. Bioassays for anticancer activities are also available [33,34]. The MTT (3-(4,5-dimethylthiazol-2-yl)-2,5-diphenyltetrazolium bromide) in vitro cell proliferation is used widely for evaluating anticancer activity of synthetic derivatives and natural products. The MTT test uses oxidoreductase enzymes which reduce the tetrazolium dye MTT. The assay indicates whole cell cytotoxicity, but does not identify the molecular target. This can be done by using the kinase inhibition assay. These enzymes play an important role in physiological reactions and their inhibitors show anticancer activity against a number of human cancer cell lines.

Carcinogenic bioassays have also been developed to identify air pollutants [35]. Epidemiological data indicate that air pollution plays a decisive role in causing lung cancer. Fiberglass membrane filters have been installed in eight American cities to collect particulate pollutants over a period of one month. By this method crude benzol, aromatic, aliphatic, and oxygenated fractions were identified. These substances were either injected or externally applied to the skin of black mice.

However, the treated mice did not develop a higher frequency of tumors than the control animals that had been treated with the suspending fluid. Pollutants from different cities induced different numbers of tumors, but there was no correlation to the lung cancer death rates in these cities.

Pharmacological and physiological studies of drugs, neurotransmitters, and hormones often result in a bunch of similar dose-response curves of sigmoidal shape [36]. Therefore it is desirable to analyze these data simultaneously rather than fitting the EC_{50} curves individually. This technique allows both for rigorous statistical analysis and the pooling of the information from individual experiments by comparing the characteristics of each curve. This method is applicable to bioassays, radioligand assays, and physiological dose-response curves. A further extension is the development of an add-on package for the language R which permits simultaneous calculation of several regression models for nonlinear fitting [37]. In addition to the analysis of dose-response curves the method can be applied to other nonlinear regression models.

Cyanobacterial toxins in drinking water may result in poisoning of humans and animals [38]. A fatal intoxication by microcystin of 50 dialysis patients has been reported in Bahia, Brazil. 140 children have been hospitalized after drinking water from a dam containing *Cylindrospermopsis raciborskii* in Armidale, Australia in which blooms of toxic *Microcystis* has been monitored for several years. Subsequently the water was treated with copper sulfate at 1 part per million (ppm) which killed the bloom. Gastroenteritis has occurred in North and South America, Africa, and Europe, due to drinking water after the appearance of cyanobacteria blooms. Rodents and larger animals have been used in bioassays to determine the potential hazard for humans and animals [39]. In these tests the cyanotoxin is injected intraperitoneal into a rodent. But a simple and effective plant test has also been used to identify the risk of cyanobacterial toxicity [40]. In this test mustard (*Sinapis alba* L.) seedlings are used for quantitative assessment of the toxin with an IC_{50} of 3 $mL\,L^{-1}$. *Daphnia* has also been shown to be a sensitive organism to detect cyanobacterial toxins [41] and has been used in an acute toxicity test to detect saxitoxins.

High concentrations of cyanotoxins can cause death from liver hemorrhage or liver failure and have been discussed to promote liver tumors following chronic exposure to low doses. Some of the cyanotoxins are alkaloid neurotoxins such as anatoxins and saxitoxins, which produce acute effects in mammals, in addition to the fact that they can accumulate in high concentrations in marine and freshwater animals which may serve as food for humans [42]. Cylindrospermopsin is known to cause histopathological damage after long-term exposure at low concentrations [43]. Cyanobacterial lipopolysaccharides can induce allergic and toxic responses in humans resulting in gastrointestinal and respiratory symptoms [43].

Dinoflagellates can also produce paralytic shellfish toxins [44]. Since these algae can form blooms (*red tides*) they are taken up in large quantities by filter feeders such as mussels and oysters [45]. Human consumption of these mollusks has repeatedly resulted in illness and death [46]. Higher temperatures due to global climate change as well as increasing fertilization of estuaries and coastal areas due to terrestrial runoff foster the occurrence and severity of these *red tides* [47]. Bioassays have been developed to determine the hazards of the toxins, such as the *Rhodomonas salina* (Cryptophyte) bioassay to quantify extracellular allelochemicals of the marine dinoflagellate *Alexandrium tamarense*, which are lytic compounds, stable over wide temperature and pH ranges, and resistant to bacterial degradation [48].

References

[1] Munch JC. Bioassay of capsicums and chillies I. J Am Pharm Assoc 1929;18 (12):1236–46.
[2] Munch JC, Crosbie H. Bioassay of aconite and its preparations. 2. The pharmacology and pharmacognosy of various species of aconitum. J Am Pharm Assoc 1929;18 (10):986–92.
[3] Abbott W. A method of computing the effectiveness of an insecticide. J Econ Entomol 1925;18(2):265–7.
[4] Munch JC, Grantham R. Bioassay of aconite and its preparations. 3. The comparative toxicity of tincture and futidextract of aconite to guinea-pigs and rats. J Am Pharm Assoc 1929;18(10):993–5.
[5] Fleming A. On the antibacterial action of cultures of a penicillium, with special reference to their use in the isolation of *B. influenzae*. Br J Exp Pathol 1929;10(3):226.
[6] Dornbusch K, Nord C-E, Olsson B. Antibiotic susceptibility testing of anaerobic bacteria by the standardized disc diffusion method with special reference to *Bacteroides fragilis*. Scand J Infect Dis 2015;59–66. Published online: 02 Jan 2015.
[7] McIntyre N. Sir Alexander Fleming (1881–1955). J Med Biogr 2007;15(4):234.
[8] Owen W, Stuckey D, Healy J, Young L, McCarty P. Bioassay for monitoring biochemical methane potential and anaerobic toxicity. Water Res 1979;13(6):485–92.
[9] Meyer B, Ferrigni N, Putnam J, Jacobsen L, Nichols Dj, McLaughlin JL. Brine shrimp: a convenient general bioassay for active plant constituents. Planta Med 1982;45 (05):31–4.
[10] McLaughlin JL. Bench-top bioassays for the discovery of bioactive compounds in higher plants. Brenesia 1990;34:1–14.
[11] Jensen A, Rystad B, Melsom S. Heavy metal tolerance of marine phytoplankton. I. The tolerance of three algal species to zinc in coastal sea water. J Exp Mar Biol Ecol 1974;15(2):145–57.
[12] Skidmore J. Resistance to zinc sulphate of the zebrafish (*Brachydanio rerio* Hamilton-Buchanan) at different phases of its life history. Ann Appl Biol 1965;56(1):47–53.
[13] Denslow JS, Vitousek PM, Schultz J. Bioassays of nutrient limitation in a tropical rain forest soil. Oecologia 1987;74(3):370–6.
[14] Sprague JB. Measurement of pollutant toxicity to fish I. Bioassay methods for acute toxicity. Water Res 1969;3(11):793–821.
[15] Bliss CI. The statistics of bioassay: with special reference to the vitamins. Elsevier; 2014.
[16] Persoone G, Marsalek B, Blinova I, Törökne A, Zarina D, Manusadzianas L, et al. A practical and user-friendly toxicity classification system with microbiotests for natural waters and wastewaters. Environ Toxicol 2003;18(6):395–402.

[17] Isidori M, Parrella A, Piazza C, Strada R. Toxicity screening of surface waters in southern Italy with Toxkit microbiotests. New microbiotests for routine toxicity screening and biomonitoring. Springer; 2000. p. 289–93.
[18] Vandenbroele M, Heijerick D, Vangheluwe M, Janssen C. Comparison of the conventional algal assay and the Algaltoxkit F™ microbiotest for toxicity evaluation of sediment pore waters. New microbiotests for routine toxicity screening and biomonitoring. Springer; 2000. p. 261–8.
[19] Ulm L, Vrzina J, Schiesl V, Puntaric D, Smit Z. Sensitivity comparison of the conventional acute *Daphnia magna* immobilization test with the Daphtoxkit F™ microbiotest for household products. New microbiotests for routine toxicity screening and biomonitoring. Springer; 2000. p. 247–52.
[20] Nalecz-Jawecki G, Persoone G. Toxicity of selected pharmaceuticals to the Anostracan Crustacean *Thamnocephalus platyurus*-Comparison of sublethal and lethal effect levels with the 1h Rapidtoxkit and the 24h Thamnotoxkit microbiotests. Environ Sci Pollut Res 2006;13(1):22–7.
[21] Törökné AK, László E, Chorus I, Sivonen K, Barbosa FA. Cyanobacterial toxins detected by Thamnotoxkit (a double blind experiment). Environ Toxicol 2000;15 (5):549–53.
[22] Radix P, Severin G, Schramm K-W, Kettrup A. Reproduction disturbances of *Brachionus calyciflorus* (rotifer) for the screening of environmental endocrine disrupters. Chemosphere 2002;47(10):1097–101.
[23] Persoone G, Wadhia K. Comparison between Toxkit microbiotests and standard tests. Ecotoxicological Characterization of Waste. Springer; 2009. p. 213–20.
[24] Blok C, Persoone G, Wever G. (Eds.). A practical and low cost microbiotest to assess the phytotoxic potential of growing media and soil. In: International Symposium on Growing Media 779; 2005.
[25] Czerniawska-Kusza I, Ciesielczuk T, Kusza G, Cichoń A. Comparison of the Phytotoxkit microbiotest and chemical variables for toxicity evaluation of sediments. Environ Toxicol 2006;21(4):367–72.
[26] Ritz C. Toward a unified approach to dose–response modeling in ecotoxicology. Environ Toxicol Chem 2010;29(1):220–9.
[27] Thompson J, Walker R, Faulkner D. Screening and bioassays for biologically-active substances from forty marine sponge species from San Diego, California, USA. Mar Biol 1985;88(1):11–21.
[28] Bergquist P, Bedford J. The incidence of antibacterial activity in marine Demospongiae; systematic and geographic considerations. Mar Biol 1978;46(3):215–21.
[29] Hamburger AW, Salmon SE. Primary bioassay of human tumor stem cells. Science 1977;197(4302):461–3.
[30] Wu G, Datar RH, Hansen KM, Thundat T, Cote RJ, Majumdar A. Bioassay of prostate-specific antigen (PSA) using microcantilevers. Nat Biotechnol 2001;19(9):856–60.
[31] Guengerich FP, Shimada T. Oxidation of toxic and carcinogenic chemicals by human cytochrome P-450 enzymes. Chem Res Toxicol 1991;4(4):391–407.
[32] Ennever FK, Noonan TJ, Rosenkranz HS. The predictivity of animal bioassays and short-term genotoxicity tests for carcinogenicity and non-carcinogenicity to humans. Mutagenesis 1987;2(2):73–8.
[33] Lieberman MM, Patterson GM, Moore RE. In vitro bioassays for anticancer drug screening: effects of cell concentration and other assay parameters on growth inhibitory activity. Cancer Lett 2001;173(1):21–9.

[34] McCauley J, Zivanovic A, Skropeta D. Bioassays for anticancer activities. Methods Mol Biol 2013;1055:191–205.
[35] Hueper W, Kotin P, Tabor E, Payne WW, Falk H, Sawicki E. Carcinogenic bioassays on air pollutants. Arch Pathol 1962;74(2):89–116.
[36] DeLean A, Munson P, Rodbard D. Simultaneous analysis of families of sigmoidal curves: application to bioassay, radioligand assay, and physiological dose-response curves. Am J Physiol 1978;235(2):G97–102.
[37] Ritz C, Streibig JC. Bioassay analysis using R. J Stat Softw 2005;12(5):1–22.
[38] Kuiper-Goodman IF., Fitzgerald J. Human health aspects. 1999. Chapter 4. In: Toxic Cyanobacteria in Water: A guide to their public health consequences, monitoring and management Edited by Ingrid Chorus and Jamie Bartram © 1999 WHO ISBN 0-419-23930-8.
[39] Li R, Carmichael WW, Brittain S, Eaglesham GK, Shaw GR, Liu Y, et al. First report of the cyanotoxins cylindrospermopsin and deoxycylindrospermopsin from *Raphidiopsis curvata* (Cyanobacteria). J Phycol 2001;37(6):1121–6.
[40] Kos P, Gorzo G, Suranyi G, Borbely G. Simple and efficient method for isolation and measurement of cyanobacterial hepatotoxins by plant tests (*Sinapis alba* L.). Anal Biochem 1995;225(1):49–53.
[41] Ferrão-Filho AdS, Soares MCS, de Magalhães VF, Azevedo SM. A rapid bioassay for detecting saxitoxins using a *Daphnia* acute toxicity test. Environ Pollut 2010;158 (6):2084–93.
[42] Codd GA, Bell SG, Kaya K, Ward CJ, Beattie KA, Metcalf JS. Cyanobacterial toxins, exposure routes and human health. Eur J Phycol 1999;34(04):405–15.
[43] Gutierrez-Praena D, Jos A, Pichardo S, Moyano R, Blanco A, Monterde JG, et al. Time-dependent histopathological changes induced in Tilapia (*Oreochromis niloticus*) after acute exposure to pure cylindrospermopsin by oral and intraperitoneal route. Ecotoxicol Environ Saf 2012;76:102–13.
[44] Teegarden GJ, Cemballa AG. Grazing of toxic dinoflagellates, *Alexandrium* spp., by adult copepods of coastal Maine: implications for the fate of paralytic shellfish toxins in marine food webs. J Exp Mar Biol Ecol 1996;196:145–76.
[45] Nagai K, Matsuyama Y, Uchida T, Yamaguchi M, Ishimura M, Nishimura A, et al. Toxicity and LD 50 levels of the red tide dinoflagellate *Heterocapsa circularisquama* on juvenile pearl oysters. Aquaculture 1996;144(1):149–54.
[46] Burkholder JM. Implications of harmful microalgae and heterotrophic dinoflagellates in management of sustainable marine fisheries. Ecol Appl 1998;8(sp1):S37–62.
[47] Hallegraeff GM. Ocean climate change, phytoplankton community responses, and harmful algal blooms: a formidable predictive challenge. J Phycol 2010;46:220–35.
[48] Ma H, Krock B, Tillmann U, Cembella A. Preliminary characterization of extracellular allelochemicals of the toxic marine dinoflagellate *Alexandrium tamarense* using a *Rhodomonas salina* bioassay. Mar Drugs 2009;7(4):497–522.

Further Reading

Häder D-P, Erzinger GS. Advanced methods in image analysis as potent tools in online biomonitoring of water resources. Recent Pat Top Imaging 2015;5(2):112–18.

Further Reading

Regulations, political and societal aspects, toxicity limits

4

Gilmar S. Erzinger[1] *and Donat-P. Häder*[2]
[1]University of Joinville Region—UNIVILLE, Joinville, SC, Brazil, [2]Friedrich-Alexander University, Erlangen-Nürnberg, Germany

4.1 Introduction

According to Rattner and Barnett [1], the field of ecotoxicology arose in the late nineteenth and early twentieth centuries, with the challenge of predicting the estimation of the effects of exposure of animals and their habitats to chemicals related to anthropogenic activities.

Toxicology, as a modern science from the 19th century onwards, aims to understand toxicities, their existence, occurrence, behavior, mechanism, and action, focusing on prevention through its application, for the recognition, identification, and quantification of risks [2].

Economic globalization has resulted in a marked increase in waste discharges across the planet. This phenomenon resulted in an exponential economic growth and, in parallel, in the increase of discharge of toxic products that resulted in serious damages to the biosphere and the ecosystems that compose it. Traditional single-species toxicity tests provide crucial evidence on death, reproductive processes, and growth rates, but fail to assess impacts on the whole ecosystem, such as community structure and function [3].

Ecotoxicology is one of the areas covered by toxicology; the term was coined in 1969 and focuses on the study of the contaminants in the biosphere and their effects on the components—including humankind—through the use of bioassays from different living beings. Thus, ecotoxicology is the science responsible for evaluating acute, chronic, and sublethal effects on organisms in the environment [1].

Covello and Mumpower [4], in a historical approach to risk analysis and management, especially in the United States, pointed out nine important factors, which were distributed between four groups, to understand the transformations that led to the contemporary way of thinking about and facing risks in the central countries of the world economy.

The first group of factors involves those related to the change in the very nature of the risks. Changes in the profile of the main causes of death, which progressively ceased to be attributed to infectious diseases in favor of chronic degenerative diseases, increase in the average life expectancy, and the growth of new risks (radioactive, chemical, and biological, all generated by the development of science and

Bioassays. DOI: http://dx.doi.org/10.1016/B978-0-12-811861-0.00004-8

technology) that became part of the daily lives of millions of people, in the form of accidents or other.

The second group is related to the scientific and technological development itself. On the one hand, the development of laboratory tests, epidemiological methods, environmental modeling, computer simulations, and engineering risk assessment has enabled advances in the ability of scientists to identify and measure risks. In parallel, there was a growth in the number of scientists and analysts who started to focus on health, safety, and environmental risks.

The third group concerns the regulatory and decision-making processes. Scientific and technological development has contributed to the growth in the number of formal quantitative analyses produced and used for decision-making processes on risk management, associated with the broadening of the role of the federal government in risk assessment and risk management. This growth was achieved through: (1) the development of legislation in the fields of health, safety, and the environment, (2) the growth of public agencies responsible for managing these risks, and (3) the increased number of cases related to the subject that reached the judicial realm.

The fourth group involves the responses of the organized society. The broadening of interest and concern for risks by the general public, demanding more and more protection, contributed substantially to the growth of social movements and interest groups that sought to participate more and more in social risk management. This process has made the activities of analysis and management of risks to health, safety, and the environment very politicized, with intense participation of those segments representing industry, workers, environmentalists, and scientific organizations among others [4].

The issue of environmental toxicity has been causing concern in the scientific community. Recent studies in this area have enabled the achievement of fundamental results for the direction of other works, as well as the development of processes for the decontamination of aggressive products in the environment. Different actors from research, economics, and politics discussed the relationship between sustainable development and economic growth. This implies a scenario that requires the revision and reorientation of existing laws that deal with environmental issues, both nationally and internationally [5].

4.2 Regulations

Most countries currently have legislation and regulations on accepted values of toxicity for the marketing of agricultural and industrial chemicals, biocides, cosmetics, food additives, medicines and other substances and use bioassays for the protection of human health and the environment. Some regulations are explained below and an overview and multisectoral overview of the regulatory data requirements for acute systemic toxicity is presented.

The Organization for Economic Co-operation and Development (OECD) is an international organization of 35 countries (largely comprising European countries, USA, Canada, Chile, South Korea, Japan, Australia, and New Zealand) which accept the principles of representative democracy and the free market economy, which seeks to provide a platform for comparing economic policies, solving common problems, and coordinating domestic and international policies. These policies include the use of similar environmental standards [6].

The OECD guidelines are a unique tool for assessing the potential effects of chemicals on human health and the environment. Internationally accepted as standard methods for safety testing, the guidelines are used by professionals in industry, academia, and government who are involved in the testing and assessment of chemicals (industrial chemicals, pesticides, personal care products, etc.). These guidelines are regularly updated with the assistance of thousands of national experts from OECD member countries. OECD test guidelines are covered by the Mutual Acceptance of Data, implying that data generated in the testing of chemicals in an OECD member country, or a partner country having adhered to the decision, in accordance with OECD Test Guidelines and Principles of Good Laboratory Practice (GLP), be accepted in other OECD countries and partner countries having adhered to the decision, for the purposes of assessment and other uses relating to the protection of human health and the environment [7].

In November 2012, the Joint Meeting of the Chemicals Committee and Working Party on Chemicals, Pesticides, and Biotechnology decided on a transition period of 18 months, between the Council Decision and the effective deletion, for Test Guidelines that have been updated or deleted [7].

4.2.1 Europe

Europe adopts a model determined by the European Commission. The European Commission is the executive of the European Union and promotes its general interest. Environmental regulations are made by the Scientific Committee on Health, Environmental, and Emerging Risks (SCHEER), which provides advice on issues relating to health, environmental, and emerging risks. In particular, the Committee shall formulate opinions on issues relating to emerging or newly identified health and environmental risks and on broad, complex, or multidisciplinary issues requiring a comprehensive risk assessment for consumer safety or public health and related matters not covered by other members. SCHEER analyzes opinions on the risks related to pollutants in environmental media, and other biological and physical factors or changes in physical conditions that may have a negative impact on health and the environment, for example in relation to air quality, water, waste, and soil, as well as in the environmental assessment of the life cycle [8].

Table 4.1 shows some examples of legislation and regulation of Acute Toxicity Testing in Sectors in different sectors of the economy in Europe.

Table 4.1 Regulatory drivers for acute toxicity testing across agrochemicals, biocides, chemicals, cosmetics, and medicinal product sectors in Europe

Agrochemical	Biocides	Cosmetics	Chemicals
Regulation (EC) No. 1107/2009 [10]; setting out scientific criteria for the determination of endocrine disrupting properties and amending Annex II to Regulation (EC) 1107/2009	Regulation (EC) No. 528/2012 of the European Parliament [11] and of the Council of May 22, 2012 concerning the making available on the market and use of biocidal products. This regulation is in the process of being replaced by the Biocidal Products Regulation [12]	Directive 2008/112/EC of the European Parliament and of the Council of December 16, 2008 [13] amending Council Directives 76/768/EEC, 88/378/EEC, 1999/13/EC and Directives 2000/53/EC, 2002/96/EC and 2004/42/EC of the European Parliament and of the Council in order to adapt them to Regulation (EC) No. 1272/2008 on classification, labeling and packaging of substances and mixtures	Regulation (EC) No. 1907/2006 of the European Parliament and of the Council of December 18, 2006 concerning the Registration, Evaluation, Authorisation and Restriction of Chemicals (REACH), establishing a European Chemicals Agency, amending Directive 1999/45/EC and repealing Council Regulation (EEC) No. 793/93 and Commission Regulation (EC) No 1488/94 as well as Council Directive 76/769/EEC and Commission Directives 91/155/EEC, 93/67/EEC, 93/105/EC and 2000/21/EC

Source: Adapted from Seidle T, Robinson S, Holmes T, Creton S, Prieto P, Scheel J, et al. Cross-sector review of drivers and available 3Rs approaches for acute systemic toxicity testing. Toxicol Sci 2010;116:382–396 [9].

4.2.2 United States of America

In the United States laws are created by Congress that governs the United States, but Congress also authorizes the United States Environmental Protection Agency (EPA) and the Food and Drug Administration (FDA) to help put those laws into effect by creating and applying regulations for the protection of health and the environment. Below you will find a basic description of how laws and regulations are developed, what they are, and where to find them, with an emphasis on health and environmental laws and regulations.

The EPA is an independent agency of the United States government that has jurisdiction over existing and developing chemicals (such as pesticides) that regulate and control their use in the environment. It regulates the manufacture, processing, distribution, and use and establishes tolerance levels of their presence in food and feed. The EPA has broad punitive powers and also filters out all chemicals before they are marketed to assess their effects on health and the environment. The director of the EPA is appointed by the President of the United States and approved by the Senate, along with an assistant administrator and nine deputy administrators, an inspector general, and a general counselor. The inspector general is responsible for investigating environmental crimes, and the general council provides legal advice. The structure of the EPA is distributed in offices that are: (1) air and radiation; (2) water; (3) pesticides and toxic substances; and (4) solid waste and emergency response. There is also a Research and Development office that works in coordination with each of the four offices of the program. The research is conducted through the EPA's main office and its regional field laboratories. Ten regional EPA offices and field laboratories work directly with state and local governments to coordinate pollution control efforts. The EPA uses a portion of its federal funding to provide grants and technical assistance to states and local government units seeking to prevent pollution [14].

Table 4.2 shows some examples of legislation and regulation of Acute Toxicity Testing in Sectors in different sectors of the economy in the United States.

4.2.3 Brazil

In Brazil the management and regulation of toxicity limits are determined by the National Council for the Environment [18] created in 1982 by Law No. 6.938/81—which establishes the National Environmental Policy. CONAMA (National Council for the Environment) is responsible for the National Environmental System—SISNAMA (National System of the Environment). In other words, [19] exists to advise, study, and propose to the government the guidelines that should be taken by government policies for the exploration and preservation of the environment and natural resources. In addition, it is also up to the body, within its competence, to create norms and standards compatible with the ecologically balanced environment and essential to a healthy quality of life [20].

As provided in article 4 of Decree 99.274/90, CONAMA [21] is made up of Plenary, Special Chamber of Recourse, Committee for Integration of Environmental

Table 4.2 Regulatory drivers for acute toxicity testing across agrochemicals, biocides, chemicals, cosmetics and medicinal products sectors in the United States

Agrochemical	Biocides	Cosmetics	Chemicals
USC. Federal Insecticide, Fungicide, And Rodenticide Act – 2014. As amended through P.L. 113–79, enacted February 7, 2014	USC. Federal Insecticide, Fungicide, and Rodenticide Act – 2012. 1. It uses a biocide or generates a biocide that is a pesticide, as defined in section 2 of the Federal Insecticide, Fungicide, and Rodenticide Act (7 U.S.C. 136), unless the biocide is registered under that Act or the Secretary, in consultation with the Administrator, has approved the use of the biocide in such treatment technology; or 2. It uses or generates a biocide the discharge of which causes or contributes to a violation of a water quality standard under section 303 of the Federal Water Pollution Control Act (33 U.S.C. 1313).	Federal Food, Drug, and Cosmetic Act [16]. Cosmetics are not subject to specific testing requirements or premarket approval in the United States. However, the Federal Food, Drug and Cosmetic Act broadly prohibits the marketing of adulterated or misbranded cosmetics, including any product (other than a hair dye) that "bears or contains any poisonous or deleterious substance which may render it injurious to users under the conditions of use prescribed in the labeling thereof, or under conditions of use as are customary and usual." Companies are encouraged to register their establishments and file Cosmetic Product Ingredient Statements with FDA's Voluntary Cosmetic Registration Program	On June 22, 2016, the Frank R. Lautenberg Chemical Safety for the 21st Century Act which amends the Toxic Substances Control Act (TSCA), the Nation's primary chemicals management law was signed [17]

Source: Adapted from Seidle T, Robinson S, Holmes T, Creton S, Prieto P, Scheel J, et al. Cross-sector review of drivers and available 3Rs approaches for acute systemic toxicity testing. Toxicol Sci 2010;116:382–396; Federal Insecticide, Fungicide, and Rodenticide Act, accessed from: <https://www.agriculture.senate.gov/imo/media/doc/FIFRA.pdf>; 2012 [15].

Policies, Technical Chambers, Working Groups, and Advisory Groups. The Technical Chambers are bodies responsible for developing, examining, and reporting to the Plenary the matters within its competence, so that it may deliberate. Under the Internal Regulations [10] there should be 11 Technical Chambers, composed of 10 Councilors, who elect a President, a Vice-President, and a Rapporteur. The Working Groups are created for a fixed period to analyze, study, and present proposals on matters within their competence [20].

Chaired by the Minister of the Environment, CONAMA holds regular meetings every three months in Brasília-DF and can hold extraordinary meetings outside the Federal District, if so convened by the president of the Council or by request of 2/3 of the members. These meetings are open to the public and there is at least one Ordinary Meeting every month. The sessions must be attended by at least an absolute majority of its members, and decisions must be reached by a simple majority of the members entitled to vote, and the chairman of the session. In the case of a tie, the president shall have the voting right [20].

The CONAMA plenary is a representative body of federal, state, and municipal bodies, the business sector, and civil society. In addition to the Minister of the Environment, who presides over it, the plenary consists of: the Executive Secretary of the Ministry of Environment, who will be its Executive Secretary; 1 member representative of IBAMA (Brazilian Institute of Environment and Renewable Natural Resources); 1 member representative of the National Water Agency (ANA); 1 member representative of each of the Ministries, Secretariats of the Presidency of the Republic, and Military Commands of the Ministry of Defense, indicated by their respective holders; 1 member representative of each of the State Governments and the Federal District, appointed by the respective governors; 8 member representatives of the Municipal Governments that have structured environmental members and an Environment Council with a deliberative character; 22 representatives of workers' organizations and civil society; eight representatives of business entities; and an honorary member nominated by the Plenary [20].

Table 4.3 shows some examples of legislation and regulation of Acute Toxicity Testing in Sectors in different sectors of the economy in Brazil.

4.2.4 China

The Ministry of Environmental Protection of the People's Republic of China (MEP), as an institution to formulate environmental protection guidelines, policy, and law, is an important government authority to deal with global affairs of environmental protection. The Ministry of Environmental Protection of the People's Republic of China was established in March 2008 to replace the State Environmental Protection Administration of China (SEPA) in accordance with the Government Reshuffle Plan of the State Council approved by the First Session of the 11th National People's Congress (NPC). It is in charge of environmental protection work under direct administration of the State Council [22–24].

Table 4.3 Regulatory drivers for acute toxicity testing across agrochemicals, biocides, chemicals, cosmetics and medicinal products sectors in Brazil

Agrochemical	Biocides	Cosmetics	Chemicals
Resolution in CONAMA 420, of December 28, 2009, published in the DOU n° 249, of 12/30/2009, pp. 81–4. It provides criteria and guiding values of soil quality for the presence of chemical substances and establishes guidelines for the environmental management of areas contaminated by these substances as a result of anthropic activities	Resolution in CONAMA 420, of December 28, 2009, published in the DOU n° 249, of 12/30/2009, pp. 81–4. It provides criteria and guiding values of soil quality for the presence of chemical substances and establishes guidelines for the environmental management of areas contaminated by these substances as a result of anthropic activities	The Ministry of Health (ANVISA) still requires tests on animals for almost all new products and ingredients that come into contact with, are ingested, or used by humans. Insecticides, detergents, waxes, disinfectants, pesticides, sanitary napkins, medicines (including phytotherapics), surgical equipment (bandages, etc.), laboratory equipment, prostheses, implants, food additives (preservatives, acidulants), chemical substances for industrial use, health research. Brazilian legislation, following the example of many countries, does not provide for the mandatory testing of animals for cosmetics	Resolution in CONAMA 420, of December 28, 2009, Published in the DOU n° 249, of 12/30/2009, pp. 81–4. It provides criteria and guiding values of soil quality for the presence of chemical substances and establishes guidelines for the environmental management of areas contaminated by these substances as a result of anthropic activities

Source: Adapted from Seidle T, Robinson S, Holmes T, Creton S, Prieto P, Scheel J, et al. Cross-sector review of drivers and available 3Rs approaches for acute systemic toxicity testing. Toxicol Sci 2010;116:382–396.

Table 4.4 Regulatory drivers for acute toxicity testing across agrochemicals, biocides, chemicals, cosmetics and medicinal products sectors in China

Agrochemical	Biocides	Cosmetics	Chemicals
Regulation on pesticide administration (SC, 2001); regulation on pesticide administration (revised in 2012); enforcement measures of the regulation on pesticide administration (MOA Decree 9, 2007); data requirement on pesticide registration (MOA Decree 010, 2007); measures for the administration of pesticide labels and manuals; measures on disinfectant administration issued by the Ministry of Health (MOH) in 2002; guidance on application of administrative approval; license of disinfectants and disinfecting apparatuses [2006]; administrative licensing procedure for health-related products [2006]	Regulated as industrial chemicals	Regulations concerning the hygiene supervision over cosmetics (MPH, 1989); regulations concerning the hygiene supervision over cosmetics [1990]; detailed rules for the implementation of the regulation on the hygiene supervision over cosmetics [2005]; hygienic standard for cosmetics [2007]; the measures for the administration of hygiene license for cosmetics (revised in 2010); guideline for risk evaluation of substances with the possibility of safety risk in cosmetics [2010]; standard Chinese names of international cosmetics ingredients inventory [2010]; cosmetics technical requirement standard [2011]; guidelines for the registration and evaluation of new cosmetic ingredients [2011]	Measures for the environmental regulation on pesticide administration (SC, 2001); regulation on pesticide administration (revised in 2012); enforcement measures of the regulation on pesticide administration (MOA Decree 9, 2007); data requirement on pesticide registration (MOA Decree 010, 2007); measures for the administration of pesticide labels and manuals; measures on disinfectant administration issued by the Ministry of Health (MOH) in 2002; guidance on application of administrative approval license of disinfectants and disinfecting apparatuses [2006]; administrative licensing procedure for health-related products [2006]

Source: Adapted from Seidle T, Robinson S, Holmes T, Creton S, Prieto P, Scheel J, et al. Cross-sector review of drivers and available 3Rs approaches for acute systemic toxicity testing. Toxicol Sci 2010;116:382–396.

Table 4.5 Regulatory drivers for acute toxicity testing across agrochemicals, biocides, chemicals, cosmetics, and medicinal products sectors in Japan

Agrochemical	Biocides	Cosmetics	Chemicals
Agricultural chemicals regulation law [31]; appendix to data introduction of the positive list system for agricultural chemical residues in foods. Department of Food Safety, Ministry of Health, Labour and Welfare (June 2006)	Regulated as industrial chemicals	Pharmaceutical affairs law; standards for cosmetics [29]. Cosmetics are not subject to specific testing requirements. However, they "shall not contain anything that may cause infection or that otherwise makes the use of the cosmetics a potential health hazard"	Act on the evaluation of chemical substances and regulation of their manufacture, etc. Two components: a premanufacturing evaluation of new chemical substances and monitoring/ regulations based on the properties of chemical substances [27]

Source: Adapted from Seidle T, Robinson S, Holmes T, Creton S, Prieto P, Scheel J, et al. Cross-sector review of drivers and available 3Rs approaches for acute systemic toxicity testing. Toxicol Sci 2010;116:382–396.

Table 4.4 shows some examples of legislation and regulation of Acute Toxicity Testing in Sectors in different sectors of the economy in China.

4.2.5 Japan

The risk assessment and regulation of the levels of product toxicity in Japan are carried out jointly by the Ministry of Health, Labor, and Welfare, the Minister for Economic Affairs, Trade, and Industry, and the Minister for the Environment, to determine the need for The "Designation as Class II Specified Chemicals" and the "Termination of Designation as Priority Assessment Chemicals" under the Japanese Chemicals Control Act [25].

The Chemical Substances Control Act of Japan defines the risk assessment for Priority Assessment Chemicals as the assessment to determine whether there is a risk of damaging human health or population and/or growth of flora and fauna in the human environment due to environmental pollution attributable to chemical substances [26–31]. Table 4.5 shows some examples of legislation and regulation of Acute Toxicity Testing in Sectors in different sectors of the economy in Japan.

4.3 Political and societal aspects

Environmental toxicology is the study of environmental causes in populations and how these risks vary in relation to the intensity and duration of exposure and other factors, such as genetic susceptibility. It is undoubtedly the basic science on which government regulatory agencies depend on the definition of standards to protect the population and ecosystems against environmental risks.

Bertoletti [32] described that the assumption that water quality for humans corresponds to that for aquatic organisms can be easily challenged based on one of the foundations of ecology, the concept of an ecological niche. This concept is expressed as "the physical space occupied by an organism, including the functional role of that organism in the community and its position in environmental gradients that give it conditions of existence." Thus, because man does not belong to an ecological niche characteristic of the aquatic environment, it can be said that it is impossible to admit that human perception and requirements are the appropriate indicators for assessing the quality of water resources as a whole.

These arguments allow us to recognize that aquatic environments are, for humankind, primarily only natural resources to be used for the purpose of permitting watering, hygiene, feeding, agro-pastoral irrigation, industrialization, and sewage disposal. Given this, it is reasonable to conclude that human perception is far from being able to understand that the natural balance of aquatic ecosystems goes beyond the quality of water required for everyday use. Therefore, it is necessary to recognize that humankind is limited, in the noblest terms, to the establishment of conditions of water quality (expressed in the standards of potability) necessary for the exclusive purpose of its watering which is known not to occur within the water resources [32].

In this sense, it can be deduced that human beings perceive the environmental imbalances where they live and, depending on the compartment where the imbalances occur, they present and express their concerns through demands of recovery of the negatively altered environments. In this way, the regulation and diffusion of public policy in terms of the limits of toxicities of different products is necessary [33].

According to Hartung [34], consumers continually tend to raise their expectations about product safety, but nearly all experimental protocols for chemicals have remained almost unchanged for more than 40 years. The European Union introduced the regulation known as Registration, Evaluation, Authorization and Restriction of Chemicals (REACH) by legislation in 2007. New products have been systematically evaluated in the European Union and the United States that includes

97% of the main chemicals in use and more than 99% of the chemicals produced by volume, but were not necessarily adequately addressed. Hartung [34] estimates that data for 86% of chemicals are missing and the REACH process aims to correct this. The regulation affects 27,000 companies, which are required to provide information on the toxic properties and uses of 30,000 chemicals, following a pre-registration phase in 2008. Expecting the REACH form to be an important tool to emphasize toxicology [35].

4.4 Toxicity limits

In this section, the description of the approach to toxicity bioassays of different wastes is essentially complete—at least as far as the measurements themselves are concerned. However, there are certain aspects of toxicity data and associated test procedures which, because they have a significant influence on the establishment of residue toxicity limits, merit inclusion. This illustrates the complexity of defining regulatory milestones between sectors and countries, and the challenges this poses for those seeking to reduce the number of animals used in acute or chronic toxicity studies, while generating acceptable data globally.

Toxicity tests determine the concentration at which 50% of the test organisms respond (i.e., the median response) within a specified period. As some of the more sensitive individuals will respond with different sensitivity over the same period of time, this Median Effect Concentration (EC_{50}) cannot be considered safe. The same applies even in the case of the Median Effective Concentration threshold—which may require longer exposures [36].

According to Seidle et al. [9], precise LC_{50} values are not required when the tests are conducted solely to meet the classification and labeling requirements. According to these authors, since the test for the upper limit of a hazard category (e.g., dose limit) is sufficient to establish a regulatory classification, therefore there is no scientific need to establish a dose-response curve for mortality.

In Table 4.6, we provide suggestions on combinations of test species that are well suited to routine compliance monitoring grouped according to the salinity of the receiving system and status of the waste. The species for use with established waste corresponds well (but not entirely) with current practice. With regard to new or modified wastes, we suggest that selections from the broader lists of species be used for initial screening to help identify the most appropriate species for subsequent routine monitoring.

One way to assess the degree of toxicity of a given product is demonstrated in Table 4.7. This work is the result of a group of EPA scientists who developed a Data Assessment Record (DER) system after reviewing an individual toxicity or ecological effect study of a pesticide. The DER summarizes the toxicity for certain groups of species that is expected when exposed to the pesticide. The conclusions of all DERs of individual ecotoxicity for pesticides are integrated and summarized in a stress-response profile, the final product of characterization of ecological

Table 4.6 **Suggested species for monitoring compliance with waste toxicity limits**

Group	Species	Response	Exposure	Remarks
Bacteria				
Light-emitting bacterium	*Vibrio fischeri* (Microtox)	Reduced light emission (inhibition of normal metabolism)	5–30 min	Commercially produced test package; acute test; bacteria reactivated from frozen culture; rapid results; useful in conjunction with other species or for multiple comparative tests
Plants/algae				
Marine alga	*Skeletonema costatum*	Reduced growth (cell division)	72 h	Standard indicator of phytotoxicity
Freshwater alga	*Pseudokirchneriella subcapitata*	Reduced growth (cell division)	72 h	Standard indicator of phytotoxicity
Freshwater alga	*Euglena gracilis*	Physiological parameters (motility, *r*-value, velocity, alignment)	10 min	System of image analysis in real time used in the equipment Ecotox; algae preserved and kept in air-conditioned greenhouse; rapid results; useful in conjunction with other species or for multiple comparative tests for both acute and chronic toxicity determination
Freshwater plant	*Lemna minor* (common duckweed)	Number of fronds	96 h	Small, very common, floating; represents vascular plants; useful for colored wastes
Crustaceans				
Freshwater cladoceran	*Daphnia magna* (water flea)	Immobilization	48 h	Small crustacean, very sensitive, international standard for acute toxicity testing; cultivated in the laboratory

(*Continued*)

Table 4.6 (Continued)

Group	Species	Response	Exposure	Remarks
Freshwater cladoceran	*Ceriodaphnia dubia*	Mortality and reproduction	7 days	Small, very sensitive crustacean, international standard for chronic toxicity testing; cultivated in the laboratory
Estuarine shrimp	*Crangon crangon* (Common shrimp)	Mortality	96 h	Widely used species, exposed in 50% seawater; represents commercially important crustaceans
Estuarine mysid shrimps	*Mysidopsis juniae*	Mortality	96 h	Small, very sensitive shrimp, exposed in 100% seawater, international standard for acute toxicity testing in seawater; cultivated in the laboratory
Marine copepod	*Tigriopus brevicornis*	Mortality	48 h	Very small, sensitive, used as test for acute toxicity of 12 metal chlorides in brackish water
Marine copepod	*Tisbe battagliai*	Mortality	48 h	Short-term bioassays based on lethal and reproductive responses; internationally recognized test species for assessing water and sediment quality
Fish				
Juvenile freshwater salmonid	*Oncorhynchus mykiss*	Mortality	96 h–36 days	Internationally used test species for acute or chronic tests; represents valuable sport and commercial fish
Juvenile marine flatfish	*Scophthalmus maximus*	Mortality	96 h	Represents common commercial fish species; juveniles available throughout the year; fish are observed for mortality and sub-lethal symptoms at specified periods of time
Tropical fish of the family of cyprinids	*Danio rerio* (zebrafish)	Mortality, reproduction, and genotoxicity		Small fish, cultivated in the laboratory, used for acute, chronic, and genotoxicity tests; model organism widely used in genetic studies and toxicity in vertebrates

Source: Adapted from Boelens R. Guidelines for the use of laboratory tests with aquatic organisms in the control of liquid waste discharges. Dublin: IIRS; 1980 [36].

Table 4.7 Ecotoxicity categories for terrestrial and aquatic organisms to classify pesticides based on toxicity data [9]

Toxicity category	Avian: acute oral concentration (mg kg^{-1} bw)	Avian: dietary concentration (mg kg^{-1} diet)	Aquatic organisms: acute concentration (mg L^{-1})	Wild mammals: acute oral concentration (mg kg^{-1} bw)	Nontarget insects: acute concentration (μg bee^{-1})
Very highly toxic	<10	<50	<0.1	<10	
Highly toxic	10–50	50–500	0.1–1	10–50	<2
Moderately toxic	51–500	501–1000	>1–10	51–500	2–11
Slightly toxic	501–2000	1001–5000	>10–100	501–2000	
Practically nontoxic	>2000	>5000	>100	>2000	>11

effects. The profile presents the set of effects for various animals and plants and an interpretation of available incident information and monitoring data. The agency compares the stress-response profile with potential exposure levels to determine the risk of exposure-related effects [14].

References

[1] Rattner BA. History of wildlife toxicology. Ecotoxicology 2009;18(7):773–83.

[2] Fukushima AR, de Azevedo FA. História da Toxicologia. Parte I—breve panorama brasileiro.. Rev Intertox Risco Ambientale Sociedade 2015;1(1):2–32.

[3] Cairns J. Ecotoxicology risk assessment for a changing world. Sci Soc 2008;6(2):113–22.

[4] Covello VT, Mumpower J. Risk analysis and risk management: an historical perspective. Risk Anal 1985;5(2):103–20.

[5] Köck-Schulmeyer M, Ginebreda A, Postigo C, López-Serna R, Pérez S, Brix R, et al. Wastewater reuse in Mediterranean semi-arid areas: the impact of discharges of tertiary treated sewage on the load of polar micro pollutants in the Llobregat river (NE Spain). Chemosphere 2011;82(5):670–8.

[6] OECD Members and Partners, accessed from <http://www.oecd.org/about/membersandpartners/>; 2017.

[7] OECD Guidelines for the Testing of Chemicals, accessed from: <http://www.oecd.org/chemicalsafety/testing/oecdguidelinesforthetestingofchemicals.html>; 2017.

[8] European Communities. Health and Food Safety, accessed from: <http://ec.europa.eu/health/scientific_committees/scheer_en>; 2017.

[9] Seidle T, Robinson S, Holmes T, Creton S, Prieto P, Scheel J, et al. Cross-sector review of drivers and available 3Rs approaches for acute systemic toxicity testing. Toxicol Sci 2010;116:382–96.

[10] Setting out Scientific Criteria for the Determination of Endocrine Disrupting Properties and Amending Annex II to Regulation (EC) 1107/200, accessed from: <http://ec.europa.eu/health/sites/health/files/endocrine_disruptors/docs/2016_pppcriteria_en>; 2009.

[11] The EU Biocides Regulation 528/2012 (EU BPR), accessed from: <http://www.hse.gov.uk/biocides/eu-bpr/>; 2012.

[12] Proposal for a Regulation of the European Parliament and of the Council Concerning the Placing on the Market and Use of Biocidal Products, accessed from: <http://eur-lex.europa.eu/legal-content/EN/TXT/PDF/?uri=CELEX:52012PC0542&from=EN>; 2009.

[13] Directive 2008/112/EC of the European Parliament and of the Council Amending Council Directive 76/768/EEC on the Approximation of the Laws of the Member States Relating to Cosmetic Products, accessed from: <http://ec.europa.eu/transparency/regdoc/?fuseaction=list&coteId=1&year=2007&number=611&langua>; 2008.

[14] EPA-United States Environmental Protection Agency. About EPA. 2017.

[15] Federal Insecticide, Fungicide, and Rodenticide Act, accessed from: <https://www.agriculture.senate.gov/imo/media/doc/FIFRA.pdf>; 2012.

[16] Federal Insecticide, Fungicide, and Rodenticide Act, accessed from: <https://legcounsel.house.gov/Comps/Federal%20Insecticide,%20Fungicide,%20And%20Rodenticide%20Act.pdf>; 2014.

[17] Frank R. Lautenberg chemical safety for the 21st century act. Public Law. 2016;114–82.

[18] LEI N° 6.938, accessed from: <http://www.oas.org/dsd/fida/laws/legislation/brazil/brazil_6938.pdf>; 1982.

[19] Portaria No. 452/2011. Regimento Interno do Conselho Nacional do Meio Ambiente, accessed from: <http://www.mma.gov.br/port/conama/legiabre.cfm?codlegi=656>; 2011.

[20] Resoluções, accessed from: <http://www.mma.gov.br/port/conama/legiano.cfm?codlegitipo=3>; 2017.

[21] CONAMA-Portaria 99.274/90. Estações Ecológicas e Áreas de Proteção Ambiental e sobre a Política Nacional do Meio Ambiente, 1990.

[22] Requirements of the Pesticide Registration Document, accessed from: <http://www.chinapesticide.gov.cn/ywb/index.jhtml>; 2014.

[23] Guideline for the Hazard Evaluation of New Chemical Substances: HJ/T 154−2004, accessed from: <http://www.mep.gov.cn/image20010518/4342.pdf>; 2004.

[24] Regulations on Pesticide Administration, accessed from: <http://www.gov.cn/english/laws/2005-08/24/content_25760.htm>; 2001.

[25] About Us, accessed from: <https://www.env.go.jp/en/aboutus/pamph/pdf.html>; 2017.

[26] Appendix to Data Requirements for Supporting Registration of Pesticides, accessed from <http://www.acis.famic.go.jp/eng/shinsei/8147main.pdf>; 2000.

[27] Act on the Evaluation of Chemical Substances and Regulation of Their Manufacture, etc, accessed from: <http://www.meti.go.jp/english/policy/mono_info_service/kagaku/chemical_substances/chemical_substances03_1.html>; 2005.

[28] Standards for Cosmetics, accessed from: <http://www.mhlw.go.jp/file/06-Seisakujouhou-11120000-Iyakushokuhinkyoku/0000032704.pdf>; 2000.

[29] The Guideline for Assessment of the Effect of Food on Human Health Regarding Food Additive, accessed from: <http://www.fsc.go.jp/english/standardsforriskassessment/guideline_assessment_foodadditives_e2.pdf>; 2010.

[30] Basic Concepts of the Risk Assessment of Priority Assessment Chemical Substances under the Japanese Chemical Substances Control Act (Draft), accessed from: <https://www.env.go.jp/en/chemi/chemicals/chemical_substances_control_act.pdf>; 2002.

[31] Agricultural Chemicals Regulation Law, accessed from: <https://www.env.go.jp/en/chemi/pops/Appendix/05-Laws/agri-chem-laws.pdf>; 1948.

[32] Bertoletti E. A presunção ambiental e a ecotoxicologia aquática. Revista das Águas 2012;1−7.

[33] Zagatto PA. Ecotoxicologia aquática: princípios e aplicações. São Carlos: Rima; 2008.

[34] Hartung T. Toxicology for the twenty-first century. Nature 2009;460(7252):208−12.

[35] Regulation (EC) No. 1907/2006 of the European Parliament and of the Council of 18 December 2006 concerning the Registration, Evaluation, Authorisation and Restriction of Chemicals (REACH), establishing a European Chemicals Agency, amending Directive 1999/45/EC and repealing Council Regulation (EEC) No. 793/93 and Commission Regulation (EC) No. 1488/94 as well as Council Directive 76/769/EEC and Commission Directives 91/155/EEC, 93/67/EEC, 93/105/EC and 2000/21/EC, 396, accessed from < http://www.dowcorning.com/content/about/OR_coverage_terms_7_Feb_2011.pdf > ; 2006.

[36] Boelens R. Guidelines for the use of laboratory tests with aquatic organisms in the control of liquid waste discharges. Dublin: IIRS; 1980.

Image analysis for bioassays – the basics

Donat-P. Häder
Friedrich-Alexander University, Erlangen-Nürnberg, Germany

5.1 Introduction

Many bioassays are based on computerized image analysis. Modern hardware and software developments allow automated, real-time analysis of online or recorded images or video sequences [1]. Applications include medicine [2], microbiology [3], plant growth and development [4], animal behavior [5], microscopy [6], oceanography [7], ecology [8], and earth sciences [9]. The hardware depends on the size of the objects, which may range from subcellular to galactic dimensions. The software is independent of the object's size, which can range from viral particles to celestial objects. Image analysis can be used to quantify dimensions and shapes of embedded objects in images such as cells [10], grains [11], or organisms [12]. One important task is to identify cellular details e.g., in the automated recognition of cancer cells [13]. Tracking of motile organisms in video sequences is an important tool in medical research [14,15], physiology [16,17], ecology [8], zoology [18], and in modern bioassays [19,20]. One interesting field is the analysis of movement patterns of microorganisms on Earth and under microgravity conditions such as in sounding rocket experiments, on parabolic airplane flights, and on satellites or the International Space Station [21–24].

The rapid development of video technology and computer hardware facilitates real-time image analysis and object tracking even with a PC [25], while in the past dedicated hardware with bulky digitizers and fast processors was required. Also the software applications have been improved over the past decades. While in the infancy of PC software it was mandatory to develop cell tracking software in Assembly language [26] because of the low CPU speed, later approaches were based on higher level languages such as Visual Basic or C++ [27]. Early systems were capable of tracking one organism at a time, while modern applications allow simultaneous analysis of quasi unlimited numbers (hundreds or thousands) of moving objects in real time [28,29]. In addition to the movement velocity and percentage of motile objects, a number of other parameters can be extracted from the moving objects such as size, position, directedness of movement, and form factors such as circularity [30,31].

Even with modern sophisticated image analysis hardware and software systems the human visual apparatus and complex neural circuits are far superior in object recognition and track analysis. In contrast, dedicated machine vision has the advantage of high objectivity and speed of analysis. In a lab course the author assigned the task

Bioassays. DOI: http://dx.doi.org/10.1016/B978-0-12-811861-0.00005-X

to a group of students of determining the direction of movement of a population of motile microorganisms in a video sequence. The students were instructed that the cells moved phototactically towards the actinic light beam entering from the right in the video sequence and they had to draw arrows on an acetate overlay on the video screen indicating the direction of movement of individual cells. The subsequent analysis showed a pronounced movement of the cells toward the light source with a high statistical significance. The next day another group had the same task with the same video sequence, but with the instruction that the light was entering from the left. The results showed a statistically significant movement of the cell population toward the light entering from the left. This little episode underlines the bias and subjective analysis of human observers who subconsciously eliminate tracks in the "wrong" direction. Even in double blind experiments an objective analysis is sometimes difficult [32]while machine image analysis is free of a subjective bias.

5.2 Hardware

While in the beginning of video processing analog cameras were used, almost all modern systems use digital CCD (charged coupled device) video cameras, the resolution of which has improved significantly over the years (Fig. 5.1).

Depending on the size of the objects under analysis the video camera can be fitted with either a standard objective or a macroobjective or it can be attached to a microscope or stereomicroscope with adjustable magnification [33]. A built-in zoom allows to focus onto a selected area within the image scene. Simple object recognition and cell tracking can be performed using a black and white (b/w)

Figure 5.1 Color and b/w digital cameras
Source: Image courtesy of Point Grey.

camera. For more sophisticated analyses color cameras are required. E.g., when a change in the photosynthetic pigments in a microorganism, an alga, or a higher plant in response to environmental stress has to be quantified, a color camera is indispensable [34]. Infrared sensitive cameras are used for cell tracking of visible light-sensitive organisms, the behavior of which may be modified by the monitoring light [20,35]. If only the tracks of motile organisms are being analyzed cellular details may not be of importance. They may even be excluded during the process of binarization (see below). For very small motile microorganisms such as bacteria, dark-field microscopy is used. In this case the organisms are irradiated perpendicular to the camera axis, so that they appear as bright objects over a black background. However, due to the halo effect the size analysis may be misleading and the objects may appear larger than in reality. This effect is avoided by phase contrast microscopy [36]; but this may lead to problems identifying the objects over the background since depending on the adjustment of the microscope the objects may appear darker or brighter than the background.

The video frequency is key for cell tracking purposes [37]. The normal video frequency of 25 Hz (Europe) or 30 Hz (e.g., in the United States) is sufficient for most movement analyses. However, for fast moving objects, such as moving flagella, swimming bacteria, flying birds, or running animals, a higher frequency may be necessary. High video frequencies may pose problems during analysis, if the digitization and analysis by the hardware and software cannot keep up. In this case the video sequence can be recorded with a high speed video camera and analyzed later offline. In contrast, for slow moving objects a lower frequency may be preferable. In these cases the frame rate is often reduced by selecting video frames from a sequence spaced over adjustable intervals. The spatial resolution of the video source should be as high as that of the digitizer. One inherent problem in object tracking is the distance traveled between subsequent video frames with respect to the object density: if an organism moves a longer distance between two frames than the statistical distance between the objects in the scene, the software may lose the identity of an individual particle and erroneously include different objects in a single track [38].

In real world video analysis the dynamic range of the video device may be of importance in order to reveal the objects in dark and bright parts of the image. This often causes problems in the analysis since the software will either regard the bright areas as background and be unable to detect objects in these areas or it will consider the dark areas as large particles. A human observer recognizes an object by its contrast to the adjacent background. One solution to this problem is the mathematical subtraction of a background image from the recorded scene, which leaves the objects visible [39]. This strategy is especially useful for detecting moving objects in front of an uneven background. Another algorithm is based on adaptive thresholding [40], also called dynamically adjusted background, which determines the brightness difference between an object and its surroundings independently of the overall brightness of the area in the image [41]. In this approach the algorithm searches for an abrupt change in the gray value as it scans the image. Then it determines the average gray value in the vicinity of the object and calculates a threshold to identify an object in this area. This threshold is adaptively changed as the software scans the image. Before applying this strategy, the evenness of a region can be improved by a smoothing operation. This technique is employed e.g., in meteorology to identify clouds in front of an unevenly lit sky.

Before using analog video devices the image has to be digitized. This is not necessary for digital CCD cameras, which already produce images with discrete spots, called pixels. The spatial resolution depends on the size of the objects to be identified. Obviously, objects with a size on the order of individual pixels will not be recognized. Typical spatial resolutions range from 512×512 to 2048×2048 pixels, but higher resolutions and different height to width ratios are also used. In 3D analysis the image represents a three-dimensional space (with axes in *X*, *Y*, and *Z* directions. The individual spots constituting the 3D image are called voxels. To digitize a 2D image of 512×512 pixels at a video frequency of 25 Hz a digitization frequency of 6.5 MHz is necessary and at 2048×2048 pixels a frequency of over 100 MHz is required.

In a b/w image with gray values each pixel has a distinct brightness. This is often an 8-bit value ranging from 0 (*black*) to 255 (*white*) [42]. In order to store the images in memory the required space for each image can be calculated from the spatial resolution; e.g., an image with a resolution of 1024×1024 pixel requires 1 MB of memory. In color images each pixel is represented by three bytes, each of which codes for the intensity of the three color channels red, blue, and green [43]. When you go very close to a TV screen you see that each pixel is constituted by a color triple. Consequently a color image requires a three-times larger memory at the same spatial resolution as a b/w image.

Early image analysis systems used bulky stand-alone digitizers, which in addition needed several minutes to digitize an image which was naturally prohibitive for real-time applications. Later on, video digitization was carried out with framegrabbers which could be inserted as cards into a slot of a standard PC. Today digitization can be done within the CPU of a modern PC and the video camera is attached via a USB, Gigabit Ethernet, or FireWire connector [44]. Commercial image analysis systems for biomonitoring are available, e.g., from Real Time Computers (Möhrendorf, Germany), LemnaTec GmbH (Aachen, Germany) and Clemex Technologies (Quebec, Canada).

5.3 Image manipulation

In contrast to the human eye, which can detect objects in noisy images and over a wide dynamic range, computerized image analysis requires high quality images with a clear distinction between objects and background. The objects may be either dark over a bright background or vice versa.

5.3.1 Pixel manipulation

As indicated above an image consists of a number of pixels organized in rows and columns. For the sake of simplicity let us consider a b/w image. Fig. 5.2 shows the numerical representation of a part of an image in memory.

	1	2	3	4	5	6	7	8	9	10	11	12	13	14	15
1	52	66	78	69	75	70	75	69	74	69	57	70	77	77	77
2	51	47	67	53	65	69	61	69	81	71	64	66	78	67	72
3	79	74	56	75	75	92	60	82	78	57	67	90	97	91	79
4	69	80	58	50	56	75	57	68	75	82	81	90	82	72	69
5	70	63	48	71	56	66	59	53	61	57	72	81	72	66	72
6	72	71	50	44	52	70	48	59	59	59	55	61	86	87	85
7	70	72	52	50	255	65	48	53	70	75	64	48	59	58	42
8	62	56	72	64	49	36	82	68	55	43	48	51	68	60	34
9	71	64	68	71	58	51	72	51	54	70	76	77	56	24	1
10	82	91	81	89	71	69	63	54	68	47	25	20	35	8	2
11	93	57	73	64	53	69	67	60	39	2	1	1	2	5	1
12	91	71	81	65	59	56	63	50	25	0	0	1	0	3	3
13	87	84	76	59	45	47	84	54	20	1	3	1	4	2	0
14	89	64	52	60	74	79	61	38	3	2	2	1	1	0	0
15	90	76	73	97	103	84	48	22	1	1	1	2	2	1	4
16	94	77	88	106	122	98	20	17	0	0	0	0	0	0	0
17	58	53	72	86	67	51	9	3	1	1	2	2	1	1	1
18	22	30	61	75	57	24	1	2	3	0	1	0	1	1	1
19	44	55	62	67	33	20	3	1	3	2	3	59	53	2	3
20	54	45	62	48	25	66	5	0	0	0	74	48	59	1	2
21	37	81	89	71	69	55	0	0	0	0	57	48	53	3	4
22	43	40	67	57	56	47	0	0	0	0	92	82	68	3	4
23	52	67	45	24	32	25	11	0	0	0	2	72	2	1	4
24	45	69	69	50	32	43	7	19	8	0	0	0	2	2	1
25	44	45	70	49	23	33	14	16	0	3	2	4	3	2	4
26	48	59	59	59	55	61	59	59	59	55	61	86	67	55	79

Figure 5.2 Section of a digitized image represented by gray values in rows and columns (*yellow*). Low values indicate a dark object (*green*) on a brighter background (*white*) with a brighter inclusion (*orange*). Note the pixel error in row 7, column 5.

The image can easily be manipulated by mathematical functions for each pixel. In order to make the image brighter a constant number is added to each pixel. However, one has to be careful not to exceed a maximal value of 255; in a larger number only the lower eight bits will be considered, so that a smaller number than the original value may occur (e.g., $74+200=274$ which will be cut to $274-255=19$, which corresponds to an even darker gray). Equally, an image can be mathematically rendered darker by subtracting a constant value, but the number may not become negative. Such manipulations are being performed by commercial photo enhancement and editor programs such as Adobe Photoshop Elements [45]. The contrast can be enhanced by multiplying the pixel values with a certain factor, again making sure that no value exceeds 255 for an 8-bit image. If an image shows a gradient in brightness a horizontal or vertical ramp can be defined which is subtracted from or added to each pixel column by column or row by row. Likewise, specific dark areas can be made brighter by adding an offset only to numbers below a predefined threshold. Similarly too bright areas in an image can be darkened by subtracting a constant value. Often in bioassay software systems a background manipulation is carried out: the camera records a frame of the scene (or several—from which the mean is calculated). This background is stored in memory and subtracted from each video frame to be analyzed [39]. By this algorithm only the objects of interest are retained while the nonuniform background is eliminated.

Often a number of subsequent manipulations are carried out to improve the image. For a given video sequence these steps can be programmed in the software. Fig. 5.3A shows a raw image from a video sequence taken from swimming *Daphnia* in front of a bright background. After background subtraction the dark objects are seen on a bright background (Fig. 5.3B). In order to further improve the visibility of the objects a contrast enhancement is performed by multiplying each pixel value with a constant number (Fig. 5.3C).

In order to clearly separate the objects of interest from the background the technique of thresholding is applied [46]. Let us assume we have a b/w image with dark objects on a bright background. Image analysis programs often provide the possibility to view a histogram of the brightness distribution of all pixels in an image [1,47]. Fig. 5.4 shows the gray level histogram derived from Fig. 5.3B.

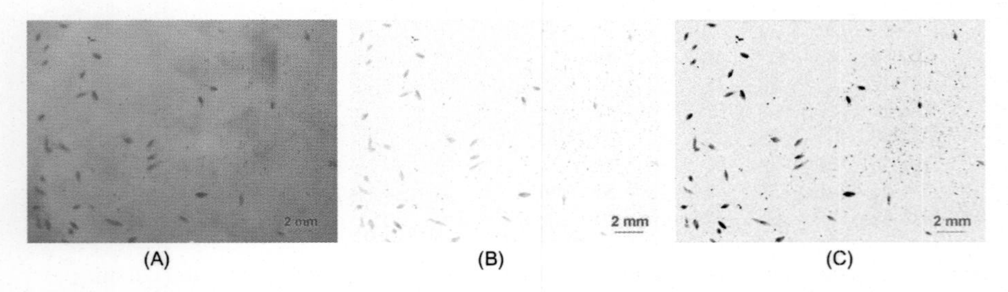

Figure 5.3 Enhancement of a frame from a video stream showing (A) moving *Daphnia magna*; (B) after background subtraction; and (C) contrast enhancement.

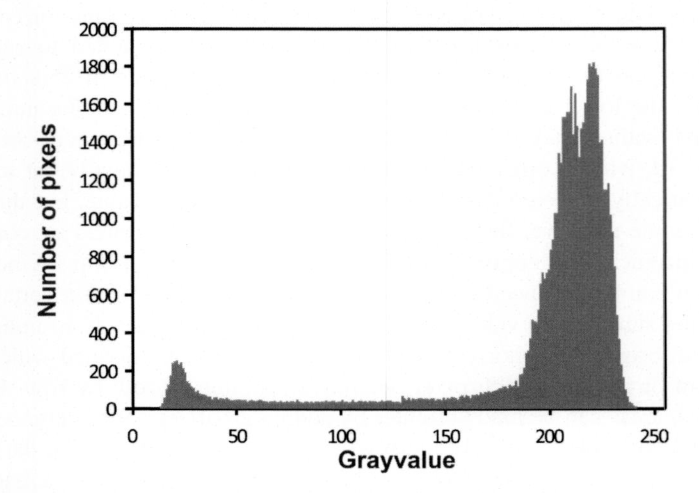

Figure 5.4 Gray level histogram of Fig. 5.3B. The abscissa shows the gray level increasing from 0 (black) to 255 (bright) and the ordinate number of pixels with the corresponding gray values. The left peak represents the organisms and the large right peak the background.

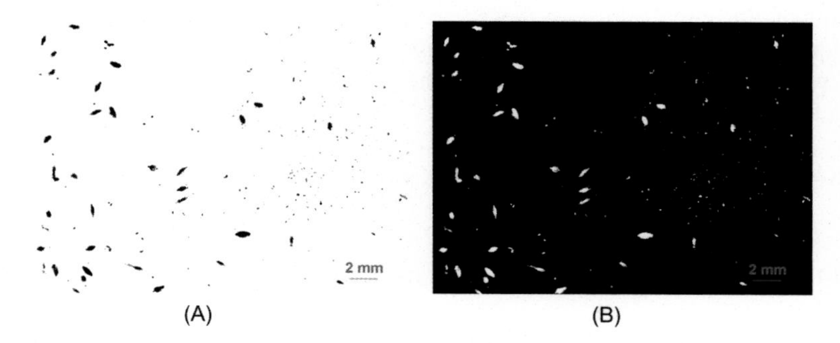

Figure 5.5 Image derived from Fig. 5.3C (A) after binarization; and (B) after inversion.

The small peak at low gray levels indicates the organisms seen in Fig. 5.3B and the large maximum at high gray levels signifies the bright background. The next step is binarization: We define a cutoff between the bright and dark values in the image and convert all values below this threshold to 0 (*black*) and all above the threshold to 255 (*bright*). The resulting image shows black organisms on an evenly bright background (Fig. 5.5A). Image analysis packages often have a slider to adjust the threshold; the resulting effect on the image is shown online [47]. Depending on the algorithm for object identification the image may have to be inverted: all dark pixels are set to 255 and all bright pixels to 0 (Fig. 5.5B). Binarization has the additional advantage of removing noise from an image.

5.3.2 Mathematical filters

A virtually unlimited number of mathematical manipulations has been developed to alter the image structure and improve the quality. CCD cameras are prone to produce pixel errors [48]. Some pixels may be constantly black or white. Also during transmission of the digital values or during image manipulation in the computer memory artifacts can occur. Noise reduction is often performed on a local basis. Let us consider a pixel X in an image which is surrounded by eight neighbors. During the manipulation this pixel X is replaced by X' which is the mean gray value of the 3×3 pixel matrix

$$X' = (X + a + b + c + d + e + f + g + h)/9 \tag{5.1}$$

$$\begin{array}{ccc} a\ b\ c & & a\ b\ c \\ d\ X\ e & > & d\ X'e \\ f\ g\ h & & f\ g\ h \end{array} \tag{5.2}$$

This manipulation is repeated for all pixels in the image row by row with the exception of the marginal rows and columns, the values of which are retained in the derived image. One side effect is that this algorithm smoothens sharp boundaries, since this algorithm constitutes a low pass filter. Related algorithms divide the sum of the gray values of the adjacent pixels by, e.g., 10. A variant of this technique is a hat filter, which weights the pixels by multiplying each pixel value with a factor before smoothing. The value of the new central pixel is then calculated by:

$$X' = (a * 1 + b * 2 + c * 1 * d * 2 + X * 4 + e * 2 + f * 1 + g * 2 + h * 1)/16 \tag{5.3}$$

All these manipulations can also be performed on a larger matrix such as 4×4 or 5×5.

Object identification is a key target in image processing. Enhancement of object contours can be obtained using a Laplace algorithm [49]. Let us assume again a 3×3 pixel matrix. The new value in the derived image is calculated by multiplying its original value with a specific kernel (usually 8) and subtracting the gray values of the eight adjacent pixels.

$$X' = 8 * X - (a + b + c + d + e + f + g + h) \tag{5.4}$$

This procedure also glides column by column and row by row over the image. The result of a Laplace filtering is an overall reduction in brightness in all areas with similar gray values [50]. Sharp boundaries between objects and background show up as bright lines in the image. Using different kernel numbers such as 4 and weighting the adjacent pixels can produce completely different results. In addition, Laplace filtering reduces pixel noise in the image.

The Sobel algorithm is more robust in relation to noise. It determines the difference between the upper and lower neighboring pixels as well as the difference between the left and right neighbors and in addition the difference for the two diagonals. Then the central pixel in the matrix is replaced by the highest difference obtained by the above calculations. The result is that the Sobel algorithm highlights not only horizontal and vertical but also diagonal boundaries in the image. In essence these methods of edge enhancement use high-pass filters. Other examples are [51]

$$\begin{matrix} 0 & -1 & 0 \\ -1 & 5 & -1 \\ 0 & -1 & 0 \end{matrix} \qquad \begin{matrix} -1 & -1 & -1 \\ -1 & 9 & -1 \\ -1 & -1 & -1 \end{matrix} \qquad \begin{matrix} 1 & -2 & 1 \\ -2 & 5 & -2 \\ 1 & -2 & 1 \end{matrix} \tag{5.5}$$

5.3.3 Look-up tables

Look-up tables can be used to manipulate images in many ways [29,52]. The gray value of each pixel in the image is looked up in a table and replaced with the value found there. In Fig. 5.6 the 256 possible gray values in an 8-bit b/w image are listed in column A. If each gray value is substituted by the corresponding value listed in

A	B	C	D	E	
0	255	0	0	0	
1	254	0	0	3	
2	253	0	0	6	
3	252	0	0	9	
4	251	0	0	12	
5	250	0	0	15	
6	249	0	0	18	
7	248	0	0	21	blue
8	247	0	0	24	
9	246	0	0	27	
10	245	0	0	30	
11	244	0	0	33	
12	243	0	0	36	
13	242	0	0	39	
14	241	0	0	42	
.	.	.	.	.	
119	136	0	0	102	
120	135	0	0	105	
121	134	0	0	108	
122	133	0	0	111	
123	132	0	0	114	
124	131	0	31	117	
125	130	0	63	120	
126	129	0	95	123	green
127	128	255	127	126	
128	127	255	159	129	
129	126	255	191	132	
130	125	255	223	135	
131	124	255	255	138	
132	123	255	255	141	
133	122	255	255	144	
.	.	.	.	.	
241	14	255	255	213	
242	13	255	255	216	
243	12	255	255	219	
244	11	255	255	222	
245	10	255	255	225	
246	9	255	255	228	
247	8	255	255	231	
248	7	255	255	234	red
249	6	255	255	237	
250	5	255	255	240	
251	4	255	255	243	
252	3	255	255	246	
253	2	255	255	249	
254	1	255	255	252	
255	0	255	255	255	

Figure 5.6 Use of look-up tables. The possible gray levels in an 8-bit b/w image are substituted by the values looked up in the table. The substitution in column A converts a positive image into its negative. Column B illustrates the process of binarization. In column C a small band of gray values is spread over the whole range of gray values, while all values below this band are set to 0 and all above to 255. Column D shows one example for pseudocolor conversion; certain gray value ranges are sent to one of the three color channels.

column B an inversion of the image is achieved (see above). In chemical photo processing this equals the transformation from a photographic negative to a positive. Column C demonstrates the binarization process used for segmentation of an image (see above). In this case all dark gray values up to 126 are replaced by 0 and all brighter values by 255. However, as indicated above, the threshold can be set to any value. Column D shows the expansion of a central band of gray levels (124–130) over the available range of gray levels. All values below this band are set to dark and all values above to white. This procedure is used when the details of an image are hidden in a small gray value band that a human observer has difficulties detecting.

Using look-up tables one can convert a b/w image into a color image. For this purpose the gray values from 0 to 85 are stretched to 0–255 by multiplying them by 3. These values are sent to the blue channel of an image. Likewise the central segments (gray values 86–170) are expanded over the range from 0–255 and sent to the green channel. Finally, the remaining 85 gray values (171–255) are expanded and sent to the red channel. Of course any other color assignment can be chosen. The color ranges can even overlap. The human eye can distinguish colors more easily than gray levels. This procedure produces pseudocolor images [53] which are used e.g., in thermal imaging [54], medicine [55], or environmental images taken by satellites [56]. Fig. 5.7 shows the image of a *Euglena* cell taken in bright field microscopy and converted to a pseudocolor image. In the front end of the cell there is an accumulation of calcium which can be visualized by applying a fluorescence dye which lights up when it binds to calcium [57]. This brightness difference is barely visible in a b/w image but clearly shows up after pseudocolor conversion.

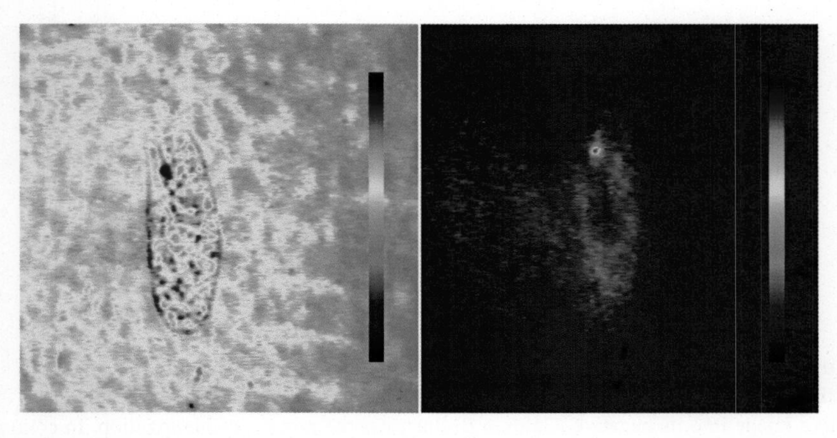

Figure 5.7 b/w bright field image of the colorless flagellate *Euglena longa* converted to a pseudocolor image (left). When bound to the fluorescent dye Calcium Crimson the high concentration of calcium in the front end of the cell clearly lights up (right). The bars indicate increasing brightness from the bottom toward the top.

5.4 Object detection by segmentation

A human observer can easily detect the position, area, and form of an object on a uniform background. How can a computer tackle this task? The algorithm is based on the segmentation of the image having two regions attributed to objects and background, respectively. Again imagine the binarized image in bright field modus (only values of 0 = object and 255 = background) being organized in columns and rows. The software starts at the top left corner and reads the gray value of each pixel in this row from left to right, then switches to the next row (Fig. 5.8). As long as it finds values of 255, it "knows" that these pixels belong to the background.

As soon as the software hits a pixel with a value of 0 it "knows" that it hit a valid object pixel. The following pixel in this row may be either again an object pixel, so these two belong to the same object. Otherwise it hits a background pixel. The basic question is then: when it scans the next row and finds an object pixel, does that belong to the same object or another one? Therefore the strategy changes at this point. The *x* and *y* coordinates of the first object pixel are stored in an array in memory and the gray value is set to a reserved value or a different color, so that it "knows" that this pixel has already been processed and is stored in the array. Now all adjacent pixels (to the right and below, since left and above have already been analyzed) are scanned. If they correspond to object pixels, they are stored in the same array and their gray value changed to the value of already identified pixels (100). Subsequently it uses adjacent pixels. Every newly found object pixel is treated in the same manner until only background pixels are detected subsequently.

From the number of pixels in the array the area can be deduced. From the minimal and maximal *x* and *y* coordinates of the object pixels in the array the center of mass can be calculated [29], which will be of importance when the movement of an object through a video sequence is analyzed (see below). Subsequently, the search

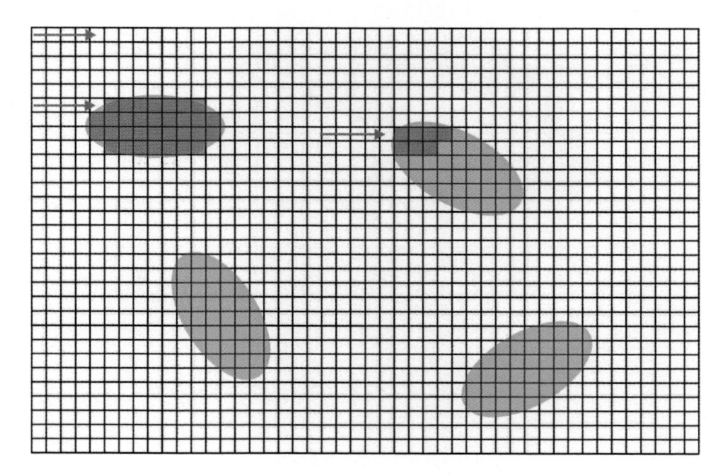

Figure 5.8 Object recognition in a binarized image by filling the whole area (for details see text).

for new objects is continued line by line until the lower right corner of the image is reached. For each object an array has been constructed from which the position and area can be deduced.

This algorithm is rather slow especially when the organisms are large and cover a large section of the image. The speed of the algorithm can be improved by skipping one or more pixels in a row or every other or more rows during the scan before the software finds a new object.

Faster options utilize edge detection algorithms ignoring the center of the object [58]. One fast strategy is the chain code algorithm introduced by Freeman [59]. Again the image is scanned pixel by pixel and line by line (perhaps with skipped interlaced pixels and lines to speed up the process) until the first edge pixel of an object has been found (Fig. 5.9).

The x and y coordinates of this pixel are stored. By definition an edge pixel has a direct neighbor unless the object consists of only one pixel. In addition, at least one background pixel has to be located outside the object. Then the algorithm searches the next edge pixel in a clockwise direction around the first found pixel. The direction is coded by a number from 0 to 7 (inset in Fig. 5.9). This number is recorded for each newly found edge pixel until the first outline pixel is found again.

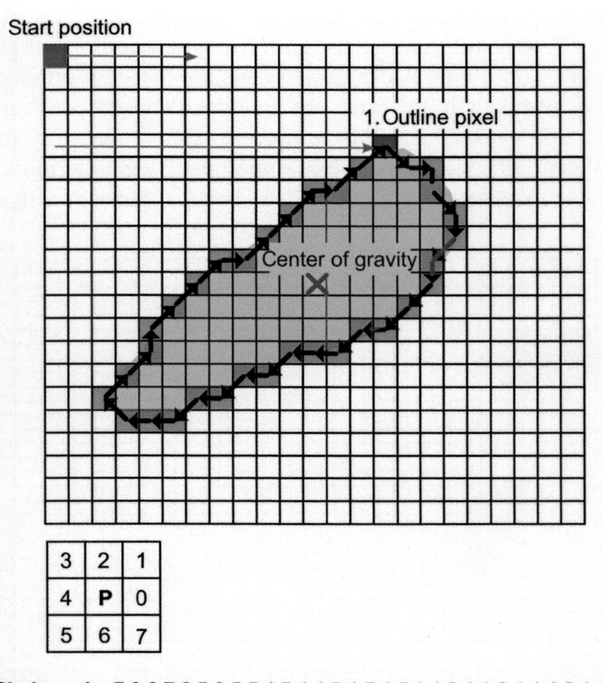

Figure 5.9 Chain code algorithm to identify the outline of an object.

All the found directions to the next edge pixel are stored in a chain code [60] which uniquely defines the outline of the found object [29]. All edge pixels are set to a predefined value so that the object is marked as being analyzed. Subsequently the search is continued in the whole image until all objects have been found. In addition to being much faster than the brute-force filling method, this robust algorithm reduces the amount of stored information considerably. However, it should be noted that it cannot detect holes in the object.

The length of the outline, the area of the object, and the coordinates of the center of gravity can be determined from the chain code. The length of the outline is calculated by:

$$L = T(n_e + n_o\sqrt{2}) \tag{5.6}$$

where n_e is the number of even links in the chain code and n_o the number of odd links and T a scaling factor which is given by the optical magnification. The area A of the two-dimensional projection of an object is calculated from the integration of the chain code elements with respect to the x axis

$$A = \sum_{i=1}^{n} a_{ix}(y_{i-1} + 0.5a_{iy}) \tag{5.7}$$

The x and y coordinates of the center of gravity are calculated from the first moments about the x and y axes; about the x axis:

$$M_1^x = \sum_{i=1}^{n} 0.5a_{ix}[(y_{i-1})^2 + a_{iy}(y_{i-1} + 1/3a_{ix})] \tag{5.8}$$

The chain is then rotated by $\pi/2$ about the origin in order to obtain the first moment about the y axis and then using the same equation. The x and y coordinates of the center of gravity are then calculated by using the first moments and the area:

$$x_c = -M_1^y/A \tag{5.9}$$

$$y_c = M_1^x/A \tag{5.10}$$

The algorithms detailed above can be used to calculate the areas and number of the objects in an image. One application is cell counts in microscopy [61]. For this purpose a glass cuvette is used under a microscope. Provided that a constant magnification is chosen, the area seen by the camera is constant and can be determined. If, in addition, the height of the liquid layer is known and kept constant, the camera sees a constant volume. This can be achieved by using a Thoma [62] or Neubauer [63] chamber or a haematocytometer [64] (to count blood cells) made of glass with etched fine lines (Fig. 5.10).

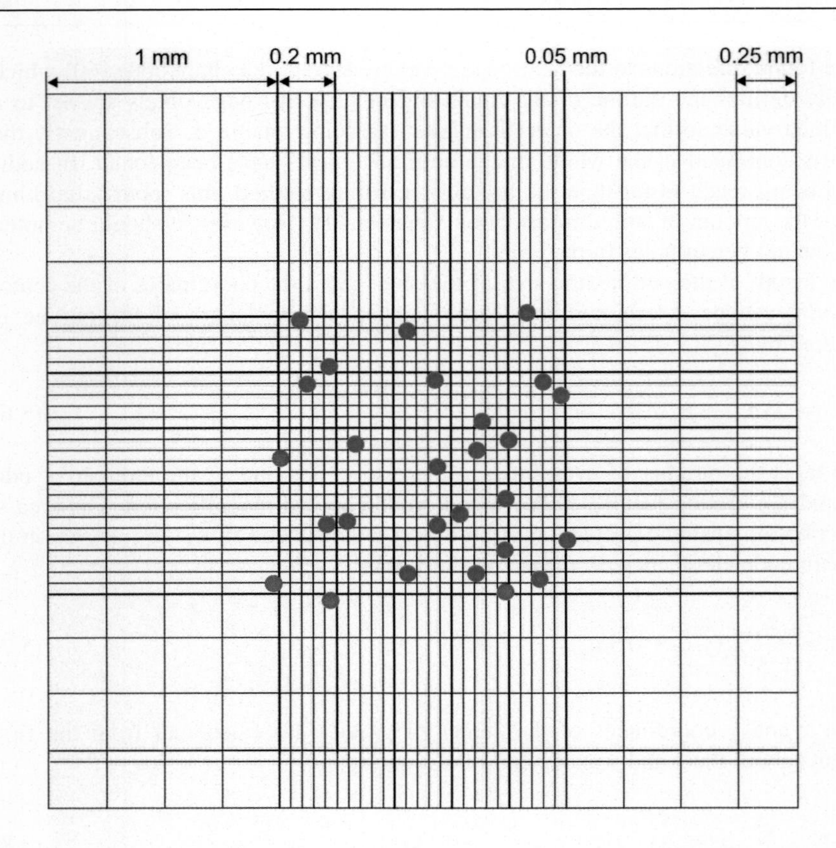

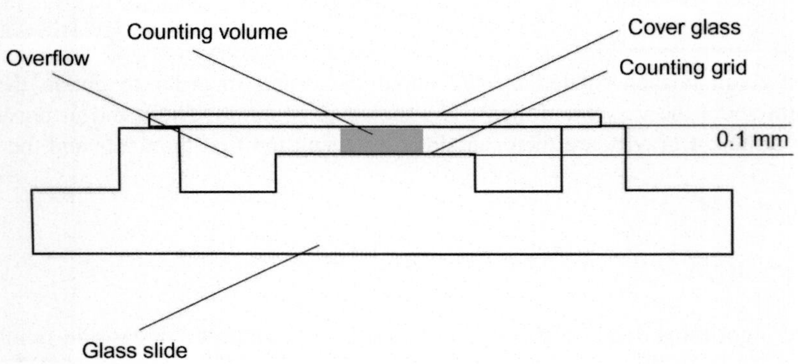

Figure 5.10 Improved Neubauer chamber for microscopic counting of microorganisms (bacteria, plankton). Each large square is 1 × 1 mm, each intermediate square in the central large square 0.2 × 0.2 mm and the smallest squares measure 0.05 × 0.05 mm. The height between the measuring grid and the cover glass is 0.1 mm.

In the example in Fig. 5.10 23 cells are counted in an area of 1 × 1 mm. By convention all cells touching the right and top border of the central large square are counted but the ones touching the left and lower boundaries are not. Since the depth is 0.1 mm the volume seen by the camera is 0.1 mm^3 assuming that the camera magnification is adjusted in such a way that the field of view corresponds exactly with the central field. In that case the cell density is 230,000 cells per mL. However, in most cases the camera field of view is not quadratic but rectangular, e.g., 1288 × 964 pixels. For this purpose a microscopic scaling device is placed in the camera field of view to find out how many pixels correspond to, e.g., 1 mm in both x and y directions. Both directions are important since in some cameras the pixels are not square. Once we know the size of the field of view in real physical units (e.g., mm^2) we can use the count function of the software and then calculate the cell density as number of cells/mL, provided the depth of field is constant and known. This function is implemented in the new Ecotox system [20] (cf. Chapter 10: Ecotox, this volume). At the beginning of a measurement cycle the concentration of cells is determined automatically and recorded in the results. One advantage of this method is that because of the short time in which a frame is recorded even motile cells can be counted automatically, while for manual microscopy motile cells need to be deactivated by adding formaldehyde or iodine. In order to improve the accuracy, multiple images can be taken from different locations and the results averaged. One disadvantage of automatic area detection is that touching objects are counted as one area while a human can easily count these as multiple objects. Therefore the cell density should not be too high in order to avoid artifacts. One way around this problem is to count areas with about the double area as two objects.

This may still result in counting errors. To solve this problem, the erosion technique can be applied. Shells one pixel wide are removed from each area and the removed pixels are set to a reserved gray value. This procedure is repeated until connecting areas are separated at the isthmus. Then the shells are dilated again to their original value. However, a gap of at least one pixel wide is left between previously touching objects. Objects can also be separated manually before analysis using the cursor controls, a mouse, a digitizing tablet, or a touch sensitive screen.

Other strategies introduce upper and lower limits for the size of objects; areas smaller or larger than the limits are excluded from the analysis [65]. This method is also useful for nonuniform cell populations, e.g., a phytoplankton sample with may include small bacteria or larger zooplankton. Setting different limits allows for measurement of different populations in the same suspension separately. Likewise, limits can be defined for the form factors in order to select more or less round cells and ignore elongated fibrous objects. One open-source platform for biological image analysis implementing these features is Fiji, which is a distribution of the popular open-source software ImageJ [66]. This software also allows the user to fill holes in the areas or leave them open. Recognized areas are highlighted on screen and a size distribution can be documented by binning the object sizes in size classes. In addition, a table is generated which records the area, x and y coordinates of the center of gravity in the image frame, the perimeter, circularity, roundness, and solidity of all objects. The form parameters allow the user to distinguish between objects of different forms such as rounded or elongated. They are defined as:

$$\text{Circularity} = 4\pi \frac{\text{area}}{\text{perimeter}^2} \tag{5.11}$$

$$\text{Roundness} = \frac{\text{area}}{\pi \times \text{major axis}^2} \tag{5.12}$$

$$\text{Solidity} = \frac{\text{area}}{\text{convex area}} \tag{5.13}$$

In order to determine the length and width of elongated objects the technique by Feret is implemented [67]. Let us look again at Fig. 5.7. The length of the uppermost object can be determined by the difference between the lowest and highest x coordinates, and the width by that of the y coordinates. However, this does obviously not work for objects oriented at an oblique angle in the image. One could use a Pythagoras calculation for the longest axis. Another strategy developed by Feret is based on the idea of turning the image repeatedly by a small angle and recording the length and width calculated from the maximal and minimal x and y coordinates of each object. The length of an object is found when a maximal value is reached for the difference between the left and right boundary x coordinates and similar for the width when a minimum for the upper and lower boundary of the y coordinates is found. In addition, the Feret angle is recorded with respect to the horizontal by which the image is turned when the maximal and minimal values have been found. This feature is implemented in the Fiji software [66].

The technology to count objects and determine their sizes is also used in plant science where leaf areas or extension of a whole plant or a root system are evaluated [64]. In geology the size distribution of particles and sediments can be determined automatically [68]. In addition, the form of the particles can be analyzed using form parameters such as roundness, circularity, and solidity defined above [20]. Other areas for automatic area analysis include medicine [69], and satellite imagery [70]. In marine ecology the cell density and their vertical distribution in the water column as well as their vertical migrations has been evaluated using pumps spaced at even distances in the water column to extract water samples at predetermined time intervals [71,72].

5.5 Organism tracking

The movement of objects with time, also called optical flow [73,74], can be extracted from a sequence of video frames. For this purpose a stack of a defined number of frames is recorded or extracted from a video stream (Fig. 5.11). The frames in the stack can be separated by 40 ms for a 25 Hz video repetition rate. For faster organisms a high frequency video recording has to be used. For slow

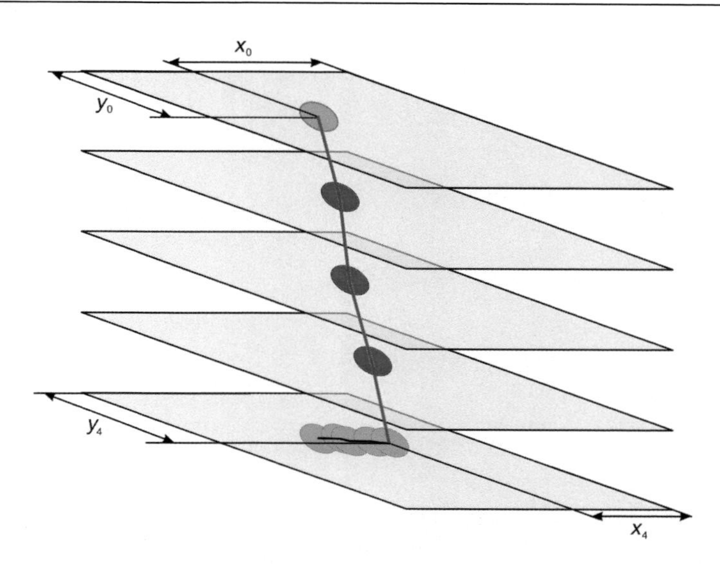

Figure 5.11 Stack with five video frames recorded at predefined time intervals (e.g., 40 ms at a 25 Hz video frequency, from top to bottom). One object is shown at an initial x_0 and y_0 position in the topmost frame, which subsequently travels through the 2D space in the following frames. The projections of this object are shown in the lowest (latest) frame [75]. From this the direction of movement is determined in x and y directions.

moving organism such as amoebae, slugs, or animal cells, one or more interlacing frames can be skipped, so that e.g., only every 10th, 20th, 30th, etc. frame is used in the stack.

Each frame in the stack is a two-dimensional representation of the three-dimensional space in which the objects are traveling with time. Thus their projections into the frame plane results in objects with specific x and y coordinates, areas, form factors and centers of gravity (centroids) [76]. The software starts by identifying the position of the object(s) in the first frame of the stack using the techniques outlined in the previous section. The basic parameters for each object are recorded in an array.

The x and y coordinates of the center of gravity of all objects are known in the first frame, so that the software does not have to scan the whole next image. Rather it starts searching at these coordinates in the second frame. If it finds an object there it assumes that this corresponds to the same object in the first frame. It determines the new position and all the other parameters and stores them in the array. Then it continues with all other objects in the second frame. This procedure is repeated for the remaining frames in the stack. If the software does not find a corresponding object at the position where it found one in the previous frame several strategies are possible. Either the software gives up on that object or it assumes that it is still in the vicinity of the previous position. Therefore the search is continued in an Archimedes spiral with a widening radius until an object is found [77]. But this strategy may result in an erroneus identification

Figure 5.12 Paths of objects (*Daphnia*) tracked through a sequence of 10 subsequent frames. The centers of gravity of each object in each frame are connected by straight lines. Only paths which have been followed over a predefined number of frames are considered for further analysis (*red*).

especially in a dense population. Therefore the search must be limited to the close vicinity. Another consideration is that the object may have left the view of the camera. For that reason all tracks with objects touching one of the boarders of the frame are excluded from further analysis. Fig. 5.12 shows the paths of objects through a stack of 10 frames. The centers of gravity of each object found in the subsequent frames are connected by straight lines. Only tracks which include object positions in a minimum predefined number of frames are considered for further analysis.

Once the position of all trackable objects has been determined in the stack the movement parameters can be extracted. The length l of the track is given by:

$$l = \sqrt{(y_2 - y_1)^2 + (x_2 - x_1)^2} \tag{5.14}$$

Often the track is not straight but the organism may follow a meandering path. In that case the true covered distance is added up by measuring the distance traveled from one frame to the next. Once we know the length dl of the track and the time interval dt between the first and the last frame we can calculate the velocity v:

$$v = \frac{dl}{dt} \tag{5.15}$$

For meandering paths the real velocity v_r is larger than the direct velocity v_d.

The movement direction can be defined as an angular deviation α in a clockwise direction from the 0 degrees direction which can be defined as toward the top of the image.

$$\alpha = tn^{-1}\frac{y_2 - y_1}{x_2 - x_1} \tag{5.16}$$

However the movement angle is only defined for the first (top right) quadrant. Therefore all other angles (in the three remaining quadrants) need to be calculated accoding to the following adjustments

$$\begin{aligned}
&\text{if } x_2 - x_1 <= 0 \text{ and } y_2 - y_1 > 0 \text{ then } \alpha = \alpha \\
&\text{if } x_2 - x_1 <= 0 \text{ and } y_2 - y_1 < 0 \text{ then } \alpha = 180^\circ - \alpha \\
&\text{if } x_2 - x_1 > 0 \text{ and } y_2 - y_1 < 0 \text{ then } \alpha = 180^\circ + \alpha \\
&\text{if } x_2 - x_1 > 0 \text{ and } y_2 - y_1 > 0 \text{ then } \alpha = 360^\circ - \alpha
\end{aligned} \tag{5.17}$$

Sometimes it is of interest to know the percentage of upward swimming cells in a chamber represented by those cells swimming into the upper quadrants ($\pm$ 90 degrees around the zero direction) or in a narrower sector such as $\pm$ 30 degrees or $\pm$ 45 degrees around 0 degrees. From this the k value can be extracted which is calculated by (Σ upward moving organisms - Σ downward moving organisms) $\times$ 100/*n*. The k value runs between 100 (all individuals move toward the upward semicircle) and -100 (all cells swim downward; 0 indicates random orientation or the same number of organisms move upward and downward. Another parameter extracted from a motile population of organisms is the alignment which indicates how well the tracks are aligned with the *x* or *y* axis. It is calculated from (Σ|sin α|-Σ| cos α|)/*n*, where α is the angular deviation of each track and *n* the total number of tracks. The alignment value ranges between $+1$ (all tracks are parallel to the *x* axis) and -1 (all tracks are parallel to the *y* axis); a value of 0 indicates that the organisms deviate by 45 degrees from the vertical and horizontal axis or are randomly distributed.

The concentration of moving objects in the field of view should be low enough so that cells rarely touch each other during tracking. But when this happens the sudden apparent increase in area will signal that these objects need to be excluded from further considerations. Likewise organisms leaving the field of view are no longer considered. One remedy is to request that a track has to be followed through a certain number of frames in order to be qualified. In any case, this is not really a problem since all these calculations are performed in real time and very large numbers of tracks can be analyzed in order to obtain a high statistical relevance, even when a few tracks are excluded.

A more advanced strategy is based on the assumption that organisms will continue more or less in the same direction even after having hit another one. The organisms do not have to physically hit each other but swim in a different layer in the chamber. In this case a look-ahead sector can be defined (e.g., a $\pm$15 degrees deviation from the previous movement direction), and the organism is searched for in that sector after the two objects are no longer in optical contact with each other. For certain applications is may be necessary to track organisms for a longer period of time, e.g., to detect if they are moving toward a specific target such as a source of a chemical [78].

A completely different strategy to identify moving objects is based on the subtraction of subsequent images from each other. This approach has been applied to study movement of people in street scenes [79] but can be applied to all situations where moving objects are to be detected in front of a rather complex background. A frame from a video sequence is digitized and stored in memory. Then the next frame is digitized and the first is subtracted from the subsequent pixel by pixel [80]. If there is no movement in the video scene the result will be a black image except for a few gray pixels resulting from thermal fluctuations and pixel noise [81]. If, in contrast, there is a change in image information between two subsequent frames due to the movement of an object in the scene this object shows up as a bright area. The algorithm for this approach begins with an image matrix (rows and columns of pixels) in which all pixels are set to a gray value of zero. When pixels at the same x and y coordinates in subsequent frames differ by a significant value the corresponding matrix pixel is incremented. The derived first-order difference picture (FOPD) combines the moving objects through the sequence, which looks like a sausage. The tracks can be calculated by the algorithm of skeletoning: shells one pixel wide are erased iteratively from the outer rim of the sausage until a line one pixel wide remains. A second order difference picture (SODP) can be derived from the FODP by setting those elements in the initial matrix to 1 at positions where elements with different values have been found in the FODP. By this method the start and end positions of the track can be determined. One disadvantage of this strategy is that it is rather time consuming and can therefore only be performed in real time using dedicated fast hardware.

A similar but faster approach determines the position and form of each object in image segments in subsequent frames and stores them for further analysis [27,82]. To identify each object in subsequent frames is based on a complex search algorithm in order to extract the movement vectors [83]. The cross-correlation method has been used to identify individual objects [84] but again this is rather time consuming. The computational velocity can be increased by reducing the correlation matrix to a small field (e.g., 32×32 pixels) centered around the centroids of the object found in the first frame [85]. The displacement of an object between two subsequent frames is indicted by a maximum in the correlation matrix, and the movement vectors are derived by an algorithm which identifies the nearest neighbor [86]. Another approach is based on analyzing moving polygons which may partially overlap by determining the corresponding vortices [87]. Also the translation, rotation, and dilation of moving objects have been used to determine the movement of objects [88]. Analysis of the movement of long overlapping filaments such as cyanobacteria is especially complicated and has been solved only by interactive image analysis [89].

Many bioassays rely on movement analysis of organisms. In modern computerized analysis movement of organisms in the time domain is extracted from video sequences [90–94]. In others the growth of an organism or organelle is being tracked [95–98]. The objects can range from individual organelles [99] and unicellular organisms such as flagellates or ciliates [100–103] to multicellular organisms such as algae or higher plants [104,105].

5.6 Oriented movement

Motile organisms are capable of moving with respect to external stimuli [106]. Not only photosynthetic unicellular algae or colonies, but also heterotrophic organisms such as flagellates or ciliates move with respect to the light direction either toward the light source (positive) or away from it (negative phototaxis) [107,108]. The gravitational field of the Earth is another important external clue that motile microorganisms use to find suitable niches for survival and reproduction in their environment [109,110]. Other environmental factors controlling the directional movement of motile organisms include chemicals [111], electrical fields [112], and the magnetic field of the Earth [3]. Several commercial bioassays monitor the oriented movement of microorganisms with respect to some of these external stimuli to monitor toxicity and pollution in aquatic ecosystems [113,114]. Motility, movement velocity, and directionality with respect to external factors deteriorate in the presence of toxic substances, and often the organisms employed in these bioassays show a high sensitivity to these toxins [115–119].

The direction of movement with respect to an external stimulus is defined by its deviation from a given direction (e.g., upward, cf. Section 5.5). Now we need to determine the precision of orientation in a population of microorganisms. This is based on the work on circular statistics established by Batschelet and Mardia [120,121]. Imagine a population of microorganisms such as the photosynthetic flagellate *Euglena* swimming in a vertical chamber and viewed from the side. The cells orient themselves with the gravitational field of the Earth and swim upward (defined as 0 degrees). This directional orientation is called negative gravitaxis [122], Not all cells swim precisely upward but deviate by an angle α from 0 degrees. The precision of orientational movement can be quantified by:

$$r = \frac{\sqrt{(\Sigma \sin \alpha)^2 + (\Sigma \cos \alpha)^2}}{n} \tag{5.18}$$

where n is the number of tracks and a the angular deviation [123] (cf. Eqs. 5.16 and 5.17). The r-value is a statistical measure and runs between 0 (the flagellates swim in random directions) and 1 (all cells swim in the same direction). However the r-value does not indicate the direction of movement. This can be deduced by calculating the mean value of the angular deviations of all tracks θ

$$\theta = \arctan\left(\frac{\Sigma \sin \alpha}{\Sigma \cos \alpha}\right) \tag{5.19}$$

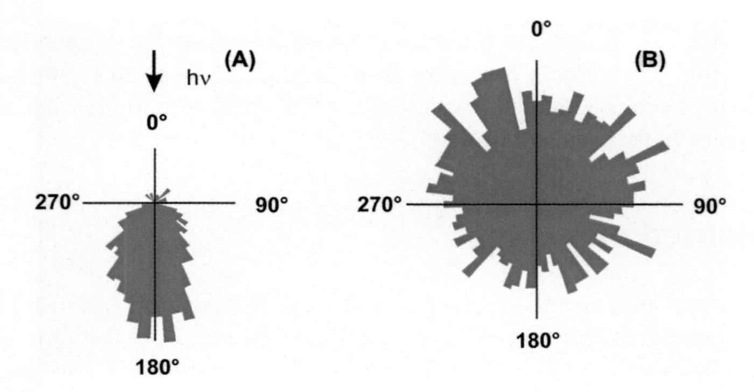

Figure 5 13 Circular histogram of movement directions of the flagellate *Euglena gracilis* (A) swimming away from a bright light beam (negative phototaxis) impinging from 0 degrees; and (B) in darkness.

Also in this case the angle θ is defined only for the first (top right quadrant) and an adjustment has to be made:

$$\begin{aligned} &\text{if } \Sigma \sin \alpha < 0 \text{ then } \theta = \theta + 360^\circ \\ &\text{if } \Sigma \cos \alpha < 0 \text{ then } \theta = \theta + 180^\circ \end{aligned} \tag{5.20}$$

The distribution of angular deviations can be visualized by a circular histogram. The angular deviation of all tracks is binned in a predetermined number of bins. The length of the sectors indicates the percentage of organisms moving in this specific direction normalized to the largest value. Fig. 5.13 shows the orientation of movement of a green flagellate in darkness and oriented with respect to a strong light source.

For most cases movement analysis in two-dimensional projections of the three-dimensional real world is sufficient, but sometimes movement analysis has to be performed in a real 3D environment. This requires recording the scene with two cameras oriented perpendicularly to each other aiming at the same volume. Therefore the computational expenses are higher [124,125]. In one approach the two images produced by the two cameras are combined in a split screen. The software starts by identifying an object in the image of one camera and then tries to find the same object in the parallel image at the same y coordinate in order to evaluate the z axis. The direction of movement in 3D is then performed in a van Mises space expanding the two-dimensional analysis into three dimensions [126].

5.7 Pattern recognition

For some applications it is not sufficient to identify objects in an image or a video sequence and follow one object over time. In some cases it is necessary to

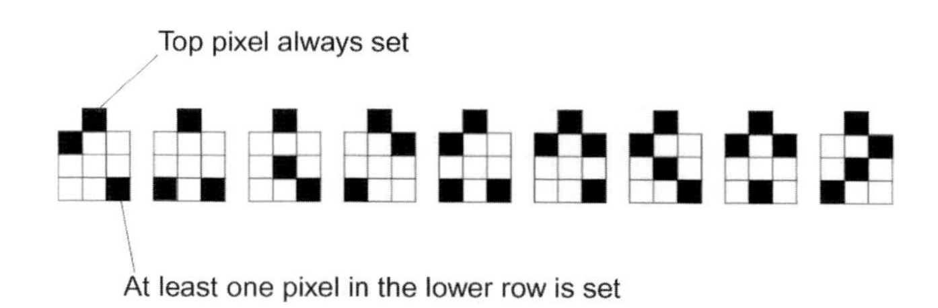
Top pixel always set
At least one pixel in the lower row is set

Figure 5.14 Patterns for recognizing moving organisms such as bees in a hive. The top square is marked in each pattern. In addition, at least one square is marked in the bottom row. During recognition the pattern is turned until the top square points upwards. The image is than expended or compressed until the pattern is recognized.

recognize a specific object in a scene. This can be achieved by pattern recognition [127,128]. One application is optical character recognition [129] in which the software has to identify the image of a written character and store it as the digital equivalent in the form of an ASCII code. This task is complicated by the representation of a character in various sizes and fonts as well as styles such as italics or boldface.

Pattern recognition is also being used in industrial production for quality control [130]. E.g., an image analysis system monitors freshly machined screws transported on a conveyor belt under the camera. The image of each individual screw is turned until it is aligned with the stored image of a perfect screw. As soon as a difference is found with respect to the stored image due to a defect of a screw this is removed from the passing objects. Pattern recognition is also used in automated face recognition [131]. This technique allows detecting persons in a crowd of moving people whose images are stored. Modern software uses a holistic matching approach in order to detect potential terrorists regardless of lightning, expression, aging, or pose [132]. However this requires immense computational power. Applications in bioassays include analysis of immunoassays [129] or analysis of nestmate recognition in ants [133]. Another example is the analysis of motile organisms in a three-dimensional space such as bees in a hive [134]. In order to assess the effect of external parameters, individual bees have to be identified e.g., to analyze the dance of a specific bee reporting a newly found food source. Since the insects are moving in a 3D space they can be hidden from view of the camera for a time and then reappear again. In order to identify individuals each animal carries a small plate on its thorax with a uniform pattern (Fig. 5.14). Since the bees can be oriented in any direction the image detected by the camera is rotated in memory until the black front square points upwards. As a second marker one square in the lowest row is always set. This arrangement warrants that the pattern is recognized even when it appears smaller or larger since the animal can be closer to or farther away from the camera. By the unique pattern for each marked bee it can be identified as long as the pattern is in the view of the camera.

References

[1] Sonka M., Hlavac V., Boyle R. Image processing, analysis, and machine vision: Cengage Learning; Boston Ma 2014.

[2] Niazi M, Chung J, Heaton-Johnson K, Martinez D, Castellanos R, Irwin M, et al. Advancing clinicopathologic diagnosis of high-risk neuroblastoma using computerized image analysis and proteomic profiling. Pediatr Dev Pathol 2017; 1093526617698603.

[3] Katzmann E, Eibauer M, Lin W, Pan Y, Plitzko J, Schüler D. Analysis of magnetosome chains in magnetotactic bacteria by magnetic measurements and automated image analysis of electron micrographs. Appl Env Microbiol 2013;79(24):7755–62.

[4] Sozzani R, Busch W, Spalding EP, Benfey PN. Advanced imaging techniques for the study of plant growth and development. Trends Plant Sci 2014;19(5):304–10.

[5] de Chaumont F, Coura RD-S, Serreau P, Cressant A, Chabout J, Granon S, et al. Computerized video analysis of social interactions in mice. Nat Methods 2012;9 (4):410–17.

[6] Zohary T, Shneor M, Hambright KD. PlanktoMetrix—a computerized system to support microscope counts and measurements of plankton. Inland Waters 2016;6 (2):131–5.

[7] Wu Q. Region-shrinking: a hybrid segmentation technique for isolating continuous features, the case of oceanic eddy detection. Remote Sens Environ 2014;153:90–8.

[8] Dell AI, Bender JA, Branson K, Couzin ID, de Polavieja GG, Noldus LP, et al. Automated image-based tracking and its application in ecology. Trends Ecol Evol 2014;29(7):417–28.

[9] Uemura M, Yamashita M, Tomikawa M, Obata S, Souzaki R, Ieiri S, et al. Objective assessment of the suture ligature method for the laparoscopic intestinal anastomosis model using a new computerized system. Surg Endosc 2015;29(2):444–52.

[10] Osman OS, Selway JL, Kępczyńska MA, Stocker CJ, O'Dowd JF, Cawthorne MA, et al. A novel automated image analysis method for accurate adipocyte quantification. Adipocyte 2013;2(3):160–4.

[11] Nyathi MS, Mastalerz M, Kruse R. Influence of coke particle size on pore structural determination by optical microscopy. Int J Coal Geol 2013;118:8–14.

[12] Ghazali KH, Hadi RS, Mohamed Z. Automated system for diagnosis intestinal parasites by computerized image analysis. Mod Appl Sci 2013;7(5):98.

[13] Kreiter S, Vormehr M, van de Roemer N, Diken M, Löwer M, Diekmann J, et al. Mutant MHC class III epitopes drive therapeutic immune responses to cancer. Nature 2015;520(7549):692–6.

[14] Hanna G, Fontanella A, Palmer G, Shan S, Radiloff DR, Zhao Y, et al. Automated measurement of blood flow velocity and direction and hemoglobin oxygen saturation in the rat lung using intravital microscopy. Am J Physiol Lung Cell Mol Physiol 2013;304 (2):L86–91.

[15] Imani Y, Teyfouri N, Ahmadzadeh MR, Golabbakhsh M. A new method for multiple sperm cells tracking. J Med Signal Sensors 2014;4(1):35–42.

[16] Azizullah A, Jamil M, Richter P, Häder D-P. Fast bioassessment of wastewater and surface water quality using freshwater flagellate *Euglena gracilis*—a case study from Pakistan. J Appl Phycol 2014;26(1):421–31.

[17] Richter PR, Schuster M, Lebert M, Häder D-P. Gravitactic signal transduction elements in *Astasia longa* investigated during parabolic flights. Micrograv Sci Technol 2003;14 (3):17–24.

[18] Kimura T, Ohashi M, Crailsheim K, Schmickl T, Okada R, Radspieler G, et al. Development of a new method to track multiple honey bees with complex behaviors on a flat laboratory arena. PLoS One 2014;9(1):e84656.
[19] Baldacchino F, Tramut C, Salem A, Liénard E, Delétré E, Franc M, et al. The repellency of lemongrass oil against stable flies, tested using video tracking. Parasite 2013;20:21.
[20] Häder D-P, Erzinger GS. Ecotox–Monitoring of pollution and toxic substances in aquatic ecosystems. J Ecol Environ Sci 2015;3(2):22–7.
[21] Hemmersbach R, Strauch SM, Seibt D, Schuber M, Häder D-P. Comparative studies on gravisensitive protists on ground (2D and 3D clinostats) and in microgravity. Micrograv Sci Technol 2006;18:257–9.
[22] Miller MS, Keller TS. *Drosophila melanogaster* (fruit fly) locomotion during the microgravity and hypergravity portions of parabolic flight. J Gravit Physiol 2006;13:35–48.
[23] Švegždienė D, Koryznienė D, Raklevičienė D. Comparison study of gravity-dependent displacement of amyloplasts in statocytes of cress roots and hypocotyls. Microgravity Sci Technol 2011;23(2):235–41.
[24] Nasir A, Strauch S, Becker I, Sperling A, Schuster M, Richter P, et al. The influence of microgravity on *Euglena gracilis* as studied on Shenzhou 8. Plant Biol 2014;16 (s1):113–19.
[25] Wang X, Zhang D. A high quality color imaging system for computerized tongue image analysis. *Ex*pert Syst Appl 2013;40(15):5854–66.
[26] Häder D-P, Lebert M. Real time computer-controlled tracking of motile microorganisms. Photochem Photobiol 1985;42:509–14.
[27] Häder D-P. Real-time tracking of microorganisms. Binary 1994;6:81–6.
[28] Meijering E, Dzyubachyk O, Smal I. Methods for cell and particle tracking. Methods Enzymol 2012;504(9):183–200.
[29] Häder D-P, Lebert M. Real-time tracking of microorganisms. In: Häder D-P, editor. Image analysis: methods and applications. Boca Raton, FL: CRC Press; 2000. p. 393–422.
[30] Tahedl H, Häder D-P. The use of image analysis in ecotoxicology. In: Häder D-P, editor. Image *anal*ysis: methods and applications. Boca Raton, FL: CRC Press; 2001. p. 447–58.
[31] Kröner S, Carbó MTD. Determination of minimum pixel resolution for shape analysis: proposal of a new data validation method for computerized images. Powder Tech 2013;245:297–313.
[32] Umbaugh SE. Digital image processing and analysis: human and computer vision applications with CVIP tools. Boca Raton, FL: CRC Press; 2016.
[33] McIntyre F, Neat F, Collie N, Stewart M, Fernandes P. Visual surveys can reveal rather different 'pictures' of fish densities: comparison of trawl and video camera surveys in the Rockall Bank, NE Atlantic Ocean. Deep Sea Res Part 1 Oceanogr Res Pap 2015;95:67–74.
[34] Goggin FL, Lorence A, Topp CN. Applying high-throughput phenotyping to plant–insect interactions: picturing more resistant crops. Curr Opin Insect Sci 2015;9:69–76.
[35] Moldrup M, Moestrup Ø, Hansen PJ. Loss of phototaxis and degeneration of an eyespot in long-term algal cultures: Evidence from ultrastructure and behaviour in the dinoflagellate *Kryptoperidinium foliaceum*. J Eukar Microbiol 2013;60 (4):327–34.

[36] Webb KF. Condenser-free contrast methods for transmitted-light microscopy. J Microsc 2015;257(1):8–22.
[37] Spalding EP, Miller ND. Image analysis is driving a renaissance in growth measurement. Curr Opin Plant Biol 2013;16(1):100–4.
[38] Jähne B. Digitale Bildverarbeitung. Berlin: Springer Verlag; 1989.
[39] Sobral A, Vacavant A. A comprehensive review of background subtraction algorithms evaluated with synthetic and real videos. Comput Vis Image Underst 2014;122:4–21.
[40] Cazorla A, Husillos C, Antón M, Alados-Arboledas L. Multi-exposure adaptive threshold technique for cloud detection with sky imagers. Solar Energy 2015;114:268–77.
[41] Korzynska A, Roszkowiak L, Lopez C, Bosch R, Witkowski L, Lejeune M. Validation of various adaptive threshold methods of segmentation applied to follicular lymphoma digital images stained with 3, 3'-diaminobenzidine & haematoxylin. Diagn Pathol 2013;8(1):1.
[42] Russ JC. The image processing handbook. Boca Raton, FL: CRC Press; 2016.
[43] Plataniotis K, Venetsanopoulos AN. Color image processing and applications. Berlin: Springer Science & Business Media; 2013.
[44] Jackson W. Capturing digital video: digital camera concepts. Digital video editing fundamentals. Heidelberg: Springer; 2016. p. 75–86.
[45] Ren Z-X, Yu H-B, Shen J-L, Li Y, Li J-S. Preprocessing with Photoshop software on microscopic images of A549 cells in epithelial-mesenchymal transition. Anal Quant Cytopath Histopath 2015;37(3):159–68.
[46] Dirami A, Hammouche K, Diaf M, Siarry P. Fast multilevel thresholding for image segmentation through a multiphase level set method. Signal Process 2013;93 (1):139–53.
[47] Hartig SM. Basic image analysis and manipulation in Image J. J Curr Protoc Mol Biol 2013;14:5.1–2.
[48] Pan B. Bias error reduction of digital image correlation using Gaussian pre-filtering. Opt Lasers Eng 2013;51(10):1161–7.
[49] Bässmann H, Besslich PW. Konturorientierte Verfahren in der digitalen Bildverarbeitung. Berlin Heidelberg: Springer-Verlag; 2013.
[50] Edges Peters JF. Lines, Ridges, and Nearness Structures. Topology of Digital Images. Berlin Heidelberg: Springer; 2014. p. 173–97.
[51] Marangoni R, Colombetti G, Gualtieri P. Digital filters in image analysis. In: Häder D-P, editor. Image analysis: methods and applications. Boca Raton, FL: CRC Press; 2000. p. 93–106.
[52] Pan B, Li K, Tong W. Fast, robust and accurate digital image correlation calculation without redundant computations. Exp Mech 2013;53(7):1277–89.
[53] Doi R. Red-and-green-based pseudo-RGB color models for the comparison of digital images acquired under different brightness levels. J Mod Opt 2014;61(17):1373–80.
[54] Deng F, Tang Q, Zeng G, Wu H, Zhang N, Zhong N. Effectiveness of digital infrared thermal imaging in detecting lower extremity deep venous thrombosis. Med Phys 2015;42(5):2242–8.
[55] Rizzardi AE, Johnson AT, Vogel RI, Pambuccian SE, Henriksen J, Skubitz AP, et al. Quantitative comparison of immunohistochemical staining measured by digital image analysis versus pathologist visual scoring. Diagn Pathol 2012;7(1):1.
[56] Liu X, Duan Z, Yang X, Xu W. Vector quantization method based on satellite cloud image. Int J Signal Proces, Image Process Pattern Recogn 2015;8(11):27–44.
[57] Eberhard M, Erne P. Calcium binding to fluorescent calcium indicators: calcium green, calcium orange and calcium crimson. Biochem Biophys Res Comm 1991;180:209–15.

[58] Shrivakshan G, Chandrasekar C. A comparison of various edge detection techniques used in image processing. IJCSI Int J Comput Sci Issues 2012;9(5):272–6.

[59] Annapurna P, Kothuri S, Lukka S. Digit recognition using freeman chain code. Int J Appl or Innov Eng Manage 2013;2(8):362–5.

[60] Kincaid DT, Schneider RB. Quantification of leaf shape with a microcomputer and Fourier transform. Can J Bot 1983;61:2333–42.

[61] Carpenter AE, Jones TR, Lamprecht MR, Clarke C, Kang IH, Friman O, et al. CellProfiler: image analysis software for identifying and quantifying cell phenotypes. Genome Biol 2006;7(10):R100.

[62] Georget E, Kushman A, Callanan M, Ananta E, Heinz V, Mathys A. *Geobacillus stearothermophilus* ATCC 7953 spore chemical germination mechanisms in model systems. Food Control 2015;50:141–9.

[63] Nordhoff V., Mallidis C., Kliesch S. Handling gametes and embryos: sperm collection and preparation techniques. In: M. Montag (Ed), *A practical guide to selecting gametes and rmbryos*, Boca Raton, FL: CRC Press; 2014:1.

[64] Biozzi G, Stiefel C, Mouton D. A study of antibody-containing cells. Immunity, cancer, and chemotherapy: basic relationship on the cellular level. New York: Academic Press Inc. 1967. p. 103–9.

[65] Häder D-P. Automatic area calculation by microcomputer-controlled video analysis. EDV in Medizin Biol 1987;18:33–6.

[66] Schindelin J, Arganda-Carreras I, Frise E, Kaynig V, Longair M, Pietzsch T, et al. Fiji: an open-sourcee platform for biological-image analysis. Nat Methods 2012;9 (7):676–82.

[67] Goldoni A, Pandolfo L, Gomes A, Folle D, Martins M, Pandolfo A. Evaluation of a method based on image analysis to obtain shape parameters in crushed sand grains. Rev IBRACON Estrut Mater 2015;8(5):577–90.

[68] Szczuciński W, Kokociński M, Rzeszewski M, Chagué-Goff C, Cachão M, Goto K, et al. Sediment sources and sedimentation processes of 2011 Tohoku-oki tsunami deposits on the Sendai Plain, Japan—Insights from diatoms, nannoliths and grain size distribution. Sediment Geol 2012;282:40–56.

[69] Hall AR, Tsochatzis E, Morris R, Burroughs AK, Dhillon AP. Sample size requirement for digital image analysis of collagen proportionate area in cirrhotic livers. Histopathology 2013;62(3):421–30.

[70] Necsoiu M, Dinwiddie CL, Walter GR, Larsen A, Stothoff SA. Multi-temporal image analysis of historical aerial photographs and recent satellite imagery reveals evolution of water body surface area and polygonal terrain morphology in Kobuk Valley National Park, Alaska. Environ Res Lett 2013;8(2):025007.

[71] Häder D-P, Griebenow K. Orientation of the green flagellate, *Euglena gracilis*, in a vertical column of water. FEMS Microbiol Ecol 1988;53:159–67.

[72] Eggersdorfer B, Häder D-P. Phototaxis, gravitaxis and vertical migrations in the marine dinoflagellates, *Peridinium faeroense* and *Amphidinium caterea*. Acta Protozool 1991;30:63–71.

[73] Honegger D., Meier L., Tanskanen P., Pollefeys M., editors. An open source and open hardware embedded metric optical flow CNOS camera for indoor and outdoor applications. In: Robotics and Automation (ICRA), 2013 IEEE International Conference; 2013.

[74] Jain R. Extraction of motion information from peripheral processes. IEEE Trans Pattern Anal Mach Intell 1981;3:489–504.

[75] Häder D-P, Erzinger GS. Advanced methods in image analysis as potent tools in online biomonitoring of water resources. Recent Pat Topics Imag 2015;5(2):112–18.

[76] Roach JW, Aggarwal JK. Determining the movement of objects from a sequence of images. IEEE Trans Pattern Anal Mach Intell 1980;2:554–62.
[77] Miralles F, Tarongi S, Espino A. Quantification of the drawing of an Archimedes spiral through the analysis of its digitized picture. J Neurosci Methods 2006;152(1):18–31.
[78] Wang X, Atencia J, Ford RM. Quantitative analysis of chemotaxis towards toluene by *Pseudomonas putida* in a convection-free microfluidic device. Biotechnol Bioeng 2015;112(5):896–904.
[79] Jain R, Nagel HH. On the analysis of accumulative difference pictures from image sequences or real world scenes. IEEE Trans Pattern Anal Mach Intell 1979;1:206–14.
[80] Kenny PA, Dowsett DJ, Vernon D, Ennis JT. A technique for digital image registration used prior to subtraction of lung images in nuclear medicine. Phys Med Biol 1990;35:679–85.
[81] Gualtieri P, Coltelli P. A digital microscope for real time detection of moving microorganisms. Micron Microsc Acta 1989;20:99–105.
[82] Häder D-P. Tracking of flagellates by image analysis. In: Alt W, Hoffmann G, editors. Biological motion, proceedings Königswinter 1989. Berlin, Heidelberg: Springer Verlag; 1990. p. 343–60.
[83] Häder D-P. Use of image analysis in photobiology. In: Riklis E, editor. Photobiology the science and its applications. New York: Plenum Press; 1991. p. 329–43.
[84] Smith EA, Phillips DR. Automated cloud tracking using precisely aligned digital ATS pictures. IEEE Trans Comput 1972;C-21:715–29.
[85] Takahashi T, Kobatake Y. Computer-linked automated method for measurement of the reversal frequency in phototaxis of *Halobacterium halobium*. Cell Struct Funct 1982;7:183–92.
[86] Noble PB, Levine MD. Computer assisted analysis of cell locomotion and chemotaxis. Boca Raton, FL: CRC Press, Inc; 1986.
[87] Aggarwal JK, Duda RO. Computer analysis of moving polygonal images. IEEE Trans Comput 1975;C-24:966–76.
[88] Schalkoff RJ, McVey ES. A model and tracking algorithm for a class of video targets. IEEE Trans Pattern Anal Mach Intell 1982;PAMI-4:2–10.
[89] Häder D-P, Vogel K. Interactive image analysis system to determine the motility and velocity of cyanobacterial filaments. J Biochem Biophys Methods 1991;22:289–300.
[90] Amos L. Movements made visible by microchip technology. Nature 1987;330:211–12.
[91] Allen RD. New directions and refinements in video-enhanced microscopy applied to problems in cell motility. In: Cowden RR, editor. Process in clinical biological research: advances in microscopy. New York: Alan R. Liss, Inc; 1985. p. 3–11.
[92] Coates TD, Harman JT, McGuire WA. A microcomputer-based program for video analysis of chemotaxis under agarose. Comput Methods Programs Biomed 1985;21:195–202.
[93] Rikmenspoel R, Isles CA. Digitized precision measurements of the movement of sea urchin sperm flagella. Biophys J 1985;47:395–410.
[94] Burton JL, Law P, Bank HL. Video analysis of chemotactic locomotion of stored human polymorphonuclear leukocytes. Cell Motil Cytoskel 1986;6:485–91.
[95] Omasa K, Onoe M. Measurement of stomatal aperture by digital image processing. Plant Cell Physiol 1984;25:1379–88.
[96] Jaffe MJ, Wakefield AH, Telewski F, Gulley E, Biro R. Computer-assisted image analysis of plant growth, thigmomorphogenesis, and gravitropism. Plant Physiol 1985;77:722–30.
[97] Popescu T, Zängler F, Sturm B, Fukshansky L. Image analyzer used for data acquisition in phototropism studies. Photochem Photobiol 1989;50:701–5.

[98] Häder D-P, Lebert M. The photoreceptor for phototaxis in the photosynthetic flagellate *Euglena gracilis*. Photochem Photobiol 1998;68:260–5.
[99] Sanderson MJ, Dirksen ER. A versatile and quantitative computer-assisted photoelectronic technique used for the analysis of ciliary beat cycles. Cell Motil 1985;5:267–92.
[100] Häder D-P, Hemmersbach R. Graviperception and graviorientation in flagellates. Planta 1997;203:7–10.
[101] Häder D-P. Gravitaxis and phototaxis in the flagellate *Euglena* studied on TEXUS missions. In: Cogoli A, Friedrich U, Mesland D, Demets R, editors. Life science experiments performed on *so*unding rockets (1985–1994). ESTEC. Noordwijk, The Netherlands: ESA Publications Division; 1997. p. 77–9.
[102] Hemmersbach R, Voormanns R, Häder D-P. Graviresponses in *Paramecium biaurelia* under different accelerations: studies on the ground and in space. J Exp Biol 1996;199:2199–205.
[103] Dusenbery DB. Using a microcomputer and videocamera to simultaneously track 25 animals. Comput Biol Med 1985;15:169–75.
[104] Gordon DC, MacDonald IR, Hart JW, Berg A. Image analysis of geo-induced inhibition, compression, and promotion of growth in an inverted *Helianthus annuus* L. seedling. Plant Physiol 1984;76:589–94.
[105] Omasa K, Aiga I, Hashimoto Y. Image instrumentation for evaluating the effects of air pollutants on plants. Acta Imeko 1982;303–12.
[106] Häder D-P. Ecological consequences of photomovement in microorganisms. J Photochem Photobiol B Biol 1988;1:385–414.
[107] Häder D-P. Strategy of orientation in flagellates. In: Riklis E, editor. Photobiology the science and its applications. New York: Plenum Press; 1991. p. 497–510.
[108] Lenci F, Ghetti F, Song P-S. Photomovement in ciliates. In: Häder D-P, Lebert M, editors. Photomovement. comprehensive series in photosciences. 1. Amsterdam: Elsevier; 2001. p. 475–503.
[109] Hemmersbach R, Häder D-P. Graviresponses of certain ciliates and flagellates. FASEB J 1999;13:S69–75.
[110] Häder D-P. Gravitaxis in unicellular microorganisms. Adv Space Res 1999;24:843–50.
[111] Sourjik V, Wingreen NS. Responding to chemical gradients: bacterial chemotaxis. Curr Opin Cell Biol 2012;24(2):262–8.
[112] Korohoda W, Mycielska M, Janda E, Madeja Z. Immediate and long-term galvanotactic responses of *Amoeba proteus* to dc electric fields. Cell Motil Cytoskel 2000;45:10–26.
[113] Azizullah A, Murad W, Adnan M, Ullah W, Häder D-P. Gravitactic orientation of *Euglena gracilis*—a sensitive endpoint for ecotoxicological assessment of water pollutants. Front Env Sci 2013;1:4.
[114] Tahedl H, Häder D-P. Fast examination of water quality using the automatic biotest ECOTOX based on the movement behavior of a freshwater flagellate. Water Res 1999;33:426–32.
[115] Ahmed H, Häder D-P. Short-term bioassay of chlorophenol compounds using *Euglena gracilis*. *SR*X Ecol 2009;2010:9.
[116] Ahmed H, Häder D-P. Rapid ecotoxicological bioassay of nickel and cadmium using motility and photosynthetic parameters of *Euglena gracilis*. Environ Exp Bot 2010;69:68–75.
[117] Ahmed H. Biomonitoring of aquatic ecosystems [PhD thesis]. Erlangen-Nürnberg: Friedrich-Alexander-Universität; 2010.

[118] Ahmed H, Häder D-P. A fast algal bioassay for assessment of copper toxicity in water using *Euglena gracilis*. J Appl Phycol 2010;22(6):785–92.
[119] Ahmed H, Häder D-P. Monitoring of waste water samples using the ECOTOX biosystem and the flagellate alga *Euglena gracilis*. J Water Air Soil Pollut 2011;216 (1-4):547–60.
[120] Batschelet E. Circular statistics in biology. London, New York: Academic Press; 1981.
[121] Mardia KV. Statistics of directional data. London: Academic Press; 1972.
[122] Häder D-P. Gravitaxis in flagellates. Biol Bull 1997;192:131–3.
[123] Batschelet E. Statistical methods for the analysis of problems in animal orientation and certain biological rhythms. Washington, DC: American Institute of Biological Sciences; 1965. p. 1–59.
[124] Kühnel-Kratz C, Häder D-P. Real time three-dimensional tracking of ciliates. J Photochem Photobiol B Biol 1993;19:193–200.
[125] Kühnel-Kratz C, Häder D-P. Light reactions of the ciliate *Stentor coeruleus*—a three-dimensional analysis. Photochem Photobiol 1994;59:257–62.
[126] David H.A., Nagaraja H.N. Generized order statistics. Hoboken NJ: Wiley; 1981.
[127] Bezdek JC. Pattern recognition with fuzzy objective function algorithms. Berlin: Springer Science & Business Media; 2013.
[128] Fukunaga K. Introduction to statistical pattern recognition. Cambridge, MA: Academic Press; 2013.
[129] Zhang Y, Qiao L, Ren Y, Wang X, Gao M, Tang Y, et al. Two dimensional barcode-inspired automatic analysis for arrayed microfluidic immunoassays. Biomicrofluidics 2013;7(3):034110.
[130] Liu D, Sun D-W, Zeng X-A. Recent advances in wavelength selection techniques for hyperspectral image processing in the food industry. Food Bioprocess Technol 2014;7 (2):307–23.
[131] Klontz JC, Jain AK. A case study of automated face recognition: The Boston marathon bombings suspects. Computer 2013;46(11):91–4.
[132] Parmar D.N., Mehta B.B. Face recognition methods & applications. arXiv preprint arXiv:14030485. 2014.
[133] Esponda F., Gordon D.M., editors. Distributed nestmate recognition in ants. Proc R Soc B 282, (1806), 2014, 28–38.
[134] Xanthopoulos P, Razzaghi T. A weighted support vector machine method for control chart pattern recognition. Comput Ind Eng 2014;70:134–49.

Growing algal biomass using wastes

Félix L. Figueroa[1], *Nathalie Korbee*[1], *Roberto Abdala-Díaz*[1], *Félix Álvarez-Gómez*[1], *Juan Luis Gómez-Pinchetti*[2] and *F. Gabriel Acién*[3]
[1]Malaga University, Malaga, Spain, [2]University of Las Palmas de GC, Las Palmas, Spain, [3]Almeria University, Almeria, Spain

6.1 Introduction

Algae are photoautotrophic organisms with high morphofunctional diversity due to their extensive evolution on the Earth from the first specimens in the ocean 1500 million years ago. They present a size from 0.2 to 200 μm in the case of microalgae, and in macroalgae from a few centimeters up to 30 m—as in the giant kelp *Macrocystis pyrifera*. The algae can live in a great variety of habitats such as the extremophiles growing in acid and hot waters, on ice, in very low irradiances as in deep water or caves (0.1% of incident light), and at all latitudes and climates. Algae the initial colonizers in ecological successions together with cyanobacteria or associated fungi (lichens). They can be found not only in waters but also in soils as biofilms, or in air bubbles or marine sprays. There are about 10,500 and 50,000 macroalga and microalga species, respectively. However, the physiological and biochemical patterns are known in a few algal species, and the biotechnological uses are still less known. The high diversity of habitats where algae are growing results in a high diversity of bioactive compounds for biotechnological uses. Although the full genome of several algal species is known, and genetic transformations are applied, the biodiversity of algae is so high that there is still a large amount of research to be conducted, and it is expected to find species with bioactive substances of interest for pharmacology, cosmetic, nutraceutic, agriculture and aquaculture, and energy applications.

The use of algae for biotechnology requires at least two key points: (1) high growth capacity in outdoor culture systems, i.e., high biomass productivity; and (2) optimal level of the biocompounds of interest in the culture conditions and high bioactive compound productivity. The combination of these two points is not easy to achieve since the biocompounds are generally accumulated under stress conditions in which the energy for growth is reduced, and consequently the biomass productivity is also reduced. Thus, the research in algal biotechnology is focused on the increase of biomass productivity maintaining good to high productivity of high value compounds. This research field has an interdisciplinary approach since it is a necessary expertise in engineering (hydraulic, mass interchange, chemical), biology, and biochemistry, and also in social sciences such as economy, commercial, and marketing, among others.

Bioassays. DOI: http://dx.doi.org/10.1016/B978-0-12-811861-0.00006-1

Microalgae present higher growth rates and photosynthetic efficiency than vascular terrestrial plants. Certain macroalgal species can also reach high growth rates under certain culture conditions. However, the costs of production are still high, and depending on the use of biomass, the production can be profitable or not.

To reduce the costs of production it is necessary to reduce the costs of hydrodynamics, nutrients, and harvesting. In this chapter we will focus on the reduction of production costs of cultured algae by using effluents of different origin such as sewage, pig farms, or fish aquaculture. Different bioassays during the whole process of algal production are presented:

1. Biofiltration assays and biomass productivity in the aquaculture grown in effluents
2. Biochemical assays in the analysis of the harvested biomass

6.2 Biofiltration assays and biomass productivity in aquaculture grown in effluents

In the last years, a great research effort has been conducted to culture algal species in effluents with two objectives: (1) Cost reduction through the production of algal biomass; and (2) the depuration of the waters by biofiltration of nutrients and the adsorption of metal (bioremediation), contributing to the increase of the ecological status of fresh and marine water environments according to the Water Framework Directive (Directive 2000/60/EC).

6.2.1 *Fishpond effluents. Integrated multitrophic aquaculture*

One of the main limitations for the development of modern aquaculture is the generation of particulate and dissolved wastes associated with animal feeding and metabolism. These residues could be the cause of eutrophication problems when dispersed to the environment. However, they might be used as "resources" at different levels. In the so-called "Integrated Multi-Trophic Aquaculture" (IMTA) approach, besides diversifying different aquatic animals and algal species that use "wastes" generated by higher trophic levels, the effects of fishpond effluents on the environment are reduced [1,2]. In particular, dissolved nutrients are mainly $N\text{-}NH_4^+$ and $P\text{-}PO_4^{3-}$ up to 70% and 20%, respectively.

In general, plants and algae (photoautotrophs) favor a stable equilibrium with respect to animals and microbials (heterotrophs) in the system, not only in relation to nutrients but also to O_2, CO_2, and pH of the medium. The role played by algae (micro- and macroalgae) is based on the use of the dissolved nutrients generated by higher trophic level organisms (fish or crustaceans grown at different carrying capacities and feed conversion rates, depending on the system), in addition to the CO_2 produced by the respiration process, being converted to biomass and O_2 through the photosynthetic process.

Macroalgae, in particular, have been described as efficient traps for the dissolved nutrients in IMTA systems. Different cultured genera (*Ulva, Gracilaria, Codium,*

Gracilariopsis, Palmaria, Hypnea, Chondrus, Kappaphycus, Porphyra, Falkenbergia, Hydropuntia, or *Laminaria*) have been successfully assayed as biofilters for effluents with dissolved nutrients—mainly ammonium—under many different factors and conditions, presenting high capacities for both uptake efficiencies (UE) and uptake rates (UR) [3]. These concepts should be very well understood when evaluating this approach. The uptake efficiency for a nutrient, i.e., nitrogen (NUE), is defined as the reduction (%) of this nutrient concentration in the effluent, while the uptake rate (NUR for nitrogen) is the nutrient concentration eliminated/assimilated from the effluent by culture surface/volume and time units. Both parameters vary depending on the environmental conditions affecting the cultivation system in a specific period (irradiance or water temperature), but also depending on important culture variables such as the system depth, algal density, or effluent turnover rate (nutrients flow).

NUE is inversely related to the effluent turnover rate and directly related to the biomass inoculation density in tanks. This was also demonstrated for NUR in the same unit, indicating that if the objective is to discharge clean waters to the environment, the NUE is the parameter to maximize; however, if the challenge is to increase biomass productivity resulting in a lower nutrients removal, the NUR is the parameter to control [4], according to the dynamic observed in Fig. 6.1.

Other multiple possibilities have been assayed to evaluate the use of algae in IMTA, not only related to the control of factors affecting biomass growth and productivity under culture conditions, but also those to improve biofiltration capacities (NUE and NUR). Some examples have shown possibilities for tanks distributed as a cascade or the co-cultivation of two different macroalgae species in the same tank

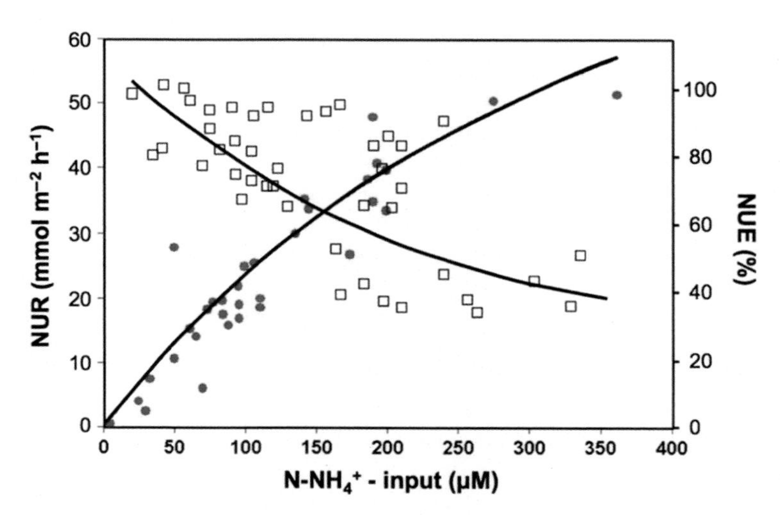

Figure 6.1 Dynamic for NUE expressed as % (*squares*) and NUR expressed as mmol $m^{-2}\,h^{-1}$ (*dots*) in relation to nutrient concentration (also for nutrient flow) in a macroalgae biofilter unit.

Table 6.1 N-ammonium biofiltration capacities in IMTA systems with different combinations of macroalgae species grown in tanks distributed in cascades

Tank distribution (TR)	Macroalgal species	$[N\text{-}NH_4^+]$ in effluent (μM)	NUE (%)	NUR (mmol $m^{-2} h^{-1}$)
Input	–	190.3 ± 63.9	–	–
1 CPT (4)	*Halopithys incurva*	9.32 ± 10.8	94.1 ± 7	45.3 ± 17.2
2 SCT (8)	*Hypnea spinella*	0	100	1.9 ± 2.3
Input	–	190.3 ± 63.9	–	–
1 CPT (4)	*Hypnea spinella*	1.4 ± 1.9	99.36 ± 0.9	47.3 ± 15.8
2 SCT (8)	*Ulva rigida*	0	100	0.3 ± 0.4
Input	–	125.9 ± 93.9	–	–
1 CPT (4)	*Halopithys incurva*	72.3 ± 41.9	38.2 ± 12.7	13.1 ± 2.5
2 SCT (8)	*Halopithys incurva*	47.9 ± 27.8	58.9 ± 6.2	5.3 ± 3.3

CPT, Polyethylene tanks with a volume of 1500 L and an aerial surface of 1.5 m^2; SCT, Semi-circular glass-fiber tanks with a volume of 750 L and an aerial surface of 1.8 m^2; TR, Turnover rate (vol d^{-1})
Source: Adapted from Gómez Pinchetti JL, Suárez Álvarez S, Güenaga Unzetabarrenechea L, Figueroa FL, García Reina G. Posibilidades para el desarrollo de sistemas integrados con macroalgas en las Islas Canarias y su entorno. In: Macroalgas en la Acuicultura Multitrófica Integrada Peninsular. Centro Tecnológico del Mar – Fundación CETMAR: Vigo; 2011. p. 75–93 [6].

[5]. Combination of species and tanks distributed in a cascade, where effluents were transferred by gravity from 1500 -L tanks with turnover rate at 4 vol d^{-1} to 750-L tanks with twice the turnover rate (8 vol d^{-1}), increased NUE to values near 100% and NUR higher than 40 mmol $m^{-2} h^{-1}$ (Table 6.1). Similar capacities were obtained when two macroalgae species, the Rhodophytes *Halopithys incurva* and *Jania adhaerens*, were cultivated in the same tank, at the same time, increasing NUE to 100% and showing NUR values higher than 30 mmol $m^{-2} h^{-1}$, values higher than those obtained when grown separately.

In relation to biofiltration capacities for P-phosphate, values for PUE and PUR are generally lower when compared to those for N-ammonium, mainly by considering that phosphorus is not a limiting element in seawater and macroalgae phosphorus requirements are generally lower [7].

The use of macroalgae in these IMTA systems has been mainly focused on their capacity as biofilters for dissolved nutrients removal and not on a potential market and added value as an additional organism produced in the same cultivation unit or water flow. Even in cases where algal biofiltration efficiency is not high and market price is low, biomass can be useful as feed for high value species such as abalone and other mollusk and echinoderm species [8]. Considering possibilities for recirculating aquaculture systems based on these described principles, many algal species show antimicrobial properties, being considered as a beneficial factor for the system [9].

6.2.2 Pig farm effluents

The utilization of manure as culture media to produce microalgae allows for treatment of this highly contaminant effluent and saves costs derived from the treatment of these residues [10]. In addition to an increase in the sustainability of the process by saving CO_2 emission, the environmental advantages result from avoiding spreading manure on land and related ammonia stripping. Nitrogen from livestock slurries contributes largely to environmental pollution through ammonia and nitrogen oxides emission to the atmosphere and nitrate leaching to the ground and surface water bodies. Moreover, the recovery of nutrients from wastes (especially N, P, and K), is becoming a research priority as they are finite resources concentrated in a few countries, or they require high energy input to be produced and/or transported, causing high greenhouse gas emissions [11].

Digested or undigested manure can be used to produce microalgae biomass, although the large concentration of contaminants in this type of effluent requires adequate design and operation of the final process to be used. Undigested manure contains the highest concentration of dissolved organic carbon (COD), nitrogen, and phosphorous, up to $20\ g\ L^{-1}$, $8\ g\ L^{-1}$ and $200\ mg\ L^{-1}$, respectively. After anaerobic digestion, the COD is partially reduced to $5\ g\ L^{-1}$ but nitrogen and phosphorous remain in the digestate. These values are one or two orders of magnitude larger than urban wastewater and whatever other wastewaters. Moreover, the nitrogen occurs mainly as ammonium that is toxic for most of microalgae strains at concentrations higher than $100\ mg\ L^{-1}$ [12]. To produce microalgae using this effluent it is mandatory to dilute it previously to adding it to the reactors, the operation in continuous mode being recommendable to minimize the adverse effect of higher concentrations of ammonia on the performance of the cultures. The optimization of the percentage of manure into the culture medium allows for maximization of the biomass productivity at the same time as the recovery of nutrients from the culture medium, with values up to $1.7\ g\ biomass \cdot L^{-1}\ day^{-1}$ equivalent to $0.17\ g\ N \cdot L^{-1}\ d^{-1}$ and $0.02\ g\ P \cdot L^{-1}\ d^{-1}$ being reported by Ledda et al. [13].

The optimal percentage of manure in the culture medium ranges from 0.5% to 20% according to the right composition of manure and strain/photobioreactor to be used. Robust strains of green microalgae such as the genera *Scenedesmus* and *Chlorella* have been reported to be most tolerant to this type of effluent, also the utilization of highly productive reactors as thin-layer cascades is recommendable to enhance the biomass productivity and nutrient consumption by the cells. Anyway, the high COD contained in manure, including after dilution, favors the presence of bacteria in the cultures. Thus, in the manure treatment process developed by microalgae-bacteria consortia, the large proportion of bacteria in the cultures is related to a higher COD or to a lower hydraulic retention time. No precise values exist regarding the contribution of bacteria to this consortia, but it can be as low as 5%, minimizing the COD supply to the reactors, or higher than the microalgae density when providing excessive COD to the systems. In any case, with adequate management and design, these systems allow for the development of robust and valuable treatment processes for these highly contaminant effluents at the same time as producing useful microalgae biomass for low-value applications such as biofertilizers or feed in aquaculture.

6.2.3 Sewage and leachate

Sewage or urban wastewater treatment is a highly nonsustainable process on a which large amount of energy (>0.5 kWh m^{-3}) and costs (>0.2 € m^{-3}) are used to remove nutrients (C, N, P) from the water and to dissipate them to the environment. The utilization of microalgae-based processes for wastewater treatment allows not only for the reduction of the energy requirement (<0.2 kWh m^{-3}) and costs (<0.1 € m^{-3}), but to produce useful microalgae biomass for low-value applications. Thus, up to 1 kg of microalgae biomass can be produced from 1 m^3 of wastewater, this proportion increasing ten times when using centrate from anaerobic digestion. The coupling between wastewater treatment and microalgae production was first reported by Oswald in the 1960s, but now it is being redesigned [14].

In conventional wastewater treatment processes three main types of wastewater can be defined: (1) after primary treatment once large particles have been removed; (2) after secondary treatment once the organic matter has been removed; and (3) centrate after anaerobic digestion of sludge produced mainly during secondary treatment. The nutrient contents of these three different effluents are largely different, thus the coupling of microalgae production with each one of them must be adequately designed. Thus, secondary treated wastewater contains low concentrations of COD, nitrogen, and phosphorous of 100, 15, and 2 mg L^{-1} respectively, that result in a limiting culture medium to produce microalgae. In either case, microalgae can be produced using this wastewater, thus allowing for a tertiary treatment to be performed, removing N and P to avoid eutrophication problems [15]. The utilization of wastewater from primary treatment is the application of greater interest because microalgae can be produced at the same time as treating wastewater, thus saving energy and costs related with wastewater treatment. Wastewater from primary treatment contains COD, nitrogen, and phosphorous of 800, 50, and 10 mg L^{-1} respectively, which are adequate for microalgae production. However, due to the presence of high concentrations of COD in these cultures microalgae-bacteria consortia always exist. The proportion of bacteria can be reduced by optimizing the conditions to favor the growth of microalgae cells as using fast-growing strains and optimal light availability [14]. Finally, centrate from anaerobic digestion is the richer effluent containing COD, nitrogen, and phosphorous of 150, 400, and 40 mg L^{-1}, respectively. The large concentration of ammonium imposed the requirement that centrate must be diluted prior to being used as culture medium, with the optimal percentage of centrate ranging from about 30%–50% [16]. Centrate is thus a nutrients-rich effluent containing low COD that can be used not only to produce freshwater but also seawater strains; at the same time the proportion of bacteria into the cultures is very low [17].

Whichever wastewater type is used the final biomass productivity is approximately the same with using fertilizers as with culture media, thus values higher than 40 $g \cdot m^{-2} d^{-1}$ can be obtained using *Scenedesmus* in thin-layer reactors [18].

Concerning removal capacity, under optimal conditions more than 90% of COD, nitrogen, and phosphorous can be removed.

6.3 Biochemical assays in the analysis of harvested biomass

The algal biomass produced in the cultures can be used in different applications depending on the effluent applied and the quality of biomass. Thus, macroalgae grown in fishpond effluents could be used both as food and feed, and as cosmetics, however the biomass produced with urban sewage can be used for the production of biostimulants and biofuels for agriculture, but not for food (Fig. 6.2).

6.3.1 Biostimulant assays

Microalgae contain organic matter, especially amino acids, and phytohormones in addition to polysaccharides, useful as biostimulants of the growth in higher plants. Thus, microalgae have been proposed as biofertilizers due to their amino acid profile and content in other algae-derived natural substances [19]. Amino acid-based fertilization supplies plants with the necessary elements to develop their structures, saving metabolic energy since the nitrogen does not have to be reduced, as in the case of nitrate, which must be reduced to ammonium prior to its incorporation and

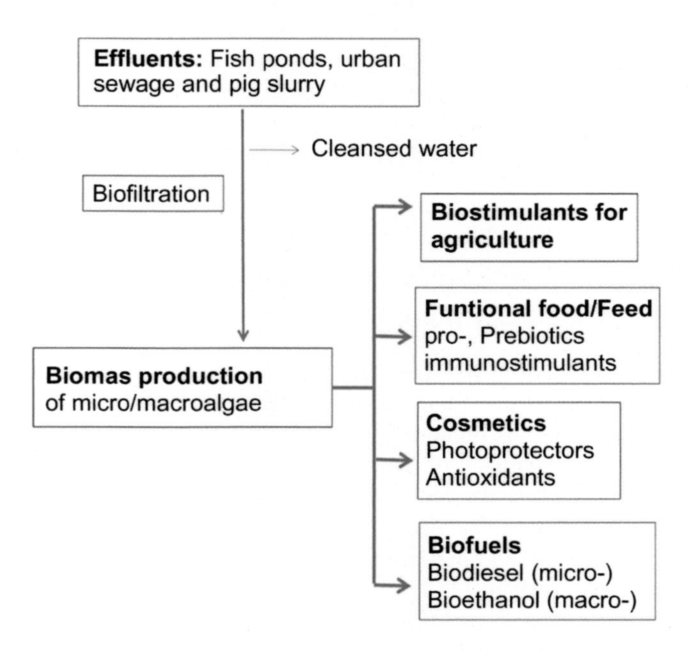

Figure 6.2 Potential uses of the biomass produced by the biofiltration of diverse effluents.

conversion to α-ketoacids in order to synthesize amino acids. It has been demonstrated that the application of amino acids to leaves [20] and soils [21] has a favorable effect on plants. Additionally, microalgae contain phytohormones including auxin, abscisic acid, cytokinine, ethylene, and gibberellins [22]. These phytohormones are small molecules that serve as chemical messengers to coordinate cellular activities in higher plants. It has been suggested that the hormone system in microalgae was the precursor to the existing hormone system in higher plants. The role of these hormones in microalgae is still unknown but there is much evidence about the positive effects of these compounds in higher plants. Thus, the improvement of maize anther culture through the supplement of microalgal and cyanobacterial biomass has been analyzed in depth [23].

To be used as biofertilizer, the microalgae biomass must be adequately produced and processed in order to avoid toxicity problems and efficient use. Thus, microalgae biomass free of pathogens and heavy metals must be produced, although it can be obtained using the inclusion of wastewater if the final composition of the biomass accomplishes legal limits for usage. Concerning processing, the microalgae biomass must be processed using soft methods that allow making available the bioactive compounds contained in the biomass but without denaturalizing it. In this sense soft enzymatic methods are preferred to intense energy or chemicals methods, including the utilization of supercritical fluid extraction (SFE) being proposed [24,25]. Enzymatic methods consist of using cocktails of enzymes with cellulolytic and proteolytic activities to perform cell disruption and protein breakdown to obtain peptides and free amino acids. Controlling the extension of protein breakdown, the quality of the final product is modified; the total breakage of proteins into free amino acids is not desired. The quality of the final product largely depends on the quality/composition of the microalgae biomass used and performance of enzymes used, in addition to the control of the processing step. The utilization of chemical methods also enables the obtaining biofertilizers from microalgae, but at the expense of reducing the quality of the final products by decomposing sensible phytohormones and L-amino acids.

6.3.2 Functional feed and food from algae

6.3.2.1 Functional feed for fish

Aquaculture is an expanding industry, and it is the largest overall user of fishmeal because this product is the major dietary ingredient used in aquafeeds [26]. However, its production has remained relatively static over last 30 years, while its demand has increased rapidly [26]. Due to the high protein requirements of teleost carnivorous species, alternative vegetable protein sources are being considered [27] to improve sustainability and reduce cost. Algae have been proposed as a valuable alternative protein source, and partial replacement of fishmeal with algae in diets of farmed fish has been assayed [28]. Algal species with a fast growth rate and low-cost production are required, and those that can be successfully cultured in an integrated way using fish farm effluents. They must be biocompatible, biodegradable, and safe for the environment and human health. However, it is important to ensure that new ingredients do not affect to factors implied in the functionality of the intestine, such as the intestinal microbiota, which is sensitive to a wide range of factors including diet [29]. In addition to this, the structure of the intestinal population of bacteria hinders the establishment of pathogenic microorganisms in the intestinal tract [30].

Substitution of fishmeal for algae meal

Partial substitutions of fishmeal by 5%, 15%, or 25% of *Gracilaria cornea* or *Ulva rigida* in experimental diets were evaluated to study their effects on biodiversity of intestinal microbiota composition of gilthead sea bream (*Sparus aurata*) [31]. The diets were offered to triplicate groups of 15 juvenile fish (15.4 g). After 70 days of feeding, results showed that the substitution of fishmeal with algae meal induced important modifications in the biodiversity of the intestinal microbiota, as a high reduction of the biodiversity when the highest percentage (25%) of *U. rigida* was included. On the contrary, an increase on the number of species was detected when the dietary inclusion of algae were 15%. Various *Lactobacillus delbrueckii* subspecies were selectively stimulated when *G. cornea* was included in the feed, and other bacterial species, such as those included in the *Vibrio* genus, were reduced.

Techniques used for evaluation of fish feed

Ingredients, experimental diets, and fish carcasses were analyzed for proximate composition according to Association of agricultural chemists (AOAC) [32].

Hematology and other metabolic analysis Plasma samples are separated from cells by centrifugation of whole blood (3 min, 10,000 g, 4°C), and stored at −80°C. The intestinal microbiota are analyzed by Denaturing Gel Gradient Electrophoresis (DGGE).

Respiratory burst activity The generation of intracellular superoxide radicals by sole phagocytes is determined by the reduction of nitro-blue tetrazolium (NBT) according to the method of Boesen et al. [33].

6.3.2.2 Functional food for human

For decades, seaweeds have been an important source of metabolites for both food and nonfood industries. Phycocoloids such as agar and carrageenan, for example, have been used as texturing agents in the food industry, while phenolic compounds have been used as antimicrobial agents in the pharmaceutical industry. Recently, polysaccharides from seaweeds have also been reported as having antitumoral properties. Algal cell-wall sugars have been shown to promote biochemical and physiological responses in a number of vertebrates. Algae are called to be the food of the future as a source of proteins.

Polysaccharides of algae with immunodulatory activity

Polysaccharides from *Porphyridium cruentum*, for example, inhibit viral replication and are a potent hypocholesterolaemic agent in rats and chickens [34]. In addition, these polysaccharides promote antioxidant activity suggesting a cell-protecting mechanism against reactive oxygen species [35]. Furthermore, the in vivo administration of the exocellular polysaccharide of *P. cruentum* to mice has resulted in an increase of the macrophage population as well as in an increase of the acid phosphatase enzyme [36]. Macrophages are the first line of defense against microbial infection. Modulation of the macrophage function by polysaccharides can help increase phagocytosis, microbicidal activity, chemotaxis, and antigen presentation to T cells thus helping in preventive and therapeutic strategies against diseases [37].

Macrophage defense against pathogens includes cytokine secretions like interleukin (IL6), tumor necrosis factor α (TNF-α), and inflammation mediators like nitric oxide (NO). The polysaccharides isolated from algae have been reported to modify macrophage activity, inducing the production of cytokines (TNF-α, IL1, and IL6). Inflammatory responses are essential but need to be tightly regulated. If this regulation fails, it can lead to chronic inflammatory diseases, such as rheumatoid arthritis (RA). In RA and other autoimmune diseases, over-expression of inflammatory cytokines, such as interleukin (IL-6) and TNF-α, has been implicated in their pathogenesis [38]. On the other hand, increased expression of the antiinflammatory cytokine IL-10 reduces inflammation [39]. Therefore, the modulation cytokine production and activity is one focus of treating these diseases.

Techniques used to assess immunological activity

Determination of cytokines In order to evaluate the immunological effect of algal polysaccharides, murine macrophage cell line RAW 264.7 (ATCC, USA) is used. RAW cells are cultured in the presence of different concentrations of the potential bioactive compounds, i.e., polysaccharides in a 24-well microtiter plate (5×10^5 cells $well^{-1}$) in a total volume of 1 mL. Bacterial LPS (50 ng mL^{-1}) is used as positive control for macrophage activation. The production of TNF-α and IL6 is measured by sandwich enzyme-linked immunosorbent assays (ELISA) as described by Martínez et al. [40]. Briefly, a purified rat antimouse monoclonal TNF-α or IL6 antibody (0.5 mg, BD Pharmingen) was used for coating at 2 μg mL^{-1} at 4°C for 16 h. After washing and blocking with PBS containing 3% bovine serum albumin, culture supernatants are added to each well for 12 h at 4°C. Unbound material is washed off and a biotinylated monoclonal antimouse TNF-α or IL-6 antibody (0.5 mg, BD Pharmingen) is added at 2 μg mL^{-1} for 2 h. Bound antibody is detected by addition of avidin peroxidase for 30 min followed by addition of the 2,2′-azinobis (3-ethylbenzothiazoline-6-sulfonic acid) (ABTS) substrate solution. Absorbance at 405 nm is measured 10 min after the addition of the substrate.

6.3.3 Cosmetics

Seaweeds provide an excellent source of bioactive compounds such as carotenoids, polysaccharides, proteins, lipids, fatty acids pigments, vitamins, polyphenols, and microelements. Algal biochemical compounds are used as cosmeceutical, derived from cosmetics with potential pharmaceutical use and referring to specific products containing active ingredients. Recently, there is increased interest in naturally produced active compounds as an alternative to synthetic substances. In recent years, there has been increased interest in the search for natural antioxidants, which usually consist of mixtures of compounds with diverse biological functions [41,42]. Of particular interest are antioxidants present in algae and their extracts, as synthetic antioxidants have been restricted because of toxicity and health risks.

The main goals in the production of algal extracts are to achieve a high yield of desired compounds, to preserve co-products, minimize energy consumption, optimize the process, increase the scale, and to be environmentally friendly [43]. Depending on the applied extraction technique, different bioactive compounds are extracted from the algal biomass. Therefore, solvents are an important factor for extracting antioxidants from these natural sources. In general, four solvents are more frequently used for antioxidant studies, in the order of frequencies, ethanol, water, methanol, and aqueous methanol [44,45]. All of these have good polarity and could be used to extract polar compounds, such as phenolic compounds, carotenoids, mycosporine-like amino acids, and flavonoids that have antioxidant properties [46,47] Among them, ethanol—being organic and nontoxic—has the highest frequency of use [48]. Although until now methanol is the third most frequently used, its toxicity must limit its use, as the extracts potentially would be used in cosmeceuticals. The use of solvents is a traditional extraction technique that requires high volumes and long extraction times. Moreover, low extraction yields of bioactive compounds with low selectivity are obtained [49].

An alternative to the traditional extraction is to use novel extraction techniques, such as SFE, microwave-assisted extraction (MAE), ultrasound-assisted extraction (UAE), enzyme-assisted extraction (EAE), pressurized liquid extraction (PLE; also known as pressurized fluid extraction, enhanced solvent extraction, high pressure solvent extraction, or accelerated solvent extraction techniques), which have been successfully used for the extraction of bioactive compounds for the food and pharmaceutical industries [43].

Antioxidant activity cannot be measured directly, but can be determined by the effects of the antioxidant compound in a controlled oxidation process. According to this, measuring a sample oxidizer, intermediate or end products may be used to evaluate antioxidant activity. Moreover, antioxidant activity should not be concluded based on a single antioxidant method. In fact, several in vitro assays are carried out for evaluating antioxidant activities with the samples of interest. Those methods based on free radical traps are relatively straightforward to perform. The 2,2-diphenyl-1-picrylhydrasyl (DPPH) method is the most rapid, simple, and inexpensive in comparison to other assays. On the other hand, the ABTS assay is also widely used, as it is applicable to both hydrophilic and lipophilic antioxidants, as the radical ABTS can be dissolved in organic or aqueous medium. By contrast, the radical DPPH can only be measured in an organic medium, which is a limiting interpretation of the antioxidant capacity of hydrophilic compounds evaluated [50]. In addition to those two methods, superoxide radical scavenging activity, hydrogen peroxide scavenging assay, and the β-carotene bleaching method (BBM) are also described in this chapter.

6.3.3.1 DPPH assay

The DPPH (2, 2-diphenyl-1-picrylhydrasyl) free-radical scavenging assay is carried out according to the method of Brand-Willliams et al. [51]. Briefly, 1.35 mL of 90% methanol (v/v) and 150 μL of each algal methanolic extract are mixed with 150 μL of a 90% methanolic DPPH solution prepared daily at 1.27 mM. The reaction is complete after 30 min in the dark at room temperature, and the absorbance is read at 517 nm.

6.3.3.2 ABTS assay

This method is based on the decrease in the absorbance (decolorization) of the radical cation of 2,2′-azinobis (3-ethylbenzothiazoline 6-sulfonate; $ABTS^+$) generated by oxidation of ABTS with potassium persulfate [52]. The cation has characteristic absorption peaks at 413, 660, 734, and 820 nm, and the addition of antioxidants to the preformed radical determines its reduction and can be followed by a decrease in absorbance. In a final incubation volume of 1 mL, $ABTS^+$ solution is diluted in phosphate buffer (50 mM) to an initial absorbance of 0.7–0.75 at 413 nm (control). After the extracts are added the absorbance at 413 nm is measured spectrophotometrically after 1 min [47].

6.3.3.3 Superoxide radical scavenging activity

The superoxide anion scavenging activity can be measured as described by Robak and Gryglewski [53]. The superoxide anion radicals are generated in 3.0 mL of Tris–HCl buffer (16 mM, pH 8.0), containing 0.5 mL of NBT (0.3 mM), 0.5 mL nicotinamide adenine dinucleotide phosphate (NADH) (0.936 mM) solution, 1.0 mL extract, and 0.5 mL Tris–HCl buffer (16 mM, pH 8.0). The reaction is initiated by adding 0.5 mL phenazine methosulfate (PMS) solution (0.12 mM) to the mixture, incubated at 25°C for 5 min and then the absorbance measured at 560 nm against a blank sample.

6.3.3.4 Hydrogen peroxide scavenging assay

This method is based on the decrease in the absorbance of H_2O_2 upon its oxidation. The method is simple, reproducible, and cost effective. The ability of algal extracts to scavenge hydrogen peroxide is determined according to the method of Ruch et al. [54]. A solution of hydrogen peroxide (40 mM) was prepared in phosphate buffer (50 mM pH 7.4). Extracts (20–60 μg mL^{-1}) in distilled water were added to a hydrogen peroxide solution (0.6 mL, 40 mM). Absorbance of hydrogen peroxide at 230 nm was determined 10 min later against a blank solution containing the phosphate buffer without hydrogen peroxide.

The calculations for these four antioxidant methods (DPPH, ABTS, superoxide radical scavenging activity and H_2O_2 scavenging assay) could be done following two procedures:

1. Calculate the antioxidant activity or percentage of scavenging of the extracts by the following equation: $AA\% = (A_c - A_s)/A_c \times 100$, where A_c is the absorbance of the control and A_s is the absorbance in the presence of the sample extracts or standards.

2. Calculate the EC_{50} value (oxidation index), which represents the concentration of the extract (mg mL^{-1}) required to scavenge 50% of the radical present in the reaction mixture (DPPH, $ABTS^+$, or H_2O_2). For this purpose it is necessary to make a calibration curve with DPPH to calculate the remaining concentration of DPPH in the reaction mixture after incubation. Then, values of DPPH concentration (μM) are plotted against plant extract concentration (mg DW mL^{-1}) in order to obtain the EC_{50} value.

A standard Trolox (6-hydroxy-2,5,7,8-tetramethylchroman-2-carboxylic acid) solution (0−15 μM), a water-soluble analog of vitamin E, is used as antioxidant of reference. That allows the comparison with other studies of the percentage of scavenging of free radicals [55], since the protocols followed by different investigators to determine antioxidant activity vary under different conditions (proportions of the reactants, concentrations, and times of measurement).

6.3.3.5 β-carotene bleaching method (BBM)

This method is based on the bleaching of β-carotene via heat-induced oxidation to discoloration being inhibited or diminished by antioxidants that donate electrons [56]. β-carotene (1.6 mg), 200 mg of Tween-20 (polyoxyethylene sorbitan monopalmitate) and 37.8 mg of linoleic acid are mixed in 0.8 ml chloroform under nitrogen atmosphere to avoid oxidation. The limitation is that the discoloration of β-carotene at 470 nm can occur through multiple pathways, thereby complicating the interpretation of results.

6.3.4 Biofuel

Production of biodiesel from microalgae has been repeatedly proposed and although it is technically feasible the actual production capacity and cost make this process unrealistic at a commercial scale. Thus, it has been reclaimed that microalgae are the most efficient biological systems to transform undesirable compounds such as CO_2 and free solar energy into biofuels [57]. However, the actual production capacity worldwide is lower than 100,000 tons per $year^{-1}$ at a production cost higher than €10 per kg^{-1} thus being not comparable with the biofuel industry requiring millions of tons of raw materials at prices lower than €0.5 per kg^{-1}. Moreover the actual microalgae production systems are focused on human applications of the biomass, thus with energy consumption being excessive for accomplishing the requirements of the biofuels market. To solve these problems it is clear that new developments must be performed in the biology and engineering of the process, new processes that are much more efficient and at a much larger scale have to be developed.

To produce almost free and sustainable microalgae biomass is only the first challenge, the second one being the adequate processing of this biomass. Thus, different processes have been proposed to obtain biofuels from microalgae biomass. The most conventional implies to extract lipids from microalgae using solvents after drying the biomass, the lipids then being used as vegetable oil to obtain biodiesel by transesterification using methanol and catalyzers. This route is scarcely efficient because microalgae produced in continuous mode contain only a maximal lipids content of 30%, half of them being nonsaponifiable lipids, thus only 15% of the biomass is suitable for producing biodiesel. Moreover, to dry the biomass requires large amounts of energy that make the process unsustainable. In order to resolve the last problem, direct transesterification of wet biomass has been proposed, but in this case larger volumes of methanol and solvents are required [58]. Alternatively, different thermochemical processes have been proposed, the most promising being the hydrothermal liquefaction [59]. In this process the biomass is subject to high pressure and temperature and different fractions are obtained as carbon-rich gases, bio-oil, and char; up to 40% of the raw biomass is transformed into bio-oil that later must be refined and processed. The quality of bio-oil can be improved by prior removal of nitrogen and oxygen-rich compounds from the raw biomass [60]. Alternatively, the production of bioethanol and bio-hydrogen has been also proposed from microalgae biomass mainly using the carbohydrate fraction, however, none of them are today performed at large scale due to the low production capacity [61,62]. The most promising alternative to producing biofuels from microalgae is coupling the production of microalgae biomass with wastewater treatment processes. Then the produced biomass is anaerobically digested to produce biogas [63]. In this case sustainable processes can be performed with up to 60% of the produced biomass being efficiently converted to biomethane, useful for transport, heat, and electricity supply.

6.4 Biorefinery and integration of the production and biochemistry processes

Microalgae biomass produced using both chemicals/fertilizers as culture medium or alternatively wastewater from human or animal sources have similar biochemical composition, thus technically they can be used for the same final purpose although regulations avoid the use of microalgae biomass obtained from wastes for direct human uses. Microalgae biomass mainly contains proteins, lipids, and carbohydrates thus different biorefinery schemes have been proposed to valorize the complete biomass [64]. High value compounds contained into the biomass must be first extracted and then the waste biomass can be used for low value applications. Valuable compounds contained in the biomass include carotenoids, polyunsaturated fatty acids, and sterols, among others. In terms of global valorization of the biomass, different schemes can be proposed, from the single production of biogas to the complete fractionation of the major components of the biomass. Whatever the proposed scheme is, two major criteria must be accomplished: (1) most valuable compounds must be first extracted to avoid losses, and (2) mild conditions must be used in each step to prevent decomposition of the remaining compounds. Fig. 6.3 shows a scheme of a biorefinery process for the complete valorization of microalgae biomass (Patent PCT/ES2013/070064).

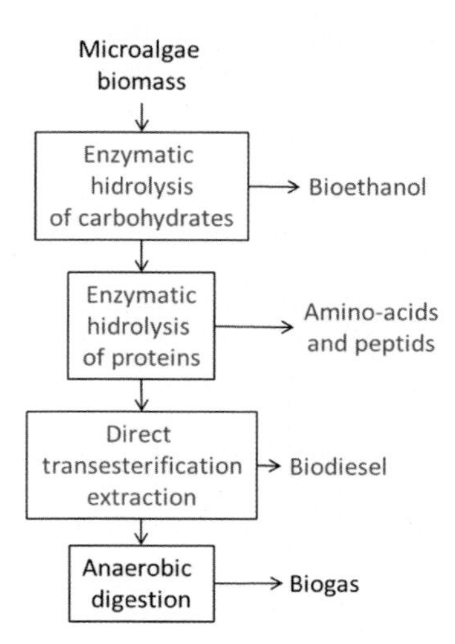

Figure 6.3 Scheme of a biorefinery process proposed for the complete valorization of microalgae biomass.

According to this scheme the most valuable proteins are first extracted as free amino acids and peptides through enzymatic hydrolysis to produce amino acid concentrates, next the carbohydrates are extracted through thermochemical hydrolysis under mild conditions, to be later fermented to produce bioethanol. The removal of proteins and carbohydrates from the initial biomass increases the percentage of lipid content in the waste biomass, enhancing the yield of the lipids valorization step. Saponificable lipids can be transformed into biodiesel through direct transesterification and the remaining waste biomass is suitable to be used to produce biogas by anaerobic digestion. The optimization of each step allows determining the maximal yield achievable, but it must be taken into account that in each step losses of other compounds can take place. Thus, only a precise optimization of the entire process can define the maximal amount of final products to be obtained. Under optimal conditions the patented process allows for utilization of 66% of the raw material; thus 34% is lost during the processing. This value is higher than when producing biogas from direct biomass; in that case a maximal biomass use efficiency of 62% can be achieved. If only single products are obtained, maximal biomass use efficiency is 30% in production of amino acid concentrates.

Although a large number of biorefinery schemes have been proposed, not all of them are technically or economically feasible. The economy of processing the biomass to obtain valuable products is greatly related to the price of the products to be obtained, and the biomass production costs. Microalgae biomass production costs are largely dependent on the technology used and the final application of the biomass [65]. The coupling of microalgae production with wastewater treatment is the only way to reduce the production costs below €1 per kg^{-1} [14]. In this scenario the production of low value products such as amino acid concentrates, bioethanol, biodiesel, and biogas is suitable. However, these type of processes must still be validated at the pilot scale to confirm reliability of the technology proposed and overall economy and sustainability of the entire process.

References

[1] Neori A. Essential role of seaweed cultivation in integrated multi-trophic aquaculture farms for global expansion of mariculture: an analysis. J Appl Phycol 2008;20 (5):567–70.

[2] Troell M, Joyce A, Chopin T, Neori A, Buschmann AH, Fang J-G. Ecological engineering in aquaculture—potential for integrated multi-trophic aquaculture (IMTA) in marine offshore systems. Aquaculture 2009;297(1):1–9.

[3] Neori A, Chopin T, Troell M, Buschmann AH, Kraemer GP, Halling C, et al. Integrated aquaculture: rationale, evolution and state of the art emphasizing seaweed biofiltration in modern mariculture. Aquaculture 2004;231(1):361–91.

[4] Buschmann AH, Troell M, Kautsky N. Integrated algal farming: a review. Cah Biol Mar 2001;42(1):83–90.

[5] Güenaga L. Influence of solar radiation and ammonium on the accumulation of antioxidant substances in seaweeds grown in a biofiltration system. Malaga: University of Málaga; 2011.

[6] Gómez Pinchetti JL, Suárez Álvarez S, Güenaga Unzetabarrenechea L, Figueroa FL, García Reina G. Posibilidades para el desarrollo de sistemas integrados con macroalgas en las Islas Canarias y su entorno. In: Macroalgas en la Acuicultura Multitrófica Integrada Peninsular. Centro Tecnológico del Mar – Fundación CETMAR: Vigo; 2011. 75–93.

[7] Harrison PJ, Hurd CL. Nutrient physiology of seaweeds: application of concepts to aquaculture. Cah Biol Mar 2001;41:71–82.

[8] Neori A, Troell M, Chopin T, Yarish C, Critchley A, Buschmann AH. The need for a balanced ecosystem approach to blue revolution aquaculture. Environ Sci Pol Sustain Dev 2007;49(3):36–43.

[9] Bansemir A, Blume M, Schröder S, Lindequist U. Screening of cultivated seaweeds for antibacterial activity against fish pathogenic bacteria. Aquaculture 2006;252(1):79–84.

[10] Ledda C, Schievano A, Scaglia B, Rossoni M, Fernández FGA, Adani F. Integration of microalgae production with anaerobic digestion of dairy cattle manure: an overall mass and energy balance of the process. J Clean Prod 2016;112:103–12.

[11] Wood S, Cowie A, editors. A review of greenhouse gas emission factors for fertiliser production. In: IEA bioenergy task; 2004;38:20.

[12] Collos Y, Harrison PJ. Acclimation and toxicity of high ammonium concentrations to unicellular algae. Mar Pollut Bull 2014;80(1):8–23.

[13] Ledda C, Villegas GR, Adani F, Fernández FA, Grima EM. Utilization of centrate from wastewater treatment for the outdoor production of Nannochloropsis gaditana biomass at pilot-scale. Algal Res 2015;12:17–25.

[14] Acién FG, Gómez-Serrano C, Morales-Amaral M, Fernández-Sevilla J, Molina-Grima E. Wastewater treatment using microalgae: how realistic a contribution might it be to significant urban wastewater treatment?. Appl Microbiol Biotechnol 2016;100(21): 9013–22.

[15] Gómez-Serrano C, Morales-Amaral M, Acién F, Escudero R, Fernández-Sevilla J, Molina-Grima E. Utilization of secondary-treated wastewater for the production of freshwater microalgae. Appl Microbiol Biotechnol 2015;99(16):6931–44.

[16] del Mar Morales-Amaral M, Gómez-Serrano C, Acién FG, Fernández-Sevilla JM, Molina-Grima E. Production of microalgae using centrate from anaerobic digestion as the nutrient source. Algal Res 2015;9:297–305.

[17] Sepúlveda C, Acién F, Gómez C, Jiménez-Ruíz N, Riquelme C, Molina-Grima E. Utilization of centrate for the production of the marine microalgae *Nannochloropsis gaditana*. Algal Res 2015;9:107–16.

[18] del Mar Morales-Amaral M, Gómez-Serrano C, Acién FG, Fernández-Sevilla JM, Molina-Grima E. Outdoor production of Scenedesmus sp. in thin-layer and raceway reactors using centrate from anaerobic digestion as the sole nutrient source. Algal Res 2015;12:99–108.

[19] Ördög V, Pulz O. Diurnal changes of cytokinin-like activity in a strain of *Arthronema africanum* (Cyanobacteria), determined by bioassays. Algol Stud 1996;82:57–67.

[20] Jie M, Raza W, Xu YC, Shen Q-R. Preparation and optimization of amino acid chelated micronutrient fertilizer by hydrolyzation of chicken waste feathers and the effects on growth of rice. J Plant Nutr 2008;31(3):571–82.

[21] Mitchell D, Fullen M. Soil-forming processes on reclaimed desertified land in north-central China. In: Millington AC, Pye K, editors. Environmental Change in Drylands. J. Wiley, Chichester. 1994. p. 393–412.

[22] Lu Y, Xu J. Phytohormones in microalgae: a new opportunity for microalgal biotechnology? Trends Plant Sci 2015;20(5):273–82.

[23] Jäger K, Bartók T, Ördög V, Barnabás B. Improvement of maize (Zea mays L.) anther culture responses by algae-derived natural substances. S Afr J Bot 2010;76(3):511–16.

[24] Garcia JR, Fernández FA, Sevilla JF. Development of a process for the production of L-amino-acids concentrates from microalgae by enzymatic hydrolysis. Bioresour Technol 2012;112:164–70.

[25] Michalak I, Dmytryk A, Wieczorek PP, Rój E, Łęska B, Górka B, et al. Supercritical algal extracts: a source of biologically active compounds from nature. J Chem 2015;2015:14.

[26] FAO. Fishery and aquaculture statistics. Rome 2012. ISBN 978-92-5-0028293-6, p. 76

[27] Bendiksen EÅ, Johnsen CA, Olsen HJ, Jobling M. Sustainable aquafeeds: progress towards reduced reliance upon marine ingredients in diets for farmed Atlantic salmon (Salmo salar L.). Aquaculture 2011;314(1):132–9.

[28] Stadtlander T, Khalil W, Focken U, Becker K. Effects of low and medium levels of red alga Nori (*Porphyra yezoensis* Ueda) in the diets on growth, feed utilization and metabolism in intensively fed Nile tilapia, *Oreochromis niloticus* (L.). Aquacult Nutr 2013;19 (1):64–73.

[29] de Paula Silva FC, Nicoli JR, Zambonino-Infante JL, Kaushik S, Gatesoupe F-J. Influence of the diet on the microbial diversity of faecal and gastrointestinal contents in gilthead sea bream (*Sparus aurata*) and intestinal contents in goldfish (*Carassius auratus*). FEMS Microbiol Ecol 2011;78(2):285–96.

[30] Ringø E, Myklebust R, Mayhew TM, Olsen RE. Bacterial translocation and pathogenesis in the digestive tract of larvae and fry. Aquaculture 2007;268(1):251–64.

[31] Rico Blanco R. Efecto fisiológico y microbiológico del uso de algas en la alimentación de peces como sustitutivo de harina de pescado. Hacia una acuicultura sostenible. Málaga: University of Málaga; 2012.

[32] AOAC. Official methods of analysis. Washington, DC: Association of Official Analytical Chemists; 1995.

[33] Boesen HT, Larsen MH, Larsen JL, Ellis AE. In vitro interactions between rainbow trout (Oncorhynchus mykiss) macrophages and Vibrio anguillarum serogroup O2a. Fish Shellfish immunol 2001;11(5):415–31.

[34] Talyshinsky MM, Souprun YY, Huleihel MM. Anti-viral activity of red microalgal polysaccharides against retroviruses. Cancer Cell Int 2002;2(1):8.

[35] Tannin-Spitz T, Bergman M, van-Moppes D, Grossman S, Arad SM. Antioxidant activity of the polysaccharide of the red microalga Porphyridium sp. J Appl Phycol 2005;17 (3):215–22.

[36] Quevedo HJM, Manrique CEM, Díaz RTA, Pupo GC. Evidencias preliminares de la actividad inmunomoduladora de la fracción polisacárida de origen marino PC-1. Rev Cubana Oncol 2000;16(3):171–6.

[37] Desai VR, Ramkrishnan R, Chintalwar GJ, Sainis K. G1-4A, an immunomodulatory polysaccharide from Tinospora cordifolia, modulates macrophage responses and protects mice against lipopolysaccharide induced endotoxic shock. Int Immunopharmacol 2007;7(10):1375–86.

[38] Brennan FM, McInnes IB. Evidence that cytokines play a role in rheumatoid arthritis. J Clin Invest 2008;118(11):3537–45.

[39] Rossol M, Heine H, Meusch U, Quandt D, Klein C, Sweet MJ, et al. LPS-induced cytokine production in human monocytes and macrophages. Crit Rev Immunol 2011;31 (5):379–446.

[40] Martinez C, Delgado M, Pozo D, Leceta J, Calvo JR, Ganea D, et al. Vasoactive intestinal peptide and pituitary adenylate cyclase-activating polypeptide modulate endotoxin-induced IL-6 production by murine peritoneal macrophages. J Leukoc Biol 1998;63(5):591–601.

[41] Stengel DB, Connan S, Popper ZA. Algal chemodiversity and bioactivity: sources of natural variability and implications for commercial application. Biotechnol Adv 2011;29(5):483–501.

[42] Park J-Y, Kim JH, Kwon JM, Kwon H-J, Jeong HJ, Kim YM, et al. Dieckol, a SARS-CoV 3CL pro inhibitor, isolated from the edible brown algae Ecklonia cava. Bioorg Med Chemi 2013;21(13):3730–7.

[43] Michalak I, Chojnacka K. Algal extracts: technology and advances. Eng Life Sci 2014;14(6):581–91.

[44] Yuan YV, Bone DE, Carrington MF. Antioxidant activity of dulse (*Palmaria palmata*) extract evaluated in vitro. Food Chem 2005;91:485–94.

[45] O'sullivan A, O'Callaghan Y, O'Grady M, Queguineur B, Hanniffy D, Troy D, et al. In vitro and cellular antioxidant activities of seaweed extracts prepared from five brown seaweeds harvested in spring from the west coast of Ireland. Food Chem 2011;126 (3):1064–70.

[46] Pérez-Rodríguez L, Mougeot F, Alonso-Alvarez C. Carotenoid-based coloration predicts resistance to oxidative damage during immune challenge. J Exp Biol 2010;213 (10):1685–90.

[47] De la Coba F, Aguilera J, Figueroa F, De Gálvez M, Herrera E. Antioxidant activity of mycosporine-like amino acids isolated from three red macroalgae and one marine lichen. J Appl Phycol 2009;21(2):161–9.

[48] Alam MN, Bristi NJ, Rafiquzzaman M. Review on in vivo and in vitro methods evaluation of antioxidant activity. Saudi Pharm J 2013;21(2):143–52.

[49] Ibañez E, Herrero M, Mendiola JA, Castro-Puyana M. Extraction and characterization of bioactive compounds with health benefits from marine resources: macro and micro

algae, cyanobacteria, and invertebrates. In: Hayes M, editor. Marine Bioactive compounds:sources, characterization and applications. SpringerScience, Bussiness Media; 2012. p. 55–98.
[50] Surveswaran S, Cai Y-Z, Corke H, Sun M. Systematic evaluation of natural phenolic antioxidants from 133 Indian medicinal plants. Food Chem 2007;102(3):938–53.
[51] Brand-Williams W, Cuvelier M-E, Berset C. Use of a free radical method to evaluate antioxidant activity. LWT-Food Sci Technol 1995;28(1):25–30.
[52] Re R, Pellegrini N, Proteggente A, Pannala A, Yang M, Rice-Evans C. Antioxidant activity applying an improved ABTS radical cation decolorization assay. Free Radic Biol Med 1999;26(9):1231–7.
[53] Robak J, Gryglewski RJ. Flavonoids are scavengers of superoxide anions. Biochem Pharmacol 1988;37(5):837–41.
[54] Ruch RJ, Cheng S-j, Klaunig JE. Prevention of cytotoxicity and inhibition of intercellular communication by antioxidant catechins isolated from Chinese green tea. Carcinogenesis 1989;10(6):1003–8.
[55] Arts MJ, Haenen GR, Voss H-P, Bast A. Antioxidant capacity of reaction products limits the applicability of the Trolox Equivalent Antioxidant Capacity (TEAC) assay. Food Chem Toxicol 2004;42(1):45–9.
[56] Hidalgo M, Ferna E, Quilhot W, Lissi E. Antioxidant activity of depsides and depsidones. Phytochemistry 1994;37(6):1585–7.
[57] Chisti Y. Biodiesel from microalgae. Biotechnol Adv 2007;25(3):294–306.
[58] Grima E, Fernández F, Medina A. Downstream processing of cell-mass and products. In: Richmond A, editor. Handbook of microalgal culture: biotechnology and applied phycology. Oxford: Blackwell; 2004. p. 215–52.
[59] Elliott DC, Hart TR, Schmidt AJ, Neuenschwander GG, Rotness LJ, Olarte MV, et al. Process development for hydrothermal liquefaction of algae feedstocks in a continuous-flow reactor. Algal Res 2013;2(4):445–54.
[60] Ramos-Suárez JL, Cuadra FG, Acién FG, Carreras N. Benefits of combining anaerobic digestion and amino acid extraction from microalgae. Chem Eng J 2014;258:1–9.
[61] Miranda J, Passarinho PC, Gouveia L. Pre-treatment optimization of Scenedesmus obliquus microalga for bioethanol production. Bioresour Technol 2012;104:342–8.
[62] Torzillo G, Seibert M. Hydrogen production by *Chlamydomonas reinhardtii*. In: Richmond A, Hu Q, editors. Handbook of microalgal culture: applied phycology and biotechnology. Chichester, England: Wiley Online Library; 2013. p. 417–32.
[63] Passos F, Solé M, García J, Ferrer I. Biogas production from microalgae grown in wastewater: effect of microwave pretreatment. Appl Energy 2013;108:168–75.
[64] Prieto CG, Ramos F, Estrada V, Díaz MS. Optimal design of an integrated microalgae biorefinery for the production of biodiesel and PHBs. Chem Eng Trans 2014;37:319–24.
[65] Acién F, Fernández J, Magán J, Molina E. Production cost of a real microalgae production plant and strategies to reduce it. Biotechnology Adv 2012;30(6):1344–53.

Toxicity testing using the marine macroalga *Ulva pertusa*: method development and application

7

Jihae Park[1], Murray T. Brown[2], Hojun Lee[3], Soyeon Choi[3], Stephen Depuydt[1], Donat-P. Häder[4] and Taejun Han[3,5]
[1]Ghent University Global Campus, Incheon, South Korea, [2]Plymouth University, Plymouth, United Kingdom, [3]Incheon National University, Incheon, South Korea, [4]Friedrich-Alexander University, Erlangen-Nürnberg, Germany, [5]Ghent University Global Campus, Incheon, South Korea

7.1 Introduction

It was once thought that the solution to pollution was dilution and that aquatic environments provided convenient, effective, and resilient locations for the disposal of anthropogenic-derived wastes. We now know this not to be the case, with rivers, estuaries and coastal waters constantly under threat from the thousands of chemicals derived annually from industrial and municipal sources. To formulate effective protective legislation for these habitats the ecological risks posed by toxicants need to be evaluated by assessing the biological responses of the biota [1,2].

The global significance of detecting toxicants in the environment and assessing their impacts on biota has resulted in the development and adoption of a range of methodologies. Direct chemical analysis has several drawbacks that include the complexity of the procedures for preparing samples, the need for expensive analytical equipment, and interference from secondary pollutants during analysis. In addition, this approach fails to account for temporal changes in exposure or the interactive effects of pollutants, nor does it provide ecologically significant information [3]. To compensate for these limitations, various biological assays have been developed to provide information on pollutant-induced toxic effects that can be used to assess environmental risks [4].

Single species toxicity tests can demonstrate causal relationships between the presence of a pollutant and the adverse effects on biota. Such testing protocols are used for monitoring and predicting the effects of chemical discharges and to derive chemical-specific water quality guidelines.

A wide range of microorganisms, invertebrates, fish, plants, and algae are increasingly being used in toxicity testing [5,6]. Of these groups of organisms, fewer tests employ algae than any other, which is at least in part due to the misconception that algal species are less sensitive to toxicants than animals [5,7]. Of the algal-based tests most employ microalgae, with macroalgae poorly represented, despite their importance as primary producers and providers of habitat and nursery grounds for fish and

Bioassays. DOI: http://dx.doi.org/10.1016/B978-0-12-811861-0.00007-3

invertebrates in marine and freshwater ecosystems. With the exception of reproductive tests on the red algae, *Ceramiun strictum*, *Champia parvula*, and the brown algae, *Laminaria Saccharina latissima* (formerly *Laminaria saccharina*), and *Fucus spiralis*, and growth tests with the red algae *Gracilaria tenuistipitata*, and *Ceramium* spp., there are few standardized toxicological testing procedures using marine macroalgae [7–12]. As for green macroalgae, apart from a test on the spore germination of *Ulva fasciata* [13], reproduction tests have not been developed as defining a measurable endpoint was considered to be extremely difficult [11].

Eklund and Kautsky [7] reported that the number of macroalgal species used for toxicity testing was 65, of which only 11 species are green algae. Yet, the species for which most information on chemical toxicants has been gathered is the green macroalga (seaweed) *Ulva intestinalis* (formerly *Enteromorpha intestinalis*) [14]. Species of the genus *Ulva* have great potential as sentinel organisms due to their wide geographical distribution, tolerance to a wide salinity range, ease of collection, simple morphology, and capacity for bioaccumulation of metals [15–17]. *Ulva* species have also been used as biofilters of waste inorganic nutrients [18–21]. For these reasons, *Ulva* spp. may have been regarded as highly tolerant and thus unsuitable for toxicological tests [22]. However, when data obtained from *Ulva* toxicity tests are compared with other commonly used test protocols they reveal that *Ulva* species are more sensitive to some toxicants than other standardized test organisms such as fluorescent bacteria, daphnids, duckweeds, microalgae, rainbow trout, and sea urchins, indicating that *Ulva* species are indeed sensitive organisms appropriate for toxicity testing [13,23].

To justify the inclusion of seaweeds in the arsenal of toxicity test organisms, it is necessary to develop bioassays that are (1) sufficiently sensitive to detect low, and harmful, concentrations of individual and mixtures of toxic agents; (2) simple to execute and inexpensive to run (not requiring costly analytical equipment); and (3) having ecological relevance, i.e., the test species is an integral component of near-shore ecosystems and the measured endpoint has relevance at biological levels higher than the individual.

Ulva spp. (sea lettuce) are some of the most abundant representatives of macroalgal communities threatened by the frequent inundations of toxic chemicals derived from human activities. They constitute important primary producers of marine and brackish coastal ecosystems and are ubiquitous in coastal benthic communities around the world. They are also important natural sources of food and pharmaceutical products and are increasingly important in coastal ecosystem management as a consequence of nutrient-driven green tides [24,25]. Reproduction is a critical process by which populations perpetuate, and disturbance of this process can cause failure in recruitment, leading to the disappearance of populations and ultimately to modifications in community structure and dynamics.

7.2 Development and application of the *Ulva pertusa* reproduction bioassay

Ulva spp. have an interesting reproductive pattern involving direct transformation of vegetative cells into reproductive cells via generative divisions. One of the most striking features during progression of reproduction in *Ulva* is a visible change in color of the thallus from yellow-green when in a vegetative state, to dark olive when reproductive, to white after release of reproductive cells (Fig. 7.1). Such changes in the color of reproductive thalli have been made use of in developing a simple, sensitive, and ecologically relevant bioassay of the reproductive processes of *Ulva pertusa*, a species recorded from the Mediterranean Sea and Pacific and Indian Oceans. Since all *Ulva* spp. undergo similar reproductive processes, the test methods developed with *U. pertusa*, and described below, should be applicable to other species of the genus occurring in most parts of the world, with only minor modifications.

A 5-day aquatic toxicity test based on inhibition of sporulation of *U. pertusa* was developed by Han and Choi [23] using a computer-assisted image analyzer to quantify color change. First, the optimal environmental conditions for undertaking the test were determined for photon irradiance, salinity and temperature and found to be 60–200 μmol photons $m^{-2}\,s^{-1}$, 25–35 psu and 15°C–20°C, respectively (Table 7.1). Then, to test the protocol, disks of *U. pertusa* were exposed to the reference toxicant sodium dodecyl sulfate (SDS), several metals (Cd^{2+}, Cu^{2+}, Pb^{2+}, Zn^{2+}), and elutriates of sludge collected from nine different locations (industrial waste, livestock waste, leather waste, urban sewage, food waste, mixed waste, industrial sewage, filtration bed, and rural sewage). From the test results the derived EC_{50} values of the individual toxicants were: 5.35 mg L^{-1} for SDS, 0.326 mg L^{-1} for Cd^{2+}, 0.061 for Cu^{2+}, 0.877 mg L^{-1} for Pb^{2+}, and 0.738 mg L^{-1} for Zn^{2+} [23]. The bioassay indicated also that the sporulation endpoint was a sensitive indicator of the toxic effects of elutriates of sludge as reflected from the no observed effect concentration (NOEC) values equal to or lower than the lowest concentration employed (6.25%).

The NOEC ranges represent the highest concentration that did not result in a significant difference in sporulation percentage from the controls, whereas the EC_{50} value is the effect concentration at which 50% inhibition of sporulation occurs.

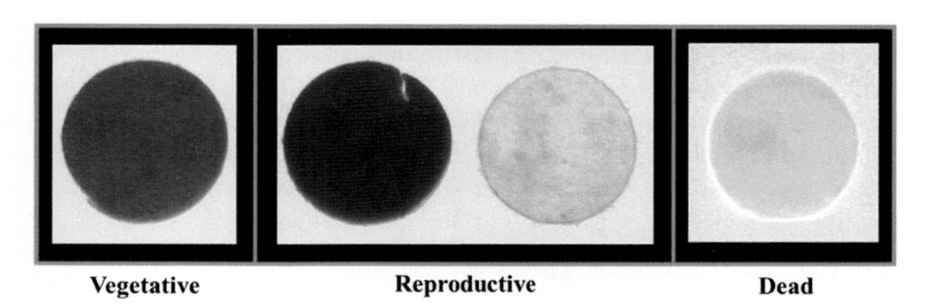

Figure 7.1 Indices of reproductive percentage of *Ulva* based on order classification of the proportion of reproductive area with thallus color change. It is noticeable that the color ("creamier white") of dead disks, due to compounds of high toxicity is different from that ("ivory white") of thallus having discharged reproductive cells.

Table 7.1 Summary of test conditions and test acceptability criteria for marine seaweed *Ulva pertusa* reproduction inhibition toxicity tests

Type of test	Static; 96-hour test
Lighting	80–100 μmol m^{-2} s^{-1}
Photoperiod	L:D 12:12
Temperature	15 ± 2°C
Salinity	25‰–35‰
pH	8.0 ± 0.2
Test vessel	24-well plate (85.4 × 127.6 mm; well dimension 15.6 mm diameter)
Growth medium	Artificial seawater
Volume of test solution	2.5 mL/well
Test organisms	Algal disk (Ø 9.4 mm)
Number of replicates	12
Endpoint	Reproduction rate

Sporulation was significantly inhibited in all elutriates with the greatest and least effects observed in elutriates of sludge from industrial waste (EC_{50} 6.78%) and filtration bed (EC_{50} 15.0%), respectively. The results of a Spearman rank correlation analyses between EC_{50} data and concentration of toxicant in the sludge gave significant results for toxicity of the four metals (Cd^{2+}, Cu^{2+}, Pb^{2+}, Zn^{2+}). Introduction of the concept of toxicity unit (TU) according to the expression ($TU = 100 \times 1/EC_{50}$) showed that these metals were the main cause of toxicity in elutriates of at least four of the nine sludge samples.

Subsequently, a bioassay using visual inspection, rather than a computer-aided analysis, of thalli during reproduction was developed [26]. This involved quantifying percentage reproduction (sporulation) rated on the scale: 0% denoted vegetative state (*yellow-green* color), 25%, 50%, 75%, and 100% denoted reproductive state (*dark olive* and *white color*) in less than one quarter, a quarter to a half, a half to three quarters, and more than three quarters of the total thallus disk area using a 16 grid measurement template (Fig. 7.2). The reliability of this simple visual inspection was validated by testing the ability of 97 first year university students, with no biology background, to evaluate reproduction by visual observation after only 30 min training before taking a test with 25 pictures of thalli with different proportions of reproductive tissue. The high test score (80.4) achieved provided good support for the method. The sensitivity of the method was then assessed using a reference toxicant (SDS; $EC_{50} = 7.1$ mg L^{-1}), the metals Cu (0.063 mg L^{-1}), Cd (0.217 mg L^{-1}), Pb (0.840 mg L^{-1}), Zn (0.966 mg L^{-1}), formalin (1.458 mg L^{-1}), diesel fuel (3.7 mg L^{-1}), and was found to be similar to the previous test using the image analyzer. Toxicity data for elutriates of sludge collected from nine different locations were directly compared with the commercially available Microtox test (https://en.wikipedia.org/wiki/Microtox_bioassay). *Ulva* reproduction was

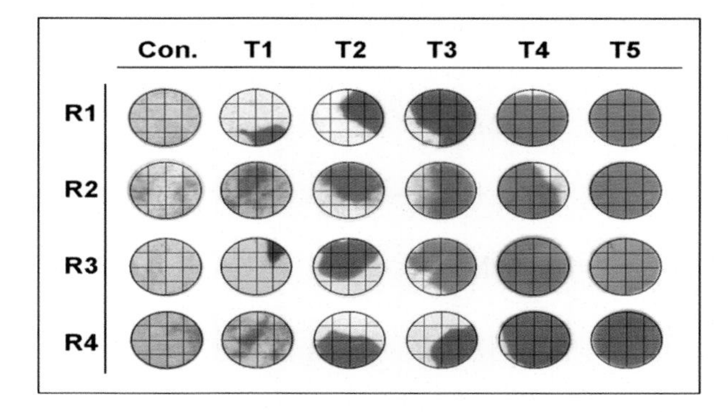

Figure 7.2 A 16-grid measurement film placed over the surface of 4 replicate disks (R1-4) after exposure to different concentrations of toxicant (Con: control, T1–5: concentrations of toxicants from lower to higher concentrations, respectively). A 16-grid measurement film on the surface of the 4 replicate disks (R1–R4), A schematic drawing of the percentage of disk areas with color change on the 4 replicate disks (R1–R4) obtained by comparing the number of grids with color change with the corresponding percent values.

significantly inhibited in all elutriates with the greatest and least toxic effects, estimated by TU observed in elutriates from industrial waste (13.1 TU) and a filtration bed (4.8 TU), whereas values ranging from 1 to 4.5 TU were obtained from the Microtox test, confirming the *Ulva* bioassay to be more sensitive. Correlation analyses between EC_{50} data and concentrations of toxicants in the sludge indicated a significant relationship between toxicity and the four meals (Cd, Cu, Pb, Zn) for the *Ulva* bioassay but no such correlation was detected from the data obtained using the Microtox test (Table 7.2).

The test was then further modified to quantify release of spores rather than sporulation. The reason for the change was that during test development it was discovered that in sporulating thalli, spores were not necessarily released. Thus, the new approach differs fundamentally from that based on sporulation as it is the area of empty cells as a proportion of sporulated areas that is measured and used to quantify spore release. The new version of the test was again evaluated by assessing the toxicity of different organic and inorganic chemicals and elutriates of sewage or waste sludge. The toxic ranking of the four metals was: Cu (EC_{50} of 0.040 mg L^{-1}) > Cd (0.095 mg L^{-1}) > Pb (0.489 mg L^{-1}) > Zn (0.572 mg L^{-1}). The EC_{50} for TBTO (tributyltin oxide) ranged from 24 to 63 μg L^{-1}. The most toxic VOC (volatile organic compound) was formalin (EC_{50} of 0.788 μL L^{-1}) and the least toxic was acetone (EC_{50} of 20,565 μL L^{-1}). Spore release was significantly inhibited in all elutriates; the greatest and least toxic effects were for industrial sewage (3.29%) and filtration bed (10.08%), respectively [26].

Further studies have since been carried out, using both sporulation and spore release as endpoints, to investigate a wider range of metals (silver, arsenic, cadmium, cobalt, chromium, copper, iron, mercury, manganese, nickel, lead, and zinc) that are prevalent in most industrial wastewaters [27]. The effect parameter of spore

Table 7.2 A comparison of EC_{50} values (mgL^{-1}) obtained from various test methods and the *Ulva* bioassay for four metals

Toxicants ($mg L^{-1}$)	Test organism					
	***Ulva* (72–120 h)**	***Selenastrum capricornutum* (96 h)**	***Lemna minor* (96 h)**	***Daphnia* (48 h)**	**Sea urchin (48 h)**	**Rainbow trout (96 h)**
Cd	0.217 [3]	0.341 [4]	0.200 [2]	0.071 [1]	9.240 [5]	1.160 [5]
Cu	0.063 [3]	0.400 [5]	1.100 [6]	0.054 [2]	0.300 [1]	0.103 [4]
Pb	0.840 [2]	2.655 [3]	8.000 [6]	4.450 [5]	0.509 [1]	3.295 [4]
Zn	0.966 [2]	0.178 [1]	10.000 [6]	1.590 [4]	1.080 [3]	3.725 [5]
/mean rank	2.50	3.25	5.00	3.00	2.75	4.50

Number inside [ ]: Rank value, Superscripts indicate reference numbers.

release was found to be more sensitive to that of sporulation. The EC_{10} and EC_{50} values derived from the data were also compared with toxicity norms of wastewater quality criteria in Korea and of global permissible limits. Based on the EC_{10} values, it can be concluded that both assays met the required standards of Korea toxicity norms for Cu, Mn, and Zn; and EC_{10} values for Cu and Zn met global effluent guidelines. According to Bruno and Eklund [28] the coefficient of variation (CV) in lab-based biological tests should range between 20 and 40%. Using this criterion, both the sporulation and spore release test could be considered to be reliable.

In a more recent study on metals, using inhibition of spore release as the measured endpoint, the interactive effects of salinity on the toxicity of cadmium (Cd), copper (Cu), lead (Pb), and zinc (Zn) was investigated [29]. The optimal salinity for maximal spore release for *U. pertusa* was between 20 and 40 psu. Under all salinity treatments the mean rank order for the relative toxicity of metals, as measured by EC_{50}, was: Cu > Cd > Pb = Zn, the same ranking as commonly found for most algae [30]. When salinity was decreased from 30 to 20 psu, the EC_{50} values for Cd toxicity decreased from 261 to 103 $g\,L^{-1}$, whereas increased salinity from 30 to 40 psu increased the EC_{50} from 261 to 801 $g\,L^{-1}$. Similarly, EC_{50} values increased with salinity for both Cu (52 $g\,L^{-1}$at 20 psu, 99 $g\,L^{-1}$ at 30 psu, and 225 $g\,L^{-1}$ at 40 psu, and Zn (720 $g\,L^{-1}$, 1,074 $g\,L^{-1}$ and 1,520 $g\,L^{-1}$, at 20, 30 and 40 psu). Only for Pb, no salinity dependent change in EC_{50} values was apparent. These findings enable us to predict that any additional increase in the metal pollution status of brackish and estuarine waters would result in a pronounced reduction in the distribution of *U. pertusa*, with consequences for the ecology of transitional waters.

The developed *Ulva* sporulation test has also recently been used to assess the acute toxicity of two new organic algicides, derivatives of thiazolidinediones (TD53 and TD49) were [31]. The compounds have been synthesized to selectively control red tides and released to the marine ecosystem. EC_{50} value of TD49 and TD53 was examined by 96-h exposure together with Solutol, which is a TD49 dispersing agent, and DMSO, as a solvent for TD53. EC_{50} value of TD53 and TD49 were 1.65 μM and 0.18 μM, respectively. From the values of NOEC, predicted no-effects concentration (PNEC) of TD53 and TD49, TD49 showed 9 times stronger toxicity than TD53.

7.3 Applying the *Ulva* reproduction test for toxicity identification evaluation (TIE)

Toxicity identification evaluation (TIE) methods have been developed by the USEPA and are generally used to characterize and identify toxicity-causing substances and ultimately reduce toxicity in industrial effluents.

A TIE based on the release of spores was conducted in 3 phases for the identification of the major toxicants in effluent from a wastewater treatment plant (WTP) and the receiving water in an adjacent stream [32]. The toxicity of the final effluent

(FE), as compared with raw wastewater (RW), and the primary and secondary effluent, showed a greater change over 12-monthly sampling events and appeared to have impacts on the toxicity of the downstream water (DS) with a significant correlation ($r^2 = 0.89$, $p < .01$). In Phase I, toxicity characterization indicated that cations were likely to be the responsible toxicants for the FE. In Phase II, cations such as Cu, Ni, and Zn were found in the FE at higher concentrations than the EC_{50} concentrations determined for the standard corresponding metals. When the concentrations of each metal in the FE samples were plotted against the respective TUs, only zinc showed a statistically significant correlation with toxicity ($r^2 = 0.86$, $p < .01$). In Phase III, using spiking and mass balance approaches, it was confirmed that Zn was indeed the major toxicant in the effluent from the WTP. Following a change in the Fenton reagent used to one with a lower Zn content, the toxicity of the FE greatly decreased in subsequent months (Fig. 7.3). The TIE developed here enabled the toxicity of FEs of the WTP to be tracked and for Zn, originating from a reagent used for Fenton treatment, to be successfully identified as the key toxicant. Fenton's reagent is a mixture of ferrous iron (catalyst) and hydrogen peroxide (an oxidizing agent) and is considered to be one of the most promising oxidative techniques for the abatement of refractory and/or toxic organic pollutants in water and wastewater, and as a powerful oxidant for organic contaminants [33–35].

The TIE method based on *U. pertusa* demonstrated utility as a low cost and simple tool to identify the risk factors for industrial effluents and provide information on regulatory control and management.

7.4 Inter-laboratory comparison test

There are important prerequisites to enable bioassays to be standardized and used as regulatory tools, which are good reproducibility and acceptable inter-laboratory variability. Various factors including handling and sensitivity of the test organisms, the quality of the water samples, and the training and experience of laboratory personnel would be possible causes of bioassay variability. An inter-laboratory evaluation elucidates these factors.

Inter-laboratory comparison tests were conducted to validate the *Ulva* test method with seven participating institutes, most of which were government institute laboratories based in Korea. In the ring test all participating laboratories used the copper standard solutions of the same CAS number [7440508]. The toxicity of Cu was tested using the following concentrations: $0\ mg\ L^{-1}$, $0.025\ mg\ L^{-1}$, $0.05\ mg\ L^{-1}$, $0.1\ mg\ L^{-1}$, $0.2\ mg\ L^{-1}$, and $0.4\ mg\ L^{-1}$. The test material package sent to participants contained the following items: an operational procedure manual, four cell plates, three small petri dishes, two plastic tweezers, six plastic Pasteur pipettes, three 16-grid films, one conical tube with artificial salts + nutrients, copper sulfate (1,000 ppm), parafilm, two conical tubes (50 mL) each containing one *Ulva* thallus, with 24 disks (Ø 9.4 mm) for visual inspection. One self-study photograph plate was provided so that participants could familiarize themselves with the

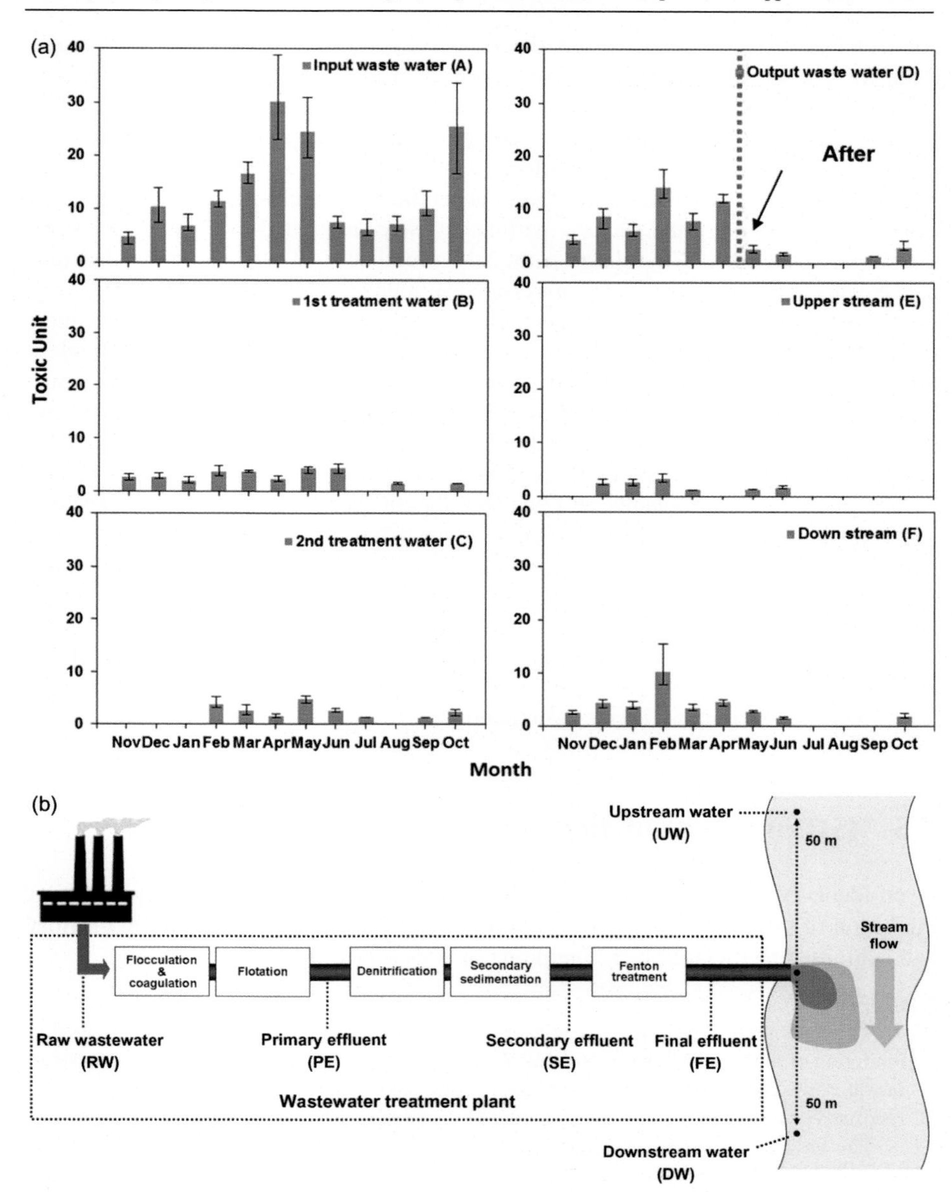

Figure 7.3 Toxicity monitoring of WTP effluents (a) and adjacent stream waters (b). (A–F) indicate toxic units measured in each sampling point: (A) raw wastewater (RW); (B) primary effluents (PE); (C) secondary effluents (SE); (D) final effluents (FE); (E) upstream water (US); (F) downstream water (DS) monthly. The error bars represent 95% confidence intervals. The vertical line between April and May indicates the time when the Fenton reagent was changed from low to a higher quality.
(Adapted from Kim et al., 2015).

color change in *Ulva* disks. Result sheets, one for the image analysis method and three for the visual inspection method, were also included. Algal disks were provided in vials labeled as No 1 for that of Ø 9.4 mm. A reference substance ($CuSO_4$, CAS No. 7440-50-8) was provided in solution form (1,000 ppm). For preparation of the test solutions, a geometric series (the ratio between 1.5 and 2.0, e.g., 2, 4, 8, 16, 32, and 64 mg L^{-1}) of at least five concentrations, exclusive of controls was designed. On receiving the package participants were requested to immediately store the 5 vials containing *Ulva* disks in the fridge until use. It was recommended that participants perform the test as soon as possible after receiving the test materials.

An assessment of color change should become less "subjective" if pretraining (mock-test), using the photographs provided in the operational procedure, is undertaken and if microscope slides and a suitable background illumination device are both used. Percentage reproduction values of each replicate, mean values, standard deviations, and CV were calculated. Determination of EC_{50} values with their confidence intervals was made based on regression analysis.

The effects of Cu toxicants on *Ulva* sporulation showed similar toxicity excluding 3 institutes. The data > 0.400 mg L^{-1} have been excluded from the calculations of mean, SD, and CV. Therefore, data from 4 out of 7 laboratories only have been used. The inter-laboratory comparison tests showed a mean EC_{50} value of 0.117 mg L^{-1} for Cu toxicity to *U. pertusa* sporulation and a CV value of 19.34% which is far below the internationally recommended levels of variation (20%–40%). A reliable *Ulva* test has therefore been developed after having been fully validated via inter-laboratory comparison exercises.

7.5 ISO protocol of the *Ulva* test

A protocol "Water quality - ISO/DIS 13308 Toxicity test based on reproduction inhibition of the green macroalga *Ulva pertusa*" has been created in the ISO format and submitted to ISO147/SC5. The protocol is as follows:

1. Scope

This International Standard specifies a method for the determination of the reproduction inhibition of the multicellular green alga (*Ulva pertusa*) to substances and mixtures contained in water, wastewater, environmental water samples, and waste or sewage sludge elutriates.

2. Normative references

The following referenced documents are indispensable for the application of this document. For dated references, only the edition cited applies. For undated references, the latest edition of the referenced document (including any amendments) applies.

ISO/TS 20281, Water quality—Guidance on statistical interpretation of ecotoxicity data

3. Terms and definitions

For the purposes of this document, the following terms and definitions apply.

3.1. algal disk

disk cut from the thallus using a hole puncher

3.2. artificial sea water medium

mixture (adjusted at salinity: 35‰) of distilled water and mineral salts

3.3. control medium

artificial sea water with nutrient salts used for stock culturing of the test alga, as medium in control cultures and dilution water in test cultures

3.4. effective concentration

ECx

concentration of test sample which results in a reduction of x % in reproduction relative to the controls

3.5. extent of reproduction

percentage of reproductive area to total area

3.6.

thallus

body of an alga

4. Principle

During progression of reproduction *Ulva* shows a visible change in thallus color from yellow-green (vegetative stage) to dark olive (reproductive stage) and then to white at the terminal stage of reproduction. Thallus disks of *Ulva pertusa* are allowed to grow as uni-algal and static cultures in different concentrations of the test sample for 96 h, and the extent of reproduction is determined by:

image analysis measurements, and/or

visual inspection of the proportion of the reproductive area out of the total area.

5. Reagents and media

5.1. Artificial seawater

Prepare artificial seawater by adding the weighed chemicals according to Table 7.3 to the desired volume of distilled water for the growth medium and test compound solutions. Aerate the artificial seawater and adjust, if necessary, to pH 8.0 ± 0.2 with 1 mol HCl or 1 mol NaOH before use. Maximum storage time for the artificial seawater is 2 months.

5.2.

5.3. Test organism

Ulva pertusa Kjellman

Table 7.3 Chemical composition of artificial sea water with a salinity of 35‰

Compounds	Quantity in g L^{-1}
NaCl	21
$MgSO_4 \cdot 7H_2O$	6
$MgCl_2 \cdot 6H_2O$	5
$CaCl_2 \cdot 2H_2O$	1
KCl	0.8
NaBr	0.1

Artificial sea water with a salinity of 35‰. In principle, artificial sea water should be prepared by adding the weighed chemicals according to Table 7.3 (ISO/DIS 13308) to the desired volume of distilled water. The artificial sea water should then be aerated and adjusted, if necessary, to pH 8.0 ± 0.2 with 1 mol HCl or 1 mol NaOH before use.
Source: Adopted from Ott FD. Synthetic media and techniques for the xenic culture of marine algae and flagellates. Va J Sci 1965;16:205–18 [36].

NOTE. The strain is available in unialgal, nonaxenic plants from the following sources:

Institute of Green Environmental Research, Department of Marine Science, 5/357, Incheon National University, 119, Academy-ro, Yeonsu-gu, Incheon, Republic of Korea Telephone: +82 32 835 4613 Facsimile: +82 32 835 0806

Website: http://marine.inu.ac.kr

5.4. Storage and cultivation

U. pertusa thalli shall be kept in artificial sea water (5.1) prepared by dissolving the required chemicals in distilled water to a concentration of 35‰ as described in Table 7.4.

10 g of *U. pertusa* thalli can be held and acclimated in plastic tanks (10 l) with aerated artificial sea water. The plants shall be maintained at 12°C ± 2°C, under a light intensity of 15 μmol m^{-2} s^{-1} ± 5 μmol m^{-2} s^{-1} of white fluorescent light with a 12 h:12 h LD cycle.

To maintain the plants for testing in healthy conditions, the growth medium shall be renewed every 3–4 days while the stock storage can be kept for up to 4 weeks with no change of medium.

6. Apparatus

The test requires standard laboratory apparatus and the following.

6.1. Temperature-controlled cabinet or room, with a white fluorescent light, providing uniform illumination in accordance with the requirements specified in Table 7.2.

6.2. Light-meter, to be used to measure photon irradiance.

6.3. pH meter

6.4. Hole puncher (4 mm or 10 mm) for prepreparation of algal disks.

6.5. Tweezers

6.6. Glassware, for the preparation of different concentration series and nutrient medium.

6.6.1. Volumetric flasks

6.6.2. Graduated cylinders

Table 7.4 Routine ecotoxicity test organisms used in different countries

	Fish	Daphnia	Algae	Bacteria
USA	O	O	O	
Canada	O	O		
England		O	O	
Netherlands	O	O	O	O
Belgium		O	O	O
Denmark	O	O	O	O
France	O	O	O	O
Germany	O	O	O	O
Northern Ireland	O	O	O	
Norway	O	O	O	
Sweden	O	O	O	
Spain		O		O
New Zealand		O	O	O
South Korea		O		
Total	9	14	11	7

'O' mark means that the organism is used as a test species in ecotoxicity test.

6.6.3. Pipettes

6.6.4. Petri dishes

6.7. Cell plates, for example 24-well cell plates with 2.5 mL per well.

6.8. Microscope slide glasses, for sandwiching thallus disks for taking measurements of reproduction.

6.9. Image analysis system, for measurements of the reproductive and total thallus area.

7. Procedure

7.1. Preparation of control medium

Add the nutrients to artificial sea water as given in Table 7.3.

7.2. Preparation of stock solutions for assessment of chemicals and products

When the test material is a chemical or a chemical product a stock solution shall be prepared by dissolving the test material in control medium at an appropriate concentration.

Adjust the pH value of the stock solution to ± 0.3 of the pH value of the medium in the control by adding hydrochloric acid or sodium hydroxide solution. No later adjustment is made.

NOTE*Ulva* generally has a narrow pH limit (between 7.5 and 8.5) for reproduction. Therefore, adjustment of pH is necessary when the pH of the sample is beyond these limits.

7.3. Preparation of test solutions for assessment of chemicals and products

An appropriate test design should consist of a geometric series (the ratio between 1.5 and 2.0, e.g. $2\ mg\ L^{-1}$, $4\ mg\ L^{-1}$, $8\ mg\ L^{-1}$, $16\ mg\ L^{-1}$, $32\ mg\ L^{-1}$, and $64\ mg\ L^{-1}$) of at least five concentrations exclusive of controls. The concentrations should be chosen to include one measured inhibition value below and one above the ECx to be estimated or to have the expected EC_{50} to be bracketed.

NOTEA suitable concentration range is best determined by carrying out a preliminary range-finding test using test concentrations with several orders of magnitude of difference.

If a solvent or carrier is used to dissolve or suspend the test sample, additional controls containing the solvent or carrier should also be included in the test to determine any effect of the solvent or carrier on the reproduction of the alga.

Measure the pH of a sample of each test solution and of the controls.

7.4. Preparation of test solution for assessment of water samples

When the test material is a sample of water, e.g., wastewater, environmental water sample or an elutriate of solid material, the salinity of this sample shall be adjusted to the same salinity as the control medium by addition of appropriate amounts of the artificial sea water salts (5.1). Then add the nutrient salts as shown in Table 7.3 (5.2). Mix the water by aeration for a minimum of 16 h to stabilize the pH. Adjust, if necessary, the pH to ± 0.3 of the pH value of the control medium by addition of hydrochloric acid or sodium hydroxide solution.

Prepare a geometric concentration series of at least five concentrations of the test material by diluting the salinity adjusted and nutrient spiked water in control medium. If possible the concentrations shall be chosen to obtain several levels of inhibition ranging from less than 10% to greater than 90% inhibition of reproduction.

Measure the pH of a sample of each test solution and of the controls.

NOTE*Ulva pertusa* generally has no reproduction problems between 25‰ and 40‰. However, adjustment of salinity is usually necessary so as the effect of temperature is not to be reflected in the test results.

7.5. Preparation of algal disks

Prepare an appropriate number of algal disks for the test (at least four disks per each of the concentrations and controls). Reproduction of *U. pertusa* is restricted to the marginal area of thalli and disks (Ø 9.4 mm) should be cut from the marginal region of the thallus using a hole puncher.

7.6. Preparation of test and control cultures

Transfer 2.5 mL of control medium to each of four wells on a 24-well cell plate. Starting with the lowest concentration, transfer 2 mL of the test solutions to each of four wells on the cell plate.

Use tweezers to transfer one algal disk to each of the wells on the cell plate.

The decision on the number of replicates per concentration will depend on the purpose of the testing (e.g., effluent screening vs. effluent compliance testing): Statistical confidence in the estimation of effects would be a reason for increasing the number of replicates per concentration or for considering an unequal replication approach, see ISO/TS 20281.

7.7. Incubation

The test cell plates shall be covered with a lid to reduce evaporation. Incubate the test cell plates for 96 h under optimal conditions: photon irradiance 80 μmol $m^{-2} s^{-1}$ – 100 μmol $m^{-2} s^{-1}$, photoperiod 12 h:12 h L/D, and temperature 15°C ± 2°C.

NOTEThe light intensity specified above can be obtained using four to six warm or cool white fluorescent lamps with approximately 0.20 m from the algal culture plate. Cosine receptors which respond to light from all angles above the measurement plane are preferred to spherical receptors which respond to light from all angles above and below the plane of measurement. Cosine receptors do respond to light from all angles 90 degrees around the vertical in a half sphere, but the measured intensity follows cos (alpha).

For light-measuring instruments calibrated in lux, an equivalent range of 4,000 lx to 5,000 lx is acceptable for the test.

Measure temperature at least twice a day; however continuous temperature control is recommended.

Before a toxicity test is conducted with new test facilities, a performance test should be conducted, in which all test plates contain control medium to ensure analyst proficiency. The coefficient of variation of reproduction should be less than 10%.

The protocol for the methods using the reproduction inhibition of *U. pertusa* is summarized in Table 7.3.

7.8. Test duration

The test duration is 96 h.

7.9. Measurements and observations

Measure percentage reproduction based on the proportion of the surface area with dark olive and white colorations.

The reproductive area of the disks can be counted using computer-assisted image analysis.

Alternatively, visual evaluation of reproduction can also be rated on the scale: 0% denoted vegetative state (*yellow-green color*), 25%, 50%, 75%, and 100% denoted reproductive state (*dark olive* and *white color*) in less than one quarter, a quarter to a half, a half to three quarters, and more than three quarters of the total thallus disk area (see Fig. 7.2). However, this can be done more objectively using a

16-grid measurement template. At the end of the 96 h exposure period, determine the color change of each of the 16 grid areas of each microplate used. A definitive rating is to be determined of each grid area based on the greatest color response within each area. The scale outlined in Fig. 7.2 shall be used to quantify the percentage of color change.

Percent reproduction is assessed from the proportion of the disk area with a color change out of the total thallus disk area.

Consideration should be given to dead disks that turn white due to bleaching after exposure to highly toxic compounds; however, the color ("creamier-white") of bleached thalli should not be confused for the "ivory-white" of the thalli that have discharged reproductive cells.

NOTE For clear visual inspection, put the thallus disks between microscope slides and place them on a white background. Illuminate from the bottom of the slides.

7.10. Reference substance

Test with reference substance, copper chloride (CAS number 7440508) is required for quality assurance and to check the sensitivity of the alga.

7.11. Validity criteria

The mean reproductive percentage in the control shall be at least 80%.

7.12. Expression of results

7.12.1. Test results

Record reproduction percentage values of each replicate, mean values and standard deviations. The coefficient of variation (CV), the standard deviation expressed as a percentage of the mean, should be calculated to estimate the precision and reproducibility of the tests.

Plotting concentration response curves is highly recommended to provide the basis for determining the ECx values for the inhibition of reproduction.

7.12.2. Determination of ECx values

Apply a linear or nonlinear regression model to the experimental data points by regression analysis. Determine ECx values with their confidence intervals. Guidance on appropriate models is given in ISO/TS 20281.

If data are too few or uncertain for regression analysis, or if inhibitions appear not to follow a regular concentration-response relation, then a graphical method may be applied by drawing a smooth eye-fitted curve of the concentration response relationship and reading ECx values from this graph.

7.12.3. Expression of results

The values calculated for ECx [e.g., EC_{20}, EC_{25}, and/or EC_{80}] and the corresponding confidence intervals (95%) and CVs are displayed with the required significant precision (digits).

The values should be given as mg/L for tests with individual chemicals or as percentage for effluent, elutriate, or leachate test samples.

8. Test report

This test report shall contain at least the following information:

the test method used, together with a reference to this International Standard (ISO/CD 13308);

name of the laboratory performing the test;

date and period of test;

test organisms (e.g., scientific name, strain, source, holding conditions);
test details;
culturing apparatus and incubation procedure;
culture types (static, unialgal, etc.)
composition of medium;
preparation of test sample (e.g., pH, salinity of effluent sample) and treatments;
concentrations tested;
replicates per concentration;
number of thallus disks per replicate;
size of cell plates;
solution volume;
light intensity and quality;
pH of test solutions including the controls at start and end of test;
salinity;
temperature range during incubation;
method of measuring reproduction percentage;
results;
table outlining reproduction percentage in each cell plate at each measuring point;
mean reproduction percentage for each test concentration (and control) at each measuring point;
relationship between reproduction and concentration in table and graphical representation;
ECx values with 95% confidence intervals and the corresponding CVs values including the method of determination;
other observed effects such as bleaching of thallus disks.

7.6 The *Ulva* kit

For effective environmental management programs, development and implementation of practical effect-measurement diagnostic tools may be required. There are various commercial toxicity testing kits available, which are manufactured by e.g., Azu Enviromental, Co., Ltd. in the USA, LemnaTech, Co., Ltd. in Germany, and MicroBiotest, Co., Ltd. in Belgium. However, to date no macroalgal toxicity test kits have been developed. Therefore, a kit for evaluating toxicity using sporulation of the green alga *U. pertusa* has been produced. This invention comprises a capsule-type bioware that includes: the green alga *Ulva* thallus prepared in a coin form, a small cylinder for measuring a sample volume of a water body, an artificial salt solution for controlling growth conditions of the green alga, a measuring film or magnifying glass for measuring the change in color area of the green alga, and a standard toxicity solution. The components of the UlvaKit are shown in Fig. 7.4 and comprise: a plastic measuring film, gloves for performing toxicity evaluation, a reference diagram for aiding determination of a color change in *Ulva* thallus, introductory CD manual containing detailed information on how to use the kit, and statistical analysis template.

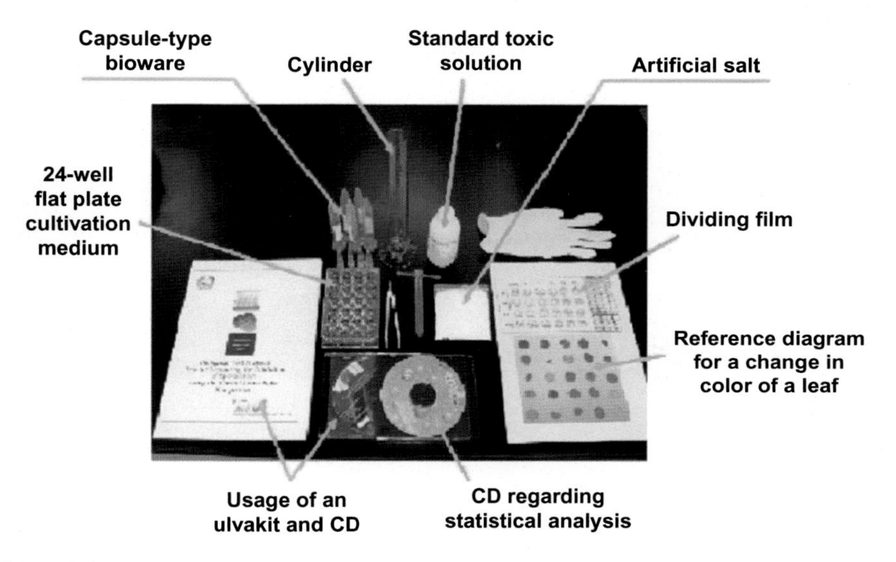

Figure 7.4 The composition of an Ulva Kit, comprising: Artificial salt, capsule-type bioware, Cd regarding statistical analysis, cylinder, dividing film, reference diagram for a change in color of a leaf, standard toxic solution, usage of an Ulva kit and CD, 24-well flat plate cultivation medium.

Since the release of the original kit we have become aware of the fact that the method based on visual inspection of thallus color changes resulting from the sporulation process is subject to bias depending on the skills of personnel and variability of instruments etc. Therefore, a more objective way of measuring the thallus color change has been tried.

During sporulation in *Ulva*, the contents of mother cells empty following spore release and die. Due to this characteristic, the emptied cells can be stained blue with Evans blue, a vital stain widely used to be taken up by dead but not living cells [37].

There is also a free image analysis program available on the Internet, called ImageJ software (National Institutes of Health, Nethesda, Maryland, USA, https://imagej.nih.gov/ij/download.html), which can be used to examine diverse morphometric characteristics based on color filtering. Most often, the gray conversion method has been used because of convenience, utilization, usefulness, economical efficiency, and objectivity of computer visions [38–41]. Image-analysis of sporulated areas stained with Evans blue has been tested to assess if this new method could complement the previous visual and/or image analysis method by eliminating subjectivity. After 96 h exposure to toxicants, *U. pertusa* disks were stained with 0.5% Evans blue for 15 min before taking measurements of stained cells. In addition, coverslips were placed on the bottom of petri dishes to ascertain whether spores had been released since some ambiguity exists in the color between fully emptied thalli after spore release and dead cells. The sporulation percentage was

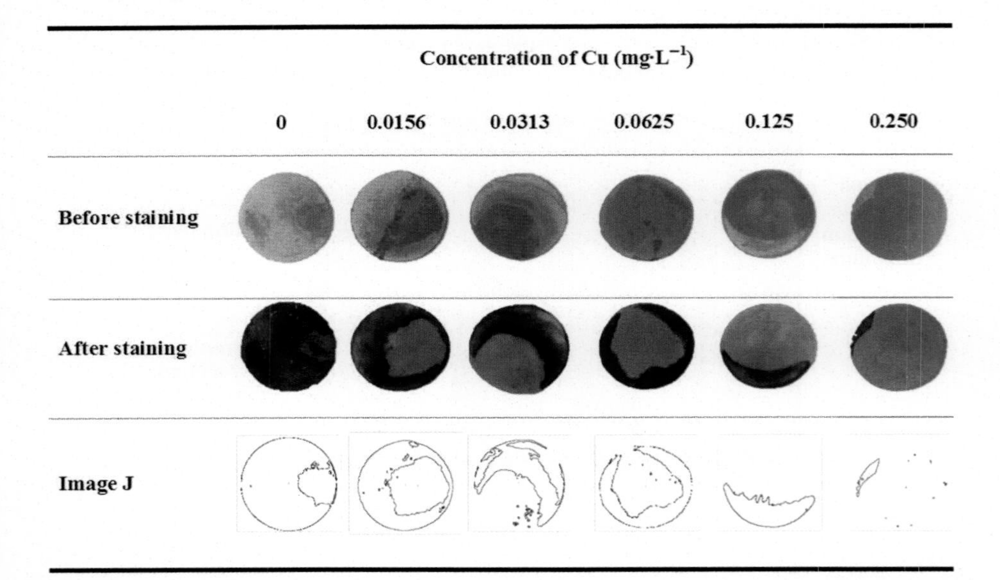

Figure 7.5 Indices of reproduction inhibition of *Ulva pertusa* exposed to different concentrations of Cu. It is noticeable that the reproductive parts of disks were stained by Evans *blue*.

then calculated by the stained area over the whole thallus disk area. Calculated EC_{50} values from the original and newly developed methods were similar: 0.1166 vs 0.1437 mg L^{-1} for Cu, 0.0368 and 0.0629 mg L^{-1} for diuron, 0.1955 and 0.2047 mg L^{-1} for atrazine, respectively. Thus, Evans blue staining followed by image analysis using the ImageJ program would eliminate the possibility of subjectivity of the *Ulva* method (Fig. 7.5).

7.7 The UlvaTox

An automated device called UlvaTox has also been developed. This is a real time image analysis program (Fig. 7.6), running under the Windows XP operating system, which allows the user to follow the development of sporulation in *Ulva* thalli segments by automatically recording images of green thalli. Thalli segments are incubated in a dedicated unit, containing a color Fire-wire camera and a multiwell plate. Algae are illuminated by 4 white Luxeon high intensity LEDs. The same LEDs are used to illuminate the field of view of the camera. The multiwell plate is temperature-controlled by an external cooler system. All measured values, i.e., thalli area, temperature, light, and supply voltage are stored in ASCII-files which can be accessed by any spreadsheet program (for example Excel). The whole system is optimized for the dedicated task of quantification of thalli development.

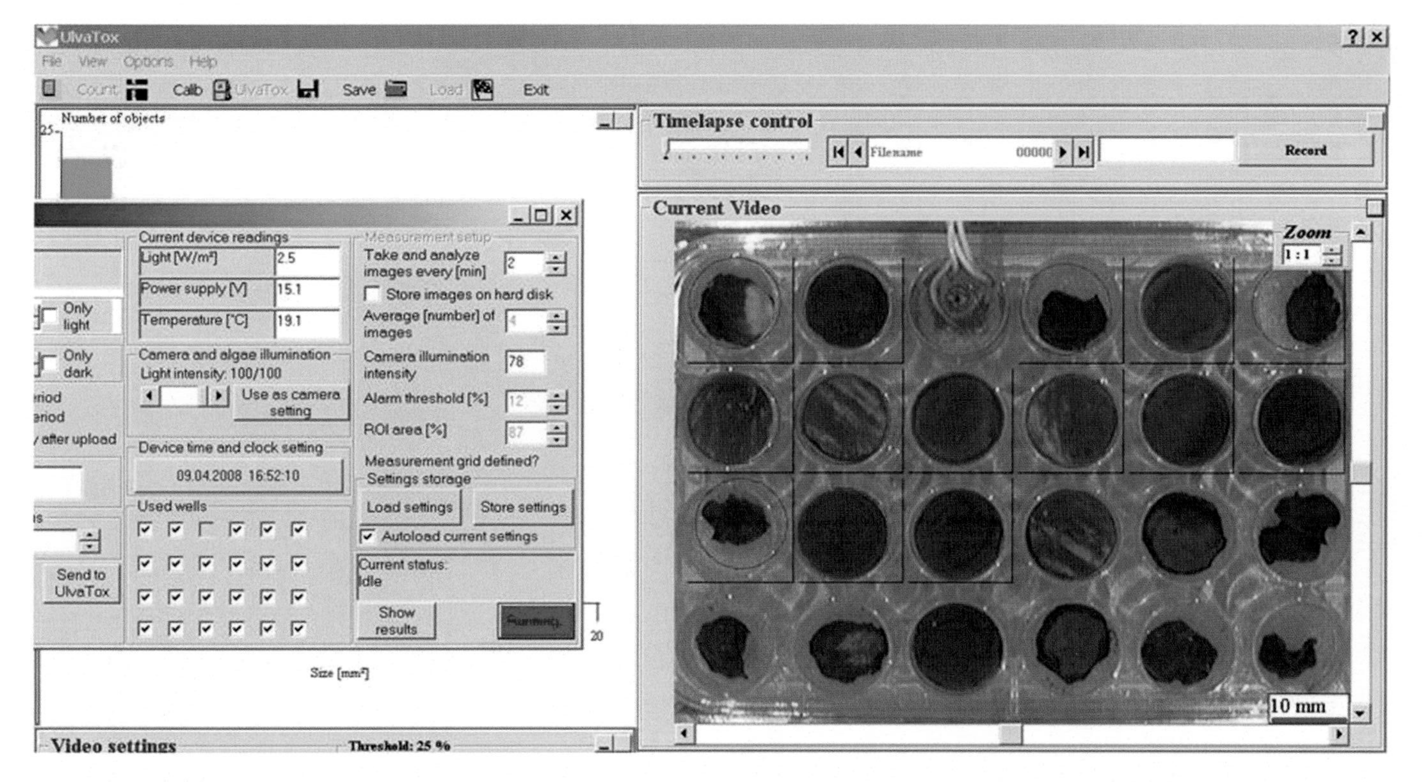

Figure 7.6 Measurement cycle image by ROI area (%) option. In this measurement cycle all wells are analyzed. According to the "ROI area (%)" settings the corresponding area part of the well is searched for the maximal thalli area. As a consequence, even if the thalli sections are significantly smaller than the well area the measurement will be accurate.

The built-in calibration unit adjusts the internal calculation algorithms to the current scale. By this means all areas, light levels, temperature, and voltage are calculated in the correct physical units. A calibration bar with the correct size is shown as an on-screen overlay.

The image quality can be adjusted by setting the brightness and contrast with slide controls under online visual verification of the result. The gray level distribution of the pixels on screen can be shown on line in a histogram. Either the stream of images is shown in real time or a snap shot is selected.

Objects are recognized by specific color via the color Fire-wire camera. Objects can be selected based on their individual parameters such as minimal or maximal size. Modern mathematical software tools are available to perform smoothing, Laplace transformations, image sharpening, and edge detection (Sobel algorithm).

Individual images can be stored and retrieved at high resolution on the hard disk of the computer. In addition an electronic time lapse video recorder is installed which allows the user to record sequences of images at high frequency (less than 100 ms per frame, depending on the hardware). These sequences can be retrieved and analyzed using the software tools described below (Fig. 7.7).

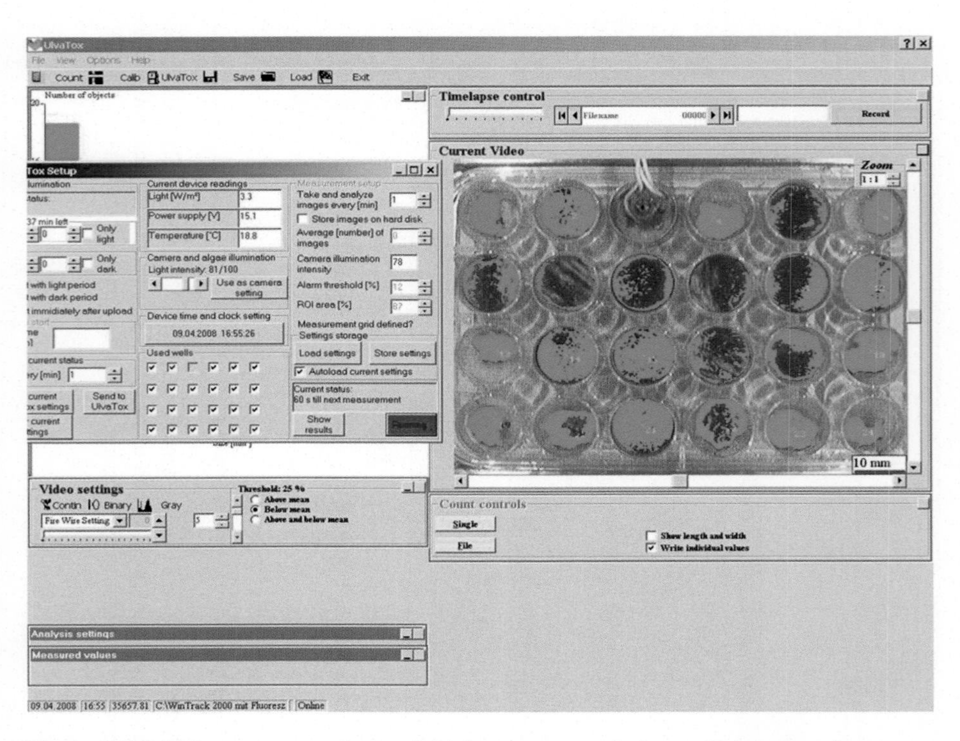

Figure 7.7 Initial measurement image. Initial measurement is done which is the reference point for all following measurements.

7.7.1 Principle of operation

No matter how the image is generated, the analog image has to be digitized in the first step. That means that the image is divided into individual pixels which are stored in a 2D (black and white camera) or 3D-matrix (color camera) of lines and columns. The number of pixels defines the spatial resolution of the digitized image and thus what size structures can be discriminated. Each pixel is square so that no horizontal or vertical distortions occur. The number of possible gray levels defines the dynamic range. The darkest black is arbitrarily given a value of 0 and the brightest white a value of 255 (in color cameras these images contain three "planes" where the same principle holds as in the black and white camera).

The video image, which consists of individual lines, is digitized at discrete times along a line, and the brightness found there is converted into a digital value by an analog/digital converter. Modern digitizers perform the digitization in real time. Since in the European video norm 25 images (50 half frames) are recorded and transmitted per second the digitization of a full frame has to be performed in 40 ms (30 images, 60 half frames, 33 ms for the American video standard). Please note: a full image consists of two half frames; i.e., every other line is updated every 20 ms.

The analog image is either constantly digitized (stream mode) or one image is digitized from the incoming sequence at a discrete point in time (snapshot mode) under software control. The digitized image is stored in memory (RAM) in the form of a matrix of digital values and mirrored into the video RAM for presentation on screen. At a resolution of 768×578 pixels one image uses about 450 Kbytes for a black and white image.

7.8 Conclusions

There have been developments of different approaches to regulating pollutants due to an urgent need for improved and effective surface water quality management, creating an increasing number of toxicity test applications. Recently, it is notable that the international community has adapted their established environmental regulations based on toxicological evaluation data from biological organisms' responses to determine the risk of water environments (Table 7.4).

The key to the process of creating eco-toxicological techniques is the selection of appropriate test species, which are ideal characteristics of the species, such as habitat fixation, broadness of distribution, ease of harvesting, and sensitivity of toxic reactions.

A novel method using the green macroalga *U. pertusa* has been developed and used for the toxicity test on more than 75 different environmental samples of contaminants, including metals, VOCs, herbicides, oils, dispersants, slag-waste, etc. [26,27,42–44]. The *Ulva* toxicity test has been proposed as a new ISO standard method. This method has several distinct advantages over other currently employed techniques. For example, no specialist expertise is required to conduct this test, and it is both cost- and time-effective, requiring only a cell plate and a small volume of water. This test can be completed within approximately 3 h following the 96 h incubation period. The sensitivity of the *Ulva* method is similar to—or in many cases exceeds—other commonly available or well established bioassays [23,26], and since reproduction is the means by which population recruitment is facilitated, the ecological significance of the measured endpoint cannot be disputed.

The development of the Ulva kit and automated UlvaTox gives additional merit of using this method in terms of the simplicity, convenience, and low-skilled requirement to operate and use over the manual method per se.

There are many advantages, but several disadvantages therein. Firstly, the *Ulva* method has not been confirmed to have applications to other *Ulva* species than *U. pertusa*. As there are taxonomic difficulties to identify *Ulva* species, it may require some experts' help to find out *U. pertusa* on the beach. Secondly, the shelf time for *Ulva* disks and thalli is not long enough, and further developments may be required for construction of a complete kit system of the *Ulva* method.

Overall, the *Ulva* system seems to be a good method of determining the toxicity of a variety of water samples including wastewaters and even freshwaters if added with salts and adjusted for salinity.

References

[1] Mallick N, Rai LC. Physiological responses of non-vascular plants to heavy metals. Physiology and Biochemistry of Metal Toxicity and Tolerance in Plants. Netherlands: Springer; 2002. p. 111–47.

[2] Bidwell JR, Wheeler KW, Burridge TR. Toxicant effects on the zoospore stage of the marine macroalga Ecklonia radiata (Phaeophyta: Laminariales). Mar Ecol Progress Ser 1998;163:259–65.

[3] Han Y-S, Kumar AS, Han T. Comparison of metal toxicity bioassays based on inhibition of sporulation and spore release in *Ulva pertusa*. Toxicol Environ Health Sci 2009; 1(1):24–31.

[4] Eullaffroy P, Vernet G. The F684/F735 chlorophyll fluorescence ratio: a potential tool for rapid detection and determination of herbicide phytotoxicity in algae. Water Res 2003;37(9):1983–90.

[5] Klaine SJ, Lewis MA. Algal and plant toxicity testing. Handbook of Ecotoxicology. Lewis Publishers, CRC Press, Inc.; 1995 163–184

[6] Williams T, Hutchinson T, Roberts G, Coleman C. The assessment of industrial effluent toxicity using aquatic microorganisms, invertebrates and fish. Sci Total Environ 1993;134:1129–41.

[7] Eklund BT, Kautsky L. Review on toxicity testing with marine macroalgae and the need for method standardization––exemplified with copper and phenol. Mar Pollut Bull 2003;46(2):171–81.

[8] Eklund B. A 7-day reproduction test with the marine red alga *Ceramium strictum*. Sci Total Environ 1993;134:749–59.

[9] USEPA. Short-term methods for estimating the chronic toxicity of effluents and receiving waters to marine and estuarine organisms. In: Environmental Protection Agency of USA, 1998 Contract No.: EPA/600/491/003, Method 1009.0.

[10] Haglund K, Björklund M, Gunnare S, Sandberg A, Olander U, Pedersén M. New method for toxicity assessment in marine and brackish environments using the macroalga *Gracilaria tenuistipitata* (Gracilariales, Rhodophyta). Fifteenth International Seaweed Symposium, Hydrobiologia. Netherlands: Springer; 1996. p. 317–25.
[11] Thursby GB, Steele RL. Sexual reproduction tests with marine seaweeds (macroalgae). Fundamentals of aquatic toxicology. Florida: Taylor & Francis; 1995. p. 171–87.
[12] Eklund BT. Aquatic primary producers in toxicity testing: emphasis on the macroalga Ceramium strictum. Stockholm: Stockholm University; 1998.
[13] Hooten RL, Carr RS. Development and application of a marine sediment pore-water toxicity test using Ulva fasciata zoospores. Environ Toxicol Chem 1998;17(5):932–40.
[14] Hayden HS, Blomster J, Maggs CA, Silva PC, Stanhope MJ, Waaland JR. Linnaeus was right all along: Ulva and Enteromorpha are not distinct genera. Eur J Phycol 2003;38(3):277–94.
[15] Lee J, Wang F. One quarter of humanity: malthusian mythology and Chinese realities, 1700–2000. Cambridge, MA: Harvard University Press; 2011.
[16] Haritonidis S, Malea P. Bioaccumulation of metals by the green alga *Ulva rigida* from Thermaikos Gulf, Greece. Environ Pollut 1999;104(3):365–72.
[17] Brown M, Hodgkinson W, Hurd C. Spatial and temporal variations in the copper and zinc concentrations of two green seaweeds from Otago Harbour, New Zealand. Mar Environ Res 1999;47(2):175–84.
[18] Mata L, Santos S. Cultivation of *Ulva rotundata* (Ulvales, Chlorophyta) in raceways, using semi-intensive fishpond effluents: yield and biofiltration. Proceedings of the 17th International Seaweed Symposium. Cape Town: Oxford University Press; 2003. p. 237–42.
[19] Chung IK, Kang YH, Yarish C, Kraemer GP, Lee JA. Application of seaweed cultivation to the bioremediation of nutrient-rich effluent. Algae 2002;17(3):187–94.
[20] Cohen I, Neori A. *Ulva lactuca* biofilters for marine fishpond effluents. I. Ammonia uptake kinetics and nitrogen content. Bot Mar 1991;34(6):475–82.
[21] Vandermeulen H, Gordin H. Ammonium uptake usingUlva (Chlorophyta) in intensive fishpond systems: mass culture and treatment of effluent. J Appl Phycol 1990; 2(4):363–74.
[22] Steele RL, Thursby GB, editors. A toxicity test using life stages of *Champia parvula* (Rhodophyta). In: Aquatic Toxicology and Hazard Assessment: Sixth Symposium; ASTM International; 1983.
[23] Han T, Choi G-W. A novel marine algal toxicity bioassay based on sporulation inhibition in the green macroalga *Ulva pertusa* (Chlorophyta). Aquat Toxicol 2005;75(3):202–12.
[24] Wang Y, Zhao X, Tang X. Antioxidant system responses in two co-occurring green-tide algae under stress conditions. Chin J Oceanol Limnol 2016;34(1):102–8.
[25] Smetacek V, Zingone A. Green and golden seaweed tides on the rise. Nature 2013;504 (7478):84–8.
[26] Han Y-S, Brown MT, Park GS, Han T. Evaluating aquatic toxicity by visual inspection of thallus color in the green macroalga Ulva: testing a novel bioassay. Environ Sci Technol 2007;41(10):3667–71.
[27] Han T, Kong J-A, Brown MT. Aquatic toxicity tests of *Ulva pertusa* Kjellman (Ulvales, Chlorophyta) using spore germination and gametophyte growth. Eur J Phycol 2009;44(3):357–63.
[28] Bruno E, Eklund B. Two new growth inhibition tests with the filamentous algae Ceramium strictum and C. tenuicorne (Rhodophyta). Environ Pollut 2003;125 (2):287–93.

[29] Oh J-j, Choi E-M, Han Y-S, Yoon J-H, Park A, Jin K, et al. Influence of salinity on metal toxicity to Ulva pertusa. Toxicol Environ Health Sci 2012;4(1):9–13.

[30] Rai LC, Gaur JP, Kumar HD. Phycology and heavy-metal pollution. Biol Rev 1981;56:99–151.

[31] Kim E, Kim S-H, Kim H-C, Lee SG, Lee SJ, Jeong SW. Growth inhibition of aquatic plant caused by silver and titanium oxide nanoparticles. Toxicol Environ Health Sci 2011;3(1):1–6.

[32] Kim Y-J, Han Y-S, Kim E, Jung J, Kim S-H, Yoo S-J, et al. Application of the *Ulva pertusa* bioassay for a toxicity identification evaluation and reduction of effluent from a wastewater treatment plant. Front Env Sci 2015;3:2.

[33] Kavitha V, Palanivelu K. Destruction of cresols by Fenton oxidation process. Water Res 2005;39(13):3062–72.

[34] Neyens E, Baeyens J. A review of classic Fenton's peroxidation as an advanced oxidation technique. J Hazard Mater 2003;98(1):33–50.

[35] Kang S-F, Liao C-H, Po S-T. Decolorization of textile wastewater by photo-Fenton oxidation technology. Chemosphere 2000;41(8):1287–94.

[36] Ott FD. Synthetic media and techniques for the xenic culture of marine algae and flagellates. Va J Sci 1965;16:205–18.

[37] Jacyn Baker C, Mock NM. An improved method for monitoring cell death in cell suspension and leaf disc assays using Evans blue. Plant Cell Tissue Organ Cult 1994;39 (1):7–12.

[38] Harris-Love MO, Seamon BA, Teixeira C, Ismail C. Ultrasound estimates of muscle quality in older adults: reliability and comparison of Photoshop and ImageJ for the grayscale analysis of muscle echogenicity. Peer J 2016;4:e1721.

[39] Mezei T, Szakács M, Dénes L, Jung J, Egyed-Zsigmond I. Semiautomated image analysis of high contrast tissue areas using Hue/Saturation/Brightness based color filtering. Acta Medica Marisiensis 2011;57(6):679–684.

[40] Igathinathane C, Pordesimo L, Batchelor W. Major orthogonal dimensions measurement of food grains by machine vision using Image. J Food Res Int 2009;42(1):76–84.

[41] Vincent L. Morphological grayscale reconstruction in image analysis: Applications and efficient algorithms. IEEE Trans Image Process 1993;2(2):176–201.

[42] Yoo J, Ahn B, Oh J-J, Han T, Kim W-K, Kim S, et al. Identification of toxicity variations in a stream affected by industrial effluents using Daphnia magna and Ulva pertusa. J Hazard Mater 2013;260:1042–9.

[43] Han T, Han Y-S, Park CY, Jun YS, Kwon MJ, Kang S-H, et al. Spore release by the green alga *Ulva*: a quantitative assay to evaluate aquatic toxicants. Environ Pollut 2008;153(3):699–705.

[44] Han Y-S, Kang SH, Han T. Photosynthesis and photoinhibition of two green macroalgae with contrasting habitats. J Plant Biol 2007;50(4):410–16.

Pigments

8

Peter Richter and Donat-P. Häder
Friedrich-Alexander University, Erlangen-Nürnberg, Germany

8.1 Introduction

Colors of bacteria, microorganisms, algae, and plants are useful indicators for the developmental and physiological status of the organisms and can therefore be used as a monitoring tool [1]. The coloration of higher plants consists mainly of the photosynthetic pigments, the chlorophylls (chl) *a* and *b*, as well as carotenoids [2]. In addition, anthocyanins can add a distinct red coloration e.g., in the purple variations of trees such as hazel or beech [3]. Flowers have developed a number of prominent colors to attract insects for pollination, which are based on a large number of carotenoids, xanthophylls, and anthocyanins [4]. In contrast, white flowers do not contain a specific dye, but the dead, air-filled epidermal cells reflect the light giving the flower a white tinge [5]. In addition, plants have developed a large number of absorbing pigments such as phytochromes, phototropins, cryptochromes, and UV-absorbing molecules (e.g., UVR8), which are utilized to control orientation in space and development [6,7]. However, the latter pigments occur at minute concentrations and are difficult to detect under the cover of the major pigments and are rarely used in bioassays.

During development the pigment contents, their ratio, and chemical nature change. E.g., during development of young foliage the concentration of the main photosynthetic pigment, chl *a*, increases, as can be seen by the lighter green in young trees, such as spruce (Fig. 8.1) [8]. Likewise, many leaves in temperate climates change their color before they are dropped before the winter (Fig. 8.2) [9]. This is mainly controlled by the internal clock of the plant cells, which sense the day length, and not by the temperature or other environmental factors to control their development [10]. Proteins and nutrients are withdrawn from the leaves as well as chlorophyll [11]. This makes the yellow and red carotenoids visible which causes the typical autumn coloration (Indian summer). However, it is not clear why many, especially American, plants synthesize and accumulate more anthocyanins during this development than European species [12]. This color change can be monitored from satellites to detect the developmental changes during autumn [13].

Changes in leaf coloration may also be induced by draught or excessive water supply [14]. Likewise, high salinity or lack of nutrients can be responsible for visible color change [15,16]. Another indication of color change in leaves may be an attack of parasites such as fungi or insects [17,18]. Fig. 8.3 shows the effect of an attack on pear leaves that results in the outgrowth of galls (or cecidia) on the lower surface. This abnormal outgrowth can be induced by various parasites such as fungi, bacteria, insects, or mites [19]. On the upper surface a red discoloration is

Bioassays. DOI: http://dx.doi.org/10.1016/B978-0-12-811861-0.00008-5

Figure 8.1 Branches of spruce (*Picea abies* L) in spring showing dark-green second year leaves and newly appearing yellow-green leaves.

Figure 8.2 Vegetative green and senescing yellow-red leaves of staghorn sumac (*Rhus typhina* L) during autumn.

visible at the position of the gall outgrowth. A recent problem is the parasitic attack of a fungal rust on soybeans which has spread to South America [20] and which is difficult to fight with fungicides [21]. Finally, changes in leaf color can also indicate the presence of pollutants or toxic substances in the soil or water. Fig. 8.4 shows the effects of K dichromate on the leaf coloration of *Wolffia* spec. at different concentrations after exposure for two weeks.

Automatic bioassays can be used to detect changes in leaf coloration (cf. Chapter 13: Image processing for bioassays, this volume). In contrast to organisms

Figure 8.3 Discoloration of pear leaves (*Prunus domestica* L) (A) induced by a parasitic attack which results in the outgrowth of galls on the lower surface (B).

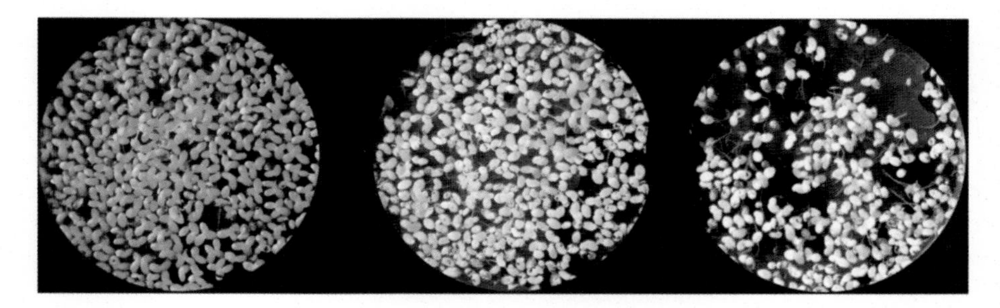

Figure 8.4 Effect of K dichromate on the leaf coloration in *Wolffia* spec. after exposure for 2 weeks. Control (left), 100 mg L^{-1} (center) and 500 mg L^{-1} (right).

tracking, in most cases color cameras are being used. In order to selectively see a color change one can select a certain color channel, e.g., red to monitor the chlorophyll concentration [22]. By this technique it is possible to monitor individual plants or a whole vegetation for developmental changes, effects of drought, or parasitic attack. To be even more specific, the fluorescence output of chl *a* can be monitored, which is induced by artificial or solar radiation [23].

Pollutants, such as high levels of heavy metals in the soil or irrigation water, can also be monitored by following the change of the photosynthetic pigments such as chlorophyll and carotenoids [24]. Photosynthetic microorganisms or algae can be employed for this purpose since they show changes in coloration due to decreases of chlorophylls or carotenoids on a faster timescale [25,26].

8.2 Effects of pollutants on photosynthetic pigments in flagellated algae

Bulk pigments of prokaryotic or eukaryotic phytoplankton or plants are useful indicators for environmental stress factors and pollutants. They respond very quickly to the presence of, e.g., heavy metal ions. Fig. 8.5A shows the concentration of chl *a* in the unicellular flagellate *Euglena gracilis* in response to exposure to Cu ions (as chloride tetrahydrate) at increasing concentrations. Even after a 24-h incubation the chl *a* concentration significantly decreased (Anova, $p < 0.05$) as determined by high performance liquid chromatography (HPLC), and at the highest concentration after 4-h exposure there was hardly any chl *a* left. Similarly, the cellular concentrations of chl *b* decreased significantly after exposure to Cu ions for up to four days (Fig. 8.5B), however, not as pronounced as chl *a*.

The photosynthetic flagellate *E. gracilis* also contains a number of carotenoids including β-carotene, diadinoxanthin, and 9-cis-neoxanthin. When exposed to increasing Cu concentrations the intracellular β-carotene concentration decreased steadily with increasing exposure times up to 4 days (Fig. 8.6A). With the exception of the concentration at 2.5 mg L^{-1} after 1 day of exposure all other treatments resulted in a statistically significant reduction (Anova, $p < 0.05$). Diadinoxanthin (Fig. 8.6B) and 9-cis-neoxanthin (Fig. 8.6C) followed a similar trend.

Similar results were obtained with cadmium. All major pigments were significantly reduced even after 1 day of incubation and even at low Cd concentrations (Table 8.1). In order to determine whether the significant decreases in the cellular pigments are due to effects on the pigments themselves or on the metabolic processes underlying destruction and resynthesis, the individual pigments were extracted from untreated cells and subsequently incubated with increasing concentrations of heavy metals. The concentration of e.g., diadinoxanthin decreased significantly when exposed to much higher Cd concentrations than those which had marked effects on the pigmentation within the cell (Fig. 8.7). The results with the other pigments and other heavy metals were similar, which confirmed the notion that the heavy metals have an effect on the physiology of the cells and do not significantly destruct the pigments.

In order to confirm that the heavy metal ions are responsible for the loss in pigmentation and not the cation (chloride), the experiments were repeated with Ca chloride. When applied at similar concentrations as for Cu chloride, application of

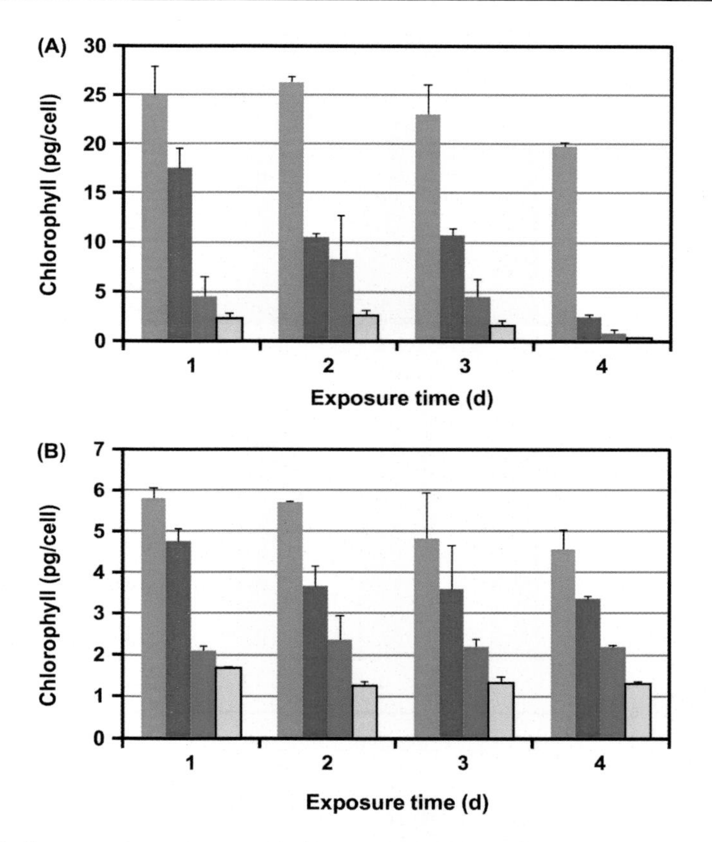

Figure 8.5 Concentrations of extracted chl *a* (A) and chl *b* (B) in *Euglena gracilis* after exposure to different copper concentrations for different incubation times. Control (*blue bars*), 2.5 mg L^{-1} (*red bars*), 5 mg L^{-1} (*green bars*) and 10 mg L^{-1} (*yellow bars*). *Source:* Redrawn after Ahmed H. Biomonitoring of aquatic ecosystems [PhD thesis]. Erlangen-Nürnberg: Friedrich-Alexander-Universität; 2010 [27].

calcium chloride did not have significant effects on the intracellular chl *a* concentrations (Fig. 8.8). Likewise it did not affect chl *b* and the carotenoids.

The composition of photosynthetic pigments is a clear characterization for the organism under analysis and has been used for fingerprinting of different organisms in phytoplankton communities [28], and using HPLC is an effective means for extraction and quantification of the main pigments [29]. Heavy metals are known to affect the metabolic processes of pigment biosynthesis and photosynthesis in higher plants [30] and phytoplankton [31]. However, the individual pigments can be affected to different degrees depending on the nature of the heavy metal ion involved; e.g., chl *b* was affected to a greater degree than chl *a* in *E. gracilis* when exposed to Ni [32]. Several mechanisms have been found for the effects of heavy metals on photosynthetic pigment synthesis: At low concentrations Cu can replace the central Mg ion in the chl molecules which also affects the photosynthetic electron

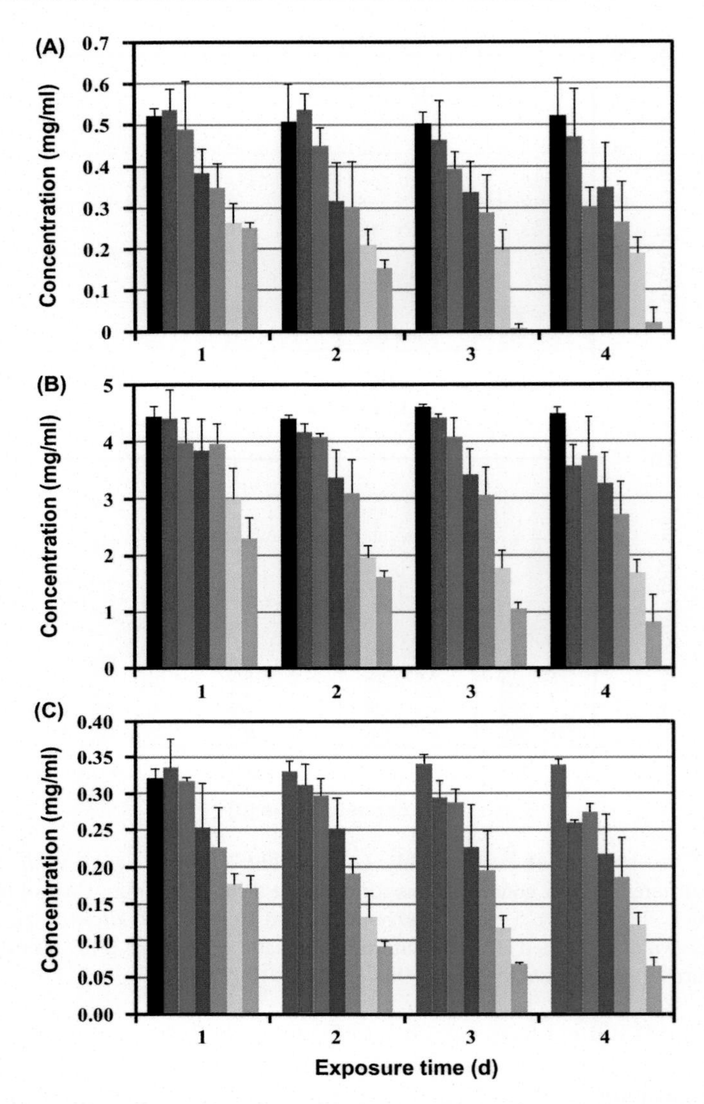

Figure 8.6 Concentration of β-carotene (A), diadinoxanthin (B) and 9-cis-neoxanthin (C) in *Euglena gracilis* cells treated with different copper concentrations after different incubation times. Control (*black bars*); 0.4 mg L^{-1} (*red bars*); 4 mg L^{-1} (*green bars*); 11 mg L^{-1} (*purple bars*); 22 mg L^{-1} (*blue bars*); 37 mg L^{-1} (*yellow bars*); and 56 mg L^{-1} (*orange bars*). *Source*: Redrawn after Ahmed H. Biomonitoring of aquatic ecosystems [PhD thesis]. Erlangen-Nürnberg: Friedrich-Alexander-Universität; 2010.

transfer [33]. Increased Cu concentrations induce oxidative stress [34] which may damage the pigment-protein complexes in the thylakoid membranes [35]. They also inhibit chlorophyll-synthesizing enzymes and induce membrane modifications

Table 8.1 Effects of the exposure of *Euglena gracilis* to increasing Cd concentration and increasing exposure times on the intracellular concentrations of chl *a*, chl *b*, β-carotene, diadinoxanthin, and 9-cis-neoxanthin (in pg/cell).

Chlorophyll *a* ($mg\ L^{-1}$)	Exposure time (d)			
	1	2	3	4
0	24.65	26.02	22.11	21.51
7	17.27	14.90	14.22	13.98
14	9.47	10.13	11.48	14.09
30	4.21	5.15	2.30	2.67
Chlorophyll *b* ($mg\ L^{-1}$)				
0	5.04	5.25	4.91	5.05
7	3.20	2.80	2.50	2.08
14	3.06	2.36	2.13	2.04
30	1.74	1.14	1.10	1.08
β-carotene ($mg\ L^{-1}$)				
0	0.521	0.526	0.501	0.490
7	0.353	0.343	0.340	0.258
14	0.265	0.255	0.270	0.104
30	0.197	0.070	0.070	0.000
diadinoxanthin ($mg\ L^{-1}$)				
0	4.10	3.50	3.60	3.40
7	2.33	1.50	1.37	1.22
14	1.87	0.90	0.95	0.59
30	1.32	0.48	0.47	0.27
9-cis-neoxanthin ($mg\ L^{-1}$)				
0	0.304	0.303	0.284	0.282
7	0.235	0.217	0.214	0.220
14	0.209	0.189	0.180	0.160
30	0.147	0.128	0.090	0.060

*Source:*After Ahmed H. Biomonitoring of aquatic ecosystems [PhD thesis]. Erlangen-Nürnberg: Friedrich-Alexander-Universität; 2010.

through lipid peroxidation [36]. Cd was found to have even stronger effects on chl *b*, β-carotene and xanthophylls which may be due to a different mechanism of inhibition. Cd affects the Calvin cycle and inhibits chlorophyll biosynthesis [33]. In addition, Cd affects protein synthesis as well as formation of mitochondria and chloroplasts [37]. Heavy metals also enhance pigment degradation through the

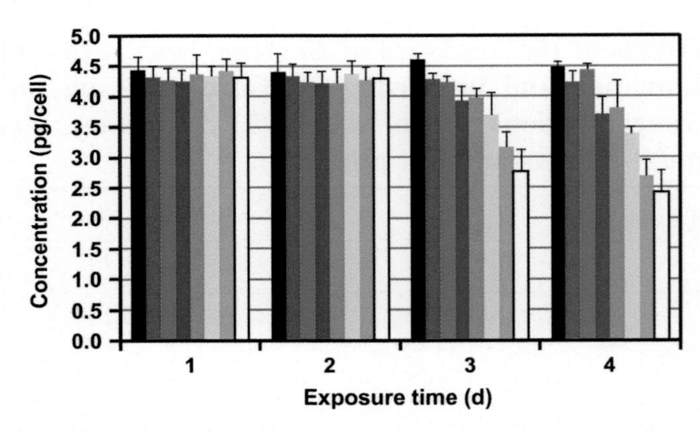

Figure 8.7 Concentration of diadinoxanthin in *Euglena gracilis* treated with different cadmium concentrations after different incubation times. Control (*black bars*); 4 mg L^{-1} (*red bars*); 10 mg L^{-1} (*green bars*); 25 mg L^{-1} (*purple bars*); 37 mg L^{-1} (*blue bars*); 50 mg L^{-1} (*yellow bars*); 100 mg L^{-1} (*orange bars*); and 200 mg L^{-1} (*white bars*).
Source: Redrawn after Ahmed H. Biomonitoring of aquatic ecosystems [PhD thesis]. Erlangen-Nürnberg: Friedrich-Alexander-Universität; 2010.

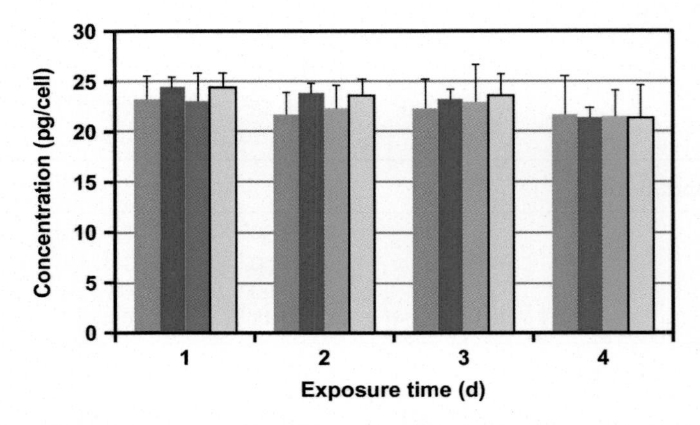

Figure 8.8 Concentration of chlorophyll *a* isolated from *Euglena gracilis* treated with increasing calcium chloride concentrations after different incubation times. Control (*white bars*); 12.5 mg L^{-1} (*red bars*); 25 mg L^{-1} (*green bars*); and 62 mg L^{-1} (*blue bars*).
Source: Redrawn after Ahmed H. Biomonitoring of aquatic ecosystems [PhD thesis]. Erlangen-Nürnberg: Friedrich-Alexander-Universität; 2010.

activation of catabolic enzymes [33] and inhibit enzymes such as protochlorophyllide reductase and δ-aminolevulinic acid dehydratase which are involved in chlorophyll synthesis [38,39].

Carotenoids play several roles in photosynthetic organisms. They are involved in the protection of the photosynthetic apparatus from environmental stresses such as excessive exposure to solar visible and UV radiation [40]. Loss of β-carotene and xanthophylls results in oxidative stress in *Euglena* [41]. Diadinoxanthin stabilizes the light-harvesting complexes in this organism [42]. In addition to heavy metals, other toxins such as norflurazon significantly reduce photosynthetic pigments in *Euglena* [43]. These and other recent studies support the notion that the concentration of photosynthetic pigments in unicellular algae such as *Euglena* can be used as a valuable marker for heavy metal stress. Thus this unicellular flagellate can be employed as a sensitive, fast, and efficient bioindicator in a monitoring system to study various environmental stress factors in aquatic ecosystems such as wastewaters, drinking water, and natural ecosystems, to detect hazards for human health and ecosystem integrity.

8.3 Delayed fluorescence as a versatile tool for measurement of photosynthetic performance

8.3.1 Delayed fluorescence – the phenomenon

Intact chloroplasts or photosynthetic organisms, respectively, which were exposed to light and then transferred to darkness, emit a low intensity light (only detectable with very light-sensitive instruments), which is termed "delayed fluorescence" (DF) [44–46]. For better understanding DF the following paragraph provides a brief overview on photosynthesis.

Photosynthesis consists of two basic reactions: the light-dependent reaction and the light-independent reaction (dark reaction). Detailed information on photosynthesis and photochemistry is provided by, e.g., [47–49]. The light-driven reaction starts with photolysis of water (where oxygen is produced as a byproduct) and results in the production of adenosin triphosphate (ATP) and nicotinamide adenine dinucleotide phosphate ($NADPH/H^+$). In the light-independent reaction (Calvin cycle) these products are utilized for carbon dioxide reduction. Elements of the light-dependent reaction are mainly located in the thylakoid membrane of the chloroplast, whereas the water-soluble enzymes of the light-independent reactions are in the stroma. The light reaction consists of various redox systems (Fig. 8.9): the water splitting complex (OEC, oxygen evolving complex), P680, pheophytin, and a bound plastoquinone (Q_A) are located in the photosystem II (PSII) complex. Plastoquinone (PQ), a benzoquinone derivate, transports electrons (two at a time) from PSII to the cytochrome b_6f protein complex (contains three redox systems: cytochrome b_6, the Rieske protein (FeSR), and cytochrom f). The small copper-containing protein plastocyanin transports a single electron from the cytochrome b_6f protein complex to the photosystem I complex (PSI). PSI consists of various redox systems (P700, A0, A1, FeSx, FeSB, and FeSA). In the excited state it reduces ferredoxin, which in turn reduces $NADP^+$ to $NAPDH/H^+$ ($NADP^+$ reductase). In addition it is an electron donor for other reducing enzymes, such as nitrite reductase (reduces NO_2^- to NH_3), sulfite reductase (reduces ${SO_4}^{2-}$ to S^{2-}), and glutamate synthase. In light a valence electron of one of the two specific chlorophyll *a* molecules (P680) in the core of the PSII complex is excited directly upon absorption of a photon or

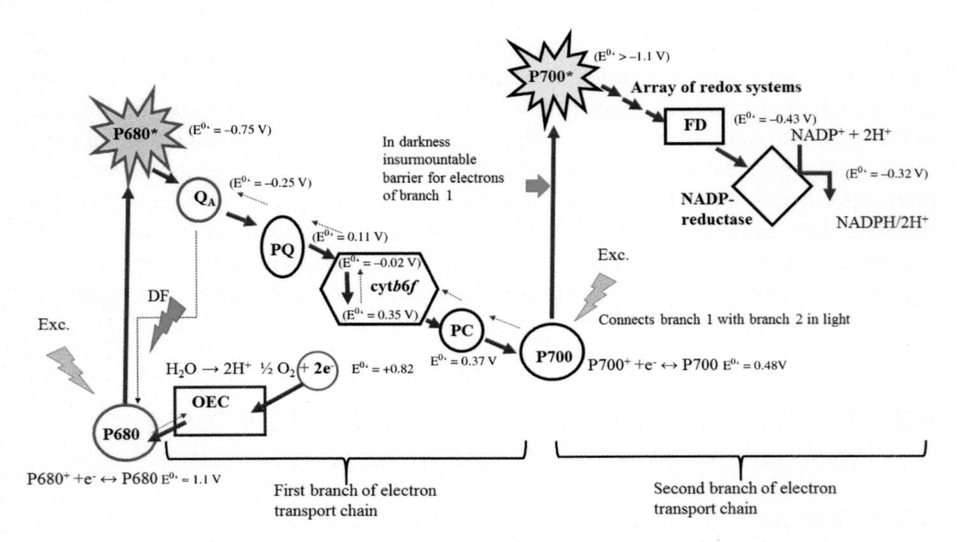

Figure 8.9 Simplified scheme of the electron transport chain of photosynthesis. Not all redox systems involved are shown. Bold arrows show the path of electrons from the water-splitting complex (oxygen evolving complex OEC) to $NADP^+$. Bold arrows: forward direction of electron transport chain, fine arrows backward reaction of electrons. *OEC water-splitting complex*, (oxygen evolving complex); *Exc.*, Excitation; *DF*, delayed fluorescence; *PQ*, plastoquinone; *cyt* b_6f, cytochrome b_6f-complex; PC, plastocyanin; FD, ferredoxin. For details see text.

indirectly via energy transfer by other accessory pigments surrounding the light-harvesting complexes. The excitation brings the electron into a higher energetic level, where the S1 state is the starting point of photosynthesis. The energy difference between the ground state (S0) and S1 is about 127.5 KJ/mol (change of standard redox potential of more than 1 V (from +0.82 to − 0.75 V, all redox potential data from [50]). Subsequently, the electron is transported away from P680 via a system of subsequent redox systems (from PSII via PQ to the cytochrome b_6f complex and from there via plastocyanin to PSI, where a second excitation by light results in the reduction of $NADP^+$, see above). The need of a second excitation step (P700 ($E^{0'}$ +0.45 V) → P700* ($E^{0'}$ > − 1.1 V)) connects the first branch of the electron transport chain P680* ($E^{0'}$ − 0.02 V) → → → plastocyanin ($E^{0'}$ +0.37 V) with the second branch: the array of redox systems starting with the special chlorophyll A_0 ($E^{0'}$ − 1.1 V) → → → $NADP^+ + 2\,H^+ \leftrightarrow NADPH^+ + H^+$ ($E^{0'}$ − 0.32 V). As we will see, this is crucial for the generation of DF. The other key elements in DF are most likely P680 and the attached water-splitting complex. After excitation and loss of the electron in the subsequent charge separation events of the electron transport chain the remaining $P680^+$ cation is an extremely strong oxidant $E_0 =$ about +1 V. This enables the oxidation of the water-splitting complex (OEC oxygen evolving complex), which delivers one of four electrons to $P680^+$. The $P680^+$cation is subsequently reduced to P680. Electron transfer from the OEC to $P680^+$ is mediated by YZ (a tyrosin residue of the D1 protein of PSII

[51]. Upon the next excitation event another electron is transferred from the water-splitting complex. When this complex has delivered four electrons to P680 it oxidizes two water molecules ($O_2 + 4\,H^+ + 4\,e^- \leftrightarrow 2\,H_2O$, $E^{0\prime}$ +0.82 V). The four electrons from water-oxidation replace the four electrons transferred into the electron-transport chain and the process starts again. In light, all redox systems of the electron transport chain are mainly in a reduced state. After sudden transition to darkness reduced redox systems of the first branch of the chain (pheophytin to P700) can no longer transfer electrons to the second branch (A0 to NADPH/H^+ or other targets of reduction). While the second branch becomes depleted, the first branch stays reduced. As the way to $NADP^+$ is blocked, the negligible back reactions (back towards PSII) in the electron transport chain become relevant. Although the "downhill" reaction is preferred, energy-rich electrons statistically move in the reverse direction. The driving force is the energy difference between P700 and the other reduced elements of the first branch and a free position of the water-splitting complex, which is about −35 kJ/mol for P700. It is believed that DF is due to the recombination of excited electrons with oxidized positions in the water-splitting complexes of PSII. The fluorescence signal originates from P680*. $P680^+$ cations have a very short lifetime in the microsecond range [52,53] which means that electrons will not encounter a $P680^+$ cation on their reverse path. It is very likely that the valence electron of P680 moves transiently to its redox partner, an open position ("electron hole") of the OEC, leaving behind a $P680^+$. An excited electron from the first branch of the electron-transport chain may take advantage of this situation and reduce the $P680^+$ to P680. Due to its high energy it will first combine with an energy-rich orbital of P680 (S1-state or T1-state), from where it relaxes to the S0-state under release of a fluorescence photon (or phosphorescence in the case of T1 → S0-relaxation). The reduction of the OEC is achieved by the former valence electron originating from P680, which cannot return because its orbital is locked by the second electron from the electron transport chain. Depending on the way of an excited electron back to the OEC, different time constants of DF can be distinguished (see Table 8.2).

As DF is only about 10^{-3} of the intensity of prompt fluorescence, detection is only feasible by means of sensitive photomultiplier-supported optical devices.

Table 8.2 Time scales of fluorescence or delayed fluorescence reactions, respectively.

Time scale	Reaction
$t < 2$ ns	Prompt fluorescence of chl *a*
$t \approx 1 - 10$ μs	p^+-I^- (from pheophytin to P680)
$t \approx 10 - 100$ μs	$Q_a - Q_b$ reactions
$t >$ ms	S_i states
$T > 100$ ms	Q_b − PQ reactions
$T > 1$ s	Discharge of PSI (P700) via PQ to PSII

Source: Modified after Watanabe M., Henmi K., Ogawa K., Suzuki T. Cadmium-dependent generation of reactive oxygen species and mitochondrial DNA breaks in photosynthetic and non-photosynthetic strains of *Euglena gracilis*. Comp Biochem Physiol C Toxicol Pharmacol 2003;134:227–34.

Intensity of light emission decreased exponentially dependent on time. Same as pulse-amplitude modulated (PAM) fluorimetry (cf. Chapter 9: Photosynthesis assessed by chlorophyll fluorescence, this volume), DF can be employed for different applications concerning photosynthetic organisms. Recording DF enables detection of environmental impact on plants and algae [54–56]. Photosynthetic organisms can be classified via excitation—DF spectra [57,58].

8.3.2 Possible DF applications

All DF data presented below were obtained with a single-photon camera (C-2400–07ER; Hamamatsu Photonics K.K., Hamamatsu, Japan). The camera is sensitive enough to detect DF single photons emitted by photosynthetic organisms. The video signals provide a spatial overview of the DF pattern of an object. Each recorded photon is represented as a bright spot (see e.g., Fig. 8.10). This enables counting and quantitative analysis by means of suitable image analysis tools. Data provided in the following were recorded and evaluated with the computer-based image analysis software Wintrack 2000 [59]. The frame grabber card of the computer running Wintrack 2000 was connected with the analog video output of the camera. Because of the high sensitivity of the camera, the samples are located in a light-tight box (Figs. 8.11 and 8.12).

8.3.3 Delayed fluorescence in ecotoxicology

The photosynthetic apparatus of plants and algae such as *Euglena* is a possible target of toxicants [35,60,61]. Changes in photosynthetic efficiency are generally determined by means of the PAM fluorometry [62–65]. Like the PAM fluorimetry, determination of photosynthesis allows detection of toxicants affecting photosynthesis [55,56,66,67].

Addition of the herbicide 3-(3,4-dichlorophenyl)-1,1-dimethylurea (DCMU) to cultures of *E. gracilis* results in a concentration-dependent decrease of DF

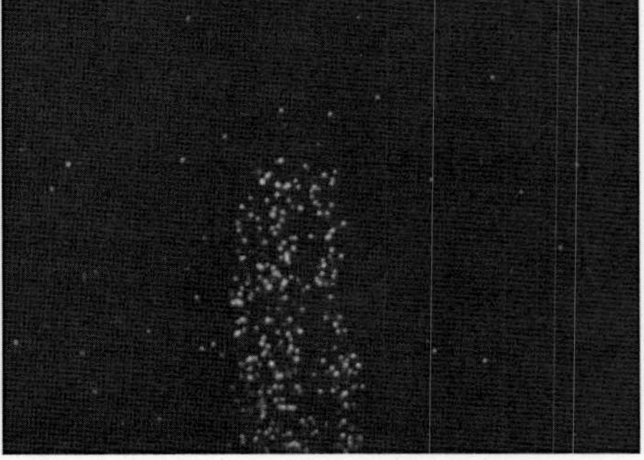

Figure 8.10 Example of delayed fluorescence. Left picture: bright field image of a cactus. Right: delayed fluorescence of the same object in darkness.

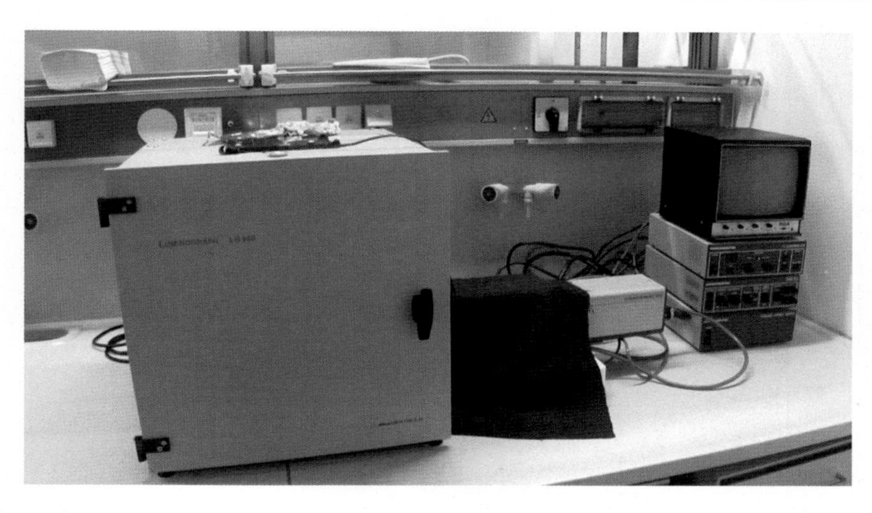

Figure 8.11 Experimental setup for detection of delayed fluorescence in plants. Left: Light-tight box; center: single-photon camera; right: panels and monitor. An analysis computer as well as the monochromator unit is not attached in the picture.

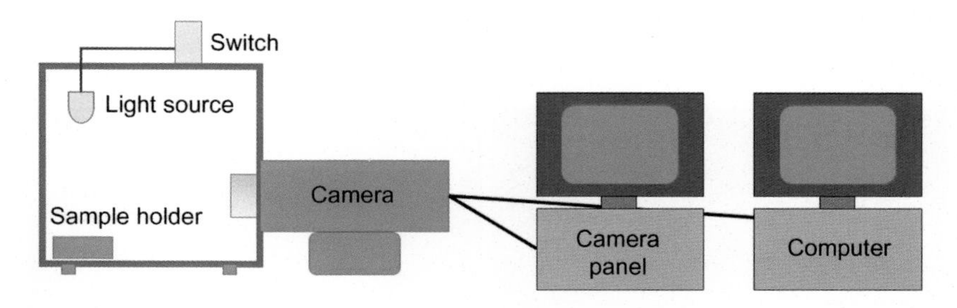

Figure 8.12 Experimental setup for spatial observation of delayed fluorescence (DF) in photosynthetic organisms. Samples were kept in a light-tight box. By means of an external trigger a desired light pulse (mostly 1 s) was applied. The subsequent DF was observed with a sensitive camera (SPC, single photon camera). The video outputs were recorded and analyzed.

(Fig. 8.13). Above all the duration of DF is impaired in the presence of lower DCMU concentrations (1 and 10 μM), while initially (until about 15 s) the signal intensity is comparable to the EtOH control (DCMU was dissolved in EtOH). DCMU blocks the electron transfer from PSII to PQ [68]. The obtained signal probably originates from electron or excition transfer, respectively, from excited light harvesting complex II (LHC II-c)- complexes. DCMU effects can also be recorded in spatial resolution. A leaf partially treated with DCMU shows decrease of DF toxicated parts of the leaf (Fig. 8.14).

Recently DF was employed to detect effects of heavy metals on photosynthesis [59]. DF was found to be as sensitive as PAM fluorimetry (Figs. 8.15 and 8.16).

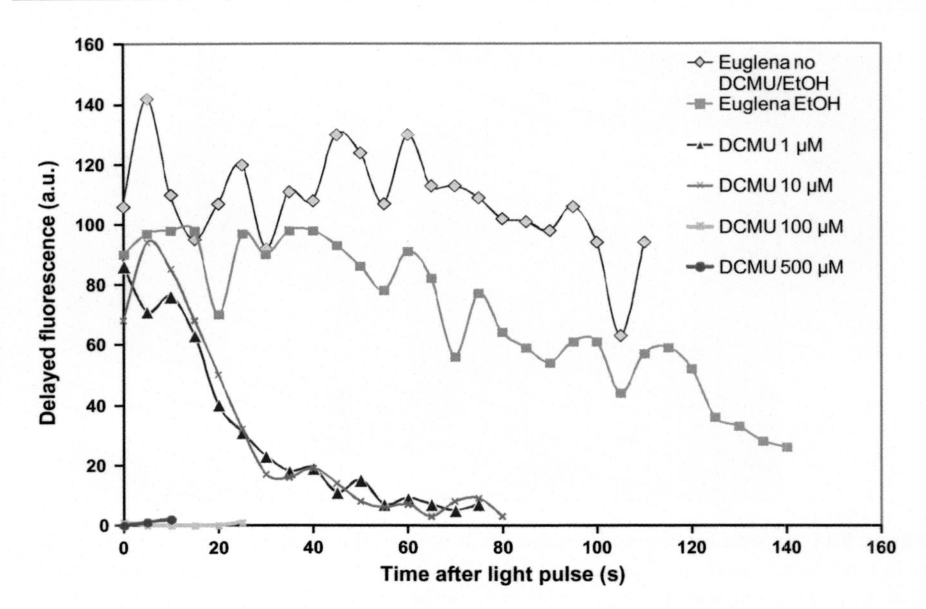

Figure 8.13 Delayed fluorescence recorded in *Euglena gracilis* after 15 min incubation at different concentrations of DCMU. Control: no DCMU or EtOH (solvent for DCMU), respectively. EtOH control: EtOH, without DCMU.

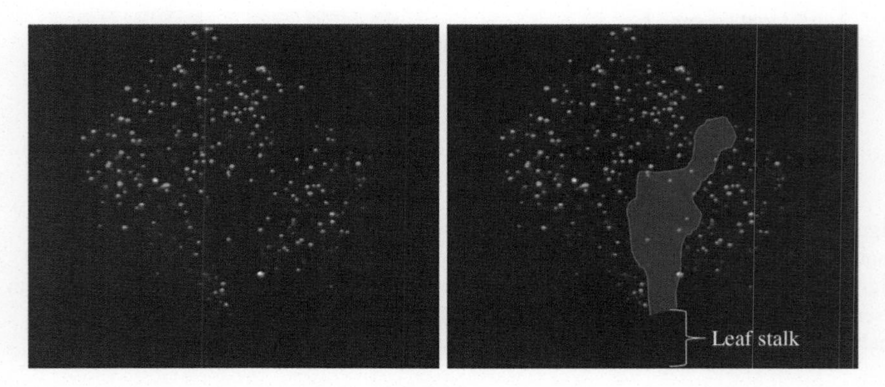

Figure 8.14 Spatial resolution of delayed fluorescence (DF) in a pea leaf, which was partially treated with DCMU. Strong reduction of DF is visible in the range of DCMU application (labeled in right photo).

The obtained EC_{50} values for Cu were 505.8 mg L^{-1} (photosystem II quantum yield, Y(PSII)), 286.5 mg L^{-1} (DF), and 491.0 mg L^{-1} (steady state fluorescence, SSF). The EC_{50} values for silver ions were 51.9 mg L^{-1} (Y(PSII)), 51.8 mg L^{-1} (DF) (values for SSF not determinable). As chloroplasts are well protected inside the cytoplasm, other parameters (motility, bioluminescence in *Vibrio fischeri*) are more sensitively impaired than photosynthesis (Table 8.3).

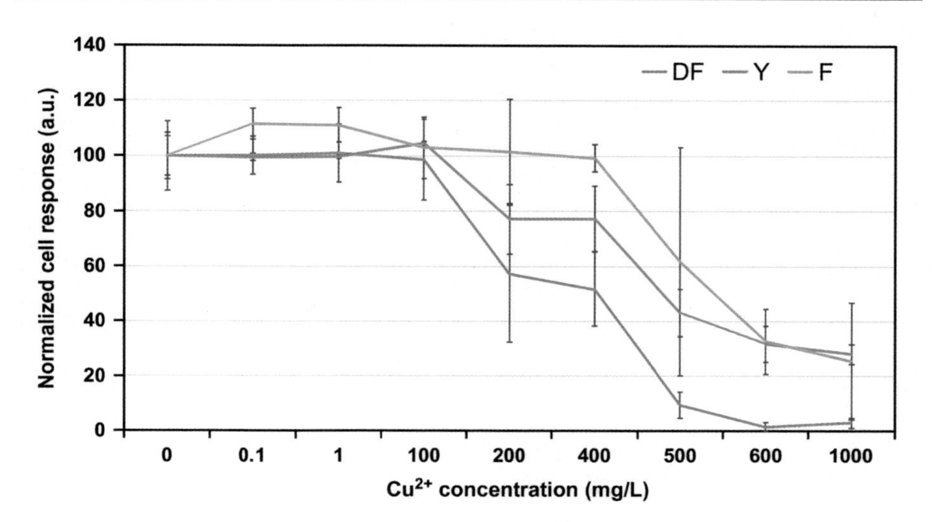

Figure 8.15 Inhibitory effect of increasing copper concentrations on photosynthesis. Comparison between the photosynthetic parameters delayed fluorescence DF, photosynthetic quantum yield Y, and chlorophyll fluorescence F.

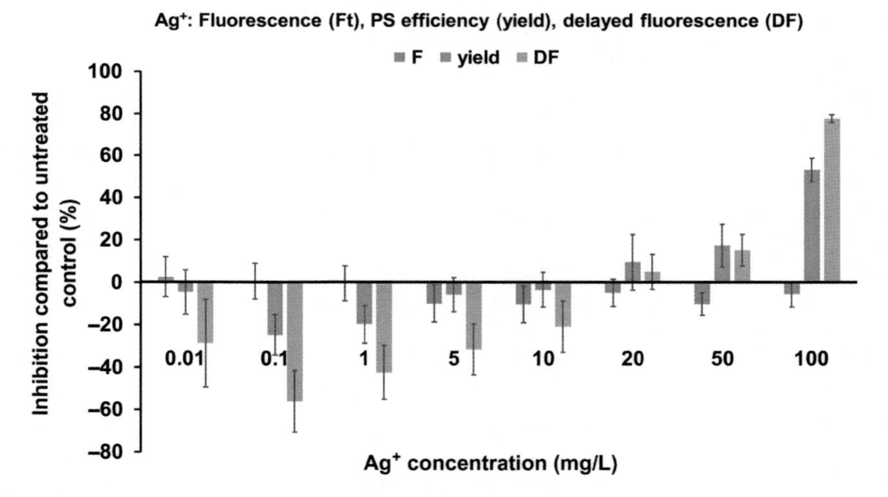

Figure 8.16 Effect of increasing silver concentrations on photosynthesis. Comparison between the photosynthetic parameters delayed fluorescence DF, photosynthetic quantum yield Y, and chlorophyll fluorescence F.

8.3.4 Recording of recovery of desiccated plants

Mainly primitive plants such as some algae and mosses are desiccation tolerant, which means that they tolerate desiccation and recover when they are again in contact with water [76,77]. Only few angiosperms have developed desiccation

Table 8.3 EC_{50} values of effects of silver and copper ions of different parameters in chosen test organisms

Metal	Parameter	EC_{50} value	Reference
Cu^{2+}	Delayed fluorescence in *Euglena gracilis*	286.5 mg L^{-1}	[59]
	Steady state fluorescence in *E. gracilis*	505.8 mg L^{-1}	[59]
	Photosystem II quantum yield in *E. gracilis*	491.0 mg L^{-1}	[59]
	Motility of *Euglena*	19.09 mg L^{-1}	[69]
	Gravitactic orientation of *Euglena*	20.4 mg L^{-1}	[69]
	Velocity of *Euglena*	23.09 mg L^{-1}	[69]
	Lemna growth inbition	0.47 mg L^{-1}	[70]
	Vibrio fischeri (bioluminescence)	34.4 mg L^{-1}	[71]
Ag^{+}	Delayed fluorescence in *E. gracilis*	51.8 mg L^{-1}	[59]
	Steady state fluorescence in *E. gracilis*	Not determinable	[59]
	Photosystem II quantum yield in *E. gracilis*	51.9 mg L^{-1}	[59]
	Orientation of *Euglena*	0.54 mg L^{-1}	[72]
	Daphnia (immobilization)	0.0023 mg L^{-1}	[73]
	Vibrio fischeri (bioluminescence)	1.856 mg L^{-1}	[74]
	Lemna minor	0.081 mg L^{-1}	[75]

tolerance [78]. In some cases, e.g., in many moss species, recovery of photosynthesis after rehydration is considerably fast, because certain mosses need to become hydrated as fast as possible before water is no longer available (e.g., mosses on a rock). Experiments with the moss *Hypnum cupressiforme* show an almost immediate recovery of photosynthesis. In the desiccated state mosses do not show DF, but immediately after rehydration significant DF signals can be recorded indicating photosynthetic activity (Figs. 8.17 and 8.18). Determination of the photosystem II quantum yield (Y(PSII)) by means of PAM fluorimetry (cf, Chapter 9: Photosynthesis assessed by chlorophyll fluorescence, this volume) confirms the results obtained with DF analysis (data not shown).

8.3.5 Determination of efficiency of accessory pigments

Accessory pigments adsorb light outside of the adsorption range of chlorophyll *a* and transmit their excitation energy to chlorophyll P680 in photosystem II increasing the wavelength range of utilizable light. By means of an excitation-dependent DF spectrum it is possible to determine the contribution of accessory pigments to photosynthesis. Only wavelengths driving photosynthetic electron transport result in detectable DF. Bodemer et al. have used a method in order to classify different alga groups in a water sample due to their different screening pigments [58]. Monochromatic excitation at certain wavelengths and subsequent recording of DF enables classification of alga groups [58,79]. Cyanobacteria possess, among other pigments, phycocyanin, which absorbs light with a peak of about 620 nm [49,80]. Rhodophytes and to a limited extent cyanobacteria use phycoerythrin as an accessory pigment, which covers the green and yellow range of the electromagnetic

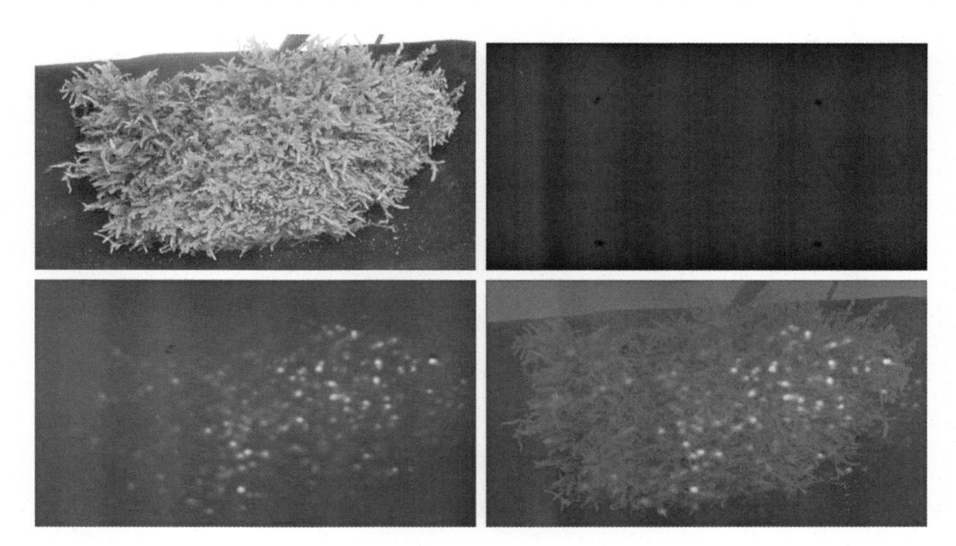

Figure 8.17 Delayed fluorescence DF in desiccation-tolerant moss *Hypnum cupressiforme* before and after rehydration. Upper left: batch of *Hypnum cupressiforme*. Upper right: DF-signal of dry moss batch. Lower left: same batch 1 min after rehydration. Lower right: combined image of moss batch and corresponding DF signal.

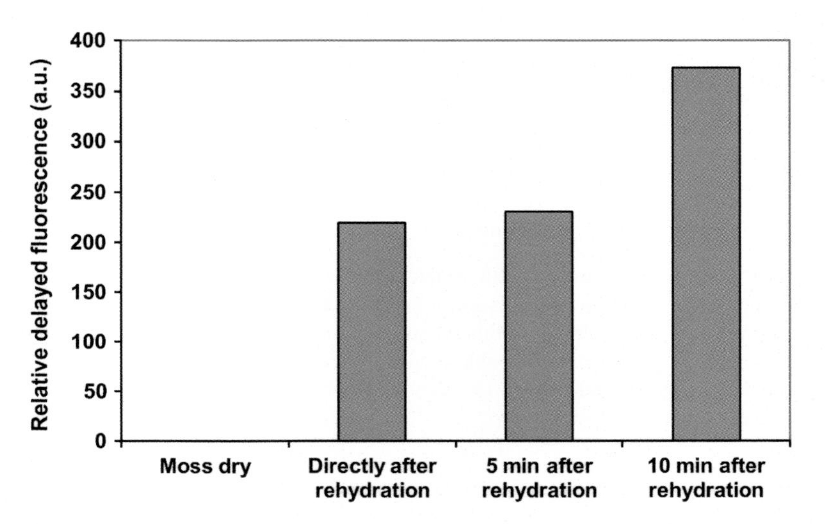

Figure 8.18 Delayed fluorescence signal of the moss *Hypnum cupressiforme* before and after rehydration.

spectrum of light [81]. Spectra of phycocyanin and phycoerythrin, respectively, are provided in Fig. 8.19. Monochromatic excitation in the range of these accessory pigments will drastically increase photosynthetic efficiency compared to plants, which do not possess corresponding accessory pigments. In addition, the method

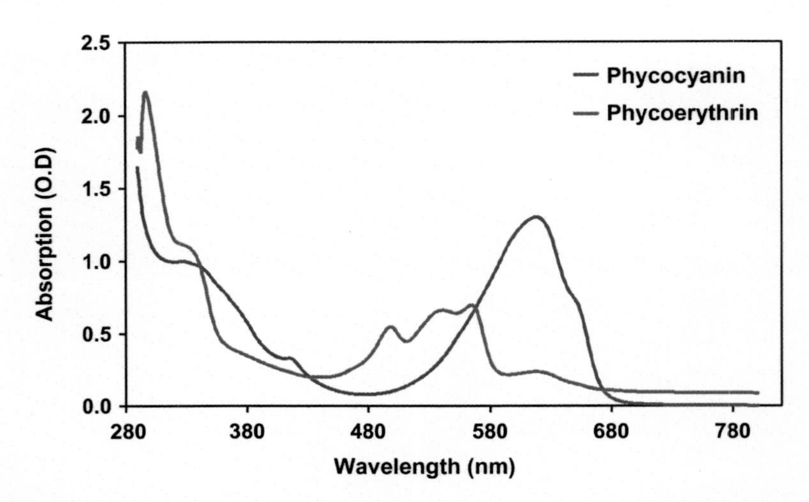

Figure 8.19 Spectra of phycocyanin and phycoerythrin, respectively, after phycobiliprotein extraction.

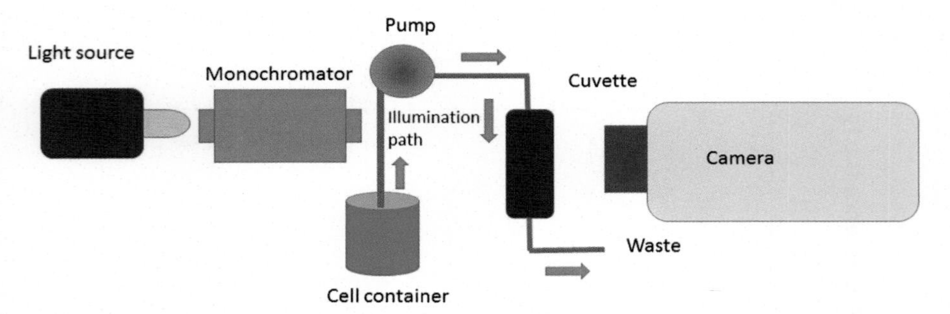

Figure 8.20 Setup for experiments with monochromatic light. Cells were kept in a light-tight container. For measurements cells were pumped into a cuvette, placed in the light-tight box shown in Fig. 8.12 (not drawn in this scheme). On their way cells were illuminated by monochromatic light provided by a double monochromator. After the cells have reached the cuvette the pump was stopped and the DF recording of cells started. Each wavelength needed to be adjusted manually.

allows to distinguish between accessory pigments and screening pigments, such as scytonemin, which do not contribute to DF after excitation [82]. In order to record the long-term DF we modified the device similar to the setup of Bodemer et al. [58] (Fig. 8.20). Cells were kept in a light-tight container. A pump transferred the cells into a cuvette located inside the observation box. On their way to the cuvette the cells passed a light transparent window of the light-tight tube, where they were irradiated with monochromatic light (from 400 to 720 nm in 20 nm intervals (400 nm, 420 nm, 440 nm, etc., Half power bandwidth: each 10 nm) provided by a monochromator in combination with a xenon arc lamp. After about 5 s the cells

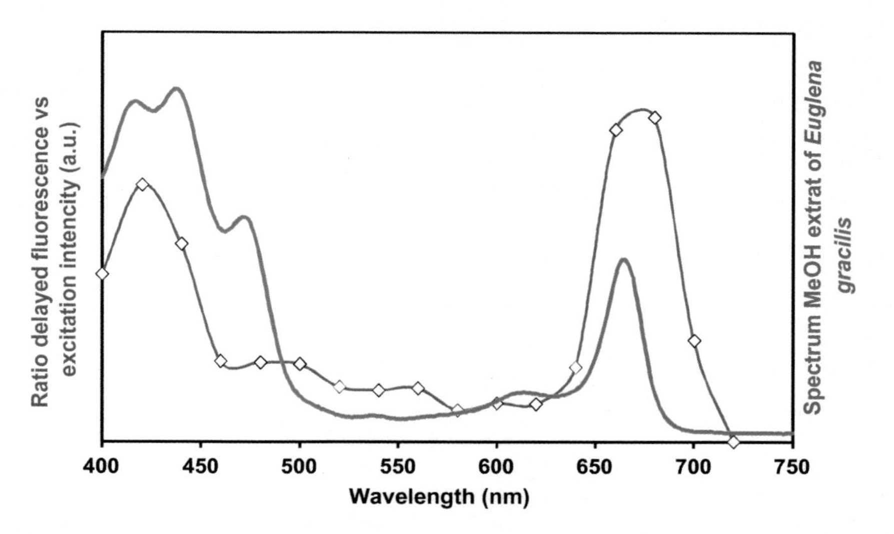

Figure 8.21 Intensity of delayed fluorescence (DF) in *Euglena gracilis* in correlation with the excitation wavelength (*diamonds*). The data are compared with the spectrum of a MeOH extract of the same cells. DF intensity was multiplied with the corresponding intensity of applied excitation (μmol/(photons m^2 s).

reached the cuvette. Excited photons were recorded with the camera and counted (60 s in 5-s intervals) with a counting module of the Wintrack 2000 software. The integral of the subsequent measurements was calculated subsequently. This was manually performed for the wavelength range 400–700 nm in 20-nm steps. The excition DF spectra show clearly which wavelength contributes to which extent to photosynthesis. In *E. gracilis* the absorption gap in the green and yellow wavelength range becomes clearly visible (Fig. 8.21). The Chrysophyceae *Ochromonas danica* possesses, among others, the screening pigment fucoxanthin, which extends the accessory range of the alga into the green range (Fig. 8.22). The cyanobacterium *Nostoc muscorum* shows significant DF in the absorption range of phycocyanin (Fig. 8.23).

8.4 Conclusions

Environmental factors, developmental stages, and pollutants have significant impacts on the pigmentation of higher and lower plants. Therefore monitoring the color of algae or plants can serve as sensitive bioassays. Pigmentation is affected by pollutants such as heavy metals. Since changes in pigmentation can be determined very fast, bioassays using this technique can be used as continuous monitoring devices and early warning systems. DF detection offers versatile possibilities in order to investigate different aspects of photosynthesis. Its sensitivity is at least in the same range as PAM fluorimetry.

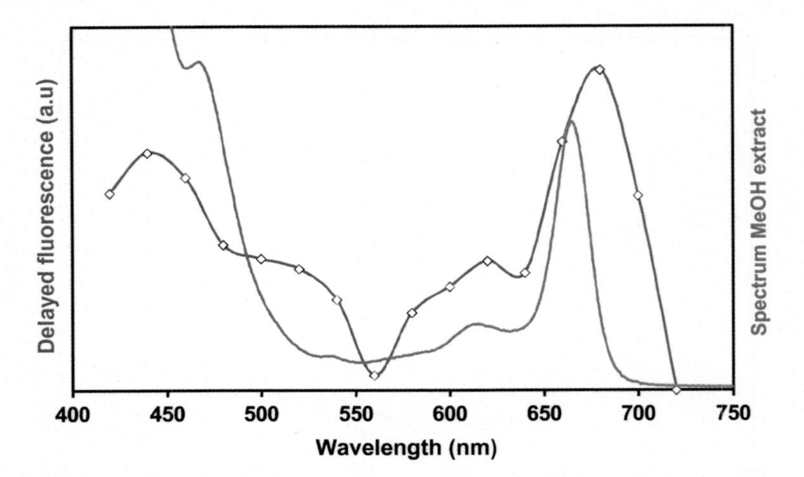

Figure 8.22 Intensity of delayed fluorescence (DF) in *Ochromonas sp.* in correlation with the excitation wavelength (*diamonds*). The data are compared with the spectrum of a MeOH extract of the same cells. DF intensity was multiplied with the corresponding intensity of applied excitation (μmol/(photons m^2 s).

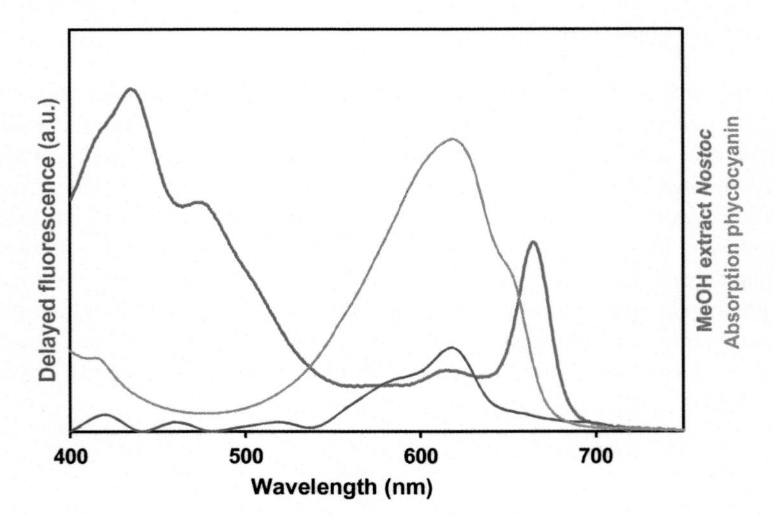

Figure 8.23 Intensity of delayed fluorescence (DF) in *Nostoc muscorum* in correlation with the excitation wavelength (*diamonds*). The data are compared with the spectrum of a MeOH extract of same cells and spectrum of phycocyanin. DF intensity was multiplied with the corresponding intensity of applied excitation (μmol/(photons m^2 s).

References

[1] Sheppard S, Evenden W, Abboud S, Stephenson M. A plant life-cycle bioassay for contaminated soil, with comparison to other bioassays: mercury and zinc. Arch Env Contamin Toxicol 1993;25(1):27–35.
[2] Gross J. Pigments in vegetables: chlorophylls and carotenoids. Berlin: Springer Science & Business Media; 2012.
[3] Close DC, Beadle CL. The ecophysiology of foliar anthocyanin. Bot Rev 2003;69 (2):149–61.
[4] Park CH, Chae SC, Park S-Y, Kim JK, Kim YJ, Chung SO, et al. Anthocyanin and carotenoid contents in different cultivars of chrysanthemum (*Dendranthema grandiflorum* Ramat) flower. Molecules 2015;20(6):11090–102.
[5] Paech K. Colour development in flowers. Annu Rev Plant Physiol 1955;6(1):273–98.
[6] Heijde M, Ulm R. UV-B photoreceptor-mediated signalling in plants. Trends Plant Sci 2012;17(4):230–7.
[7] Wade HK, Bibikova TN, Valentine WJ, Jenkins GI. Interactions within a network of phytochrome, cryptochrome and UV-B phototransduction pathways regulate chalcone synthase gene expression in *Arabidopsis* leaf tissue. Plant J 2001;25:675–85.
[8] Pallardy SG. Physiology of woody plants. London: Academic Press; 2010.
[9] Lev-Yadun S. Spring versus autumn or young versus old leaf colors: Evidence for different selective agents and evolution in various species and floras. In: Lev-Yadun S, editor. Defensive (anti-herbivory) coloration in land plants. Switzerland: Springer; 2016. p. 259–66.
[10] Arora VK, Boer GJ. A parameterization of leaf phenology for the terrestrial ecosystem component of climate models. Global Change Biol 2005;11(1):39–59.
[11] Anderson R, Ryser P. Early autumn senescence in red maple (*Acer rubrum* L) is associated with high leaf anthocyanin content. Plants 2015;4(3):505–22.
[12] Chalker-Scott L. Why do leaves turn red? Pullman, WA: Washington State University; 2016. Available from: https://research.libraries.wsu.edu/xmlui/bitstream/handle/2376/6123/FS209E.pdf.
[13] Liu L, Liang L, Schwartz MD, Donnelly A, Wang Z, Schaaf CB, et al. Evaluating the potential of MODIS satellite data to track temporal dynamics of autumn phenology in a temperate mixed forest. Remote Sens Environ 2015;160:156–65.
[14] Ahmed IM, Dai H, Zheng W, Cao F, Zhang G, Sun D, et al. Genotypic differences in physiological characteristics in the tolerance to drought and salinity combined stress between Tibetan wild and cultivated barley. Plant Physiol Biochem 2013;63:49–60.
[15] Honsdorf N, March TJ, Berger B, Tester M, Pillen K. High-throughput phenotyping to detect drought tolerance QTL in wild barley introgression lines. PLoS One 2014;9(5): e97047.
[16] Álvarez S, Sánchez-Blanco MJ. Comparison of individual and combined effects of salinity and deficit irrigation on physiological, nutritional and ornamental aspects of tolerance in *Callistemon laevis* plants. J Plant Physiol 2015;185:65–74.
[17] Gange AC, Eschen R, Wearn JA, Thawer A, Sutton BC. Differential effects of foliar endophytic fungi on insect herbivores attacking a herbaceous plant. Oecologia 2012;168(4):1023–31.
[18] Jeon CW, Hong SW, Cho H, Kwak Y-s. *Alternaria solani* causing leaf blight disease on *Aster glehni* in Korea. J Inf Serv Syst KoreaScholar 2016;50(4):249–53.
[19] Mani MS. Ecology of plant galls. Netherlands: Springer; 2013.

[20] Koga LJ, Canteri MG, Calvo ES, Martins DC, Xavier SA, Harada A, et al. Managing soybean rust with fungicides and varieties of the early/semi-early and intermediate maturity groups. Trop Plant Pathol 2014;39(2):129–33.

[21] Aguiar RA, Cunha MG, Araújo FG, Carneiro LC, Borges EP, Carlin VJ. Efficiency loss of recorded fungicides for the control of asian soybean rust in Central region of Brazil. Rev Agric Neotrop 2016;3(2):41–7.

[22] Cai J, Okamoto M, Atieno J, Sutton T, Li Y, Miklavcic SJ. Quantifying the onset and progression of plant senescence by color image analysis for high throughput applications. PLoS One 2016;11(6):e0157102.

[23] Yuan Y, Shu S, Li S, He L, Li H, Du N, et al. Effects of exogenous putrescine on chlorophyll fluorescence imaging and heat dissipation capacity in cucumber (*Cucumis sativus* L) under salt stress. J Plant Growth Regul 2014;33(4):798–808.

[24] John R, Ahmad P, Gadgil K, Sharma S. Heavy metal toxicity: effect on plant growth, biochemical parameters and metal accumulation by *Brassica juncea* L. Int J Plant Prod 2012;3(3):65–76.

[25] Ahmed H, Häder D-P, Richter P. Screening of photosynthetic pigments as an indicator of heavy metal stress in *Euglena gracilis*. In: Sinha RP, Richa, Rastogi RP, editors. Biological sciences – innovations and dynamics. New Delhi, India: New India Publishing Agency; 2015. p. 263–78.

[26] Garcia Jr J, De Llano L, Sanchez Jr A, Ynalvez R. The comparison of growth, bioaccumulation of heavy metals, and Cia7 gene expression levels between two strains of *Chlamydomonas reinhardtii*. FASEB J. 2015;29(1 Supplement):887.14.

[27] Ahmed H. Biomonitoring of aquatic ecosystems [PhD thesis]. Erlangen-Nürnberg: Friedrich-Alexander-Universität; 2010.

[28] Gieskes W. Algal pigment fingerprints: clue to taxon-specific abundance. Part Anal Oceanogr 2013;27:61.

[29] De Backer A, Van Hoey G, Hostens K. Long-term trends in the soft bottom benthic fauna of the Belgian part of the North Sea: results of 30 years environmental monitoring. In: Mees J, et al., editors. Book of abstracts – VLIZ Marine Scientist Day. Brugge, Belgium, 12 February 2016. VLIZ Special Publication, 2016, 75:33.

[30] Sytar O, Kumar A, Latowski D, Kuczynska P, Strzałka K, Prasad M. Heavy metal-induced oxidative damage, defense reactions, and detoxification mechanisms in plants. Acta Physiol Plant 2013;35(4):985–99.

[31] Cabrita MT, Gameiro C, Utkin AB, Duarte B, Caçador I, Cartaxana P. Photosynthetic pigment laser-induced fluorescence indicators for the detection of changes associated with trace element stress in the diatom model species *Phaeodactylum tricornutum*. Environ Monit Assess 2016;188(5):1–13.

[32] Manankina E, Mel'nikov S, Budakova E, Shalygo N. Effect of Ni^{2+} on early stages of chlorophyll biosynthesis and pheophytinization in *Euglena gracilis*. Russ J Plant Physiol 2003;50(3):390–4.

[33] Hou W, Chen X, Song G, Wang Q, Chang CC. Effects of copper and cadmium on heavy metal polluted waterbody restoration by duckweed (*Lemna minor*). Plant Physiol Biochem 2007;45:62–9.

[34] Watanabe M, Henmi K, Ogawa K, Suzuki T. Cadmium-dependent generation of reactive oxygen species and mitochondrial DNA breaks in photosynthetic and non-photosynthetic strains of *Euglena gracilis*. Comp Biochem Physiol C Toxicol Pharmacol 2003;134:227–34.

[35] Rocchetta I, Küpper H. Chromium-and copper-induced inhibition of photosynthesis in *Euglena gracilis* analysed on the single-cell level by fluorescence kinetic microscopy. New Phytologist 2009;182(2):405–20.
[36] Shakya K, Chettri M, Sawidis T. Impact of heavy metals (copper, zinc, and lead) on the chlorophyll content of some mosses. Arch Env Contamin Toxicol 2008;54 (3):412–21.
[37] Lamaia C, Kruatrachuea M, Pokethitiyooka P, Upathamb ES, Soonthornsarathoola V. Toxicity and accumulation of lead and cadmium in the filamentous green alga *Cladophora fracta* (OF Muller ex Vahl) Kutzing: A laboratory study. Sci Asia 2005;31(2):121–7.
[38] Aravind P, Prasad MNV. Zinc alleviates Cd induced oxidative stress in *Ceratophyllum demersum* L.: A free floating fresh water macrophyte. Plant Physiol Biochem 2003;41:391–7 PubMed PMID: 35915.
[39] Dai LP, Xiong ZT, Huang Y, Li MJ. Cadmium-induced changes in pigments, total phenolics, and phenylalanine ammonia-lyase activity in fronds of *Azolla imbricata*. Environ Toxicol 2006;21(5):505–12.
[40] Bartwal A, Mall R, Lohani P, Guru S, Arora S. Role of secondary metabolites and brassinosteroids in plant defense against environmental stresses. J Plant Growth Regul 2013;32(1):216–32.
[41] Pancaldi S, Bonora A, Dall'Olio G, Bruni A, Fasulo MP. Ageing of *Euglena* chloroplasts *in vitro*. I. Variations in pigment pattern and in morphology. J Exp Bot 1996;47:49–60.
[42] Casper-Lindley C, Björkman O. Fluorescence quenching in four unicellular algae with different light-harvesting and xanthophyll-cycle pigments. Photosynth Res 1998;56:277–89.
[43] Tschiersch H, Ohmann E, Doege M. Modification of the thylakoid structure of *Euglena gracilis* by norflurazon-treatment: consequences for fluorescence quenching. Environ Exp Bot 2002;47:259–70.
[44] Amesz J, Gorkom H. Delayed fluorescence in photosynthesis. Annu Rev Plant Physiol 1978;29(1):47–66.
[45] Goltsev V, Zaharieva I, Chernev P, Strasser RJ. Delayed fluorescence in photosynthesis. Photosynth Res 2009;101(2-3):217–32.
[46] Strehler BL, Arnold W. Light production by green plants. J Gen Physiol 1951;34 (6):809–20.
[47] Lawlor DW. Photosynthesis, productivity and environment. J Exp Bot 1995;46:1449–61.
[48] Abraham ER, Law CS, Boyd PW, Lavender SJ, Maldonado MT, Bowie AR. Importance of stirring in the development of an iron-fertilized phytoplankton bloom. Nature 2000;407:727–30.
[49] Barber DJW, Richards JT. Energy transfer in the accessory pigments R-phycoerythrin and C-phycocyanin. Photochem Photobiol 1977;25:565–9.
[50] Kadereit JW, Körner C, Kost B, Sonnewald U. Strasburger – Lehrbuch der Pflanzenwissenschaften. Springer-Verlag; Berlin, Heidelberg; 2014.
[51] Christen G, Renger G. The role of hydrogen bonds for the multiphasic $P680^+$ reduction by Y_Z in photosystem II with intact oxygen evolution capacity. Analysis of kinetic H/D isotope exchange effects. Biochemistry (USA) 1999;38:2068–77.
[52] Christen G, Seeliger A, Renger G. $P680^+$ reduction kinetics and redox transition probability of the water oxidizing complex as a function of pH and H/D isotope exchange in spinach thylakoids. Biochemistry (USA) 1999;38:6082–92.

[53] Christen G, Reifarth F, Renger G. On the origin of the '35-μs' kinetics of $P680^+$ reduction in photosystem II with an intact water oxidising complex. FEBS 1998;429:49–52.
[54] Horio T, Nishikawa K, Katsumata M, Yamashita J. Possible partial reactions of the photophosphorylation process in chromatophores from *Rhodospirillum rubrum*. Biochim Biophys Acta 1965;94:371–82.
[55] Katsumata M, Koike T, Kazumura K, Takeuchi A, Sugaya Y. Utility of delayed fluorescence as endpoint for rapid estimation of effect concentration on the green alga *Pseudokirchneriella subcapitata*. Bull Environ Contam Toxicol 2009;83(4):484–7.
[56] Zhang L, Xing D. Rapid determination of the damage to photosynthesis caused by salt and osmotic stresses using delayed fluorescence of chloroplasts. Photochem Photobiol Sci 2008;7:352–60 PubMed PMID: 35238.
[57] Scordino A, Musumeci F, Gulino M, Lanzanó L, Tudisco S, Sui L, et al. Delayed luminescence of microalgae as an indicator of metal toxicity. J. Phys. D: Appl. Phys 2008;41:155507.
[58] Bodemer U, Gerhardt V, Yacobi YZ, Zohary T, Friedrich G, Pohlmann M. Phytoplankton abundance and composition in fresh water systems determined by DF excitation spetroscopy and conventional methods. Arch Hydrobiol 2000;55:87–100.
[59] Strauch SM, Richter PR, Haag FW, Krüger M, Krüger J, Azizullah A, et al. Delayed fluorescence, steady state fluorescence, photosystem II quantum yield as endpoints for toxicity evaluation of Cu^{2+} and Ag^+. Environ Exp Bot 2016;130:174–80.
[60] De Filippis L, Ziegler H. Effect of sublethal concentrations of zinc, cadmium and mercury on the photosynthetic carbon reduction cycle of *Euglena*. J Plant Physiol 1993;142 (2):167–72.
[61] De Filippis LF, Hampp R, Ziegler H. The effects of sublethal concentrations of zinc, cadmium and mercury on *Euglena*. II. Respiration, photosynthesis and photochemical activities. Arch Microb 1981;128:407–11.
[62] Ferroni L, Baldisserotto C, Fasulo MP, Pagnoni A, Pancaldi S. Adaptive modifications of the photosynthetic apparatus in *Euglena gracilis* Klebs exposed to manganese excess. Protoplasma 2004;224(3–4):167–77.
[63] Kalaji HM, Schansker G, Ladle RJ, Goltsev V, Bosa K, Allakhverdiev SI, et al. Frequently asked questions about in vivo chlorophyll fluorescence: practical issues. Photosynth Res 2014;122(2):121–58.
[64] Ferroni L, Klisch M, Pancaldi S, Häder D-P. Chlorophyll fluorescence analysis in *Euglena gracilis*: a survey on the use of a portable PAM fluorometer for photosynthesis studies in flagellate algae. Trends Photochem Photobiol 2006;11:23–32 PubMed PMID: 34740.
[65] Baldisserotto C, Ferroni L, Medici V, Pagnoni A, Pellizzari M, Fasulo MP, et al. Specific intra-tissue responses to manganese in the floating lamina of *Trapa natans* L. Plant Biol 2004;6:578–89.
[66] Wang C, Xing D, Chen Q. A novel method for measuring photosynthesis using delayed fluorescence of chloroplast. Biosens Bioelectron 2004;20(3):454–9.
[67] Katsumata M, Koike T, Nishikawa M, Kazumura K, Tsuchiya H. Rapid ecotoxicological bioassay using delayed fluorescence in the green alga *Pseudokirchneriella subcapitata*. Water Res 2006;40(18):3393–400.
[68] Hsu BD, Lee JL. A study on the fluorescence induction curve of the DCMU-poisoned chloroplast. Biochim Biophys Acta 1991;1056:285–92.
[69] Ahmed H, Häder D-P. A fast algal bioassay for assessment of copper toxicity in water using *Euglena gracilis*. J Appl Phycol 2010;22(6):785–92.

[70] Khellaf N, Zerdaoui M. Growth response of the duckweed *Lemna minor* to heavy metal pollution. J Environ Health Sci Eng 2009;6(3):161–6.
[71] Anonymous. Reference list of chemicals for the LUMIStox luminating bacteria test (Referenzliste Chemikalien für den LUMIStox Leuchtbakterientest), 1997. Info 04. Dr. Bruno Lange GmbH, Berlin (Germany).
[72] Tahedl H, Häder D-P. Automated biomonitoring using real time movement analysis of *Euglena gracilis*. Ecotoxicol Environ Safety 2001;48(2):161–9.
[73] Asghari S, Johari SA, Lee JH, Kim YS, Jeon YB, Choi HJ, et al. Toxicity of various silver nanoparticles compared to silver ions in *Daphnia magna*. J Nanobiotechnol 2012;10(14):1–14.
[74] Binaeian E, Safekordi AA, Attar H, Saber R, Chaichi MJ, Kolagar AH. Comparative toxicity study of two different synthesized silver nanoparticles on the bacteria *Vibrio fischeri*. Afr J Biotechnol 2012;11(29):7554.
[75] Naumann B, Eberius M, Appenroth K-J. Growth rate based dose-response relationships and EC-values of ten heavy metals using the duckweed growth inhibition test (ISO 20079) with *Lemna minor* L. clone St. J Plant Physiol 2007;164:1656–64.
[76] Hoekstra FA, Golovina EA, Buitink J. Mechanisms of plant desiccation tolerance. Trends Plant Sci 2001;6:431–8.
[77] Gray DW, Lewis LA, Cardon ZG. Photosynthetic recovery following desiccation of desert green algae (Chlorophyta) and their aquatic relatives. Plant Cell Environ 2007;30 (10):1240–55 PubMed PMID: 35204.
[78] Gaff DF, Oliver M. The evolution of desiccation tolerance in angiosperm plants: a rare yet common phenomenon. Funct Plant Biol 2013;40(4):315–28.
[79] Gerhardt V, Bodemer U. Delayed fluorescence excitation spectroscopy: a method for determining phytoplankton composition. Arch Hydrobiol 2000;55:101–20.
[80] Mörschel E, Wehrmeyer W. Multiple forms of phycoerythrin-545 from *Cryptomonas maculata*. Arch Microb 1977;113:83–9.
[81] Brooks C, Gantt E. Comparison of phycoerythrins (542,566 nm) from crytophycean algae. Arch Microb 1973;88:193–204.
[82] Garcia-Pichel F, Castenholz RW. Characterization and biological implications of scytonemin, a cyanobacterial sheath pigment. J Phycol 1991;27:395–409.

Photosynthesis assessed by chlorophyll fluorescence

9

Dieter Hanelt
University of Hamburg, Hamburg, Germany

9.1 Introduction

The technique of in vivo chlorophyll fluorescence measurements is ubiquitously used in ecophysiological studies of plants. Especially under field conditions, where environmental factors cannot be controlled, fluorescence measurements can be helpful in understanding how plants react under natural, nonlaboratory conditions. For this purpose, a lot of instruments have been developed also for usage in the field, and which seem to be easy to handle. Nevertheless, even with easy handling, the underlying theory, the process behind the measurements, and the adjustment of the instrument need to be understood, otherwise the data produced will cause misleading interpretations [1]. In addition, a huge amount of data are produced and their interpretation remains quite complex, and also often leads to controversial results reported in the literature. Moreover, fluorometric analysis is not a silver bullet to explain what happens in photosynthesis or to calculate biomass production from fluorescence parameters as photosynthesis and its downstream processes are too complex to be investigated only with a single method. In the past, activity of photosynthesis was mostly determined by oxygen evolution and less frequently by carbon assimilation (C^{14}) instead of fluorescence measurements, because there was no instrument for fluorescence measurements available for the field. Many studies have been performed by measuring both oxygen evolution and carbon assimilation, and thus have, e.g., determined the apparent photosynthetic quotient ($PQ = mol\ O_2$: mol C). As an example, Hatcher [2] measured PQs of the kelp *Laminaria longicruris* being between 0.7 and 1.5, and measurements made on five brown and two red algal species from Antarctica gave PQs between 1.1 and 5.3 [3,4]. High values of the PQ are typically obtained when nitrate is the dominant nitrogen source and low values when ammonium is mainly taken up due to the difference in their reduction states [5]. Enhanced production of lipids and proteins also result in higher PQs compared to conditions when polysaccharides are predominantly synthesized [6,7]. However, although oxygen production and carbon fixation can indicate important alternative pathways of anabolism, chlorophyll fluorescence measurements have been more frequently used in recent years. The main reason is the easy and practical use of the fluorescence devices, which allows fast and repeated measurements in the laboratory as well as in the field. Moreover, fluorescence kinetics gives closer insights into the performance of the photosynthetic machinery than can be observed by oxygen or CO_2 measurements.

Bioassays. DOI: http://dx.doi.org/10.1016/B978-0-12-811861-0.00009-7

However, fluorescence measurements do only report brutto photosynthesis (neglecting respiration) and mainly give information of events directly occurring at photosystem II, often disregarding consequences of downstream processes in the metabolism. As oxygen consumption of plants depends also on the state of photosynthetic activity, e.g., reduced respiration when energy to the metabolism is supplied by photosynthesis, or high photorespiration under high light conditions or CO_2 limitation, this information is unfortunately also not obtained by fluorescence studies. Therefore, it is rather hard to estimate growth rates only by extrapolations from fluorescence-based assessments of photosynthetic activity, although this is often published in ecological studies, e.g., under temperature stress quantum efficiency of photosynthesis measured by fluorescence or CO_2 assimilation deviated strongly because photosynthetic reducing equivalents are not only used for CO_2 fixation but also to remove stress-induced harmful oxygen radicals [8]. This chapter will demonstrate why fluorescence measurements are nevertheless a preferred method to investigate photosynthetic activity and will give a short introduction to the method and its parameters, as well as several instruments available on the market and for which experimental design they are recommended for usage. A compilation of the most commonly used photosynthetic and fluorescence parameters are shown in Table. 9.1.

9.2 Principle of fluorescence measurements

Light energy which is absorbed by the photosynthetic apparatus is mainly channeled into three pathways: photochemistry, fluorescence, and heat production as a consequence of the first law of thermodynamics. Whereas absorbed energy is lost by fluorescence light and heat conversion, the part converted into photochemistry is used to power anabolism of the plant.

$$\text{Energy absorbed} = \text{energy used for photochemistry} + \text{energy dissipated as heat} + \text{energy dissipated as fluorescence}$$

In principle the fluorescence signal of photosystem II (PSII) competes for the excitation energy with the photochemical energy conversion and heat dissipation. Schreiber and co-workers [9] introduced the pulse amplitude modulation fluorescence technique (PAM method) which offers the opportunity to easily determine the quantum yield of PS II by measuring the difference of fluorescence dissipating from closed (primary electron acceptor plastoquinone A (Q_A) is reduced) and open PS II reactions centers (oxidized Q_A). As long as the photosystem is in a dark-adapted state the primary electron acceptor of the reaction center can be filled with electrons (therefore open state) whereas in light charge separation occurs so that the reaction center chlorophyll (P680) becomes temporarily ionized and the primary acceptor (Q_A) is in reduced (closed) state [10]. Under normal conditions the closed reaction centers will transfer electrons into the electron

Table 9.1 Abbreviations and definitions of the most common fluorescence parameters

Actinic light	Light which drives electron transport so that parts of reaction centers are temporarily closed, needed to estimate ETR curves (means cause an action or reaction)
A_{leaf}	Incident light conversion efficiency of photosystem II (generally 0.84 in leaves of green plants)
α	The initial slope of the PI or ETR curve
ETR	Absolute electron transport rate represents photosynthetic activity measured in μmol electrons m^{-2} s^{-1}
ETR curve	Photosynthesis-irradiance curve, showing the correlation between the photon fluence rate (PFR) to the linear electron transport rate between both photosystems
ETR_{max}	Maximal photosynthetic rate, reached when a higher irradiance cannot cause higher electron flow from the water splitting complex to $NADP^+$. It is indicative of the photosynthetic capacity or the amount of active reaction centers contributing to the electron transfer
Far red	Light with wavelength >700 nm which mainly excites photosystem I and causes a fast oxidation of the electron transport chain between both photosystems
Fm	Maximal fluorescence produced by the oversaturating light pulse (SP), in totally recovered reaction centers (PSII) and after dark adaptation. Centers are temporarily closed, all primary electron acceptors (Q_A) are reduced and thus PS II centers will not contribute to electron transport. The absorbed energy is dissipated only as heat or fluorescence. It is indicative of the amount of reaction centers which are able to close in light
Fm′	Maximal fluorescence produced by the oversaturating light pulse (SP) during actinic irradiation. It indicates the amount of reaction centers which are still able to close in light
Fo	Minimal fluorescence of totally recovered photosynthesis of dark-adapted reaction centers (PS II). It corresponds to the energy transfer efficiency within the photosystem, when no charge separation is induced
Fo′	Minimal fluorescence during actinic illumination when already a fraction of fluorescence quenching occurs. It is measured after a preceding short far-red irradiation in the dark
Ft	Instantaneous fluorescence which is caused when a fraction of the reaction centers are closed or open, normally caused during actinic light irradiation
Fv′	Variable fluorescence of the dark-acclimated reaction center; however, still lower due to long-lasting photoinhibitory quenching of not fully recovered photosynthesis

(*Continued*)

Table 9.1 (Continued)

Fv = Fm−Fo	Variable fluorescence, which depends on the light acclimation status of the dark-adapted reaction center. It is lower if heat dissipation in the photosystem is increased, mainly due to photoinhibition or photodamage
Fv/Fm = (Fm−Fo)/Fm	Optimal quantum yield (sometimes also called maximal quantum yield), ratio which gives information on photosynthetic efficiency of dark-adapted reaction centers, i.e., which amount of absorbed photons can be converted into electron transport
ΔF/Fm′ = (Fm′-Ft)/Fm′	Effective quantum yield, during actinic irradiation. It indicates the performance of photosynthesis in light, or how efficient absorbed photons are converted into electron transport during illumination
I_c	Compensation point, when photosynthetic oxygen production levels out respiration, so that no increase or decrease of oxygen production is measured
I_k	Saturation level, artificial point calculated by the intersection of the slope (α) of the PI/ETR-curve and the photosynthetic capacity at the P_{max} level. Gives information on the light acclimation status of photosynthesis
ML	Measuring light: modulated light applied with constant but very low irradiance to estimate the amount of modulated fluorescence, in competition to photochemistry and heat production. The measuring light itself should not cause photochemical quenching
NPQ	Nonphotochemical quenching ranges between 0 and infinity. It does not depend on the estimation of Fo' and indicates better heat dissipation within the antenna system
OIJP	Rapid fluorescence rise kinetics after dark adaptation during actinic light irradiation. It indicates an increasing reduction of the electron acceptor pool of PS II so that a decreasing amount of energy is used for the first photosynthetic process (reduction of the primary electron acceptor)
PAM	Pulse amplitude modulation (fluorometer)
PAR	Photosynthetically active radiation in the waveband between 400 and 700 nm measured as μmol photons $m^{-2}\ s^{-1}$, also called light
PI-curve	Photosynthesis-irradiance curve, showing the correlation between the photon fluence rate (PFR) and a product of photosynthesis, which can be oxygen production or CO_2 fixation
P_{max}	Maximal photosynthetic rate, reached when a higher irradiance cannot cause a higher oxygen production or CO_2 consumption. It is indicative for the photosynthetic capacity or the amount of active reaction centers contributing to the electron transfer
PS I	Photosystem with the reaction center chlorophyll P700 which leads to a reduction of $NADP^+$. This photosystem does not contribute to chlorophyll fluorescence under in vivo conditions
PS II	Photosystem attached to the water splitting complex with the reaction center chlorophyll P680. The in vivo chlorophyll fluorescence originates mainly from this photosystem
Q_A	Final electron acceptor plastoquinone A of PS II

(*Continued*)

Table 9.1 (Continued)

qI	Fluorescence quenching representing photoinhibition and/or photodamage. A longer time is needed for recovery than observed by the other nonphotochemical quenching mechanisms
qN	Nonphotochemical quenching, fraction of energy which is dissipated as heat or fluorescence. It ranges between 0 and 1 and gives a good resolution at lower quenching activity as well is indicative for thylakoid membrane energization.
qP	Photochemical quenching, a measure of the fraction of still open reaction centers. 1-qP indicates the fraction of centers which are continuously closed and do not contribute to charge separation or the reduction state of the final electron accepter plastoquinone A. They will dissipate absorbed energy as heat
qT	Fluorescence quenching due to different energy distribution to both reaction centers. It increases if PS I is supplied with a higher fraction of the absorbed energy by spillover in state 2
R	Respiration: oxygen consumption of the metabolism
rETR	Relative electron transport rate without units used for the purpose of comparison. There is no need to know the amount of absorbed photons in vivo
SP	Saturation pulse: high intensity light pulse which totally reduces the primary acceptor of PS II and thus causes the "closer" of the reaction centers. Under this condition no photochemical quenching (*qP*) occurs during the period of the pulse
$Y_{(NO)}$	Quantum yield of nonregulated thermal energy dissipation due to persistent closure of reaction centers in light
$Y_{(NPQ)}$	Quantum yield of regulated energy dissipation by PS II via regulated heat emission
Y (I)	Effective photochemical quantum yield of PS I
Y (II)	Effective photochemical quantum yield of PS II

transport chain and become subsequently regenerated by receiving new electrons from the water-splitting complex, so opens again. Using a very low irradiance of modulated measuring light the chlorophyll molecules only dissipate the absorbed energy by low fluorescence (Fo) or some heat, because all reaction centers remain still "open". Then, no electron transport occurs, as irradiation "pressure" is not sufficient to activate the physicochemical charge separation, so that the center remains in a quasi dark-acclimated state. As a result, the fluorescence level Fo (caused only by the measuring light intensity) is very low. The fluorescence emission of Fo should be on a stable level for a while, as if its intensity changes then the photosynthetic apparatus would acclimate by fluorescence quenching mechanisms and thus would react to the measuring light irradiance, e.g., by first activating state transitions. This would be an indication that the photosynthetic apparatus changes from a dark-acclimated into a light-acclimated state and then Ft (temporary fluorescence during light acclimation) instead of Fo is emitted. In contrast, an additional very short oversaturating light pulse produced usually by a second high intensity lamp causes a total charge separation within the reaction centers so that all centers are "closing". The reaction center chlorophylls become oxidized and do not contribute to any electron transport for a moment, thus, also no photochemical quenching of the fluorescence yield can occur. This drives the fluorescence caused by the modulated measuring light into its maximal level, which is called Fm. Prerequisite of both states (Fo, Fm) is that the chloroplasts are always dark acclimated and no photoacclimation to any preceding light level has occurred. If no dark acclimation had taken place before this might lead to higher heat dissipation (nonphotochemical fluorescence quenching). As under these both extreme conditions no energy dissipation into photochemistry takes place, absorbed light energy can be only dissipated by heat or fluorescence. The amount of fluorescence emission can be measured by a suitable fluorometer, the amount of absorbed energy by measuring the incident irradiance in dependence on the spectral absorption of the chloroplast pigments. The still unknown variable is heat dissipation which can now be easily estimated (heat dissipated energy = absorbed light energy minus dissipated energy by fluorescence). By standardization of the difference Fm − Fo to the maximal fluorescence level Fm caused by the oversaturating light pulse the ratio (Fm − Fo)/Fm = 1 − (Fo/Fm) = Fv/Fm becomes independent of the amount of absorbed energy, and thus the pigment concentration, and defines the *optimal quantum yield* of PS II in the dark-acclimated state [11]. (This is a mathematical trick as enumerator as well as denominator depend both on the same pigment concentration and its respective light absorption and, thus, the dependence on the pigment concentration is erased in the formula). The ratio is also called *yield (II)* and indicates the amount of the absorbed photons which contribute to the electron transport within photosynthesis and neither to heat dissipation nor fluorescence emission. It is correlated to the slope (α) in the beginning of a photosynthesis-irradiance curve (for explanation see below). Furthermore, Genty et al. [12] demonstrated a linear relationship between the quantum yield of CO_2 fixation and the fluorescence ratio (Fm′-Ft)/Fm′ = ΔF/Fm′ also called *effective quantum yield* of light-acclimated chloroplasts. [Fm′ is the maximal fluorescence in the light-acclimated state]. The difference to the calculation of the optimal quantum yield is that the fluorescence level of partly closed or open reaction centers in light (Ft) is subtracted from the fluorescence level of all closed reaction centers under light acclimation (Fm′). The Ft level is generally an intermediate level between Fo and Fm as the number of closed reaction centers increases the fluorescence emission over the Fo level. If all reaction centers are closed under oversaturating light conditions Ft = Fm′ then no further light energy can be used for photochemistry. Generally, the ratio ΔF/Fm′ at low irradiances is lower than Fv/Fm due to an increased concurrence of heat dissipation of a certain amount of closed reaction centers which lowers the fluorescence yield and the photosynthetic efficiency of the system. If the effective quantum yield in light appears higher than the comparable optimal quantum yield measured after dark acclimation, some preceding light acclimation has still caused fluorescence quenching in the dark-adapted state, which contributes to nonphotochemical energy dissipation, or easily an error in the measurement was done (e.g., often produced by a reduced distance to the fluorescence probe during the measurement in light compared to the dark measurement).

Knowing the irradiance of absorbed light and that at least two photons are necessary to drive both photosystems, the electron transport rate (ETR) of the photosynthetic apparatus can be calculated, which follows in most cases in green plants and green algae (e.g., Ref. [13]):

$$\begin{aligned} \text{ETR} = \text{PAR} & \left(\text{amount of incident light in } \mu\text{mol photons m}^{-2}\ \text{s}^{-1}\right) \\ & \times 0.84\ (A_{\text{leaf}},\ \text{incident light conversion efficiency in green plants}) \\ & \times 0.5\ (\text{two photons necessary to drive both photosystems}) \times \Delta\text{F}/\text{Fm}' \end{aligned}$$

ETR is the apparent amount of electrons which are transferred by photosynthesis to the electron end acceptor $NADP^+$ as a consequence of the incident light irradiance. However, often the software of commercial fluorometers automatically calculates values of ETR by assuming that green leaves have values of A_{leaf} of 0.84 and *fraction of PS II* of 0.5, often leading to substantial errors in calculations of the absolute ETR [14]. It has to be considered that this depends on several parameters which, e.g., differ with the inherent construction of the photosynthetic system (different in green, brown, and red algae) or under different ecophysiological conditions (e.g., adaptation to altered light conditions). Thus, for comparison in species acclimated to different stress conditions, it is often recommendable to calculate only the relative ETR (rETR), which does not depend on the knowledge on the fraction of absorbed light or energy distribution between both photosystems, assuming that this remains constant in a species during the course of an experiment. To study changes of the light conversion into a rETR in a certain species, e.g., after light stress, give normally already sufficient information without knowing the exact amount of spectrally absorbed light and its distribution between both photosystems. The calculated ETR versus photosynthetically active radiation (PAR) curve is often in good correlation to the corresponding photosynthesis–irradiance curves (PI-curve, Fig. 9.1) measured by oxygen production [13], at least the calculated saturation irradiance I_k showed a good fit to oxygen measurements performed as shown by unicellular algae [15]. The I_k value is a theoretical point when light irradiance is sufficient to cause saturation of photosynthesis and can be estimated by the intersection of the initial slope of the PI or ETR-curve (α, indicator of the photosynthetic activity) with the photosynthetic capacity (P_{max} or ETR_{max}, indicator of the amount of active reaction centers, respectively the maximal amount of transported electrons within photosynthesis).

Fig. 9.1 shows an example for a comparison of a PI-curve measured by oxygen production in the sublittoral low light-adapted red alga *Phycodris rubens* (Fig. 9.1A) and an ETR-curve of the more high light-adapted *Gigartina skottsbergii* measured by fluorescence (Fig. 9.1B). It is visible that the rETR is not transferable to the oxygen production as at least 4 electrons are needed to produce one oxygen molecule, which means at least 8 absorbed photons. In reality Björkman and Demmig [16] calculated a higher average value of 9.37 photons (a photon efficiency of 0.107) of 37 different plant species as there is already a thermal energy loss in the system. For the example of *G. skottsbergii*, 31 photons would be

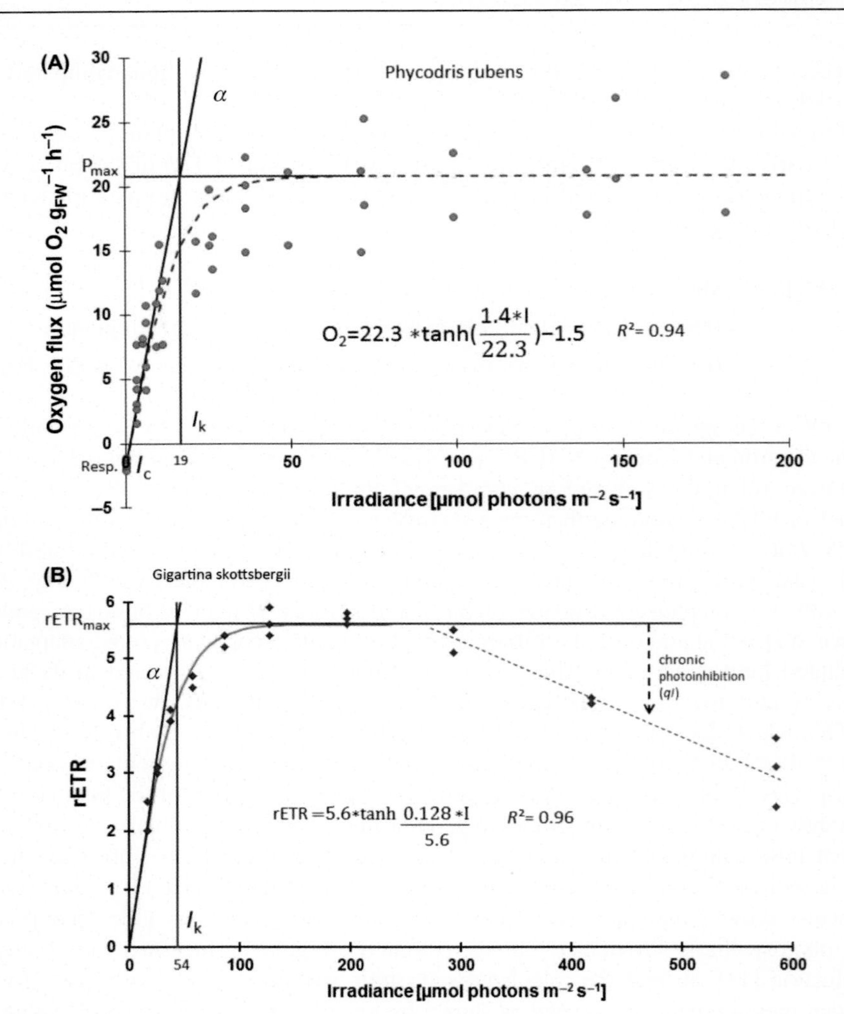

Figure 9 1 (A) Oxygen measurement: example of a PI-curve of the red alga *Phycodrys rubens* from 10 m water depth, fitted broken curve is calculated by a modified formula of Jassby and Platt. P_{max}, photosynthetic capacity; α, slope of the curve or indicator of the photosynthetic activity; I, irradiance; I_c, compensation point when photosynthetic rate levels out respiration rate; I_k, saturation irradiance. (B) Relative electron transport measured by fluorescence: example of a rETR-curve of the red alga *Gigartina skottsbergii* from shallow water, fitted green curve is calculated by a formula of Jassby and Platt, broken line extrapolated $rETR_{max}$, maximal relative electron transport rate; α, slope of the curve or indicator of the photosynthetic activity; I, irradiance; I_k, saturation irradiance.

necessary for the transport of four electrons which is much overestimated; here α is indicative for photosynthetic activity, but not efficiency. The reason for the divergence is that the rETR-curve is not standardized to a certain biomass and that the incident light conversion factor and energy distribution between both photosystems is unknown. However, the higher I_k-value, which is of course comparable between these two different methods, demonstrates that *G. skottsbergii* is much better adapted to higher light intensities than *P. rubens* as it grows in shallow water depths. In contrast, *P. rubens* performs photosynthesis much better under low light conditions in 10 m depth than *G. skottsbergii.* Exceeding 300 μmol photons m^{-2} s^{-1} $rETR_{max}$ of *G. skottsbergii* starts to decline due to photoinhibition, which will be explained below.

Furthermore, the fluorescence method gives insight into what happens at the PS II reaction center and how the absorbed energy is used. With fluorescence quenching analysis it is possible to determine the effect of photochemistry by fluorescence quenching (*qP*), of energy-dependent quenching associated with light-induced proton transport into the thylakoid lumen (*qE*), of the nonphotochemical quenching (*qN*) (i.e., fluorescence and heat dissipation), photoinhibitory quenching (*qI*), as well as quenching associated with state transitions (*qT*) (energy distribution between PSII and PSI). In recent times also the quantum yield of regulated energy dissipation by PS II via heat emission ($Y_{(NPQ)}$) or of nonregulated energy dissipation ($Y_{(NO)}$) due to persistent closure of reaction centers in light is determined. Under moderate light intensities the Kautsky effect (change of fluorescence level over time) in predark acclimated chloroplasts can be observed, which gives information about the activation of the electron flow, kinetics of closure of reaction centers, ΔpH formation, attachment of the antenna to the PSII reaction center, or even alternative oxygen-dependent electron flow where e.g., also the light protection by ascorbate peroxidase is involved [17]. By a combination of measurement of chlorophyll fluorescence at PS680 (II) and P700 (I) absorbance changes as in the Dual-PAM method (this will be explained below) a simultaneous assessment of energy conversion in PS I and PS II can be conducted. Both reaction centers need to transport electrons at a similar rate to maintain a stable flow of electrons without accumulations at the acceptor sides of the respective photosystem. Therefore, it is of interest to observe deviations in the activities of both reaction centers e.g., under stress conditions.

9.3 Method of fluorescence measurements

Prerequisite to measure the fluorescence of the photosynthetic pigments is the excitation at an appropriate wavelength. As red light (around 650 nm) already corresponds to part of the red absorption band of the chlorophylls, many fluorometers use light emitting diodes (LEDs) with their inherent red emission peak as light sources. Moreover, as LEDs can produce light pulses in high frequencies, this is often used for modulation of the measuring light (i.e., switching on/off of the light at a certain frequency). With development of high emission blue LEDs also the Soret band of chlorophyll can be excited (e.g., by 460 nm LED emission). This has the advantage that the blue excitation wavelength of the measuring light can be more easily separated by optical filters from the red chlorophyll fluorescence signal (at about 680 nm at room temperature) which is necessary to screen the fluorescence sensitive photodetector of the instrument against stray light, as the photodetector should detect fluorescence light emitted by the plant and not the measuring light intensity. Many simple instruments use only red measuring light with a nonmodulated irradiance, so that it is more difficult to separate the emitted fluorescence produced by the measuring light at the reaction center from simultaneous fluorescence signals induced by an external actinic light source. An additional radiation of an external actinic light is necessary for a stimulation of the photosynthetic electron transport and thus to adjust the photosynthetic performance to measure e.g., $\Delta F/Fm'$. As the intensity of the resulting fluorescence signal depends on the irradiance of the exciting irradiation as well as the pigment concentration, it is advisable that the irradiance of the measuring light and the pigment concentration is kept constant throughout the measurements. Otherwise, the observed alterations of the fluorescence signal induced by the measuring light would be caused by change in its irradiance, and not by changes in the performance of the photosynthetic apparatus, which is the purpose of the measurements. Some instruments use the same light source for detection of fluorescence changes as well as to produce the actinic light for induction of changes in the performance of the photosynthetic apparatus. In this case, a complex study of the kinetics of the fluorescence formation is necessary to calculate the fluorescence parameters of interest for estimation of the photosynthetic performance. In contrast, the PAM principle uses modulated measuring light at high frequencies which stimulates modulated chlorophyll fluorescence in the chloroplast. After passing an optical long pass filter ($\lambda > 700$ nm), the fluorescence light is usually detected by a PIN-photodiode. Only the pulsed signal of the photodiode is electronically amplified by a selective lock-in or pulse-selective amplifier. As only the measuring light is pulsed, any other fluorescence signal or even stray light which would also pass the blocking optical filter is electronically ignored as it is not synchronously produced with the measuring light pulses. This enables the instrument to distinguish between the fluorescence caused by the weak pulsed measuring light signal and additional high background fluorescence signals from e.g., sunlight, which would be used as external actinic light source in the field. Thus, external light sources do not disturb the measurements, so that a mixture of different light sources for ecophysiological measurements can be simultaneously applied. For example, this can be used for demonstration of the Emerson enhancement effect [18] by testing the photosynthetic efficiency of different spectra of several external light sources. It is important that any change of the pulsed fluorescence signal should only originate by a changing energy distribution or conversion within the photosynthetic apparatus and not by a change in the irradiance of any external actinic light trigger. During an intentional application of different external light irradiances needed to activate the photosynthetic machinery (e.g., to measure ETR-curves), the intensity of the measuring light is kept constant during the whole measurement. This is what the pulse amplitude modulated fluorescence principle (PAM) means. The PAM instruments can tolerate a ratio between measured fluorescence and background signal of $1:10^5$ or even higher [9], which is more than sufficient to work in full sunlight without any disturbance of the measurement.

Irradiation with increasing intensity of an actinic light will cause more and more photosynthetic centers to temporarily close, so that in consequence the fluorescence level also increases. As a result, the pulsed (PAM) fluorescence, despite the constant measuring light irradiation, also increases. When the maximal fluorescence (Fm) is reached, all centers are closed and additional energy flow by higher photon fluence rates of absorbed actinic light cannot be used for photosynthesis anymore. The surplus of energy will be dissipated by heat radiation. In consequence, the yield of photosynthesis continuously decreases with increasing fluence rate of actinic light, i.e., it is necessary to irradiate more photons to reach a similar photosynthetic effect as before using a lower photon fluence rate (e.g., demonstrated by constant oxygen production rate). As a result of increasing irradiances, an increasing portion of photon energy is converted into heat and less energy is used to drive the electron transport. The result is that the photosynthetic efficiency decreases and the increase of heat dissipation is indicated by an increased q_N. This is illustrated by the calculation of a PI-curve, where its slope (α), which is in the beginning representative for the maximal photon conversion efficiency, decreases at higher irradiances (Fig. 9.1). At saturating irradiances the photosynthetic activity reaches its maximal or saturation level (ETR_{max} or P_{max}). This represents the photosynthetic capacity, the number of electrons that can be maximally transported by the electron transport chain. After the ETR_{max} level is reached, additionally absorbed energy will be mainly dissipated as heat. In contrast, a PI-curve measured by oxygen production indicates the respiration rate in darkness (i.e., oxygen consumption by the cell, respiration in Fig. 9.1) as well as the compensation point (I_c). The intersection with the abscissa when oxygen production balances the oxygen consumption shows the minimum irradiance when primary production starts. This information is obviously not visible by fluorescence measurements. Only the photon conversion efficiency (α), the ETR_{max} level and saturation irradiance I_k can be calculated by fluorescence measurements which shows the ratio between the photon conversion efficiency and the photosynthetic capacity of photosynthesis. Accordingly, this is an indication of the light acclimation of the photosynthetic apparatus. For example, a higher I_k value can be caused by a higher capacity of the ETR, i.e., a higher number of active reaction centers, and/or a lower photon conversion efficiency (lower α) as is rather typical in strong light-adapted chloroplasts [19].

Which means the effect of the part of absorbed light energy which drives the electron transport, is in the beginning of a PI-curve rather constant, so that the incline of the resulting PI-curves depends only on the increase of the irradiance. When heat dissipation increases and reaction centers become temporarily closed at higher irradiances the effective quantum yield also starts to decrease, so that the slope of the curve starts to decline. Reaching the saturation level, the decrease of the yield is balanced by the increasing number of absorbed photons, i.e., the photosynthetic activity keeps constant with further increase of light intensity and the slope of the curve will be zero. Under very high irradiances heat dissipation surpasses the energy conversion into electron transport with the result that also the photosynthetic capacity starts to decrease, i.e. the slope becomes negative, which is called photoinhibition (Fig. 9.1B). This is a stress parameter, as the energy absorbed clearly overwhelms the energy need of the metabolism of the cell.

9.4 Fluorescence quenching parameters

The distribution of absorbed energy depends on the physiological state of the photosynthetic organism. Under normal conditions most of the absorbed energy will be channeled into the electron transport for the generation of chemically conserved energy by adenosintriphosphate (ATP) and the reducing equivalent NADPH as electron acceptor. The photochemical quenching parameter (*qP*) is related to the fraction of this energy channeling because it indicates the proportion of excitation energy trapped by open reaction centers [20]. It is a relative value ranging between 0 and 1, with the maximal value meaning a maximal usage of excitation energy for photosynthetic electron transport. It can be calculated as follows:

$$qP = (Fm' - Ft)/(Fm'\text{-}Fo')$$

where the prime stands for the parameters measured in the light-acclimated state [11,21]. A decrease of *qP* means that reaction centers of PS II are temporarily closed and therefore cannot contribute to the charge separation or electron transport, respectively. Alternatively, the equation 1-*qP* gives the proportion of closed reaction centers and are thus a measure for the excitation pressure on PS II [22] which will result in an increase in heat dissipation. A high value of 1-*qP* means PS II centers are closed and charge separation to the electron acceptor plastoquinone is hindered. In early fluorescence studies Kautsky et al. [23] observed that leafs transferred from dark to moderate light conditions show a typical change in the fast and slow fluorescence emission within a time period of several minutes. An increase of the fluorescence level primarily occurs (OIJPT-fluorescence course, for a more detailed explanation see Ref. [24]) which emerges from an increasing reduction of the electron acceptor pool of PS II so that a decreasing amount of energy is used for the first photosynthetic process (reduction of the primary electron acceptors, O-P level change). After some time re-oxidation occurs by a proper electron flow to the final acceptors, which causes a decrease in the fluorescence emission until the florescence level closely reaches again Fo (T-level), which means that the absorbed energy is used for photochemistry with the highest yield. The fast increase of the fluorescence indicates nonphotochemical energy dissipation, whereas the decline of the second phase shows that more and more energy is channeled into photosynthetic processes and is visible by an increase of the photochemical fluorescence quenching mechanism (*qP*). The Kautsky transients of the light-induced/dark-reversible changes in the chlorophyll fluorescence induction above described reflect changes in the quantum yield as well as changes in the number of Chl *a* in photosystem II (PSII; state transitions). Both are related to excitation trapping in PS II and the resultant photosynthetic electron transport and to secondary effects, such as ion translocation across thylakoid membranes and filling or depletion of post-PS II and post-PS I pools of metabolites [25].

The nonphotosynthetic fluorescence quenching mechanisms compete with photosynthetic ones, and this can be easily monitored or calculated by additional fluorescence parameters. The general term is Nonphotochemical quenching (*NPQ*) or *qN* accordingly to how it is calculated:

$$NPQ = \mathrm{Fm} - \mathrm{Fm}'/\mathrm{Fm}'$$

is related to heat dissipation and ranges from 0 to infinity [26]. Generally, values do not exceed values over 10 under saturating light, but it depends on the former light history of the organism as well as on species.

$$qN = (\mathrm{Fm\text{-}Fm}')/(\mathrm{Fm\text{-}Fo}')$$

and it ranges between 0 and 1 (often inversely correlated to *qP*) [27]. qN reflects better nonphotochemical quenching at a lower extent as well as thylakoid membrane energization, which will be further explained below. *NPQ* indicates a better heat dissipation in the antenna system, and, in addition, there is no need to measure Fo′ for this parameter, which requires a certain short interruption of the light phase for dark acclimation, which may hinder the experimental course. Both nonphotochemical parameters must be related to the dark-acclimated state determined before the experimental light phase assuming that at the start nonphotochemical quenching was minimal. The increase of the nonradiative energy dissipation has variously been suggested to be due to an increase of the zeaxanthin content in the photosystem II antenna [28], to an aggregation of the light-harvesting complexes (LHCs) [29], and/or to an increase of the amount of inactive photosystem II centers (e.g., PS II_{β} centers), which may also be able to protect active photosynthetic centers [30–32]. The nonphotochemical quenching can be divided into a few subcategories. A main contributor is the energy dependent quenching (*qE*), which requires a low pH in the lumen of the thylakoids, which means membrane energization, and the light-induced formation of the carotenoid zeaxanthin [33,34]. The main indicator for the occurrence of *qE* is its capability for a fast relaxation, when the photosynthetic apparatus is exposed back to low light conditions. This kind of fluorescence quenching declines already within seconds to minutes as it controls the naturally fast regulation of the energy conversion within the photosynthetic apparatus, and thus, the fast acclimation to the energy need of the metabolism. Here the xanthophyll cycle and hence the enzymatic conversion of carotenoids (de-epoxidation), e.g., violaxanthin in zeaxanthin and vice versa is involved. Already under low light conditions a second process causes fluorescence quenching by the change of energy distribution to both photosystems, which is called *qT*, and this depends on state transitions [35]. In darkness the LHC is normally attached to PS II, especially in green plants (State1). By reversible phosphorylation of the LHC it detaches from PS II and couples to PS I (State 2). This enables a better energy distribution between both photosystems, as the activity of both "light driven pumps" should be balanced for a smoothly running linear electron transport. If e.g., PS I receives less light energy, its activity would be lower than that of PS II which would cause an accumulation of electrons in the electron transport chain between both photosystems. This would cause a down-regulation of PS II activity as the membrane energization increases, which would increase qE and, thus, dissipation of the absorbed energy by heat. State transitions happen already under low light conditions but qT contributes only in a small amount to nonphotochemical quenching. Changes in qT can be observed by the slow change of the fluorescence level of PS II during dark-to-low light transitions of the chloroplasts. The qE as well as qT function both as regulative and photoprotective mechanisms for photosynthesis. Another significant role plays qI as photoinhibitory mechanism but with much slower relaxation kinetics. Photoinhibition causes a long lasting down regulation of the photosynthetic activity and can be observed by comparing the optimal quantum yield of dark acclimated fully recovered photosynthesis (Fv/Fm), e.g., after the night period, with the quantum yield after a short period of dark acclimated photosynthesis within a light period (Fv′/Fm′, the superscript prime means the values are measured after short dark acclimation but not with fully recovered photosynthesis).

$$qI\% = 100 - (100 * (\mathrm{Fv}'/\mathrm{Fm}')/(\mathrm{Fv}/\mathrm{Fm}))$$

qE and qT relax fast when a light period is temporarily interrupted by darkness or sometimes even at low light conditions. After photoinhibitory quenching the short dark-acclimated quantum yield (qT and qE should be relaxed) is still lower as compared to the morning value during an experimental course of a day.

$$(\mathrm{Fv}'/\mathrm{Fm}' < \mathrm{Fv}/\mathrm{Fm})$$

The recovery process of the photoinhibited quantum yield may need several hours or even the full night period to return to a similar value as in the morning at the start of the irradiation experiment. This slow recovery process often occurs when the photosynthetic apparatus was under strong high light stress, so that a long lasting down regulation or even a mechanistic photodamage of the photosynthetic apparatus due to production of oxygen radicals was caused. Then damaged proteins have to be exchanged in a slow process e.g., by the repair cycle of PS II. This can be also observed by a strong decrease of the slope (α) of a PI-curve and the decrease of the photosynthetic capacity (ETR_{max}) which is indicative of damage to the photosynthetic reaction centers (e.g., Fig. 9.1B at highest intensities). However, even after damaging light conditions still a faster regulative part ("dynamic photohinhibition") can be distinguished from the slow repair of the damaged centers due to "chronic photoinhibition" by determining the 2-phase kinetics of the recovery phase of qI after high light stress [36]. The xanthophyll cycle is also involved in the occurrence of the dynamic photoinhibitory process, whereas the repair cycle of PS II is active to recover the damaged centers after chronic photoinhibition [37].

This shows that an observation of a decrease in the quantum yield of photosynthesis alone does not give sufficient information on the mechanisms involved. It can be regulative due to temporary changes in the irradiance, e.g., by changing cloud cover in the field, but can also be caused by stress to the photosynthetic apparatus, which finally results in molecular damage. More important is to measure the following relaxation or recovery processes after down regulation of the quantum yield as the kinetics of recovery will indicate if photosynthesis was only temporarily down regulated (by e.g., *qE*), strongly affected, or even damaged (by chronic *qI*).

In the field, recovery can be artificially induced by covering the samples with a light-absorbing sheet or filter. However, in red algae it was found that low light is still necessary for recovery of *qI*, because darkness will conserve the photoinhibitory state of photosynthesis [38] and no recovery occurs. The reason might be that the energy needed for the repair process is supplied by photosynthesis itself.

The interaction between photochemical and nonphotochemical quenching controls the photosynthetic activity of the plant in the course of a day. This can be particularly well demonstrated in algae from the intertidal, which underlie strong changes in the irradiance, as the daily course of daylight is additionally modulated by tidal effects [39].

Thus, photosynthetic activity in the intertidal and upper subtidal zone is often depressed on sunny days in a typically diurnal pattern (Fig. 9.2). Strongest depression, i.e., highest degree of dynamic photoinhibition, occurs mostly in the early afternoon [40–43]. Full recovery of photosynthesis is reached at the latest in the evening in species capable of dynamic photoinhibition. Thus, photosynthetic oxygen production measured at nonsaturating fluence rates and the optimal quantum yield (Fv/Fm) show an approximately inverse course in comparison to the fluence rate of daylight measured continuously during the day (Fig. 9.2) [43,44]. However, if the tidal level of a turbid water column decreases during low tide, resulting in an increase of irradiance under water although sunlight decreases in the afternoon, the

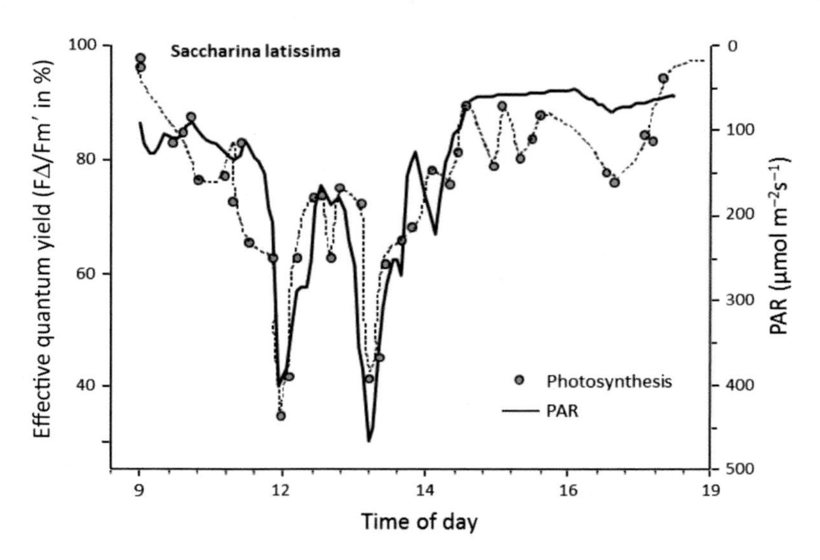

Figure 9.2 Photosynthetic performance of *Saccharina latissima* over the course of a day. Broken line shows photosynthetic activity measured by ΔF/Fm′ in percent of the control in low light (ΔF/Fm′, 0.7) and solid line the photon fluence rate of sunlight. Please note that the PFR in air is shown with an inverse ordinate, to demonstrate better how the quantum yield follows the course of natural radiation.

irradiance underwater overrules the above-described photosynthetic pattern and maximal inhibition of photosynthesis can occur also in late afternoon [39]. It is also possible that high tide during midday hours can protect against photoinhibition in the intertidal when the water column is turbid, resulting in low light transmittance. To measure such effects, operation of submersible fluorometers is advantageous, as can be carried out with a Diving-PAM. A relation between the formation of zeaxanthin and a decrease in oxygen production and optimal quantum yield in macroalgae was demonstrated early [45–47]. Most studies of photoinhibition have been performed so far with respect to PAR and to UV radiation. For the determination of photosynthetic performance under natural light conditions, field experiments are indispensable, as the results of field and laboratory experiments on photosynthetic performance carried out under comparable conditions can be inconsistent [38,43,48]. In laboratory experiments using cultured red algae, a period of 48 h was required to attain full recovery of photosynthesis after the same level of photoinhibition, whereas in the field, the same species had already fully recovered by the evening, and hence, much faster. Without a submersible fluorometer, it was only possible to measure the optimal quantum yield (Fv/Fm) in algae after collection under water and temporarily dark-exposed in a special cuvette on the shore. Thus, the short-term dynamic depression of the quantum yield could not be assessed. Nevertheless, to measure quantum yield under natural light conditions in the field, battery powered instruments are necessary, which are optimized for the field application.

9.5 Instrumentation for fluorescence measurements

The measurement of fluorescence is used to elucidate the process of physical light conversion and chemical downstream processes within photosynthesis by biophysicists, but also to investigate photosynthetic performance under natural conditions in ecophysiological studies. As the biophysical background is quite complex and would complicate the facts important for environmental studies, the instrumentation for usage in biology will be further discussed (for deeper information on the biophysical background the reviews by [14,24,25] and the literature included are recommended). Fluorometer for in vivo measurements are not comparable to standard laboratory spectrofluorometers, as the excitations wavelength cannot be adjusted by a monochromator and they also do not use high pressure lamps for production of the excitation radiation. As there is no need to perform a scan of the emission wavelengths since it is well known which wavelength ranges excite chlorophyll fluorescence best, a monochromator at the emission side is also missing. Photosynthetic reactions centers show fixed, well known emission peaks and do not differ between different photosynthetic eukaryotes. The optical parts are therefore much easier constructed, using generally only optical filters to select for the fluorescence emission peaks and an arrangement of LED with discrete emission peaks for their excitation wavelengths. However, detection velocity and time resolution of the in vivo fluorometers are much faster than those of normal spectrofluorometers as fast changes in milliseconds shall be resolved for studies of fluorescence kinetics. A computer and the respective software replace the oscilloscope function and the control of the first developed instruments [9]. Measurements are done in a normal photometer cuvette with a holder for excitation lights and the detector (e.g., Walz PAM 101, Photon System Instruments PSI FL 3500) which is suitable for microorganisms or isolated chloroplasts, or the excitation as well as the fluorescence light are guided through a flexible fiber optics to the sample and back to the detector (e.g., Walz PAM, Hansatech Instrument FMS 1). Often a leaf clip or magnetic holder fastens the fiber optics to a laminar sample, e.g., a leaf or phylloid, and it is also used for dark acclimation of the sample. An alternative is that exciting LEDs and the photodetector are placed in a small sensor head which can be directly attached to a laminar sample (e.g., Walz, Hansatech, LiCor) and which is connected by a cable to the control and detection unit. The fiber optics is light and easy to handle but has the disadvantage that light energy is lost by coupling the fiber to the light sources and to detect fluorescence from the sample surface. It is also not recommendable to fix a fiber, consisting of a mixture of fibers, directly to the surface of the sample, as the light point by an irradiating fiber does not intersect to the light field monitored by a measuring fiber. A small distance of some millimeters is necessary between the multifiber optics and the sample to ensure uniformity in the illuminated area. This can easily be done by placing a small 2 mm thick acrylic glass disc between sample and fibers, which in addition protects the end of the fiber optics against damage and dirt of a sample. A constant distance between sample and fiber optics is necessary during the measurement as this also changes excitation. Additionally, the measured fluorescence intensity, i.e., signal amplitude, will change strongly and wrong data will be produced. For irradiation with sunlight, e.g., for measurement of the effective quantum yield, it is not possible to put the sensor head or fiber optics directly onto the leaf, but a certain distance and angle relative to the sample plane (e.g., Walz distances clip 60 degrees) is necessary to avoid shading of the measuring area. Also here it is important that the distance does not change during the measurement, so the leaf needs to be fixed by a leaf holder or at least the experimenter should have a very calm hand during the short measurement. In recent times larger surface areas can be monitored by using digital CCD cameras combined with filteroptics which allow only transmittance of the fluorescence light (e.g., Walz Imaging PAM, Technologia CF Imager, PSI Fluorcam FC 800). From large areas, illuminated by LED panels for exciting fluorescence, it generates images of the fluorescence signal which are converted by special software and presented often in false color scale.

9.6 Saturation pulse method or how to measure quantum yield properly

A standard method to determine stress of the photosynthetic apparatus is the so-called saturation pulse method. Here, the fluorescence level of totally closed reaction centers are compared with the level when all centers are open in darkness (optimal quantum yield) or partly closed under actinic radiation (effective quantum yield) as described above. This procedure is easy to handle and is therefore used in many ecophysiological studies in the laboratory as well as in the field. A decrease in the respective quantum yield will indicate an increase in thermal energy dissipation, so that photosynthesis works less efficiently. Such a decrease can for example occur after high light stress, when energy supply by photosynthesis exceeds the needs of the metabolism, during high or low temperature stress, after drought or after poisoning the biochemical process with, e.g., heavy metals.

For comparison and to calculate the quenching coefficient, it is necessary to determine the fully recovered optimal quantum yield of PS II, e.g., after a certain recovery period in dim light or darkness when qT, qE, and qI are zero. This may need several hours of recovery if photosynthesis was strongly inhibited before. The first step is to determine the right level of Fo, which depends mainly on the chlorophyll concentration, the light irradiance, and the temperature. The irradiance of measuring/exciting light should not be too low to get a proper fluorescence signal much higher as the electronic noise or background signal (a good signal to noise ratio). The user may think that it may be advisable to use a high exciting irradiance to get a strong fluorescence signal, but this has the disadvantage that the measuring light causes itself premature photosynthetic acclimation to the higher light intensity. Then qT would already be induced or electrons transported from the primary acceptors, which would mean that reaction centers start to close temporarily. The result is that the level of Fo decreases during temporary measuring light irradiation mainly in green plants and brown algae, or may in contrast increase as observed in red algae [38]. It has been observed that in red algae fast changes in the balance of energy distribution to both reaction centers occur. Often also the sensitivity to the measuring light is different even within the same species coming either from laboratory culture or from samples grown in the field. The second parameter which needs to be determined is the proper irradiation of the saturation pulse and the pulse duration. If the irradiation is too low, the Fm level will not be reached so that maximal fluorescence is underestimated. This can easily be observed by increasing the irradiance of the saturation light pulse causing higher maximal fluorescence. However, if it becomes too high, fluorescence quenching mechanisms will be induced and the maximal fluorescence would be lower as the maximal possible Fm-level. Then the number of closed centers appears lower than it would really be when all centers are closing without fluorescence quenching by qN. This can also be caused when the length of the saturation pulse is too short (underestimation of Fm) or too long, because quenching mechanism can be also induced during a longer pulse length. If consecutive saturation pulses are given during an experimental time course and irradiance of the saturation pulses are high, the time interval between the successive pulses needs to be longer, otherwise the consecutive pulses cause the same effect of a too high, single pulse, and thus induce a quenching of the maximal fluorescence level. Often the reaction centers from field material close at lower irradiances of the saturation pulses than material cultured in the laboratory because of a different light adaptation status. Therefore, it is recommendable to start with lower saturation pulse irradiances and to test whether higher irradiances still cause an increase in Fm. Depending on the light source, the period the saturation lamp needs to reach its maximal light emission should be also considered. Thus, for halogen lamps a saturation pulse length below 600 ms should be not used, as the filament of the lamp is still in its heating-up phase and the light emission is not stable and the spectrum is variable. By experience a length of 800 ms is sufficiently long for the pulses and causes no technical problems (<1 s recommended by Murchie and Lawson [49]). In contrast, using flash lights or LED the multiturnover saturation pulses can be applied for very short durations, without these technical problems. In green plants or green alga an Fv/Fm value of about 0.83 indicates highest possible photosynthetic efficiency, whereas in brown algae a value of 0.7–0.8, in red algae of 0.6–0.7 and diatoms 0.5–0.6 normally present maximal photosynthetic efficiency [27,49]. The different values depend on the different structure of the respective photosynthetic apparatus and are not due to a lower photosynthetic efficiency, as is possible to be misled when values of red algae are compared to green algae. Red algae can most efficiently funnel the absorbed energy to the reaction center chlorophyll because of its special construction of the antenna system and thus are the organisms which perform still net photosynthesis under the lowest irradiances in the sea, so are most efficient in the yield of light. In comparison within one algal class, any lower value of Fv/Fm will be a hint that the organism is under stress and has not fully recovered, because there will be still photoinhibitory quenching visible, which reduces the photosynthetic efficiency.

After determination of the optimal quantum yield as a control value, the organism can be exposed to actinic or sunlight. As a result, the fluorescence-level determined by Fo will primarily increase to a Ft-level, because centers are partly closed, and will therefore contribute more to the fluorescence emission. The level depends on the ratio of how many centers are temporarily closed and thus cannot contribute to charge separation and how many are open, so they will still be able to convert light energy into chemically bound energy. With a second saturation pulse all centers will again temporarily close and the difference of the resulting Fm′ level to Ft is the variable fluorescence level (ΔF) which shows the capacity of light energy still be convertible by photosynthesis for electron transport. With this parameter the effective quantum yield of the system in light can be calculated. Simultaneously to an increase of Ft, Fm will start decreasing to a lower Fm′-level because increasing heat dissipation will lower the competing fluorescence. In consequence, a lower Ft increase and a stronger decrease of Fm′ indicate a decreasing photosynthetic efficiency during a high light phase. When the product of increasing actinic irradiance and decreasing effective quantum yield is constant, the saturation level of photosynthesis is reached, i.e., P_{max} or ETR_{max} = constant. With further increasing actinic or solar radiation ΔF will asymptotically reach zero. Then no further energy input to the system can be photochemically fixed and any further absorbed energy will be mainly dissipated as heat. Moreover, qP will also reach zero (all PSII reaction centers are closed) but NPQ will further increase (the surplus of absorbed energy is totally converted into heat). But not only the decrease of Fm′ and the reduced increase or even decrease of Ft is caused by increasing heat dissipation, also Fo will decrease to a Fo′-level.

Bilger and Schreiber [21] showed that this is a result of ΔpH-dependent nonphotochemical quenching, i.e., *qE*. Therefore, it is necessary to determine the new level Fo′ during the light illumination phase by a short dark acclimation phase with an application of a weak far-red background light, to temporarily activate solely PSI during the dark phase. This causes a fast reoxidation of the electron carriers within the transport chain between both photosystems (also plastoquinone A oxidation), which relaxes *qE* and *qT* quenching fast. Knowing Fo', the parameters *qP* and *qN* can be calculated. Modern fluorometers automatically apply a several second-lasting dark period with a weak Far-Red pulse to determine the Fo' after each saturating light pulse and calculate automatically the quenching parameters. During the course of a sunny day, the effective quantum yield of photosynthesis is often lowest during midday as the system dissipates most of its absorbed energy by heat (Fig. 9.2). However, this does not necessarily mean that there is no carbon fixation because when calculating the PI-curve, it will be visible that photosynthetic activity will still be higher than the compensation point, when respiration equals photosynthetic oxygen production.

Irrespective of the manifold parameters which can be assessed by PAM measurements, it should still be noticed that data should always be interpreted carefully or should be complemented by oxygen measurements. Using marine algae Hanelt and Nultsch [48], Beer and Axelsson [1] and Nielsen and Nielsen [50] already reported severe discrepancies between yield values determined by simultaneous fluorescence and oxygen measurements. There is underestimation of oxygen evolution derived from fluorescence yield, especially when the fluorescence yield is < 0.1 [1] or fluorescence yield becomes already reduced by *qI* whereas oxygen production rate was still constant or even increased [48]. Moreover, the action spectrum of photoinhibition differs between both methods, showing that light absorbed by R-phycocyanin or R-allophycocyanin was ineffective to cause inhibition of oxygen production in *Palmaria palmata* but contrarily caused a reduction of the fluorescence yield [48].

9.7 Description of instrumentation

9.7.1 Submersible fluorometers for algal research

Use of submersible instruments opens a new field of ecophysiological in situ research. With the underwater fluorometer (Diving-PAM) [51] or a SCUBA-based fast repetition rate fluorometer (FRR-fluorometer) [52] it is possible to perform underwater measurements of the effective quantum yield with benthic organisms in situ. This offers information on how photosynthesis is regulated under water in relation to the impinging photon fluence rate.

The FRR-fluorometer (e.g., Chelsea instruments) measures chlorophyll fluorescence transients using a controlled series of sub-saturating blue light flashes that cumulatively saturate PS II within about 100 μs, i.e., within a single photochemical turnover. The optimal quantum yield of brown algae determined with this kind of fluorometer was somewhat lower than determined with a PAM-fluorometer [52]. Measurements in corals or phytoplankton have been published by e.g., [53–55] but discrepancies between the results to ^{14}C based in situ primary production were also found [55]. The FRR-measurement is often used in oceanography attached to a CTD-device (probe for conductivity, temperature, and depth) but it is apparent that FRR-fluorescence measurements in situ can exhibit a strong taxonomic (adaptive) component, especially visible if the phytoplankton community structure changes [56]. Changes within a community can also be observed with the submersible bbe FluoroProbe (Moldaenke, Germany) as a sensitive measuring instrument for analysis of chlorophyll which promises algae class determination [57]. Individual profiles during the measurements are taken for green algae, cyanobacteria, diatoms, dinoflagellates, and cryptophytes. The different groups of pigments within the algal classes are excited by different wavebands with a consortium of different LEDs. A similar principle as used in the Phyto-PAM (Walz, Germany, Table. 9.1) which is, however, not watertight. This instrument will give information if the composition in the phytoplankton community changes. However, by experience, a correlation to the cell amount of different classes is vague and microscopic determination is still recommendable. The spectrofluorometric AlgaeTorch for determination of the water quality down to 100 m, or the BenthoTorch for measurement of the Chl *a* concentration via fluorescence in benthic green algae, cyanobacteria, and diatom communities are offered by Moldaenke.

The Diving-PAM was the first submersible and commercially available chlorophyll fluorometer. It has a robust housing based on a cylindrical Perspex tube with a touch sensitive keypad. A flexible fiberoptics is used to connect a flat algal thallus or a coral to the radiation sources and the detector mounted within the fluorometer. A water temperature and depth sensor are also integrated in the casing. An external fiber quantum sensor with a cosine similar characteristics measures vertical incident photon irradiance. The battery-powered instrument is submersible down to 50 m, thus, it enables in situ fluorescence measurements of photosystem II by Scuba diving. Data are stored in a CMOS RAM. The main application of the DIVING-PAM is in the determination of effective PSII quantum yield by the saturation pulse method. Investigations of corals, sponges, and macrophytes using a Diving-PAM (Walz, Germany) have already been performed early on (e.g., [51,58–60]. Generally, in situ the yield is often low, so that e.g., with the Diving–PAM a high measuring light intensity (to induce modulated fluorescence) and a high electronic signal gain of the instrument are required, which unfortunately decrease the signal: noise ratio and, hence, increase the variance of the achieved data. A prerequisite for an accurate measurement is that the measuring light does not affect electron transport and all reaction centers remain open, as high measuring light intensities induce electron charge separation and close reaction centers. The ambient light necessary for the calculation of the ETR should be measured with the light sensor close to the investigated specimen. However, as the variance within the fluorescence signal is often large, measurements of a high number of specimens at one location are necessary to calculate a representative average. The construction of the Diving-PAM only allows the internal halogen lamp for the induction of both the saturation pulse and the actinic light. Halogen lamps used for actinic illumination cause major problems as unfortunately the spectrum of these lamps changes when actinic light settings are varied with the power through the lamp filament as e.g., in the Diving-PAM (see as example Fig. 3 in Ref. [61], law of Wien i.e., spectral shift to shorter wavelength caused by higher lamp filament temperature). Moreover, light output of the high power consuming lamp depends strongly on the battery voltage and is not stable and changes quickly. It results in a shift in the ratio of blue to red light and it may have several effects on photosynthesis and could affect photosynthetic performance. For example, photosynthetic capacity of brown algae in red light was increased by an additional blue light due to activation of a carbon supply [62]. Whether the photosynthetic rate, and thus, the course of the P-I curve is affected by alterations in the *blue: red light* ratio during short term irradiation (e.g., by the Emerson enhancement effect) is not clear. However, the emission of a halogen lamp within the waveband of 400–550 nm can change by about 35%. This affects the quantum yield, and quenching parameters of different macroalgae, in an unpredictable manner [61]. Therefore PI-curves measured with a Diving-PAM or polychromatic light produced by halogen lamps in other fluorometers should be interpreted carefully. In contrast, measurements of the effective quantum yield under ambient light are not problematic and can be done in different light conditions, which can also be used for the calculation of a PI-curve but under real environmental conditions [61]. In situ determinations of the effective quantum yield with underwater fluorometers are more realistic than values determined only with land-based PAM fluorometers [63–65]. As normal fluorometers are not watertight, the algal sample has to be removed from its growth site. Even if the period between the collection of the alga and measurement is small (< 15 min, see Ref. [64]), changes of the light field and temperature will rapidly affect quantum yields as well as photochemical and nonphotochemical quenching mechanisms. Therefore, the results obtained with underwater devices better reflect the real photosynthetic performance of algae at their growth site than any land-based fluorometer can do.

9.7.2 Land-based fluorometers

Many investigations on algae are done in the laboratory or at the coastline so that submersible instruments are not obligatory. A high number of the available instruments are battery-powered and rain-proof so that they can also be used in the field. As long as the setup of suitable fluorometers is compact, battery-powered, and easily transportable it can be recommended for field operation. The user should consider that often the sampling side is difficult to access and the instruments have to be carried over a longer distance. The first compact fluorometer on the market was the PAM 2000 (now PAM 2500, Walz, Germany), which was first successfully tested in Antarctica [66]. The fiber optics of the instrument were watertight so that measurements of photosynthetic activity of macroalgae in the daily course were done in shallow waters. Nowadays, many different instruments by various manufacturers are on the market, which cannot all be presented here, so that only a few outstanding instruments are introduced. A compilation of the original PAM fluorometers of the company Walz with important features is presented in Table. 9.2 as this company offers the largest variety of instruments.

Table 9.2 Compilation of the features of the actual PAM fluorometers of the company Heinz Walz GmbH, Germany

Fluorometer	Typical samples	Approx. detection limit for suspensions μg Chl/L	Measuring light color	Actinic light color	Comments
Standard fiberoptics fluorometers using a photodiode detector					
JUNIOR-PAM	Leaves, suspensions	500	Blue or white	Blue or white, far red	Teaching device
DIVING-PAM-II	Macroalgae, corals	3000	Blue or red	Blue or red, far red	Portable, underwater device
MINI-PAM-II	Leaves, suspensions	3000	Blue or red	Blue or red, far red	Portable, many add-ons
PAM-2500	Leaves, suspensions	500	Red	Red and blue, far red	Portable, fast kinetics
High sensitivity fluorometers using a photomultiplier detector					
MICROFIBER-PAM	Tissues, photosynthetic layers	Single cells	470, 520, 630, 650 nm	470, 520, 630, 650 nm	For photosynthesis gradients
MICROSCOPY-PAM	Tissues, cells, chloroplasts	Single cells	470, 615 nm	470, 615 nm	Non imaging
WATER-PAM	Dilute suspensions, natural water, photosynthetic layers (fiber version)	0.03	Blue or red	Blue or red, far red	Cuvette or fiber optics version
XE-PAM	Leaves, dilute suspensions	0.05	Variable, defined by filters	470, 520, 620, 655 nm; white	UV excitation

(*Continued*)

Table 9.2 (Continued)

Fluorometer	Typical samples	Approx. detection limit for suspensions µg Chl/L	Measuring light color	Actinic light color	Comments
Multicolor fluorometers					
MULTI-COLOR-PAM	Leaves, suspensions	200	400, 440, 480, 540, 590, 625 nm	440, 480, 540, 590, 625 nm; white, far red	Functional absorption cross section of PS II
PHYTO-PAM-II	Dilute suspensions, natural water samples	0.5	440, 480, 540, 590, 625 nm	440, 480, 540, 590, 625 nm; white, far red	Differentiation of algae groups
Specialized fluorometers					
DUAL-PAM-100	Leaves, suspensions	Fluorescence: 100–1000 (photodiode), 0.1 (photomultiplier) Absorption (P700): 20,000	Blue or red	Blue, red, far red	Simultaneous measurements of PS I and PS II, many add-ons
MONITORING-PAM	Leaves, macroalgae, corals, suspensions	Data not available	Blue or white	Blue or white	Long term monitoring
IMAGING-PAM fluorometers					
MAXI	Leaves, suspensions (in well plates)	50	Blue or red	Blue or red	10 × 13 cm viewing area
MINI	Leaves, photosynthetic surfaces	Data not available	Blue or red	Blue or red	Portable, also GFP fluorescence
MICROSCOPY	Tissues, cells	Data not available	Blue or red-orange	Blue or red-orange	Tissue spatial variations
MICROSCOPY RGB	Tissues, cells	Data not available	620, 520, 460 nm	620, 520, 460 nm	Differentiation of algae groups

Using fiber optics and special holders, laminar or filamentous algal thalli can be easily investigated, whereas cuvettes are the best choice for unicellular, planktonic, or even spore and gamete suspensions. The sensitivity of the instrument should be chosen depending on the expected chlorophyll concentration. Whereas in macroalgae chlorophyll concentration and density are high, in natural phytoplankton it is rather low. Highly sensitive fluorometers use photomultipliers as detectors (e.g., Water PAM, Xe PAM, Table. 9.1) and are too sensitive for the measurement of thalli, dense algae cultures, or blooms without strong cell dilution. If the instrument sensitivity is medium, low cell concentration of open water bodies cannot be measured as the signal: noise ratio would be too low. An increase in the concentration of cells by centrifugation or filtration causes stress to the cells and strongly affects the photosynthetic status so that it is not recommendable. In contrast, if concentration of cells and thus chlorophyll density within the cuvette is too high, the self-absorption of fluorescence by high pigment concentration causes an underestimation, also if the signal is still standardized to a second fluorescence parameter (e.g., calculation of yields). The user should consider for which purpose the respective fluorometer will be used, as one instrument cannot cover all requirements.

All of the instruments discussed above, except the Imaging Maxi PAM, measure only a small area of a little piece of an algal thallus to deduce the photosynthetic performance of the whole thallus. However, it may be necessary to investigate a larger area or a multicellular community covering a rock. This can be done by using fluorescence sensitive CCD-cameras which, recently, are even connected to a 3D scanner. Such instruments are e.g., represented by the Image-PAM series (Walz) or the FluorCAM (PSI, Czech Republic), however they are rather applicable in the laboratory only. CCD cameras observe fluorescence changes only in the upper layer of a thallus or biofilm, due to technical reasons. Comparing results obtained with e.g., normal fiber optics fluorometers, the yield measured by CCD cameras often occurs much lower, as photosynthetic activity in deeper cell layers cannot be detected (own unpublished observation). A frequent application of Imaging PAM is to investigate cultures of different unicellular algae in multiwell plates simultaneously for physiological stress treatments. In such experiments or in thick algae thalli, it should be considered that the fluorescence signal only corresponds to the upper cell or chlorophyll layer, whereas deeper layers may respond differently. Excitation light does not penetrate deep enough and chlorophyll fluorescence cannot be excited and measured because only the upper cell layers are technically observed. Nielsen and Nielsen [50] found that the saturating pulse of the conventional PAM was saturating throughout the tissue of both thin as well as thick-thallus algae, while imaging PAMs tend to underestimate oxygen evolution at high super-saturating irradiances due to physical/anatomical properties of the algae. Moreover, they reported the thallus thickness may affect the relationship between O_2 evolution and ETR in all PAM methods.

Photon Systems instruments developed a small handheld device called AquaPen-C. It is a cuvette version of the FluorPen fluorometer (PSI). Due to its small size it is very convenient for quick measurements in the field or laboratory. With a blue and red LED light emitter and optical filter it delivers irradiances of up to 3,000 μmol photon m^{-2} s^{-1} in suspensions of unicellular green, red algae, or cyanobacteria. *Blue light* (455 nm) excites chlorophylls to cause chlorophyll fluorescence. *Red-orange* excitation light (620 nm) is used for the excitation of phycobilins and is suitable for measurements in cyanobacteria. Due to its high sensitivity (0.5 μg Chl/L) the AquaPen-C can measure natural water samples containing low concentrations of phytoplankton.

9.7.3 Simultaneous measurement of PS II and PSI using a Dual PAM

The DUAL-PAM-100 (Walz, Germany) is the most sophisticated instrument and combines the properties of a high performance PAM chlorophyll fluorometers and those of a dual wavelength absorbance spectrometer. In vivo chlorophyll fluorescence of PS I compared to PS II is rather low and mainly visible at deep temperatures, so it is not applicable for in vivo measurements. Therefore, the instrument measures PS I absorbance changes at 830 nm [67] and simultaneously chlorophyll fluorescence of PSII. Assuming that under donor-side electron limitation of PS I, which can be produced by a preceding Far-Red exposure, a multiturnover saturation pulse induces transiently total oxidation of the reaction center chlorophyll P700 which is subsequently followed by its full reduction. Fully oxidized PS I has a broad absorbance peak of around 800–840 nm. It becomes re-reduced by the activated electron transport from PS II by the multiturnover saturation pulse with electrons originating from the water splitting complex. The difference in the low absorbance of fully reduced P700 with its absorbance level Po and high absorbance of the fully oxidized P700 is denoted as Pm. Under actinic illumination the Pm level changes to Pm' according to the fluorescence nomenclature. This is analogous to the Chl fluorescence yield, which varies between its minimal level Fo, the maximal level Fm, and Fm′ during actinic illumination. A comparison of the absorbance of 830 nm with a reference beam at 875 nm (similar to a physiological internal standard), which is independent from the redox state of PS I, ensures that only absorbance changes by the redox state are measured. The quantum yield of PS I, denoted Y(I), is also calculated by the saturation pulse analysis. Per definition $Y(\mathrm{I}) = 0$ when P700 is fully oxidized as charge separation is only possible with reduced P700. However, reduced P700 is not necessarily capable of charge separation, as it may be still inhibited at the PS I acceptor side. Therefore, Y(I) corresponds only to the fraction of reduced P700, when the PS I acceptor side is not limiting as supposed after a preceding 10 s FR irradiation. Unicellular alga suspensions can be measured within the cuvette of the optical unit, whereas pieces of macroalgae should be measured with the enclosed leaf holder. However, a serious problem may arise with unicellular algae because the fluorescence measurement needs a lower chlorophyll concentration as the instrument is quite sensitive, whereas vice versa the P700 measurement requires a relative high chlorophyll concentration because of the need of a high absorption of exciting light (see Table. 9.1). Thus, it is not easy to find the best cell or chlorophyll concentration so that fluorescence stays within a good detectable range but is also high enough to observe the absorption changes. A successful compromise can give new insights into the performance of both photosystems as presented by Patzelt et al. [68] after chemical cell stress. However, it is advised that the dual performance of the instrument is not recommendable for inexperienced beginners, as some experience with PAM fluorometry is obviously necessary.

References

[1] Beer S, Axelsson L. Limitations in the use of PAM fluorometry for measuring photosynthetic rates of macroalgae at high irradiances. Eur J Phycol 2004;39(1):1–7.

[2] Hatcher B, Chapman AO, Mann K. An annual carbon budget for the kelp *Laminaria longicruris*. Mar Biol 1977;44(1):85–96.

[3] Thomas DN, Wiencke C. Photosynthesis, dark respiration and light independent carbon fixation of endemic Antarctic macroalgae. Polar Biol 1991;11:329–37.

[4] Weykam G, Thomas D, Wiencke C. Growth and photosynthesis of the Antarctic red algae *Palmaria decipiens* (Palmariales) and *Iridaea cordata* (Gigartinales) during and following extended periods of darkness. Phycologia 1997;36(5):395–405.

[5] Williams P, Raine R. Agreement between the c-14 and oxygen methods of measuring phytoplankton production-reassessment of the photosynthetic quotient. Oceanol Acta 1979;2(4):411–16.

[6] Laws EA. Photosynthetic quotients, new production and net community production in the open ocean. Deep Sea Res Part A Oceanogr Res Pap 1991;38(1):143–67.

[7] Williams PI, Robertson J. Overall planktonic oxygen and carbon dioxide metabolisms: the problem of reconciling observations and calculations of photosynthetic quotients. J Plankton Res 1991;13(suppl):153–69.

[8] Fryer MJ, Andrews JR, Oxborough K, Blowers DA, Baker NR. Relationship between CO_2 assimilation, photosynthetic electron transport, and active O_2 metabolism in leaves of maize in the field during periods of low temperature. Plant Physiol 1998;116 (2):571–80.

[9] Schreiber U, Schliwa U, Bilger W. Continuous recording of photochemical and nonphotochemical chlorophyll fluorescence quenching with a new type of modulation fluorometer. Photosynth Res 1986;10:51–62.

[10] Schreiber U, Bilger W. Progress in chlorophyll fluorescence research: major developments during the past years in retrospect. In: Lüttge U, Ziegler H, editors. Progress botany, vol. 54. Berlin: Springer Verlag; 1993. p. 151–73.

[11] Schreiber U, Bilger W, Neubauer C. Chlorophyll fluorescence as a nonintrusive indicator for rapid assessment of in vivo photosynthesis. In: Schulze ED, Caldwell MM, editors. Ecophysiology of photosynthesis ecological studies, vol. 100. Berlin: Springer Verlag; 1994. p. 49–70.

[12] Genty B, Briantais JM, Baker NR. The relationship between the quantum yield of photosynthetic electron transport and quenching of chlorophyll fluorescence. Biochim Biophys Acta 1989;990:87–92.

[13] Beer S, Larsson C, Poryan O, Axelsson L. Photosynthetic rates of *Ulva* (Chlorophyta) measured by pulse amplitude modulated (PAM) fluorometry. Eur J Phycol 2000;35 (01):69–74.

[14] Baker NR. Chlorophyll fluorescence: a probe of photosynthesis in vivo. Annu Rev Plant Biol 2008;59:89–113.
[15] Blache U, Jakob T, Su W, Wilhelm C. The impact of cell-specific absorption properties on the correlation of electron transport rates measured by chlorophyll fluorescence and photosynthetic oxygen production in planktonic algae. Plant Physiol Biochem 2011;49 (8):801–8.
[16] Björkman O, Demmig B. Photon yield of O_2 evolution and chlorophyll fluorescence characteristics at 77 K among vascular plants of diverse origins. Planta 1987;170:489–504.
[17] Schreiber U. New emitter-detector-cuvette assembly for measuring modulated chlorophyll fluorescence of highly diluted suspensions in conjunction with the standard PAM fluorometer. Z Naturforsch 1994;49:646–56.
[18] Emerson R. Dependence of yield of photosynthesis in long wave red on wavelength and intensity of supplementary light. Science 1957;125:746.
[19] Prezelin BB. Light reactions in photosynthesis. Can Bull Fish Aquat Sci 1981;210:1–43.
[20] Krause GH. Photoinhibition of photosynthesis. An evaluation of damaging and protective mechanisms. Physiol Plant 1988;74:566–74.
[21] Bilger W, Schreiber U. Energy-dependent quenching of dark-level chlorophyll fluorescence in intact leaves. Photosynth Res 1986;10:303–8.
[22] Maxwell K, Johnson GN. Chlorophyll fluorescence - a practical guide. J Exp Bot 2000;51:659–68.
[23] Kautsky H, Apel W, Amann H. Chlorophyllfluoreszenz und Kohlensäureassimilation. XIII. Die Fluoreszenkurve und die Photochemie der Pflanze. Biochem Unserer Zeit 1960;322:277–92.
[24] Strasser RJ, Srivastava A, Tsimilli-Michael M. The fluorescence transient as a tool to characterize and screen photosynthetic samples. In: M.Yunus, U. Pathre and P. Mohanty, (Eds.), Probing Photosynthesis: Mechanism, Regulation and Adaptation, Chap. 25, 2000, Taylor and Francis, London, UK, 443–80.
[25] Papageorgiou GC, Tsimilli-Michael M, Stamatakis K. The fast and slow kinetics of chlorophyll a fluorescence induction in plants, algae and cyanobacteria: a viewpoint. Photosynth Res 2007;94(2–3):275–90.
[26] Bilger W, Björkman O. Role of the xanthophyll cycle in photoprotection elucidated by measurements of light-induced absorbance changes, fluorescence and photosynthesis in leaves of *Hedera canariensis*. Photosynth Res 1990;25:173–85.
[27] Büchel C, Wilhelm C. *In vivo* analysis of slow chlorophyll fluorescence induction kinetics in algae: progress, problems and perspectives. Photochem Photobiol 1993;58:137–48.
[28] Adams Iii WW, Demmig-Adams B. Operation of the xanthophyll cycle in higher plants in response to diurnal changes in incident sunlight. Planta 1992;186:390–8.
[29] Ruban AV, Young AJ, Horton P. Induction of non-photochemical energy dissipation and absorbance changes in leaves. Evidence for changes in the state of the light-harvesting system of Photosystem II in vivo. Plant Physiol 1993;102:741–50.
[30] Guenther JE, Melis A. The physiological significance of photosystem II heterogeneity in chloroplasts. Photosynth Res 1990;23:105–9.
[31] Öquist G, Chow WS. On the relationship between the quantum yield of photosystem II electron transport, as determined by chlorophyll fluorescence and the quantum yield of CO_2-dependent O_2 evolution. Photosynth Res 1992;33:51–62.
[32] Critchley C, Russell AW. Photoinhibition of photosynthesis in vivo: the role of protein turnover in photosystem II. Physiol Plant 1994;92:188–96.

[33] Horton P, Ruban AV, Walters RG. Regulation of light harvesting in green plants. Annu Rev Plant Physiol 1996;47:655–84.
[34] Demmig-Adams B. Carotenoids and photoprotection in plants: a role for the xanthophyll zeaxanthin. Biochim Biophys Acta 1990;1020:1–24.
[35] Krause GH, Weis E. Chlorophyll fluorescence and photosynthesis: the basics. Annu Rev Plant Physiol 1991;42:313–49.
[36] Hanelt D. Capability of dynamic photoinhibition in Arctic macroalgae is related to their depth distribution. Mar Biol 1998;131:361–9.
[37] Osmond CB. What is photoinhibition? Some insights from comparisons of shade and sun plants. In: Baker NR, Bowyer JR, editors. Photoinhibition of photosynthesis, from the molecular mechanisms to the field. Oxford: BIOS Scientific Publ; 1994. p. 1–24.
[38] Hanelt D, Hupperts K, Nultsch W. Photoinhibition of photosynthesis and its recovery in red algae. Botanica Acta 1992;105:278–84.
[39] Bischof K, Hanelt D, Wiencke C. UV radiation and Arctic marine macroalgae. In: Hessen D, editor. *UV* radiation and arctic ecosystems ecological studies, vol. 153. Berlin, Heidelberg: Springer Verlag; 2002. p. 227–43.
[40] Huppertz K, Hanelt D, Nultsch W. Photoinhibition of photosynthesis in the marine brown alga *Fucus serratus* as studied in field experiments. Mar Ecol Prog Ser 1990;66:175–82.
[41] Henley WJ, Lindley ST, Levavasseur G, Osmond CB, Ramus J. Photosynthetic response of *Ulva rotundata* to light and temperature during emersion on an intertidal sand flat. Oecologia 1992;89:516–23.
[42] Henley W, Levavasseur G, Franklin L, Lindley ST, Ramus J, Osmond CB. Diurnal responses of photosynthesis and fluorescence in *Ulva rotundata* acclimated to sun and shade outdoor culture. Mar Ecol Prog Ser 1991;75:19–28.
[43] Hanelt D, Huppertz K, Nultsch W. Daily course of photosynthesis and photoinhibition in marine macroalgae investigated in the laboratory and field. Mar Ecol Prog Ser 1993;97:31–7.
[44] Hanelt D. Photoinhibition of photosynthesis in marine macrophytes of the South China Sea. Mar Ecol Prog Ser 1992;82:199–206.
[45] Franklin LA, Levavasseur G, Osmond CB, Henley WJ, Ramus J. Two components of onset and recovery during photoinhibition of *Ulva rotundata*. Planta 1992;186:399–408.
[46] Uhrmacher S, Hanelt D, Nultsch W. Zeaxanthin content and the degree of photoinhibition are linearly correlated in the brown alga *Dictyota dichotoma*. Mar Biol 1995;123:159–65.
[47] Schofield O, Evens TJ, Millie DF. Photosystem II quantum yields and xanthophyll-cycle pigments of the macroalga *Sargassum natans* (Phaeophycea): responses under natural sunlight. J Phycol 1998;34(1):104–12.
[48] Hanelt D, Nultsch W. Field studies of photoinhibition show non-correlations between oxygen and fluorescence measurements in the Arctic red alga *Palmaria palmata*. J Plant Physiol 1995;145:31–8.
[49] Murchie EH, Lawson T. Chlorophyll fluorescence analysis: a guide to good practice and understanding some new applications. J Exp Bot 2013;64(13):3983–98.
[50] Nielsen HD, Nielsen SL. Evaluation of imaging and conventional PAM as a measure of photosynthesis in thin- and thick-leaved marine macroalgae. Aquat Biol 2008;3:121–31.
[51] Schreiber U, Gademann R, Ralph PJ, Larkum A. Assessment of photosynthetic performance of *Prochloron* in *Lissoclinum patella* in hospite by chlorophyll fluroescence measurments. Plant Cell Physiol 1997;38(8):945–51.

[52] Gorbunov MY, Falkowski PG, Kolber ZS. Measurement of photosynthetic parameters in benthic organisms in situ using a SCUBA-based fast repetition rate fluorometer. Limnol Oceanogr 2000;45(1):242–5.
[53] Lesser MP, Gorbunov MY. Diurnal and bathymetric changes in chlorophyll fluorescence yields of reef corals measured in situ with a fast repetition rate fluorometer. Mar Ecol Prog Ser 2001;212:69–77.
[54] Lombardi MR, Lesser MP, Gorbunov MY. Fast repetition rate (FRR) fluorometry: variability of chlorophyll *a* fluorescence yields in colonies of the corals, *Montastraea faveolata* (w.) and *Diploria labyrinthiformes* (h.) recovering from bleaching. J Exp Mar Biol Ecol 2000;252(1):75–84.
[55] Raateoja MP. Fast repetition rate fluorometry (FRRF) measuring phytoplankton productivity: a case study at the entrance to the Gulf of Finland, Baltic Sea. Boreal Environ Res 2004;9:263–76.
[56] Suggett DJ, Moore CM, Hickman AE, Geider RJ. Interpretation of fast repetition rate (FRR) fluorescence: signatures of phytoplankton community structure versus physiological state. Mar Ecol Prog Ser 2009;376:1–19.
[57] Beutler M, Wiltshire KH, Meyer B, Moldaenke C, Lüring C, Meyerhöfer M, et al. A fluorometric method for the differentiation of algal populations *in vivo* and *in situ*. Photosynth Res 2002;72:39–53.
[58] Beer S, Ilan M. In situ measurements of photosynthetic irradiance responses of two Red Sea sponges growing under dim light conditions. Mar Biol 1998;131(4):613–17.
[59] Beer S, Ilan M, Eshel A, Weil A, Brickner I. Use of pulse amplitude modulated (PAM) fluorometry for in situ measurements of photosynthesis in two Red Sea faviid corals. Mar Biol 1998;131(4):607–12.
[60] Beer S, Vilenkin B, Weil A, Veste M, Susel L, Eshel A. Measuring photosynthetic rates in seagrasses by pulse amplitude modulated (PAM) fluorometry. Mar Ecol Prog Ser 1998;174:293–300.
[61] Hanelt D, Wiencke C, Bischof K. Photosynthesis in marine macroalgae. In: Larkum AW, Douglas SE, Raven JA, editors. Photosynthesis in algae. The Netherlands: Kluwer Academic Publication; 2003. p. 413–35.
[62] Schmid R, Dring MJ. Blue light and carbon acquisition in brown algae: an overview and recent developments. Sci Mar 1996;60(1):115–24.
[63] Häder D-P, Lebert M, Flores-Moya A, Jiménez C, Mercado J, Salles S, et al. Effects of solar radiation on the photosynthetic activity of the red alga *Corallina elongata* Ellis et Soland. J Photochem Photobiol B Biol 1997;37(3):196–202.
[64] Häder D-P, Figueroa FL. Photoecophysiology of marine macroalgae. Photochem Photobiol 1997;66:1–14.
[65] Häder D-P, Herrmann H, Schäfer J, Santas R. Photosynthetic fluorescence induction and oxygen production in two Mediterranean *Cladophora* species measured on site. Aquat Bot 1997;56:253–64.
[66] Hanelt D, Jaramillo JM, Nultsch W, Senger S, Westermeier R. Photoinhibition as a regulative mechanism of photosynthesis in marine algae of Antarctica. Serie Cientifica, Instituto Antartico Chileno 1994;44:67–77.
[67] Schreiber U, Klughammer C, Neubauer C. Measuring P700 absorbance changes around 830 nm with a new type of pulse modulation system. Z Naturforsch C 1988; 43(9-10):686–98.
[68] Patzelt DJ, Hindersin S, Kerner M, Hanelt D. Responses of photosystems I and II of *Acutodesmus obliquus* to chemical stress caused by the use of recycled nutrients. Appl Microbiol Biotechnol 2016;100(1):361–70.

Ecotox

10

Azizullah Azizullah[1] *and Donat-P. Häder*[2]
[1]Kohat University of Science and Technology (KUST), Kohat, Pakistan,
[2]Friedrich-Alexander University, Erlangen-Nürnberg, Germany

10.1 Introduction

The increasing pollution of aquatic environments with different classes of natural and synthetic pollutants is one of the serious environmental issues of the world. In order to prevent its further pollution and sustain healthy aquatic ecosystems, proper monitoring of the pollution level in water is essential. Traditionally, chemical methods are used to measure the load of pollutants in water, but chemical methods have several disadvantages like high expenses and high time consumption. Also chemical methods may not detect all the pollutants present in a sample. The most serious drawback is that these methods do not reflect the adverse effects of the pollutant on living organisms [1,2]. Therefore, assessment of water quality involving living organisms, i.e., bioassessment, is essential to understand the actual effects of pollutants on living organisms in aquatic systems. A number of bioassays based on different classes of organisms like fish, invertebrates, duckweeds, different algal species, and bacteria are employed to assess ecotoxicity of pollutants in aquatic environments. Different parameters of these organisms like mortality, growth, respiration, photosynthesis, motility, and many other parameters are employed as endpoints to evaluate the toxicity of a substance under test. However, most of these tests have several drawbacks. For example, many bioassays require a long time on the order of hours or even days. Tests with animals such as fish involve ethical issues. The Ecotox system was developed to overcome these issues and establish a fast and cost-effective bioassay for easy monitoring of aquatic pollution [3–6] based on an earlier test system called Erlanger flagellate test [7].

The Ecotox instrument employs the unicellular flagellate *Euglena gracilis* as a test organism [8]. It is based on the observation that motility, orientation, and cell shape of *Euglena* are very sensitive to pollution in water and are impaired by any change in the physicochemical characteristics of water. These parameters of *Euglena* are determined in real time by image analysis software of Ecotox and are used as endpoint parameters for observing toxicity. For determining toxicity of a pollutant, a cell culture of *Euglena* and the test solution are automatically mixed in a mixing chamber and passed to an observation chamber by three stepper motor pumps. A miniaturized microscope connected to a firewire camera records the behavior of *Euglena* cells in the observation cuvette. In order to determine the toxicity of a sample, the behavior of an unpolluted control is compared with that of a

Bioassays. DOI: http://dx.doi.org/10.1016/B978-0-12-811861-0.00010-3

treated sample and the results are being compared and recorded in numerical and graphical form.

All these functions are automatically controlled by the connected computer running the software of the system. A virtually unlimited number of cells are tracked by the system in parallel. A number of motility, orientation, and morphological parameters of the tracked cells are measured in real time. The system automatically makes different dilutions of the solution or wastewater under test and records its effect on the different endpoint parameters of *Euglena* as percent inhibition [5,6].

The measurement principle of Ecotox is very simple. First a control reading of a fresh *Euglena* culture not mixed with any test solution is recorded. After the control measurement, the observation cuvette is automatically rinsed with clean water and fresh cell culture is automatically mixed with the test solution and transferred to the observation cuvette, and all endpoint parameters are determined again. The reading of the test measurement is automatically compared with the control reading and the effect on various parameters is determined as percent inhibition. If there are toxic substances present in the samples, one or more parameters can be affected. The inhibition is shown graphically on the monitor and it is indicated in red when the results of the test sample deviate significantly from the control sample. All readings and results are saved in a user-friendly format that can be opened by Microsoft Excel for further analysis according to the user's requirement. Although *Euglena* is a model organism for the Ecotox instrument, other motile small organisms can also be used in the instrument [9,10]; they just need a change in the cuvette and camera settings according to the size of the organism used. For example, it was successfully tested with different species of marine unicellular algae like *Dunaliella*, *Prorocentrum*, and *Tetraselmis* [11].

The Ecotox has been successfully used to evaluate the ecotoxicological effects of a wide range of water pollutants including heavy metals, pesticides, fertilizers, detergents, wastewater, and many other organic and inorganic pollutants [4,6,12–15]. Ecotox can be applied to monitor potential water pollutants both in short- and long-term tests [15–17]. The sensitivity of Ecotox to different pollutants and toxicants has been found to be better than or comparable to other commonly used biotests such as the bioluminescence assay with *Vibrio fischeri*, the motility assay with *Daphnia*, mortality tests with fish, and the algal growth test [4,6,16]. This bioassay has several advantages over other commonly used bioassays, as described later in this chapter.

10.2 *Euglena gracilis* - model organism for Ecotox

E. gracilis (Fig. 10.1), a unicellular motile flagellate belonging to the phylum Euglenophyta, is used as a model organism by Ecotox [6,18,19]. It has two flagella but is powered by only one flagellum because the second is rudimentary and does not extend from the anterior invagination [20]. The cell lacks a cell wall but has a pellicle constructed from proteinaceous ribbons surrounding the cell [21,22]. The

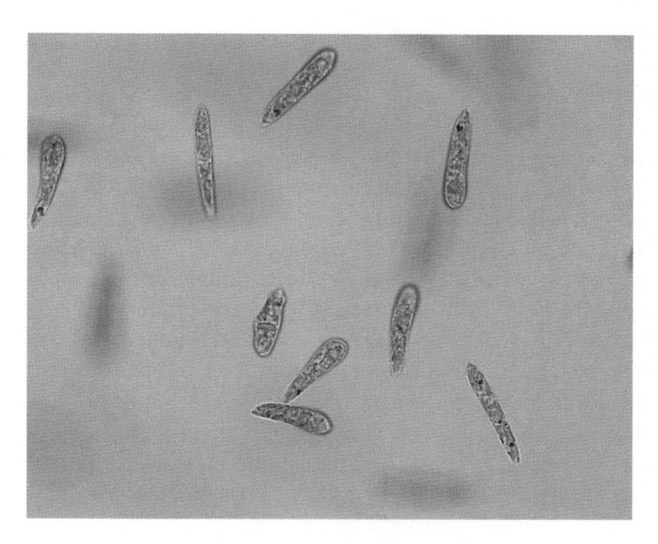

Figure 10.1 The flagellate *Euglena gracilis* used as bioindicators in the Ecotox bioassay. The unicellular organism is powered by an anterior flagellum. The *red dot* at the anterior end is the stigma, which has been shown not to be the photoreceptor for phototaxis. Photosynthesis is conducted by the numerous chloroplasts. Other species such as the colorless *E. longa* lack chloroplasts and live heterotrophically. The cells do not possess a cell wall but are enclosed by a pellicle consisting of proteinaceous strips.

organism lacks sexual reproduction and cell division is by mitosis and very fast [23,24]. *Euglena* is very sensitive to physicochemical changes and pollution in the surrounding environment and is considered among the most sensitive and widely used organisms in ecotoxicological studies [25,26]. A number of parameters based on its behavioral, biochemical, morphological, and physiological responses can be used as endpoints in assessment of water pollutants. The flagellate orients itself in the water column using light and gravity as the main clues [27–29]. Ecotox uses its motility, swimming speed, and orientation with respect to light and gravity and cell shape as endpoints. All these endpoints have been recommended as sensitive and reliable parameters for assessing water pollutants of both organic and inorganic nature as well as wastewaters [15,26,30–32].

For experiments with Ecotox, *Euglena* can be grown either in organic or inorganic medium or in a mixture of both media under continuous light of 20 W m^{-2} and at a temperature of 20–22°C. Young cultures of *Euglena* show a positive gravitaxis (so that they move downward in the water column) [33], while older cells show negative gravitaxis, which is more precise [29,34]; therefore stationary phase cultures are recommended for experiments with Ecotox. One- or two-weeks old cultures grown in organic or inorganic medium, respectively, generally show good orientation and motility and are suitable for use in experiments with Ecotox. In order to warrant homogeneity of the culture and avoid interference by the diurnal rhythm in motility and orientation of cells exposed to a light-dark cycle [35,36], cultures grown in continuous light are recommended for Ecotox.

10.3 Hardware of Ecotox

The Ecotox instrument is shown in Fig. 10.2. The hardware, including a miniaturized microscope, a firewire camera (DMK 21F04, Imaging Source, Bremen, Germany), three stepper motor pumps (motors: PK245–03 A, Oriental, Neuss; pumping heads: SPQ-048, Möller Feinmechanik, Fulda, Germany), a mixing chamber, and an observation cuvette is contained in a compact housing. The microscope has a 10 × objective lens mounted in a horizontal position and is connected to the camera. The objective of the microscope can be focused onto the swimming chamber. In order to avoid the interference of white light with the orientation of the organisms, an infrared diode ($\lambda = 875$ nm) is used as a light source. The schematic diagram of the instrument is shown in Fig. 10.3. The three stepper motor pumps are equipped with pipes and valves and are used to transport the *Euglena* culture, rinsing water, and test solution (any toxin solution to be tested) into the observation cuvette through the mixing chamber to ensure homogeneity of the solution. The number of steps of the motor pumps is automatically controlled by the software to ensure correct dilutions of the provided test solution [6,16]. The observation cuvette is made of stainless steel with glass windows (inner diameter 50 mm and thickness 0.2 mm) for observation. The cuvette is positioned vertically, so that gravitactic orientation of cells swimming in the cuvette can be assayed by the horizontally fixed microscope [6]. The device is equipped with a leakage sensor that sounds an alarm if a leakage occurs from the pumps or tubes.

The hardware is connected to a computer running the Ecotox software via a serial interface, which automatically controls all functions of the system. The test organisms (cell cultures) are automatically filled into the observation cuvette and the movement behavior of the cells is analyzed by a complex image analysis software [5]. The system operates in real time and tracks a virtually unlimited number of cells in parallel. The software of the systems determines vectors of the tracks to

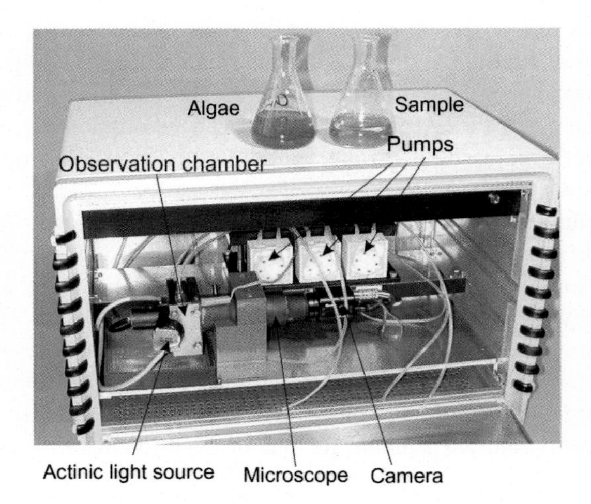

Figure 10.2 Ecotox instrument in operation showing the main hardware elements.

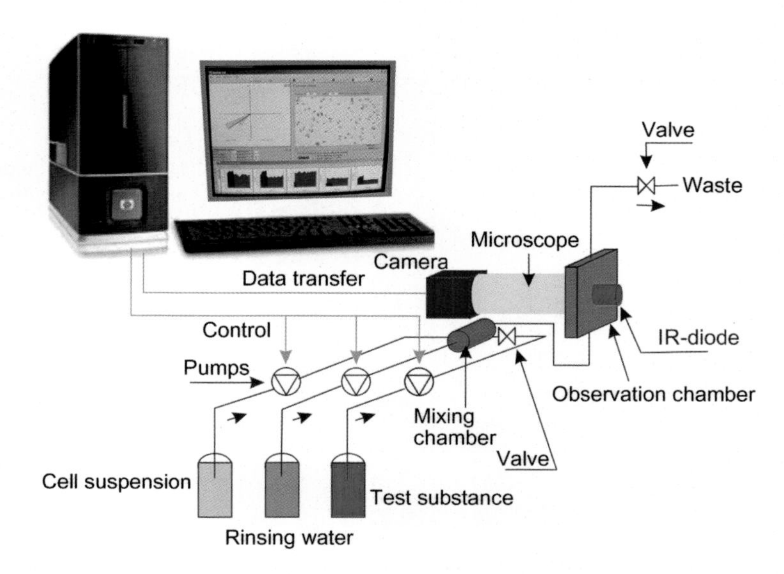

Figure 10.3 Schematic diagram of the Ecotox hardware showing the miniaturized microscope with a firewire camera attached. The cell suspension, the solubilized toxin, and the rinsing water are transported by three stepper motor pumps to a mixing chamber and forwarded from there to the observation chamber.

calculate various parameters such as percent motility, mean swimming velocity, cell compactness (cell shape), percentage of cells moving upwards, alignment, and *r*-value of cells (cf. Chapter 5: Image analysis for bioassays – the basics, this volume). The system automatically rinses the cuvette with clean water before and after each measurement as well as between each two measurements in the case of single toxin or online modes.

The screen of the Ecotox software is displayed on the computer monitor (shown in Fig. 10.4) and has different menus which can be selected by the user to set various parameters according to the experimental requirements. The user can select the mode of operation from the main screen he or she wants to use (control, single toxin, or online). Similarly, the user can use these menus to define different setting parameters like rinsing and filling time of the cuvette, dark incubation, and tracking time, etc. In addition, other settings (e.g., camera setting, program setting) can be selected using these menus.

10.4 Endpoint parameters

10.4.1 Motility

The motility parameter indicates the percentage of cells moving at a speed equal to or faster than the minimum velocity set in the program; all other parameters are

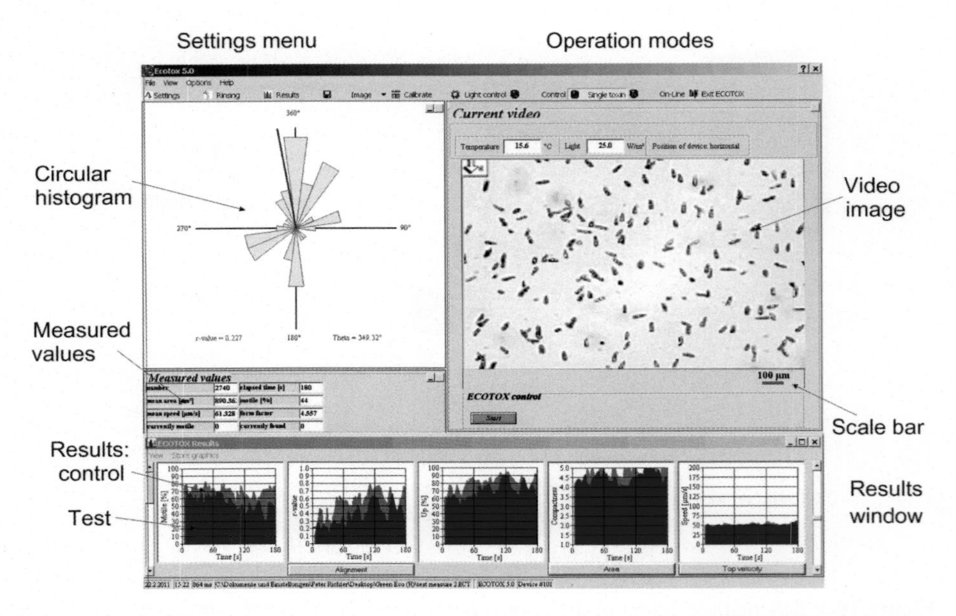

Figure 10.4 Main screen of the Ecotox software. The moving cells are shown in real time in the top right hand video image. A scale bar indicates the real length. Further information such as the ambient temperature, the irradiance of the actinic light and the position of the microscope (vertical or horizontal) are indicated. The diagram in the top left shows the movement of the cells as a circular histogram. The measured data are shown in numerical form as well as graphical output over time.

calculated only for objects which fulfill this criterion. The percent motility of cells is calculated by Eq. (10.1).

$$\text{Motility} = \frac{n_S}{n} \cdot 100\% \tag{10.1}$$

where n is the number of all calculated vectors and n_s the number of vectors with a velocity higher than a predefined threshold value [6].

10.4.2 Velocity

The velocity gives the mean speed (swimming velocity) of all motile cells in $\mu m\ s^{-1}$ and is calculated using Eq. (10.2).

$$\text{Velocity} = \frac{d}{\Delta t} \cdot fs \tag{10.2}$$

where $d = ((\Delta x^2) + (\Delta y^2))^{0.5}$ and Δx and Δy is the distance in x and y-direction, respectively, Δt the time delay the between first and fifth frame (160 ms) in the video sequence and fs is a scaling factor [6].

10.4.3 Cell compactness

The cell compactness or form factor (which is the ratio of the circumference to the area normalized to a circle) describes the shape of a cell and is calculated by the software using Eq. (10.3) [6]. The cell compactness has the lowest value of 1 when the outline of the object is a circle and increases as the cell increases in length.

$$\text{Compactness} = \frac{\sum_{i=1}^{n_s} \frac{S_i^2}{A_i \cdot 4\pi}}{n_s} \tag{10.3}$$

where S is the length of the outline and A is the area of the object and n_s is the number of all calculated vectors.

10.4.4 Upward swimming %

This parameter gives the percentage of cells which are swimming towards the upper part of the cuvette (± 90 degrees around the vertical direction upward) and is calculated by the system using Eq. (10.4) as described by Tahedl and Häder [6].

$$\text{Upward} = \frac{n_o}{n_s} \cdot 100\% \tag{10.4}$$

where n_o is the number of vectors with -90 degrees $\leq \alpha < 90$ degrees and n_s is the total number of cells. The upward direction is defined as 0 degrees.

10.4.5 R-value

It is a statistic parameter which describes the precision of gravitactic orientation of swimming cells and ranges from 0 (when the cells are moving randomly) to 1 (when all the cells are moving in the same direction). The r-value is calculated according to Eq. (10.5) (cf. Chapter 5: Image analysis for bioassays – the basics, this volume).

$$r\text{-value} = \frac{\sqrt{\left(\sum_{i=1}^{n_s} \sin \alpha_i\right)^2 + \left(\sum_{i=1}^{n_s} \cos \alpha_i\right)^2}}{n_s} \tag{10.5}$$

where α is the angle of the movement vector and n is the number of all calculated vectors.

10.4.6 Alignment

The alignment, calculated by Ecotox using Eq. (10.6) [6] (cf. Chapter 5: Image analysis for bioassays – the basics, this volume), is used to describe whether the movement of cells is horizontal or vertical; its value ranges from -1 (when there is movement of cells only in the horizontal direction) to 1 (when there is movement of cells only in the vertical direction).

$$\text{Alignment} = \frac{\sum_{i=1}^{n_s} |\sin \alpha_i| - \sum_{i=1}^{n_s} |\cos \alpha_i|}{n_s} \tag{10.6}$$

10.5 Operational modes of Ecotox

The Ecotox can be operated in three different modes. Like other settings, the desired operational mode can be selected from the menu on the main screen (Fig. 10.4).

10.5.1 Control mode

In this mode, the *Euglena* culture (diluted with water 1:1 or undiluted, depending upon the setting) is pumped into the observation chamber and all parameters are measured. This mode is usually used to examine cell behavior of control cultures without mixing with any toxin solution or cultures already treated with toxin.

10.5.2 Single toxin mode

This mode is used when the toxicity of a test solution is assessed. In this mode, a control reading of a *Euglena* culture (undiluted or diluted with water (1:1) depending on the setting of the software) is measured first, followed by measurement of a culture automatically mixed with a toxin solution. In this mode one or five (depending upon the setting) dilutions of the toxin stock solution are made automatically and the effect on various endpoint parameters is recorded. In the setting of "five dilutions" the provided toxin solution is automatically diluted with clean water in a ratio of 1:31, 1:15, 1:7, 1:3, or 1:1 (dilutions are made in ascending or descending order depending upon the setting of the software).

10.5.3 Online mode

In this mode, measurements described above in "single toxin mode" are repeated at predetermined time intervals (e.g., 60 or 120 min). This mode is generally used for monitoring ecotoxicity for a long time at regular intervals, for example monitoring the efficiency of a water treatment plant at regular intervals.

10.6 Data recording, storage and analysis

The data are recorded and saved both in graphic forms (Fig. 10.5). The data saved as ASCII files can be opened and used with Microsoft Excel for further analysis. In the single toxin and online modes, the calculated value of the control measurement is presented by a *green graph* while that of the toxin in *red* color. The area of the toxin graph (*red*) is compared to that of the control (*green*) to calculate the percent inhibition using Eq. (10.7) [6].

$$\text{Inhibition} = \frac{A_c - A_s}{A_c} \cdot 100\% \tag{10.7}$$

where A_c is the area under the data curve of the control measurement (*green*) and A_s is the area under the data curve of the sample measurement (*red*). The observed effects are expressed as percent inhibition for all parameters and shown as graphics under the kinetic representations. The concentration of toxin solution (wastewater

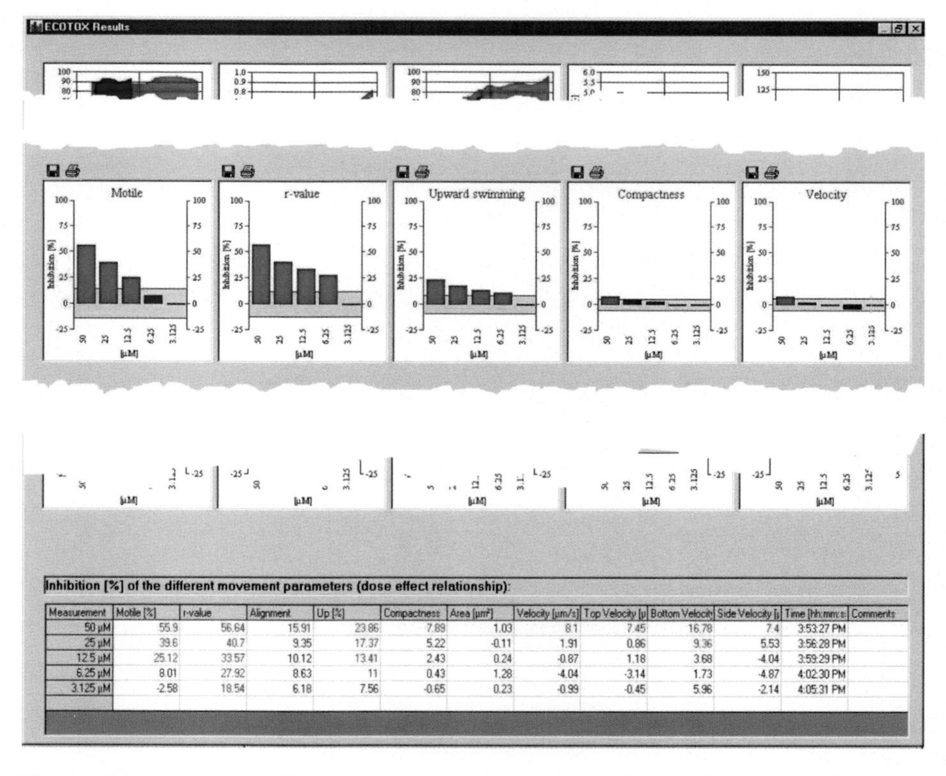

Inhibition [%] of the different movement parameters (dose effect relationship):

Measurement	Motile [%]	r-value	Alignment	Up [%]	Compactness	Area [µm²]	Velocity [µm/s]	Top Velocity [µ	Bottom Velocit	Side Velocity [	Time [hh:mm:s	Comments
50 µM	55.9	56.64	15.91	23.86	7.89	1.03	8.1	7.45	16.78	7.4	3:53:27 PM	
25 µM	39.6	40.7	9.35	17.37	5.22	-0.11	1.91	0.86	9.36	5.53	3:56:28 PM	
12.5 µM	25.12	33.57	10.12	13.41	2.43	0.24	-0.87	1.18	3.68	-4.04	3:59:29 PM	
6.25 µM	8.01	27.92	8.63	11	0.43	1.28	-4.04	-3.14	1.73	-4.87	4:02:30 PM	
3.125 µM	-2.58	18.54	6.18	7.56	-0.65	0.23	-0.99	-0.45	5.96	-2.14	4:05:31 PM	

Figure 10.5 Data recorded by the automatic Ecotox system. The calculated values for the endpoints show the results for the control in *green* and those for the toxin in *red*. The data are also available in numerical form (bottom).

or any chemical) is displayed on the *x*-axis and the percent inhibition of a given parameter on the *y*-axis (Fig. 10.5). The blue color of the graph indicates a non-significant effect while the red color shows that the inhibitory effect is significant. The effect is regarded as significant if the percent inhibition for a given parameter exceeds the threshold limits (e.g., 11.4% for motility, 12.3% for *r*-value, 3.1% for upward, 3.4% for compactness, and 6.8% for velocity). These threshold values have been calculated by a hundred measurements of control readings and are fitted in the system software. In addition, the percent inhibition of various parameters in displayed in a tabular form. In addition to motility, orientation, and form parameters of *Euglena*, the Ecotox also automatically records physical parameters like actinic light intensity (for phototaxis induction) and temperature in the observation chamber.

The data obtained from Ecotox in ASCII file form can be further processed and presented in different forms according to the user requirements. The percent inhibition can be drawn against the concentration of the tested substance using regression analysis with a suitable equation to calculate EC_{50} values for a tested substance (Fig. 10.6). The percent inhibition can also be drawn versus the concentration of the tested substance in the form of a bar chart. In addition, the actual values of different parameters for control and treatments can be obtained from the ASCII file and presented as bar graphs.

Measuren	Motile [%	r-value	Alignmen	Theta [°]	Up [%]	Compactn	Area [μm^2	Velocity [	Top veloci	Bottom ve	Side veloc	Time [s]	circular hi	Position o	Irradiance	Temperature [°C]
Control	83	0.304	0.125	304.72	57	3.423	1174.267	46.729	43.151	49.271	50.087	5.26	No	Horizonta	37.8	12.5
Control	83	0.348	0.156	330.53	63	3.331	1168.955	45.959	41.9	48.513	51.241	7.37	No	Horizonta	37.8	12.2
Control	91	0.38	0.175	336.74	67	3.321	1158.198	45.572	43.364	48.671	48.162	9.49	No	Horizonta	37.8	12.2
Control	80	0.449	0.273	341.77	73	3.553	1143.327	46.538	46.727	50.85	41.611	11.69	No	Horizonta	37.8	12.2
Control	81	0.471	0.325	351.24	74	3.589	1123.327	47.175	45.987	52.191	46.384	13.82	No	Horizonta	37.8	12.2
Control	64	0.446	0.232	357.63	73	3.546	1151.821	46.748	43.882	55.822	47.861	17.06	No	Horizonta	37.8	12.5
Control	76	0.424	0.184	346.02	74	3.587	1157.528	46.063	44.208	52.15	46.229	19.23	No	Horizonta	37.8	12.2
Control	77	0.425	0.163	326.11	69	3.532	1132.728	45.882	44.8	45.808	48.126	21.25	No	Horizonta	37.8	12.2
Control	77	0.492	0.137	326.07	70	3.51	1135.385	47.624	46.913	49.926	48.319	23.47	No	Horizonta	37.8	12.2
Control	84	0.462	0.164	341.86	74	3.728	1144.353	47.385	47.639	49.785	45.745	25.68	No	Horizonta	37.8	12.2
Control	90	0.265	0.254	357.88	67	3.906	1151.453	46.101	46.494	44.652	46.869	27.83	No	Horizonta	37.8	12.5
Control	88	0.256	0.275	9.12	66	3.935	1167.811	46.149	45.846	46.919	45.903	29.99	No	Horizonta	37.8	12.2
Control	83	0.37	0.232	10.57	71	3.837	1201.21	45.196	44.245	53.777	41.058	32.16	No	Horizonta	37.8	12.5
Control	91	0.38	0.216	10.22	70	3.712	1234.898	43.953	42.594	53.638	39.736	33.25	No	Horizonta	37.8	12.2
Control	91	0.452	0.222	8.03	76	3.709	1226.233	45.363	43.405	51.083	46.938	35.37	No	Horizonta	37.8	12.5
Sample	54	0.626	0.282	223.49	34	2.341	862.301	33.25	33.506	36.798	24.185	48.67	No	Horizonta	37.8	12.8
Sample	22	0.567	0.276	227.84	37	2.357	870.756	35.933	34.815	38.735	30.163	56.23	No	Horizonta	37.8	12.8
Sample	21	0.558	0.249	232.13	39	2.248	870.322	39.317	35.348	42.603	35.552	63.78	No	Horizonta	37.8	12.8
Sample	15	0.554	0.127	253.87	48	2.559	927.916	41.39	37.304	46.97	36.608	73.7	No	Horizonta	37.8	12.8
Sample	15	0.69	-0.174	277.42	55	2.796	983.195	40.34	38.273	48.801	37.68	88.88	No	Horizonta	37.8	12.8
Sample	19	0.755	-0.294	280.43	52	2.682	953.899	37.583	36.747	38.045	37.958	99.9	No	Horizonta	37.8	12.8
Sample	31	0.754	-0.269	271.23	48	2.757	975.095	37.351	34.347	37.873	38.808	107.58	No	Horizonta	37.8	12.8
Sample	10	0.825	-0.335	258.49	41	2.69	980.789	38.756	33.766	36.668	41.757	115.18	No	Horizonta	37.8	12.8
Sample	21	0.796	-0.24	250.41	34	2.635	992.018	42.157	36.264	41.618	43.904	129.29	No	Horizonta	37.8	12.8
Sample	9	0.597	-0.052	239.34	33	2.884	1025.26	41.31	36.596	42.514	41.441	141.22	No	Horizonta	37.1	12.8
Sample	33	0.422	0.194	216.66	32	2.874	1078.264	36.937	34.743	37.629	37.221	153.97	No	Horizonta	37.8	12.8
Sample	18	0.315	0.493	202.72	34	2.576	1081.131	32.299	31.672	32.222	34.979	162.43	No	Horizonta	37.8	12.8
Sample	15	0.024	0.617	78.21	50	2.581	1037.751	28.196	29.777	26.522	28.323	171.06	No	Horizonta	37.8	12.8
decrease in area [%]																
50%	61.63	-25.12	14.68		53.58	43.93	19.29	35.96	44.49	33.43	45.34	15:55:52		Horizonta	37.8	12.8

Figure 10.6 A representative sample of an ASCII file showing data recorded with Ecotox.

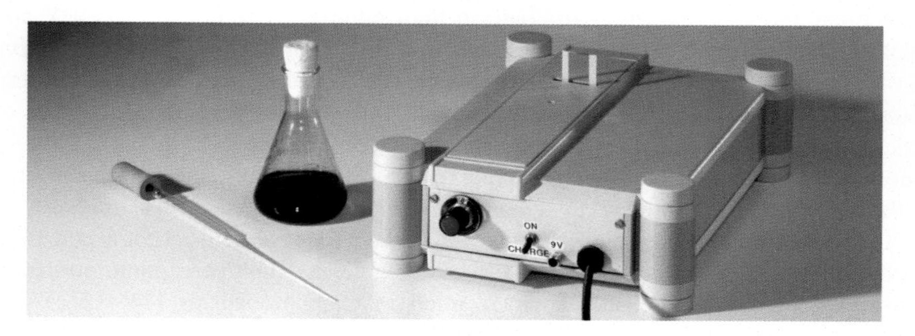

Figure 10.7 Manual version of the Ecotox instrument which houses a miniaturized microscope with 10× objective and a firewire camera. The cell suspension (with or without a toxin) is filled into observation chambers made from two slides separated by a spacer which is inserted into a slit in the top of the instrument.

10.7 Manual version of Ecotox

In order to determine toxicity of water samples outside the laboratory a manual version has been developed (Fig. 10.7). The instrument is portable and characterized by a small size. Like the automatic instrument, the manual Ecotox contains a miniaturized microscope with a 10× microscope objective and a firewire camera. Instead of the installed swimming chamber, the device is fitted with removable observation chambers, which consist of two object slides separated by a U-shaped spacer. The user fills the chamber with the control suspension or a suspension of cells with some toxin added and inserts it into a slit at the top. The instrument is connected to an external power supply, but in addition has internal batteries, which allow an independent operation for several hours. The camera is connected to the firewire input of a laptop computer, which in essence runs the same software as the automatic instrument without the motor controls.

These features allow the employment of the bioassay in the field because of the independence from an external power supply as well as the small size and low weight. The only disadvantage is that the user has to calculate and prepare the toxin dilutions and fill them manually into the observation chamber. The data produced by the device and the graphical output are identical to the ones in the automatic instrument.

10.8 Applications for Ecotox

The Ecotox instrument has been employed for numerous tasks. The control mode has been used to study gravitaxis in *Euglena*. The mechanism of this orientational behavior has been revealed by application of a number of inhibitors. As indicated above (cf. Section 10.2) young cells shortly after incubation into new medium show positive gravitaxis and swim downwards [33]. Application of heavy metals such as copper, mercury, cadmium, or lead causes the cells to change their direction of movement and swim upwards. In contrast, high salinity [37] and excessive UV and visible radiation cause a change in gravitactic orientation from negative to positive in older cells [38,39]. This reversal of movement direction is not caused by the photoreceptor of the cell which controls phototaxis [40] but rather by the generation of reactive oxygen species as shown by the fluorescent probe 2′.7′-dichlorodihydrofluorescein diacetate [38]. Also the application of reduced dithionite or flushing the cells with nitrogen suppressed the change of directional movement [41].

The gravireceptor is a membrane channel of the TRP (transient receptor potential) protein family [42] which are involved in osmoregulation, light perception, mechanosensitivity, thermosensing, and nociperception in many organisms [43–45]. Upon stimulation the protein channel allows the influx of Ca^{2+} along a previously established gradient by a Ca^{2+} ATPase which pumps Ca^{2+} out of the cell into the outer medium. This pump can be inhibited by gadolinium and the Ca^{2+} gradient can be broken down by insertion of passive Ca^{2+} gates such as A23187 (calcimycin) into the membrane, both of which abolish gravitactic orientation [46]. The influx of Ca^{2+} caused by the stimulation of the TRP channel subsequently activates one of the five calmodulins (CaM.2) found in *Euglena* as can be shown by experiments in Ecotox using RNA inhibition [47]. Likewise, inhibition of the calmodulin by fluphenazin, trifuluphenazin, or W7 abolishes gravitaxis [48]. The activated calmodulin in turn activates a specific adenylyl cyclase. The adenylyl cyclase produces 3',5'-cyclic adenosine monophosphate (cAMP) from adenosine triphosphate (ATP) necessary for the subsequent sensory transduction chain. Inhibition of the adenylyl cyclase using indomethacine decreases the gravitactic orientation; in contrast, foskulin activates the enzyme and augments the precision of orientation. The naturally produced cAMP can be replaced by externally applied 8-bromo-cAMP which cannot be metabolized by the cell and therefore enhances gravitactic orientation of *Euglena* [49]. The cAMP signal is quenched by a phosphodiesterase which can be impaired by caffeine, theophylline, or IBM-X, resulting in a higher precision of gravitactic orientation as studied by Ecotox [1,50]. The cAMP signal eventually activates a protein kinase A as indicated by the application of staurosporine [28,51]. *Euglena* possesses five isoforms of this enzyme, but only one (PK-4) is involved in graviperception as shown by RNA inhibition [52].

Ecotox is used to monitor toxicity of heavy metals [14,53] and industrial and household effluents [31,32,54,55]. Fig. 10.8 shows the effect of nickel on the motility and upward swimming of *Euglena*. It also proved to be useful in monitoring the efficiency of wastewater treatment plants comparing the toxicity of input and output streams [16] (cf. Chapter 18: Ecotoxicological monitoring of wastewater, this volume). The toxicity in very diverse natural ecosystems has been studied, such as in lakes in Egypt [16,25]. Excessive use of fertilizers and herbicides in agriculture pose serious threats for freshwater systems and drinking water as also shown by Ecotox [13,15,55].

10.9 New Ecotox

The original Ecotox instrument was based on dedicated hardware and custom-written software. New hardware and software developments have facilitated online computer-controlled image analysis and real time cell tracking. Therefore we

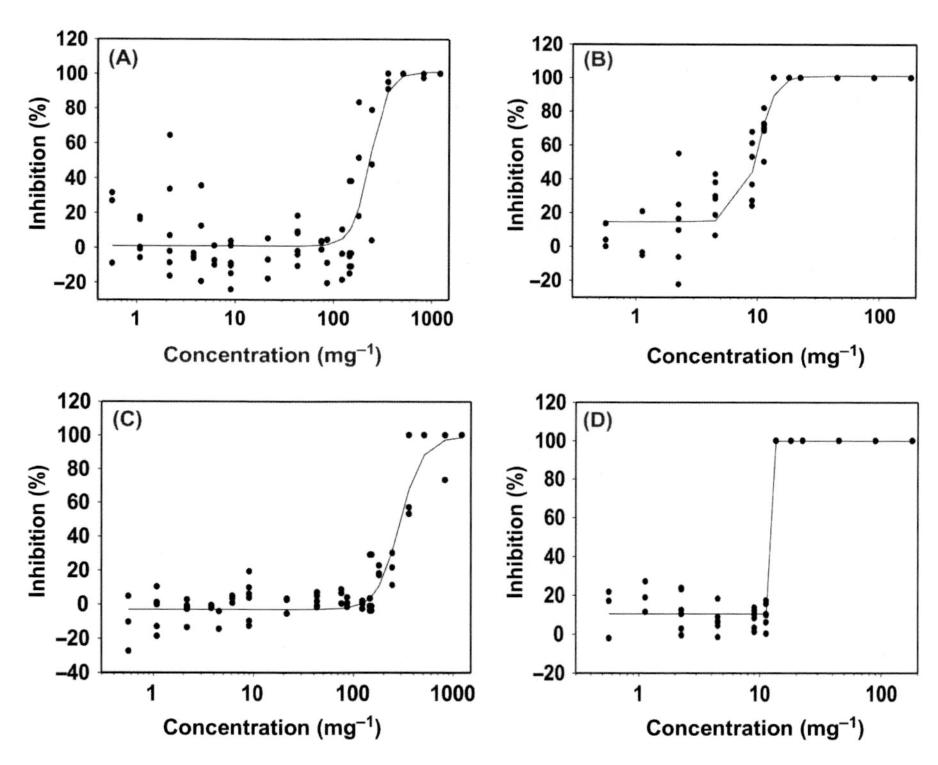

Figure 10.8 Inhibition by nickel of the motility (A,B) and upward swimming of *Euglena* (C,D) immediately after application of the heavy metal (A,C) and after 1 day of incubation (B,D). The EC_{50} values were 235.44 mg L^{-1} (A), 10 mg L^{-1} (B), 292.4 mg L^{-1} (C), and 12.7 mg L^{-1} (D).

decided to develop a new version of the instrument and develop novel software based on the freeware software package ImageJ [56]. Other considerations were ease of use, a large number of endpoints indicating toxic effects, low costs of the instrument and consumables as well as an enhanced number of output options for the results in numeric and graphical form [57].

As in the original system, typically the motile unicellular flagellate *E. gracilis* is being used [26,53]. But also other motile freshwater and marine flagellates can be employed [9,11]. The swimming chamber is oriented vertically so that the swimming flagellates show gravitaxis [28] which is an important endpoint of the bioassay since it responds very sensitively to toxic substances and other environmental stressors such as toxicants and pollutants [14,31,53], UV [58], excessive white light [39], increased salinity [37], and gamma rays and high-energy carbon rays [59].

10.9.1 Hardware

The hardware of the new Ecotox consists of a custom-made basic microscope with a standard microscope objective which is focused onto the swimming flagellates in

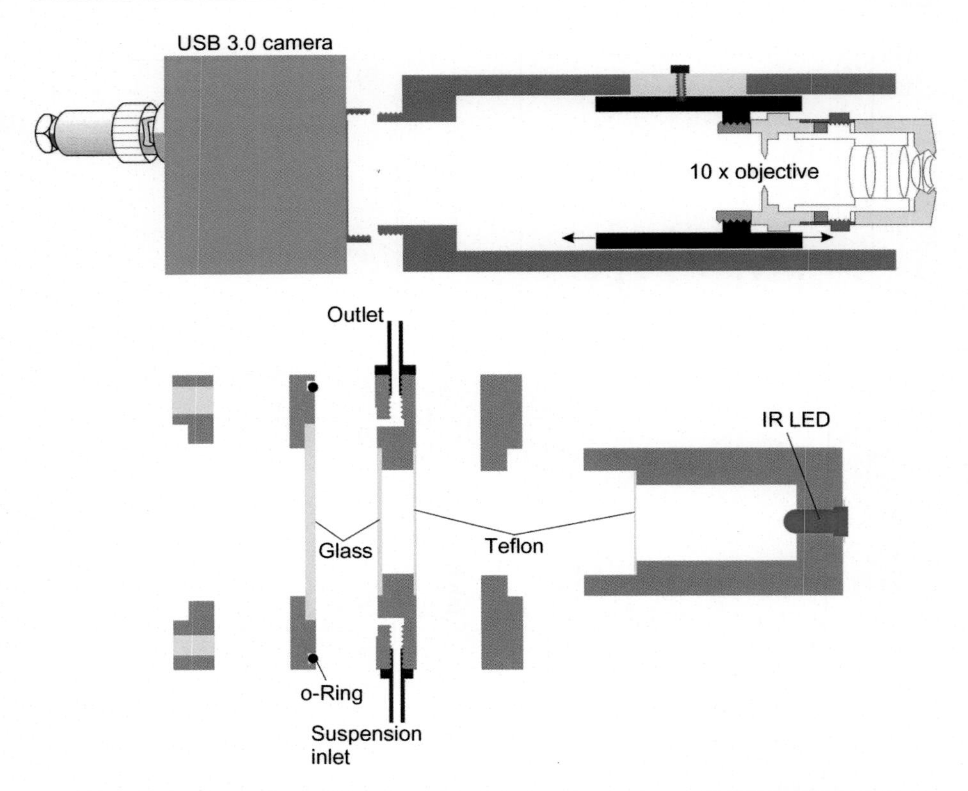

Figure 10.9 The optical instrument of the new Ecotox instrument with a Blackfly camera, 10 × microscope objective, stainless steel observation chamber, infrared monitoring LED, and two Teflon diffusors.

the observation chamber with 170-μm depth (Fig. 10.9). The actinic light beam is produced by an infrared LED in order to exclude visible radiation which might result in photoresponses of the organisms used in the bioassay such as phototaxis, photophobic reactions, or photokinesis [60]. It is evenly diffused by two Teflon disks to warrant uniform irradiation. A monochromatic USB camera (Point *Grey*, Blackfly BFLY-U3-13S2M-CS) with 1.3 mega pixel resolution (1288 × 964) is attached to the microscope. A 10× microscope objective projects the image of the moving cells onto the charged coupled device (CCD) target of the camera. Since the microscope is oriented horizontally and thus the observation chamber vertically, the organisms can orient themselves with the gravity vector of Earth and swim upward or downward by an active process called gravitaxis [47,61]. One important aspect of the system is that the moving cells are always visible in a window on the computer screen.

Two versions of the instrument are available. One is a manual instrument in which the organism suspension is mixed with the potentially polluted sample or toxic substance at a chosen concentration and filled into the swimming chamber.

After the suspension has settled and the cells have resumed normal swimming (visually controlled by the operator on screen), an analysis run is started. By changing the concentration of the toxicant and running subsequent analyses, dose-response curves can be obtained.

As in the original Ecotox the automatic system uses three stepper motor pumps to transport the flagellate suspension, the toxic sample, and the rinsing water. Before the liquids enter the observation chamber, they are mixed thoroughly in a mixing chamber. While in the original system the pumps were controlled by the PC, in the new system a dedicated Ardiuno microcomputer (Mega 2560) based on the Atmel AVR microcontroller ATmega 2560 is employed, which has many input/output (I/O) channels as well as several analog-to-digital (A/D) input and output pins [62,63]. The microprocessor sends its commands to a motor controller board (Quadstep, Sparkfun.com) which is capable of driving up to four stepper motors with 1.8 degrees resolution (NEMA 17 DamenCNC, Alphen aan den Rijn, The Netherlands), each of which are connected to an attached peristaltic pump (PP60). The motors are powered by a 3 A, 12 V switching power supply. Using stepper motors with small step sizes allows for precise dosing of the liquids to guarantee exact dilution of the toxins. When at rest the stepper motor pumps effectively block the connecting tubes so that no liquid can flow back or forth; therefore no magnetic valves are necessary.

The current action, such as which pump is active or which part of the program is currently being performed, is indicated by a 4×20 character LCD display. The hardware is integrated into a solid block housed in a box, which also holds the containers for the fluids and the waste. The temperature is measured by a sensor and displayed on the LCD screen.

10.9.2 Software for image analysis and hardware control

The image analysis and cell tracking software is based on the open source software Fiji, derived from ImageJ [56]. ImageJ (and Fiji) can process only recorded videos (.avi). As a workaround a plugin was developed to allow online video input into the imaging software instead of opening recorded videos or frames (Phase, Lübeck, Germany).

The dedicated Ecotox software is written in the ImageJ Macro Language, which allows easy modifications to adapt the analysis to different tasks.

To get started the user can select a number of parameters for the analysis, such as the subsequent number of frames collected in a stack through which the movement of organisms is tracked (cf. Section 5.5, this volume), the maximal number of organisms, and the maximal time of analysis after which the test is stopped (whichever comes first). In order to eliminate smaller or larger objects (bacteria, bubbles) minimal and maximal limits are chosen for the area of the analyzed organisms as well as the minimal and maximal velocity in order to avoid analyzing sedimenting organisms or those moved by a drift in the swimming chamber. In order to analyze using true physical units a factor is defined which converts pixels into real length (μm, mm etc.). This is used to calculate the size and velocity of the analyzed objects. Since the volume seen by the camera (length and width of the field of view and the depth of the swimming chamber) is known, another factor can be defined which calculates the cell density (cells per ml) from the number of detected objects (cf. Section 5.4, this volume).

When the analysis is initiated by clicking on a button in the taskbar the software logo is shown and then the camera control is started, which displays the life camera screen. The analysis is started with another button in the taskbar using the parameters defined above and stored in a settings file. The software uses background subtraction and contrast enhancement to improve the image quality, and then binarization to display white objects in front of a black background in a stack with a predetermined number of frames. In each frame the motile organisms are determined and a number of movement, area, and form parameters are analyzed such as percentage of motile organisms, swimming velocity (only of the swimming cells, maximally 256 per stack), as well as area, length, and perimeter. The form of the objects is described by the perimeter, circularity, aspect ratio, roundness, and solidity [57]. Meandering deviations from the direct path between beginning and end of a track can be calculated by comparing the velocity and the direct velocity.

After analysis of the current frame stack, a new one is captured from the live video stream and the analysis is repeated until the maximal chosen number of organisms is reached or the maximal analysis time has elapsed, whatever comes first. Then the user is prompted to input a description of the experiment and choose a file name for the results. The result file shows the date and time of the experiment, the temperature, and the cell density per ml, and the description of the experiment given by the user. All data from the individual frames are pooled and the mean velocity, direct velocity, the percentage of upward swimming cells (± 90 degrees around the vertical upward), the area, perimeter, circularity, length, aspect ratio, roundness, and solidity are recorded as mean values with standard deviation. In addition, the statistical value for the precision of orientation with respect to the gravity vector of Earth (r-value) and the mean direction of the whole population (θ) are given.

If the user wants to store the detailed analysis he or she can select this by checking a box in the dialog. In that case another file is stored with the user-defined name + area.txt. This includes the area, the x and y coordinates of all objects, and in addition the perimeter, circularity, Feret angle (the angular deviation of the long axis from the horizontal), the x and y coordinates of the Feret, the aspect ratio, roundness, and solidity of each individual object. Another file shows a histogram of the angular distribution of the movement directions as .png file (cf. Section 5.6, this volume). This file also gives the number of tracks, the percentage of motile organisms, the swimming velocity, the area and length as well as the r-value, theta, and the percentage of upward swimming organisms.

The results obtained with the Ecotox system can be used to calculate effect-concentration curves (Fig. 10.10). These are used to extract information on toxicity like no observed effect concentration (NOEC), lethal dose (LD), and EC_{50} values [31]. The system can be used for short-term as well as long-term (online) monitoring of pollution in surface or wastewater.

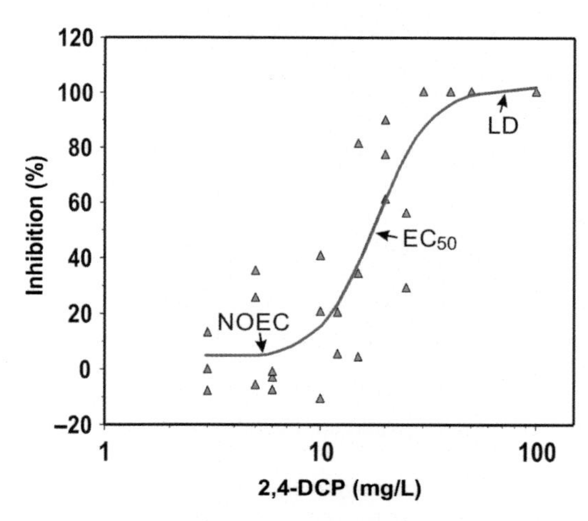

Figure 10.10 Effect-concentration curve for the inhibition of motility in *Euglena gracilis* by 2-4-dichlorophenol indicating the NOEC (no observed effect concentration). EC_{50} (concentration at which 50% inhibition occurs) and LD (lethal dose).
Source: Modified after Ahmed H. Biomonitoring of aquatic ecosystems [PhD thesis]. Erlangen-Nürnberg: Friedrich-Alexander-Universität; 2010.

10.10 Suitability and advantages of Ecotox

Low costs and short measurement times as well as high sensitivity are the main advantages of the new Ecotox instrument over other bioassays [64]. Also the operational costs are almost negligible and involve only keeping a *Euglena* culture. The image analysis and calculation of track parameter are performed online in real time so that the analysis time is as short as 6–10 min including control and sample [4,65]. The software was designed with a stress on user friendliness and does not require extensive training of personnel. Since all analyses operate fully automatically, user measurement errors and interpretation biases are avoided. The reliability and sensitivity of the Ecotox system is enhanced by using 14 movement and orientation values as well as size and form parameters as endpoints [64]. It was found that the sensitivity of the Ecotox system is higher than other bioassays using algae (growth), *Daphnia* (motility), fish (mortality), and bacteria (bioluminescence) [31]. The different form, motility, and orientation parameters show different sensitivities to various groups of toxins, which allows for the deduction of a first indication of the potential group of pollutants.

The bioassay can be used as an early warning system for pollution in surface, municipal, and industrial wastewaters as well as for monitoring efficiency of wastewater treatment plants. Since it is a low-cost development, the system can be used in developing countries where poor funding and lack of environmental experts renders efficient water quality monitoring difficult.

References

[1] Streb C, Richter P, Sakashita T, Häder D-P. The use of bioassays for studying toxicology in ecosystems. Curr Top Plant Biol 2002;3:131–42.
[2] Streb C, Richter P, Ntefidou M, Lebert M, Häder D-P. ECOTOX-biomonitoring based on real time movement analysis of unicellular organisms. J Gravit Physiol 2002;9:345–6.
[3] Tahedl H, Häder D-P. Vollautomatische Überprüfung der Wasserqualität durch Bewegungsanalyse eines Einzellers. GIT Labor-Fachz 1999;9:898–902.
[4] Tahedl H, Häder D-P. Fast examination of water quality using the automatic biotest ECOTOX based on the movement behavior of a freshwater flagellate. Water Res 1999;33:426–32.
[5] Tahedl H, Häder D-P. The use of image analysis in ecotoxicology. In: Häder D-P, editor. Image analysis: methods and applications. Boca Raton, FL: CRC Press; 2001. p. 447–58.
[6] Tahedl H, Häder D-P. Automated biomonitoring using real time movement analysis of *Euglena gracilis*. Ecotoxicol Environ Saf 2001;48(2):161–9.
[7] Häder D-P, Lebert M, Tahedl H, Richter P. The Erlanger flagellate test (EFT): photosynthetic flagellates in biological dosimeters. J Photochem Photobiol B 1997;40:23–8.
[8] Johnson LP. The taxonomy, phylogeny and evolution of the genus *Euglena*. In: Buetov DE, editor. The biology of *Euglena*. New York: Academic Press; 1968. p. 1–25.
[9] Breiter R, Streb C, Richter P, Häder D-P. Anpassung des Biotestsystems ECOTOX für den Brack- und Meerwasserbereich. In: Neeße T, editor. Jahrestagung 2002 der Fachgruppe Umwelttechnik und Ökotoxikologie der GDCh, Forschung und Entwicklung im Dienste des Umweltschutzes, vom 06–08 November 2002 in Braunschweig, Kurzreferate. Frankfurt: GDCh; 2002. p. 166.
[10] Willemann RL. Development of an application of the ECOTOX system in the Estuarine Zone of the Baía da Babitonga, SC, Brazil [Master Thesis]. Germany: Friedrich-Alexander University Erlangen-Nürnberg; 2002.
[11] Millán de Kuhn R, Streb C, Breiter R, Richter P, Neeße T, Häder D-P. Screening for unicellular algae as possible bioassay organisms for monitoring marine water samples. Water Res 2006;40:2695–703.
[12] Azizullah A, Richter P, Häder D-P. Comparative toxicity of the pesticides carbofuran and malathion to the freshwater flagellate *Euglena gracilis*. Ecotoxicology 2011;20 (6):1442–54.
[13] Azizullah A, Nasir A, Richter P, Lebert M, Häder D-P. Evaluation of the adverse effects of two commonly used fertilizers, DAP and urea, on motility and orientation of the green flagellate *Euglena gracilis*. Environ Exp Bot 2011;74:140–50.
[14] Ahmed H, Häder D-P. A fast algal bioassay for assessment of copper toxicity in water using *Euglena gracilis*. J Appl Phycol 2010;22(6):785–92.
[15] Pettersson M, Ekelund NG. Effects of the herbicides Roundup and Avans on *Euglena gracilis*. Arch Environ Contam Toxicol 2006;50(2):175–81.
[16] Ahmed H. Biomonitoring of aquatic ecosystems [PhD thesis]. Erlangen-Nürnberg: Friedrich-Alexander-Universität; 2010.
[17] Azizullah A, Richter P, Häder D-P. Toxicity assessment of a common laundry detergent using the freshwater flagellate *Euglena gracilis*. Chemosphere 2011;84 (10):1392–400.
[18] Leedale GF. Euglenida/Euglenophyta. Annu Rev Microbiol 1967;21:31–48.

[19] Kivic PA, Walne PL. An evaluation of a possible phylogenetic relationship between the Euglenophyta and Kinetoplastida. Orig Life Evol Biosph 1984;13:269–88.

[20] Wolken JJ. *Euglena*: an experimental organism for biochemical and biophysical studies. Berlin: Springer Science & Business Media; 2012.

[21] Sbrana F, Barsanti L, Passarelli V, Gualtieri P. Atomic force microscopy study on the pellicle of the alga *Euglena gracilis*. From cells to proteins: imaging nature across dimensions. Heidelberg, Berlin: Springer; 2005. p. 395–403.

[22] Leander BS, Farmer MA. Comparative morphology of the euglenid pellicle. I. patterns of strips and pores. J Eukaryot Microbiol 2000;47:469–79.

[23] Sommer JR, Blum JJ. Cell division in *Astasia longa*. Exp Cell Res 1965;39:504–27.

[24] Edmunds Jr. LN, Tamponnet C. Oscillator control of cell division cycles in *Euglena*: role of calcium in circadian timekeeping. In: O'Day DH, editor. Calcium as an intracellular messenger in eucaryotic microbes. Washington, DC: American Society for Microbiology; 1990. p. 97–123.

[25] Ekelund NGA, Aronsson KA. Assessing *Euglena gracilis* motility using the automatic biotest ECOTOX application to evaluate water toxicity (cadmium). Vatten 2004;60:77–83.

[26] Engel F, Pinto L, Del Ciampo L, Lorenzi L, Heyder C, Häder D, et al. Comparative toxicity of physiological and biochemical parameters in *Euglena gracilis* to short-term exposure to potassium sorbate. Ecotoxicology 2015;24(1):153–62.

[27] Häder D-P, Ntefidou M, Iseki M, Watanabe M. Phototaxis photoreceptor in *Euglena gracilis*. In: Wada M, Shimazaki K, Iino M, editors. Light sensing in plants. Tokyo, Berlin, Heidelberg, New York: Springer; 2005. p. 223–9.

[28] Häder D-P, Faddoul J, Lebert M, Richter P, Schuster M, Richter R, et al. Investigation of gravitaxis and phototaxis in *Euglena gracilis*. In: Sinha R, Sharma NK, Rai AK, editors. Advances in life sciences. New Delhi: IK International Publishing House; 2010. p. 117–31.

[29] Richter P, Ntefidou M, Streb C, Lebert M, Häder D-P. Physiological characterization of gravitaxis in *Euglena gracilis*. J Gravitat Physiol 2002;9:279–80.

[30] Azizullah A, Murad W, Adnan M, Ullah W, Häder D-P. Gravitactic orientation of *Euglena gracilis* - a sensitive endpoint for ecotoxicological assessment of water pollutants. Front Environ Sci 2013;1:4.

[31] Ahmed H, Häder D-P. Monitoring of waste water samples using the ECOTOX biosystem and the flagellate alga *Euglena gracilis*. J Water Air Soil Pollut 2011;216 (1-4):547–60.

[32] Azizullah A. Ecotoxicological assessment of anthropogenically produced common pollutants of aquatic environments [PhD thesis]. Erlangen, Germany: Friedrich-Alexander University; 2011.

[33] Stallwitz E, Häder D-P. Effects of heavy metals on motility and gravitactic orientation of the flagellate, *Euglena gracilis*. Eur J Protistol 1994;30:18–24.

[34] Häder D-P., Lebert, M. Graviorientation in flagellates. In: Proceedings 2nd China-Germany Workshop on Microgravity Sciences, September 1–3, 2002, Dunhuang, China; 2002; Beijing, China: National Microgravity Laboratory, Chinese Academy of Sciences.

[35] Mittag M. Circadian rhythms in microalgae. Int Rev Cytol 2001;206:213–47.

[36] Hagiwara SY, Bolige A, Zhang Y, Takahashi M, Yamagishi A, Goto K. Circadian gating of photoinduction of commitment to cell-cycle transitions in regulation to photoperiodic control of cell reproduction in *Euglena*. Photochem Photobiol 2002;76:105–15.

[37] Richter P, Börnig A, Streb C, Ntefidou M, Lebert M, Häder D-P. Effects of increased salinity on gravitaxis in *Euglena gracilis*. J Plant Physiol 2003;160:651–6.
[38] Ntefidou M, Richter P, Streb C, Lebert M, Häder D-P. High light exposure leads to a sign change in gravitaxis of the flagellate *Euglena gracilis*. J Gravitat Physiol 2002; 9(1):277–8.
[39] Richter PR, Ntefidou M, Streb C, Faddoul J, Lebert M, Häder D-P. High light exposure leads to a sign change of gravitaxis in the flagellate *Euglena gracilis*. Acta Protozool 2002;41:343–51.
[40] Ntefidou M, Iseki M, Watanabe M, Lebert M, Häder D-P. Photoactivated adenylyl cyclase controls phototaxis in the flagellate *Euglena gracilis*. Plant Physiol 2003;133 (4):1517–21.
[41] Richter P, Ntefidou M, Streb C, Lebert M, Häder D-P. The role of reactive oxygen species (ROS) in signaling of light stress. Recent Res Dev Biochem 2003;4:957–70.
[42] Häder D-P, Richter P, Schuster M, Daiker V, Lebert M. Molecular analysis of the graviperception signal transduction in the flagellate *Euglena gracilis*: involvement of a transient receptor potential-like channel and a calmodulin. Adv Space Res 2009;43 (8):1179–84.
[43] Bender FLP, Mederos Y, Schnitzler M, Li Y, Ji A, Weihe E, et al. The temperature-sensitive ion channel TRPV2 is endogenously expressed and functional in the primary sensory cell line F-11. Cell Physiol Biochem 2005;15:183–94.
[44] Barritt G, Rychkov G. TRPs as mechanosensitive channels. Nat Cell Biol 2005; 7(2):105–7.
[45] Maroto R, Raso A, Wood TG, Kurosky A, Martinac B, Hamill OP. TRPC1 forms the stretch-activated cation channel in vertebrate cells. Nat Cell Biol 2005;7:179–85.
[46] Lebert M, Häder D-P. How *Euglena* tells up from down. Nature 1996;379:590.
[47] Daiker V, Häder D-P, Lebert M. Molecular characterization of calmodulins involved in the signal transduction chain of gravitaxis in *Euglena*. Planta 2010;231(5):1229–36.
[48] Häder D-P, Richter P, Lebert M. Signal transduction in gravisensing of flagellates. Sig Transd 2006;6:422–31.
[49] Lebert M, Richter P, Häder D-P. Signal perception and transduction of gravitaxis in the flagellate *Euglena gracilis*. J Plant Physiol 1997;150:685–90.
[50] Streb C, Richter P, Ntefidou M, Lebert M, Häder D-P. Sensory transduction of gravitaxis in *Euglena gracilis*. J Plant Physiol 2002;159:855–62.
[51] Häder D.-P. Rock 'n' Roll - Wie Mikroorganismen die Schwerkraft spüren. In: Spektrum der Wissenschaft Extra: Schwerelos Europa forscht im Weltall 2010; p. 114–120.
[52] Daiker V, Häder D-P, RP R, Lebert M. The involvement of a protein kinase in phototaxis and gravitaxis of *Euglena gracilis*. Planta 2011;233:1055–62.
[53] Ahmed H, Häder D-P. Rapid ecotoxicological bioassay of nickel and cadmium using motility and photosynthetic parameters of *Euglena gracilis*. Environ Exp Bot 2010;69:68–75.
[54] Azizullah A, Jamil M, Richter P, Häder D-P. Fast bioassessment of wastewater and surface water quality using freshwater flagellate *Euglena gracilis*—a case study from Pakistan. J Appl Phycol 2014;26(1):421–31.
[55] Ahmed H, Häder D-P. Short-term bioassay of chlorophenol compounds using *Euglena gracilis*. SRX Ecol 2009;2010.
[56] Schindelin J, Arganda-Carreras I, Frise E, Kaynig V, Longair M, Pietzsch T, et al. Fiji: an open-source platform for biological-image analysis. Nat Methods 2012; 9(7):676–82.

[57] Häder D-P, Erzinger GS. Ecotox – Monitoring of pollution and toxic substances in aquatic ecosystems. J Ecol Environ Sci 2015;3(2):22–7.
[58] Streb C, Richter P, Häder D-P. ECOTOX—a biomonitoring system for UV-effects and toxic substances. In: Ghetti F, Checcucci G, Bornman JF, editors. Environmental UV Radiation: Impact on Ecosystems and Human Health and Predictive Models. IV. Earth and Environmental Sciences, Vol. 57. The Netherlands: Springer; 2006. p. 288.
[59] Sakashita T, Doi M, Yasuda H, Takeda H, Fuma S, Nakamura Y, et al. Comparative study of gamma-ray and high-ernergy carbon ion irradiation on negative gravitaxis in *Euglena gracilis* Z. J Plant Physiol 2002;159:1355–60.
[60] Lebert M, Häder D-P. Photoperception and phototaxis in flagellated algae. Res Adv Photochem Photobiol 2000;1:201–26.
[61] Richter PR, Streb C, Ntefidou M, Lebert M, Häder D-P. High light-induced sign change of gravitaxis in the flagellate *Euglena gracilis* is mediated by reactive oxygen species. Acta Protozool 2003;42:197–204.
[62] Thomas CD, Cameron A, Green RE, Bakkenes M, Beaumont LJ, Collingham YC, et al. Extinction risk from climate change. Nature 2004;427(6970):145–8.
[63] Nayyar A, Puri V. A review of Arduino board's, Lilypad's & Arduino shields. In: Computing for sustainable global development (INDIACom), 2016 3rd international conference. IEEE.ISO; 2016, p. 1485–92.
[64] Azizullah A, Richter P, Häder D-P. Effects of long-term exposure to industrial wastewater on photosynthetic performance of *Euglena gracilis* measured through chlorophyll fluorescence. Journal of Applied Phycology 2014;27:303–10.
[65] Azizullah A, Richter P, Ullah W, Ali I, Häder D-P. Ecotoxicity evaluation of a liquid detergent using the automatic biotest ECOTOX. Ecotoxicology. 2013;22(6):1043–52.

Daphniatox

11

Donat-P Häder[1] and Gilmar S. Erzinger[2]
[1]Friedrich-Alexander University, Erlangen-Nürnberg Germany,
[2]University of Joinville Region—UNIVILLE, Joinville, SC, Brazil

11.1 Introduction

As detailed in the introductory chapter of this volume, chemical analysis of potential toxicants and pollutants in aqueous habitats is neither cost efficient nor exhaustive [1]. Due to the thousands of chemicals of natural or anthropogenic origin it is not feasible to monitor aquatic ecosystems such as drinking water reservoirs, natural ecosystems, or industrial and household wastewaters by chemical analysis [2]. Not even the major groups of chemicals can be detected quantitatively.

An alternative to chemical monitoring is the employment of bioassays [3]. While these systems are not capable of identifying the chemical nature of a potential pollutant, the responses of the biomaterial being used do indicate the potential hazard for humans or ecosystem health. The organisms used for this purpose need to be sensitive to a wide variety of toxicants in the environment. In addition, they should be easy to obtain or cultivate and the running costs should be minimal. Furthermore, the analysis should be simple and not require trained personnel, and should be preferably fully automatic. The time required for analysis should be short in order to allow instant warning if a potential hazard occurs [4].

The water flea *Daphnia* is a genus of the order Cladocera in the subphylum Crustacea with numerous species (more than 200) such as *Daphnia pulex, Daphnia moina,* or *Daphnia magna,* which dwell in a wide variety of freshwater habitats such as swamps, lakes, ponds, streams, and rivers [5]. Some species are adapted to brackish or saltwater [6]. Most species have a length of 0.2 to 5 mm and are covered by a carapace with a ventral gap for the five or six pairs of legs [7]. The head is fused with the carapace and caries two large compound eyes and extended antennae (Fig. 11.1). At the rear end the organisms carry a pair of abdominal setae. The organisms are filter feeders which use their second and third pair of legs for filtering plankton, bacteria, and organic detritus out of the stream of water which is created by the other legs [8]. The exoskeleton is semitranslucent and allows viewing of the beating heart as well as the eggs in an abdominal egg pouch in females. In addition to the sexual reproduction, there is a parthenogenetic, asexual reproduction. During their life cycle females produce a brood of from 2 to over 100 eggs each time they molt [9]. The normal lifespan of the organisms is about five to six months, but in cold, oligotrophic lakes and the absence of predating fish it can extent to 13–14 months [10].

The eggs hatch after about 24 h under suitable conditions, but remain in the egg pouch for about three days before they are released into the water. During the

Bioassays. DOI: http://dx.doi.org/10.1016/B978-0-12-811861-0.00011-5

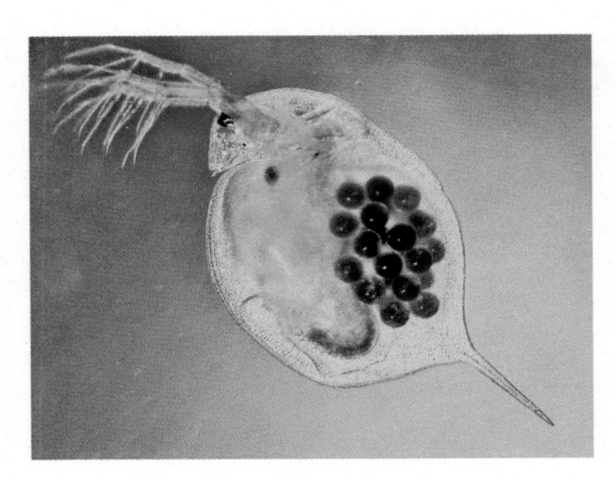

Figure 11.1 Light microscopic image of *Daphnia magna*.

following 5–10 days they pass through 4–6 instar stages before they become reproductive. At the end of the growth season the females asexually produce resistant resting eggs which normally give rise to female organisms, but can also develop into males when the environmental conditions are unfavorable [11]. These resting eggs can survive extreme cold, drought or lack of nutrients protected by an extra shell layer and hatch when the environmental conditions improve.

Because of the ease of cultivation *Daphnia* are used as life food source for fish and amphibia [12]. *Daphnia's* thin membrane allows drugs such as adrenaline or capsaicin to penetrate into the body, and their effects e.g., on the heartbeat pattern can be studied because of the semitranslucent carapace. Because of the easy uptake of substances through the carapace, *Daphnia* is a sensitive bioassay organism for aquatic pollutants [13]. It has been shown to have a high sensitivity to many toxic chemicals [14]. E.g., the organisms have been used to indicate acute and chronic stress by copper [15]. In addition, *Daphnia* responds to many organic and inorganic pollutants [16] including carbamate, pyrethroids, and organophosphorous pesticides [17]. Several *Daphnia* species, *Chaoborus crystallinus*, and *Mesocyclops leuckarti*, have been employed to detect the pesticide triphenyltin hydroxide as well as novel antibiofouling chemicals [18]. Detergent pollution was determined by quantifying the motility of *D. magna* [19,20]. Mortality of *Daphnia* is another defined endpoint used in bioassays for water quality; e.g., to monitor the pollution by the fungicide metalaxyl [21]. The heartbeat pattern was monitored in *Daphnia* to reveal exposure of the organisms to toxic pollutants [22].

D. magna as well as other copepods respond to visible light by positive phototaxis (movement toward the light source) [23]. This response toward the direction of light is involved in the control of diurnal vertical migrations which the organisms perform in the water column. In contrast, *Daphnia* was found to show negative phototaxis (downward movement in the water column) when exposed to UV radiation (260–380 nm) [24].

When exposed to heavy metals the precision of positive phototaxis deteriorates [25] so that this response can be used as another endpoint in a bioassay. Phototaxis has been found to be affected in the presence of Cu^{2+} and pentachlorophenol (PCP) [26]. Also sublethal concentrations of silver and NaCl have been found to affect phototaxis in water fleas [27]. For this reason monitoring of this endpoint has been implicated in an instrument developed by one of the author of this chapter (D.-P. H.). This instrument has been employed, e.g., in water quality monitoring in Canada [28].

Another important application as bioassay is the sensitivity toward cyanobacterial toxins such as microcystin [29]. However, different species show markedly different sensitivities toward purified toxins such as microcystin from *Microcystis aeruginosa* and nodularin from *Nodularia spumigena*. *Diaptomus birgei* was found to be the most sensitive organism with an LC_{50} of 0.45 to 1 $\mu g\ mL^{-1}$ after 48 h of exposure while *Daphnia pulicaria* was the least sensitive (LC_{50} 21.4 $\mu g\ mL^{-1}$) [30]. These differences are due to the physiological sensitivity and feeding behavior. *D. pulicaria* was found to have a low sensitivity to purified toxin as well as an inhibition of feeding when exposed to toxic organism which enables the crustaceans to coexist with toxic cyanobacteria. *Daphnia* is widely used as a standard indicator organism in bioassays monitoring exotoxicology [31]. However the experimental conditions for acute and chronic tests need to be standardized to warrant repeatability and comparability between laboratories. In addition, genotypes as well as culture condition before and during the tests need to be standardized. For these reasons the biotest using *Daphnia* has been defined and certified by several national and international agencies such as the German standard methods (DIN 38412 L-30) and the OECD test 202 to determine freshwater, wastewater, and sludge toxicity [32].

11.2 Culture conditions

Daphnia can be obtained in the local pet store. However, the quality and sensitivity of the organisms may vary between batches. Therefore it is preferable to maintain a culture for analysis in the bioassay. Culture conditions are standardized by ISO-6342 [33]. E.g., 50 *D. magna* are kept in 5-L glass beakers with 2 L M4 culture medium at room temperature [34]. The culture medium is renewed and the newly hatched organisms removed every three or four days. After 4 weeks the breeding *Daphnia* are replaced by neonatal organisms. The organisms can be fed using the unicellular green alga *Raphidocelis subcapitata* (previously *Selenastrum capricornutum*) according to ISO-8692 [35]. This can be automated using a timer-controlled pump which delivers a specified amount of algal suspension at a likewise defined cell concentration.

11.3 Bioassays using *Daphnia*

The Daphniatoximeter is a commercial instrument developed by the company bbe Moldaenke GmbH, Schwentinental, Germany, which uses a static test with *Daphnia* according to DIN 388412 (Fig. 11.2A). This instrument can be used to

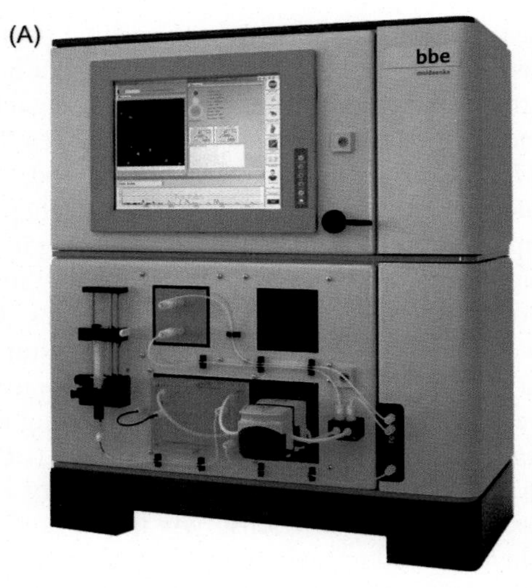

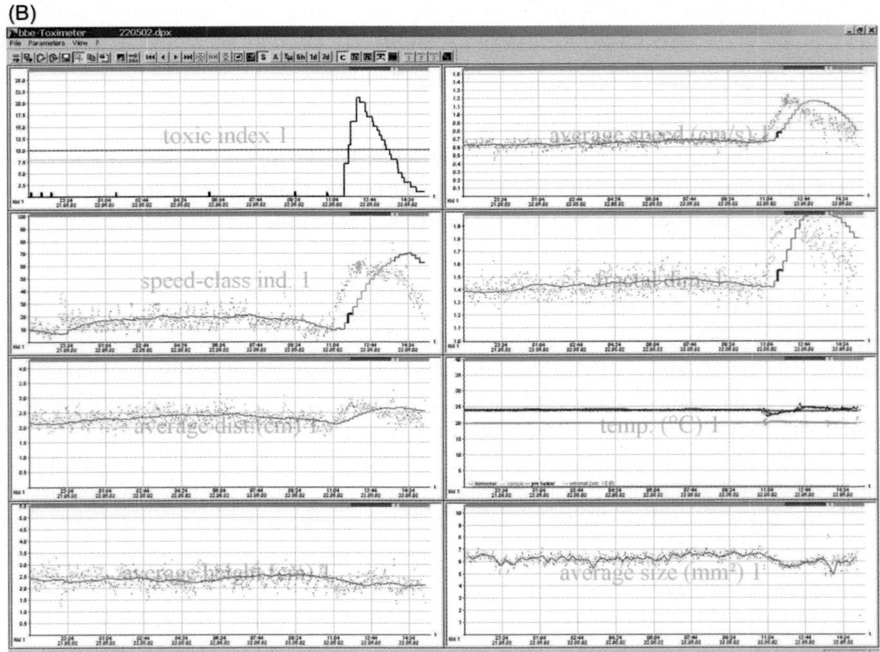

Figure 11.2 Daphniatoximeter (A) and the result screen (B).
Source: Photo courtesy bbe Moldaenke GmbH, Schwentinental, Germany.

detect insecticides, neurotoxins, and heavy metals. In this instrument a water stream is pumped continuously through a vertical chamber holding about ten *D. magna*. The movement of the organisms is recorded by a camera. In the control the organisms show a continuous movement and under the effect of a toxic substance they either display hypo- or hyperactivity. The software analyzing the movement patterns elicits an alarm when it detects a change in the monitored parameters which include mean swimming velocity, velocity distribution, and percentage of motile organisms (Fig. 11.2B). Also the orientation with respect to gravity is an important endpoint in the analysis of the motile behavior of *Daphnia*. However, this instrument does not monitor the orientation with respect to light (see below). From these parameters the software generates a toxindex indicating the degree of toxicity and potential hazard. This bioassay system is being used by the city of Hamburg, Germany, which monitors the water quality. Another application is found in Jochenstein, Bavaria, where the instrument serves as an early warning system for the water quality in the river Danube.

Another instrument was developed for laboratory conditions based on the postexposure feeding depression in *Daphnia* as a sensitive and robust endpoint [36]. This instrument was used in situ in the field where the organisms were exposed to four contaminated and control sites. About 90% of the *Daphnia* could be recovered alive from the test chambers after exposure and the feeding rates were determined. Pollution at the deployment site resulted in a significant depression of the feeding rate measured after exposure while other endpoints monitored in the benthic macroinvertebrate community did not show any effect. This indicates the feeding depression is a reliable and sensitive endpoint for monitoring pollution in aquatic ecosystems.

Barata et al. [37] also used the feeding depression as sensitive endpoint to monitor aquatic pollution and compared the sensitivity with that of the standardized bioluminescent bacteria inhibition bioassay [38], algae growth test (DOC 89/88/XI, Directive 79/831), and fish bioassays [39]. The authors determined that the *D. magna* feeding bioassay is a cost-effective and ecological relevant sublethal toxicity test for environmental risk assessment of toxic effluents.

In another bioassay the toxicity of metals, including Na, Ca, Mg, K, Sr, Ba, Fe, Mn, As, Sn, Cr, Al, Zn, Au, Ni, Pb, Cu, Pt, Co, Hg, and Cd on neonatal *D. magna* was assayed measuring the 50% lethal concentration after 48-h and 3-week exposure [40]. At concentrations which did not kill the organisms the metals inhibited reproduction and growth. Total protein and glutamic oxalacetic transaminase activity were affected to a different extent depending on the metal.

Fresh and weathered crude oils and chemical dispersants are common pollutants in aquatic ecosystems. This was measured with a new bioassay using *D. magna* [40]. The results indicate that dispersions of crude oil vesicles are more toxic than the water soluble fractions or dispersants. Since *Daphnia* feeds on particles depending on size, the toxicity was also determined by the size of the oil particles.

11.4 Daphniatox

A novel bioassay instrument has been developed using *D. magna* or other Cladocerans [41]. The aim was to create an instrument which allows real time

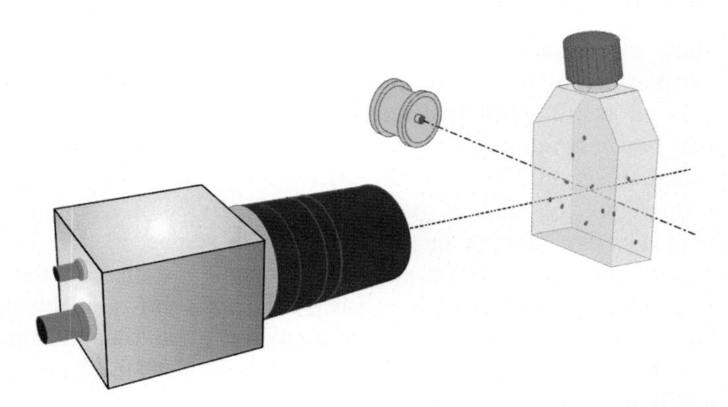

Figure 11.3 Optical setup of the Daphniatox instrument with a Blackfly USB 3.0 camera equipped with a macro zoom lens and diffuse background monitoring infrared radiation. Actinic blue light is provided by an LED aiming perpendicularly to the optical axis.

analysis of many movement parameters using state-of-the-art hardware and software technology with minimal costs in a portable housing.

11.4.1 Hardware

The underlying construction principle is based on the online monitoring of motile *Daphnia* swimming in a commercial cell culture flask (50 mL) which is irradiated from behind with a diffuse light beam so that the camera records dark organisms in front of a white background (Fig. 11.3). In order to avoid phototactic responses of the organisms to this monitoring light, infrared radiation is used which is not sensed by the organisms [42,43].

The movement tracks of the swimming organisms are being recorded by a monochromatic USB 3.0 camera (Point Grey, Blackfly BFLY-U3-13S2M-CS) with 1.3 mega pixel resolution (1288 × 964). Because of the size of the Cladocerans a macro zoom objective (Computar 2.8–12 mm, 1:1.3 IR 1/3″) is used and aimed at the cell culture flask (Aldrich), which contains about 100 *Daphnia*. When observed in a horizontally orientated flask nonmotile or dead organisms collect at the bottom but are in the view of the camera. By this means the percentage of motile organisms can be determined correctly from the movement vectors analyzed by the software (see below).

In order to induce phototactic orientation the cell culture flask is placed horizontally on a light table using infrared radiation with the camera looking vertically down. A blue LED light aims horizontally at the observation flask perpendicular to the vertical optical axis of the camera. The camera is oriented in such a way that the actinic light beam impinges from the top (0 degrees) in the image.

In addition to light, the diurnal vertical migrations in *Daphnia* are controlled by a circadian rhythm and by gravity [44,45]. Gravitaxis allows the organisms to orient within the water column even in the absence of light [46]. When monitoring the movement vectors the organisms show upward movement alternating with downward swimming. In order to evaluate the precision of gravitactic orientation the optical axis of the camera is horizontal aiming at the observation flask being in a vertical position.

11.4.2 Software for image analysis in the Daphniatox bioassay instrument

Like the Ecotox bioassay (cf. Chapter 10: Ecotox, this volume) the Daphniatox software was developed on the basis of the open source software Fiji, which is a further development from the open source ImageJ [47]. One problem of this software package is that it accepts only recorded videos in the .avi format and does not allow direct input of an online video stream from a camera. However, a software plugin has been developed (Phase, Lübeck, Germany) to circumvent this shortcoming so that the Daphniatox software accommodates real time video camera input.

Based on ImageJ the Daphniatox software has been developed in the ImageJ macro language. The program follows the software approach, which extracts important movement, orientation, as well as size and form parameters from a stack of video frames taken from the online video stream [48].

After loading the software package ImageJ into memory the user can start the camera control by an additional button in the tool bar to display the live camera image. It is one of the advantages of the software to permanently view the moving *Daphnia* and to control the validity of the parameters extracted from the analyzed video tracks in order to avoid including sedimenting organisms or other artefacts in the analysis (Fig. 11.4). A second important aspect is the use of correct physical

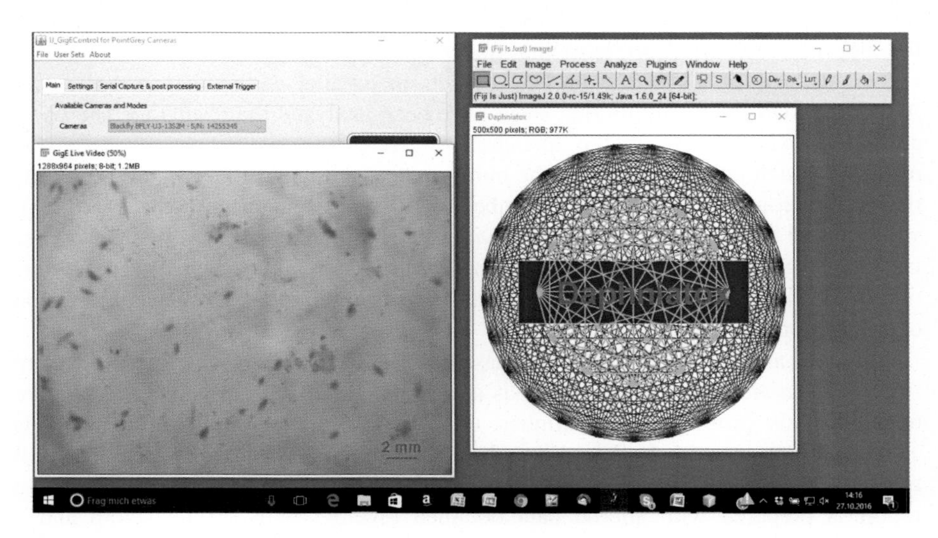

Figure 11.4 Screenshot of the Daphniatox software showing the ImageJ task bar and the live video image of the swimming organisms.

units for the analyzed parameters. A digital video image is organized in rows and columns of pixels [49]. These need to be translated into physical units by a calibration. For this purpose a microscopic ruler is viewed by the camera and the correlation between the number of pixels and real distances determined (cf. Chapter 5: Image analysis for bioassays – the basics, this volume). This is initiated by pressing another button, which allows modification of the calibration parameters. This part of the software is also used to preselect the upper and lower size limits for valid objects (*Daphnia*) to exclude e.g., food organisms to be tracked. Also a range of velocities is predefined in order to exclude drifting or sedimenting organisms. The timeframe for a complete analysis is defined by the maximal number of tracked organisms and a maximal allotted time. The analysis is terminated when the maximal number is exceeded or the maximal time reached—whichever comes first.

After defining the settings, the actual motion analysis is started by another button in the task bar. This algorithm starts with taking a predefined number of snapshots from the input video stream. One can either take every subsequent frame or skip frames to accommodate for a lower velocity of the organisms. These frames are stored in a stack together with the time interval between the frames which is later on used to determine the swimming velocity.

Even though the optical conditions have been optimized for the video recording, images can be enhanced by background subtraction [50] and contrast enhancement [51], resulting in an image of dark organisms on a bright background. In a subsequent step the image is binarized so that only two gray levels derive: 255 for the objects and 0 for the background; during this process the image is inverted showing white organisms in front of a dark background [52].

After this image improvement by preprocessing the video images in the stack, the algorithm for the analysis of the movement vectors commences. The software allows analysis of up to 256 moving objects in parallel within the current stack. After the video frames in the first stack have been analyzed the results are shown in a separate window. A circular histogram indicates the percentage of organisms moving in each respective sector, the number of which had been defined in the settings definition. Furthermore, the number of tracks recorded, the mean swimming velocity, the percentage of motile *Daphnia*, the mean area and length of the organisms, the mean angular deviation from the vertical as well as—in the case of a vertical movement—the percentage of upward swimming organisms and the precision of orientation are displayed.

After the analysis of the first stack has been finished a new set of frames is taken from the live video stream, the analysis is repeated, and the intermediate results are updated. This procedure is continued until the predefined number of tracks is exceeded or the maximal allotted time has elapsed, whichever comes first. Then all results are combined, mean values and standard deviations are calculated, and a report is prepared with all the data obtained together with a time stamp and a description of the experiment entered by the user. This file is stored and—if the user decides to—so are the raw data such as areas of individual organisms as well as the track coordinates and the angular distribution. Another file contains the data on the area and form parameters of each analyzed object.

The movement parameters extracted from the tracks include the angular deviation from a predefined 0 degrees direction such as the direction of the incident light or the upward direction in case of gravitactic orientation (cf. Chapter 5: Image analysis for bioassays – the basics, this volume) [53]. The precision of gravitactic or phototactic orientation of the whole population is determined by calculating the *r*-value, which is a statistical measure running between 0 (movement in random directions) and 1 (all organisms move in the same direction) from the angular deviation of all tracks [54]. Since the *r*-value does not indicate the direction in which the organisms move, a mean angle θ is calculated from the whole analyzed population [54]. In addition, in the case of gravitactic orientation in a vertical swimming chamber, the percentage of upward swimming cells ($\pm$ 90 degrees around the vertical upward) is calculated. Likewise, in the case of phototactic orientation the percentage of organisms swimming in the two quadrants toward the light ($\pm$ 90 degrees) is quantified.

Daphnia does not necessarily swim along straight paths; therefore two different swimming velocities are calculated. The direct velocity is determined from the distance in a straight line between the start and end of each track and the time interval recorded for this track. In addition, the total swimming velocity is calculated by using the real length of the (meandering) path and the same time interval. The swimming directedness is calculated as the ratio between the direct path length and the actually covered path length. The endpoint motility is calculated as the percentage of motile organisms from the whole population, which falls into the bracket of the predefined velocity range.

Several size and form parameters are determined for which mean values and standard errors are calculated. The area of each object—which fulfils the size definition—is determined from the number of pixels it covers in the video image and the calibration factor (cf. above). In addition, the perimeter (length of the outer boundary, the circularity, roundness, solidity, and the aspect ratio of the major and minor axes of an ellipse which fits the objects are determined (cf. Chapter 5: Image analysis for bioassays – the basics, this volume) [48]. The length of the organisms cannot be calculated from the minimal and maximal x or y coordinates, when the object is oriented at an oblique angle. Rather this is determined using the technique introduced by Feret [47]. The distance between the minimal and maximal x coordinates is determined, then the image is turned stepwise by a small angle in a clockwise direction and the distance is reassessed. This process is repeated until a maximum is found for the interval. This degree of turning (Feret angle) also indicates the orientation of the longest axis of an object as an angular deviation from the horizontal. Thus, this angle runs between 0 and 180 degrees. Using the settings information, all size and velocity parameters are calculated in real physical units (mm, mm^2, mm s^{-1}).

Table 11.1 Results from a control analysis using *Daphnia magna*

3.11.2015 13:22:58
Daphnia magna in a horizontal cuvette, control, no toxins added
84% motile
Mean velocity = 0.14 ± 0.07 mm s^{-1}
Mean direct velocity = 0.10 ± 0.05 mm s^{-1}
Mean directedness = 0.76 ± 0.24
47.00% organisms swam upward
$\theta = 177.11°$
r-value = 0.21
Mean area = 0.15 ± 0.06 mm^2
Mean perimeter = 1.94 ± 0.58 mm
Mean circularity = 0.51 ± 0.17
Mean organism length = 0.71 ± 0.22 mm
Mean aspect ratio = 2.28 ± 0.94
Mean roundness = 0.48 ± 0.17
Mean solidity = 0.79 ± 0.13

11.5 Application of the Daphniatox instrument

Table 11.1 shows a typical result file from a control experiment with *D. magna* in a horizontal observation chamber. Due to the lack of gravitational or phototactic stimuli the organisms swam in random directions as indicated by the low *r*-value. As expected, about 50% swam in the two quadrants adjacent to the 0 degrees direction; the mean direction is meaningless in this case. The swimming velocity and directedness as well as area and form factors are typical for *Daphnia* neonates. Table 11.2 shows an extract of a typical example of the area results. In a horizontal observation chamber the *Daphnia* swim in random directions (Fig. 11.5A). In a vertical chamber they preferentially swim upward or downward and less frequently in horizontal directions (Fig. 11.5B).

Results determined with the Daphniatox instrument can serve as a basis to determine effect-concentration curves from which important toxicity parameters such as no observed effect concentration (NOEC) (concentration at which no effect is observed for the studied end point), LD (lethal dose), and EC_{50} values (concentration at which 50% inhibition of the observed response is found) [55]. Both short-term and long-term responses of the organism can be monitored with this instrument in a wide variety of aquatic ecosystems including groundwater, surface, or wastewater, as well as natural ecosystems.

Potassium dichromate is a pollutant in industrial wastewaters which are often discharged into natural ecosystems and groundwater [56,57]. This chemical is used in the tanning and galvanoplastic industry and to produce chrome-sulfuric acid. In the laboratory it also serves to stain neurons and nerves [58]. For its importance as environmental pollutant it is marked as a reference to determine the sensitivity of organisms used in bioassays and to compare results among laboratories. The MAC

Table 11.2 Excerpt from a typical result file for the area and form factor parameters from a control experiment using *Daphnia magna*. The area of each object is recorded as a number of pixels before using the calibration factor. The x and y coordinates indicate the position of the object in the screen. Perimeter, circularity, roundness, aspect ratio, and solidity are form parameters. In addition, the Feret parameters are used to calculate the true length of each organism and its orientation with respect to the horizontal axis

Area	XM	YM	Perim.	Circ.	Feret	FeretX	FeretY	Feret Angle	MinFeret	AR	Round	Solidity
340	14.49	32.39	97.3	0.45	25.55	11	20	120.58	21.92	1.13	0.88	0.84
511	244.58	46.92	171.18	0.22	42.45	220	45	164.98	25.32	1.77	0.56	0.69
241	690.35	47.18	109.78	0.25	23.43	680	41	129.81	20.97	1.27	0.79	0.65
253	181.3	51.68	105.1	0.29	26.91	173	41	131.99	16.03	1.78	0.56	0.74
858	302.19	71.46	231.61	0.2	60.01	270	67	156.43	29.35	2.31	0.43	0.62
206	977.83	90.56	97.2	0.27	24.21	969	82	128.29	18.68	1.5	0.67	0.6
1180	106.61	102.34	290.76	0.18	82.01	68	114	15.56	24.35	3.84	0.26	0.8
462	156.46	169.92	138.31	0.3	35.11	142	178	19.98	22.64	1.67	0.6	0.79
1176	231.94	206.73	271.69	0.2	71.51	221	174	110.46	27.73	3.27	0.31	0.74
306	518.7	192.85	108.61	0.33	32.45	505	185	146.31	16.81	2.05	0.49	0.8
654	25.5	227.95	209.22	0.19	52.39	18	205	103.24	24.57	2.82	0.35	0.68
456	443.67	219.97	111.78	0.46	38.29	429	232	40.76	19.69	2.15	0.47	0.86
649	133.83	239.08	137.24	0.43	39.36	116	231	152.78	26.25	1.58	0.63	0.85
853	383.19	259	190.49	0.3	57.94	370	232	111.25	24.94	2.37	0.42	0.82
247	12.11	277.2	114.17	0.24	26.48	9	291	79.11	15.71	1.82	0.55	0.75
402	305.22	288.15	128.99	0.3	35.17	288	291	14.83	18.65	1.81	0.55	0.79
395	37.4	296.33	117.68	0.36	37.22	20	299	6.17	19.35	1.88	0.53	0.8

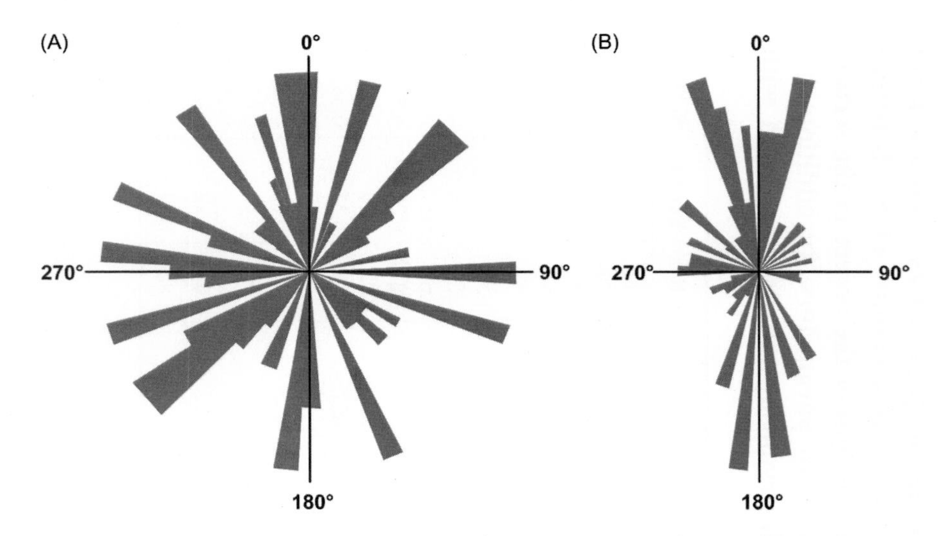

Figure 11.5 Angular histogram of tracks of *Daphnia magna* in a horizontally oriented swimming chamber showing a random distribution (A) and in a vertical chamber showing both positive and negative gravitaxis (B) binned in 60 sectors.

value in air is 5 μg m^{-3} [59]. Motility and mortality of *Daphnia* are common endpoints in tests after specified exposure times of e.g., 24 or 48 h [60]. However, chronic tests using survival and reproduction as endpoints require longer exposure times [61,62].

11.5.1 Typical results obtained with Daphniatox

Short-term tests were carried out according to the standard NBR 12713 protocol using neonates of *D. magna*, 2–26 h old [63,64]. The organisms were exposed to different concentrations of K dichromate in 50-mL cell culture flasks, which were also used for videomonitoring of the movement for up to 48 h [65] (Fig. 11.5). Because of the horizontal orientation of the swimming chamber the organisms did not show gravitactic orientation and moved in random directions, so that the precision of orientation (*r*-value) was very low (0.02). Of the 101 analyzed organisms 71% were motile, swimming at a mean velocity of 0.11 mm s^{-1}. The mean area was 0.19 mm^2 and the mean length 0.82 mm. 47.5% of the organisms swam in the two quadrants adjacent to 0 degrees direction. Because of the random orientation, the mean angular deviation θ (236.66 degrees) is meaningless.

Immediately after incubation in K dichromate the *Daphnia* remained motile, but after 4 h an increasing number of organisms stopped moving at 1 and 2 mg L^{-1} and after 24 h there were no motile organisms left at concentrations $\geq$ 1 mg L^{-1} (Fig. 11.6). At lower concentrations the percentage of motile organisms remained constant at about 85%. And the swimming velocity was almost constant at

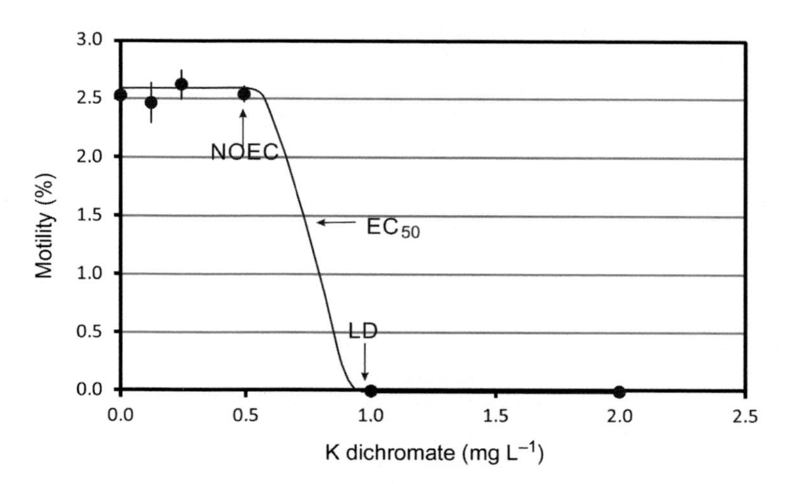

Figure 11.6 Effect-concentration curve for the inhibition of motility (percentage of motile organisms, mean values ± SD, $n = 6$) in *Daphnia magna* by K dichromate indicating the NOEC (= 0.56 mg L^{-1}, no observed effect concentration). EC_{50} (=0.78 mg L^{-1}, concentration at which 50% inhibition occurs) and LD (=1 mg L^{-1}, lethal dose).
Source: Redrawn from Häder D-P, Erzinger GS. Daphniatox—Online monitoring of aquatic pollution and toxic substances. Chemosphere 2017;167:228–235.

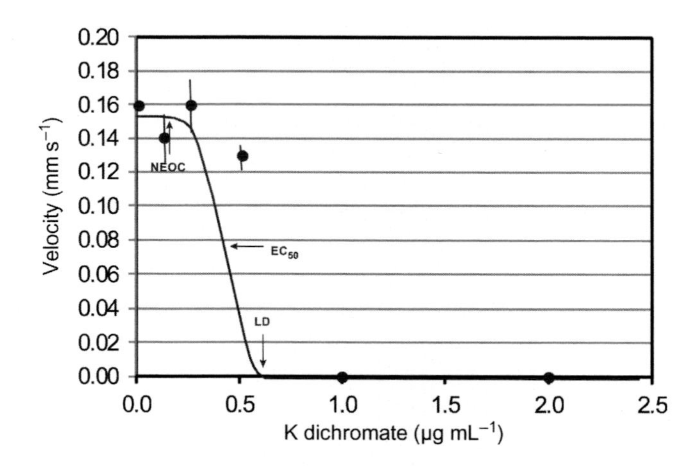

Figure 11.7 Effect-concentration curve of the mean swimming velocity (mean values ± SD, $n = 6$) of *Daphnia magna* after incubation in potassium dichromate for 24 h indicating the NOEC (=0.25 mg L^{-1}, no observed effect concentration). EC_{50} (=0.68 mg L^{-1}, concentration at which 50% inhibition occurs) and LD (=1 mg L^{-1}, lethal dose).

concentrations < 1 mg L^{-1} (Fig. 11.7). *Daphnia* has a rigid calcified exoskeleton [66]. Therefore it was not expected that the length of the organisms decreased with increasing toxin concentrations while the area increased (Fig. 11.8A,B).

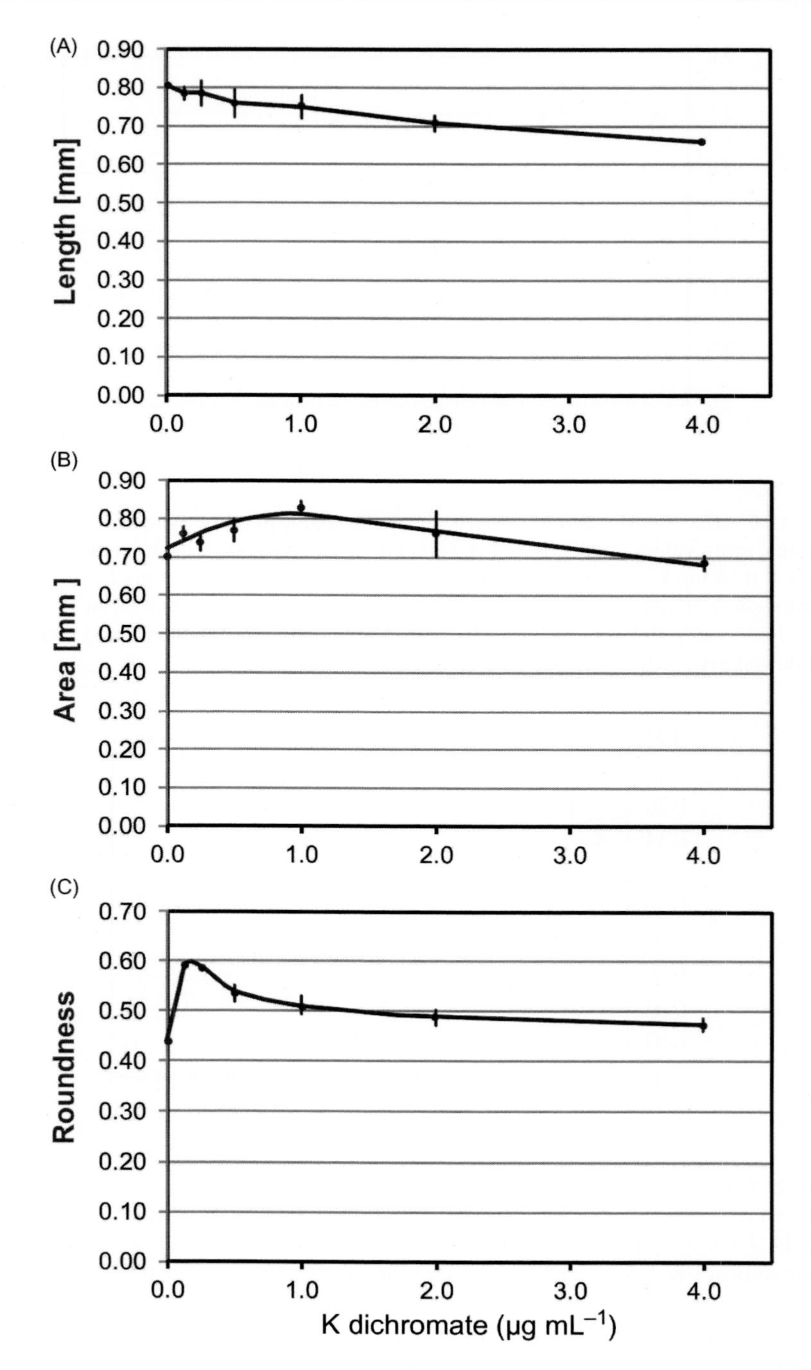

Figure 11.8 Length (A), area (B) and roundness (C) of *Daphnia magna* after incubation in potassium dichromate for 24 h. Mean values ($n = 6 \pm$ SD).

Measurements of the form factors such as roundness and aspect area indicted a swelling of the organisms (Fig. 11.8C), which reversed at higher concentrations.

11.5.2 Suitability of Daphniatox for monitoring environmental pollution

Daphnia is an established bioassay organism and the biomonitoring protocol is standardized in many countries [67]. The sensitivity of *Daphnia* to low concentrations of toxicants and pollutants makes the organism suitable for the use in the instrument Daphniatox. The tests with dichromate indicate that even low concentrations in the water are detected. After a 2 h exposure time EC_{50} values of less than 2 mg L^{-1} have been found for motility, velocity, and other endpoints, and even less after 24 h of exposure. The EC_{50} value for oral application of potassium dichromate in rats is 25 mg kg^{-1} after transdermal application, and 14 mg kg^{-1} for rabbits [68,69]. The sensitivity of *Daphnia* and related organisms to toxic pollutants has been determined to be higher than that of organisms utilized in other bioassays including algae (growth), fish (mortality), and bacteria (bioluminescence) [55].

The large number of 14 measured endpoints warrants a reliable quantification of negative effects on human health and the environment. While motility and mortality are used in other bioassays employing *Daphnia*, analysis of the size and form factors adds an additional dimension. Finding changes in length, area, and the various form factors was not expected and can be found only using precise and sensitive, computer-aided image analysis while it would escape manual evaluation.

Modern bioassays must fulfill a number of requirements. Short analysis time is of importance for online monitoring and early warning of potential hazards. Using Daphniatox, a complete measurement cycle can be performed in less than 2 min due to the fully automatic, computer-controlled image analysis being much faster than many other commercial bioassays (cf. Chapter 17: A comparison of commonly used and commercially available bioassays for aquatic ecosystems, this volume). Another requirement is high precision and repeatability, which again is warranted by computerized image analysis, which avoids human error and subjective evaluation. Using high numbers of organism tracks guarantees high statistical significance. Employing an invertebrate circumvents ethical problems that are involved in using fish, mammals, or other vertebrates [70]. Especially when used in developing countries, where poor funding and lack of environmental experts render efficient water quality monitoring difficult, the initial costs of the instrument should be low and the running costs minimal [41], which is not the case for some commercial instruments [71]. For Daphniatox it is sufficient to keep a supply of the microcrustaceans and their food source, which is not complicated. In addition to short-term tests and online real-time monitoring, modern bioassays should be capable of performing long-term tests analyzing the effects of pollutants and toxic substances over extended periods of time. Another requirement is that the operation of the instrument is user friendly and does not require extensive training of personnel.

While bioassays are not capable of revealing the chemical nature of a toxin involved in aqueous pollution, the analysis of different endpoints may give some clues, since they vary in their sensitivity toward different classes of toxic substances. Using 14 different parameters in motility, orientation, size, and form factors guarantees a large array of endpoints for evaluating pollution. Ecotox using the photosynthetic flagellate *Euglena gracilis* likewise uses a large number of endpoints [72].

In contrast to chemical analysis, bioassays can indicate potential threats of combined toxins, which may act additively or synergistically. In natural ecosystems the toxicity of pollutants is often enhanced by solar radiation, especially in the UV-B range which is detected by bioassays [73]. Modern bioassays have been used to monitor the toxicity of a wide range of pollutants such as heavy metals, herbicides, pesticides, fertilizers, detergents, and many other organic and inorganic pollutants [72,74] in both short- and long-term tests of ecotoxicity [72,75]. Daphniatox can be employed as an early warning system for pollution of surface and groundwaters, and municipal and industrial wastewaters, as well as for controlling the efficiency of wastewater treatment plants.

References

[1] Wille K, De Brabander HF, Vanhaecke L, De Wulf E, Van Caeter P, Janssen CR. Coupled chromatographic and mass-spectrometric techniques for the analysis of emerging pollutants in the aquatic environment. TrAC Trends Analyt Chem 2012;35:87–108.
[2] Keith L, Telliard W. ES&T special report: priority pollutants: Ia perspective view. Environ Sci Technol 1979;13(4):416–23.
[3] Phillips DJ, Rainbow PS. Biomonitoring of trace aquatic contaminants. Berlin: Springer Science & Business Media; 2013.
[4] Bae M-J, Park Y-S. Biological early warning system based on the responses of aquatic organisms to disturbances: a review. Sci Total Environ 2014;466:635–49.
[5] Jeong H, Kotov AA, Lee W, Jeong R, Cheon S. Diversity of freshwater Cladoceran species (Crustacea: Branchiopoda) in South Korea. J Ecol Environ 2015;38(3):361–6.
[6] Meunier CL, Boersma M, Wiltshire KH, Malzahn AM. Zooplankton eat what they need: copepod selective feeding and potential consequences for marine systems. Oikos 2016;125(1):50–8.
[7] Crittenden RN. Morphological characteristics and dimensions of the filter structures from three species of *Daphnia* (Cladocera). Crustaceana 1981;41(3):233–48.
[8] O'Sullivan P, Reynolds CS. The lakes handbook: limnology and limnetic ecology. New York: John Wiley & Sons; 2008.
[9] Ebert D. Ecology, epidemiology, and evolution of parasitism in *Daphnia*, 2005. Available from: ncbi.nlm.nih.gov.
[10] Pietrzak B, Bednarska A, Markowska M, Rojek M, Szymanska E, Slusarczyk M. Behavioural and physiological mechanisms behind extreme longevity in *Daphnia*. Hydrobiologia 2013;715(1):125–34.
[11] Sharma C, Langer S, Noorani IA. Population structure and some biological parameters of *Daphnia similis*, an important fish food organism. Int J Fish Aquat Stud 2016; 4(1):111–14.

[12] Armitage P. Chironomidae as food. In: The chironomidae. Berlin, Heidelberg: Springer; 1995. p. 423–35.
[13] LeBlanc GA. Laboratory investigation into the development of resistance of *Daphnia magna* (Straus) to environmental pollutants. Environ Pollut Ser A Ecol Biol 1982;27 (4):309–22.
[14] Chen L, Fu X, Zhang G, Zeng Y, Ren Z. Influences of temperature, pH and turbidity on the behavioral responses of *Daphnia magna* and Japanese Medaka (*Oryzias latipes*) in the biomonitor. Procedia Environ Sci 2012;13:80–6.
[15] Winner RW, Farrell MP. Acute and chronic toxicity of copper to four species of *Daphnia*. J Fish Board Can 1976;33(8):1685–91.
[16] Knops M, Altenburger R, Segner H. Alterations of physiological energetics, growth and reproduction of *Daphnia magna* under toxicant stress. Aquat Toxicol 2001;53:79–90.
[17] Ren Z, Zhang X, Wang X, Qi P, Zhang B, Zeng Y, et al. AChE inhibition: one dominant factor for swimming behavior changes of *Daphnia magna* under DDVP exposure. Chemosphere 2015;120:252–7.
[18] Gergs A, Kulkarni D, Preuss TG. Body size-dependent toxicokinetics and toxicodynamics could explain intra- and interspecies variability in sensitivity. Environ Pollut 2015;206:449–55.
[19] Pettersson A, Adamsson M, Dave G. Toxicity and detoxification of Swedish detergents and softener products. Chemosphere 2000;41:1611–20.
[20] Uc-Peraza R, Delgado-Blas V. Acute toxicity and risk assessment of three commercial detergents using the polychaete *Capitella* sp. C from Chetumal Bay, Quintana Roo, Mexico. Int Aquat Res 2015;7(4):1–11.
[21] Chen S, Liu W. Toxicity of chiral pesticide Rac-metalaxyl and R-metalaxyl to *Daphnia magna*. Bull Environ Contam Toxicol 2008;81:531–4.
[22] Kiss I, Kováts N, Szalay T. Evaluation of some alternative guidelines for risk assessment of various habitats. Toxicol Lett 2003;140-141:411–17.
[23] Ringelberg J. The positively phototactic reaction of *Daphnia magna* Straus: a contribution to the understanding of diurnal vertical migration. Neth J Sea Res 1964;2(3):319 IN1,335–334,IN2,406.
[24] Storz U, Paul R. Phototaxis in water fleas (*Daphnia magna*) is differently influenced by visible and UV light. J Comp Physiol A 1998;183(6):709–17.
[25] Zhou Q, Zhang J, Fu J, Shi J, Jiang G. Biomonitoring: an appealing tool for assessment of metal pollution in the aquatic ecosystem. Analyt Chim Acta 2008;606(2):135–50.
[26] Michels E, Leynen M, Cousyn C, De Meester L, Ollevier F. Phototactic behavior of *Daphnia* as a tool in the continuous monitoring of water quality: Experiments with a positively phototactic *Daphnia magna* clone. Water Res 1999;33(2):401–8.
[27] Kolkmeier MA, Brooks BW. Sublethal silver and NaCl toxicity in *Daphnia magna*: a comparative study of standardized chronic endpoints and progeny phototaxis. Ecotoxicology 2013;22(4):693–706.
[28] Netto I. Assessing the usefulness of the automated monitoring systems Ecotox and DaphniaTox in an integrated early-warning system for drinking water [Master thesis]. Toronto, Canada: Ryerson Univeristy; 2010.
[29] Wogram J, Liess M. Rank ordering of macroinvertebrate species sensitivity to toxic compounds by comparison with that of *Daphnia magna*. Bull Environ Contam Toxicol 2001;67(3):0360–7.
[30] DeMott WR, Zhang QX, Carmichael WW. Effects of toxic cyanobacteria and purified toxins on the survival and feeding of a copepod and three species of *Daphnia*. Limnol Oceanogr 1991;36(7):1346–57.

[31] Baird DJ, Barber I, Bradley M, Calow P, Soares AM. The *Daphnia* bioassay: a critique. Environmental bioassay techniques and their application. Berlin, Heidelberg: Springer; 1989. p. 403–6.
[32] GSM. Examination of water, waste water and sludge, bio-assays (group L), Determining tolerance of *Daphnia* to toxicity of waste water by way of a dilution series. German Standard Methods, 1989.
[33] ISO 6341. Water quality–Determination of the inhibition of the mobility of *Daphnia magna* Straus (Cladocera, Crustacea). Acute Toxicity Test, 2012.
[34] Erzinger GS, Souza SC, Pinto LH, Hoppe R, Del Ciampo LF, Souza O, et al. Assessment of the impact of chlorophyll derivatives to control parasites in aquatic ecosystems. Ecotoxicology 2015;24(4):949–58.
[35] ISO 8692. Water quality–Fresh water algal growth inhibition test with unicellular green algae, 2012.
[36] McWilliam RA, Baird DJ. Application of postexposure feeding depression bioassays with *Daphnia magna* for assessment of toxic effluents in rivers. Environ Toxicol Chem 2002;21(7):1462–8.
[37] Barata C, Alanon P, Gutierrez-Alonso S, Riva M, Fernández C, Tarazona JA. *Daphnia magna* feeding bioassay as a cost effective and ecological relevant sublethal toxicity test for environmental risk assessment of toxic effluents. Sci Total Environ 2008;405 (1):78–86.
[38] ISO 11348-3. Water quality determination of the inhibitory effect of water samples on the light emission of *Vibrio fischeri* (luminescent bacteria test) Part 3: Method using freeze-dried bacteria. Geneva: International Organization for Standardization; 1998.
[39] E1192-97 A. Standard guide for conducting acute toxicity tests on aqueous ambient samples and effluents with fishes, macroinvertebrates, and amphibians. West Conshohocken, PA: ASTM International; 2008.
[40] Bobra AM, Shiu WY, Mackay D, Goodman RH. Acute toxicity of dispersed fresh and weathered crude oil and dispersants to *Daphnia magna*. Chemosphere 1989;19(8-9): 1199–222.
[41] Häder D-P, Erzinger GS. Daphniatox–Online monitoring of aquatic pollution and toxic substances. Chemosphere 2017;167:228–35.
[42] Wolken JJ. Invertebrate photoreceptors: a comparative analysis. New York and London: Academic Press; 2013.
[43] Buchanan C, Goldberg B. The action spectrum of *Daphnia magna* (Crustacea) phototaxis in a simulated natural environment. Photochem Photobiol 1981;34:711–17.
[44] Ringelberg J. A mechanism of predator-mediated induction of diel vertical migration in *Daphnia hyalina*. J Plankton Res 1991;13(1):83–9.
[45] Ringelberg J. The photobehaviour of *Daphnia* spp. as a model to explain diel vertical migration in zooplankton. Biol Rev 1999;74:397–423.
[46] Gonçalves RJ, Barbieri ES, Villafane VE, Helbling EW. Motility of *Daphnia spinulata* as affected by solar radiation throughout an annual cycle in mid-latitudes of Patagonia. Photochem Photobiol 2007;83(4):824–32.
[47] Schindelin J, Arganda-Carreras I, Frise E, Kaynig V, Longair M, Pietzsch T, et al. Fiji: an open-source platform for biological-image analysis. Nat Methods 2012;9(7):676–82.
[48] Häder D-P, Erzinger GS. Advanced methods in image analysis as potent tools in online biomonitoring of water resources. Recent Pat Top Imag 2015;5(2):112–18.
[49] Dickinson A, Ackland B, Eid E-S, Inglis D, Fossum ER, editors. A 256/spl times/256 CMOS active pixel image sensor with motion detection. In: Solid-State Circuits

Conference, 1995 Digest of Technical Papers 41st ISSCC, 1995 IEEE International; 1995.

[50] Barnich O, Van Droogenbroeck M. ViBe: a universal background subtraction algorithm for video sequences. IEEE Trans Image Process 2011;20(6):1709–24.

[51] Shah GA, Khan A, Shah AA, Raza M, Sharif M. A review on image contrast enhancement techniques using histogram equalization. Sci Int 2015;27(2):1297–302.

[52] Sauvola J, Pietikäinen M. Adaptive document image binarization. Pattern Recogn 2000;33(2):225–36.

[53] Häder D-P, Lebert M. Real-time tracking of microorganisms. In: Häder D-P, editor. Image analysis: methods and applications. Boca Raton, FL: CRC Press; 2000. p. 393–422.

[54] Batschelet E. Circular statistics in biology. London, New York: Academic Press; 1981.

[55] Ahmed H, Häder D-P. Monitoring of waste water samples using the ECOTOX biosystem and the flagellate alga *Euglena gracilis*. J Water Air Soil Pollut 2011;216(1-4): 547–60.

[56] Ma L, Wang D, Mai X, Ren S. Research on the treatment of low concentration chromium-containing groundwater by iron filings. Environ Eng 2011;S1.

[57] Zhang B, Wu T, Qin T, Liu J-h. Research focus and green development trend on treatment of waste water containing chromium in leather industry. West Leather 2009;17:012.

[58] Fisher-Scientific. Material Safety Data Sheet Potassium dichromate. In: Sigma-Aldrich, editor. Fair lawn. Pittsburgh, New Jersey: Fisher Scientific; 2007.

[59] Robu B, Zaharia C, Macoveanu M. Environmental impact assessment for steel processing. Environ Eng Manage J 2005;4(1).

[60] Kim Y, Jung J, Oh S, Choi K. Aquatic toxicity of cartap and cypermethrin to different life stages of *Daphnia magna* and *Oryzias latipes*. J Environ Sci Health 2008;43(1):56–64.

[61] Naddy RB, Gorsuch JW, Rehner AB, McNerney GR, Bell RA, Kramer JR. Chronic toxicity of silver nitrate to *Ceriodaphnia dubia* and *Daphnia magna*, and potential mitigating factors. Aquat Toxicol 2007;84:1–10.

[62] Tong Z, Huailan Z, Hongjun J. Chronic toxicity of acrylonitrile and acetonitrile to *Daphnia magna* in 14-d and 21-d toxicity tests. Bull Environ Contam Toxicol 1996;57:655–9.

[63] ABNT- Associação Brasileira de Normas Téchnicas Ecotoxicologia aquática – Toxicidade crônica – Método de ensaio com *Ceriodaphnia*spp. (Crustacea, Cladocera). Rio de Janeiro, 2003. 12 p.

[64] Weber C. I. Short-term methods for estimating the chronic toxicity of effluents and receiving waters freshwater organisms, 1998, Vol. 89, No. 220503. DIANE Publishing.

[65] Flohr F, Brentano DM, Carvalho-Pinto CRS, Machado VM, Matias WG. Classificação de resíduos sólidos industriais com base em testes ecotoxicológicos utilizando *Daphnia magna*: uma alternativa. Biotemas 2005;18(2):7–18.

[66] Porcella DB, Rixford CE, Slater JV. Molting and calcification in *Daphnia magna*. Physiol Zool 1969;42(2):148–59.

[67] Asghari S, Johari SA, Lee JH, Kim YS, Jeon YB, Choi HJ, et al. Toxicity of various silver nanoparticles compared to silver ions in *Daphnia magna*. J Nanobiotechnol 2012;10(14):1–14.

[68] Tandon S. Organ toxicity of chromium in animals. Biol Environ Aspects Chromium 1982;5:209.

[69] Gumbleton M, Nicholls P. Dose-response and time-response biochemical and histological study of potassium dichromate-induced nephrotoxicity in the rat. Food Chem Toxicol 1988;26(1):37–44.

[70] Farré M, Barceló D. Toxicity testing of wastewater and sewage sludge by biosensors, bioassays and chemical analysis. TrAC Trends Analyt Chem 2003;22(5):299–310.

[71] Chen P, Wu P, Chen J, Yang P, Zhang X, Zheng C, et al. Label-free and separation-free atomic fluorescence spectrometry-based bioassay: sensitive determination of single-strand DNA, protein, and double-strand DNA. Analyt Chem 2016;88 (4):2065–71.

[72] Azizullah A, Richter P, Häder D-P. Effects of long-term exposure to industrial wastewater on photosynthetic performance of *Euglena gracilis* measured through chlorophyll fluorescence. J Appl Phycol 2014;27:303–10.

[73] Zepp R, Erickson III D, Paul N, Sulzberger B. Effects of solar UV radiation and climate change on biogeochemical cycling: interactions and feedbacks. Photochem Photobiol Sci 2011;10(2):261–79.

[74] Danilov R, Ekelund N. Applicability of growth rate, cell shape, and motility of *Euglena gracilis* as physiological parameters for bioassessment at lower concentrations of toxic substances: an experimental approach. Härnösand, Sweden: Mid Sweden University, John Wiley & Sons, Inc; 2000.

[75] Ahmed H. Biomonitoring of aquatic ecosystems [PhD thesis]. Erlangen-Nürnberg: Friedrich-Alexander-Universität; 2010.

Bioluminescence systems in environmental biosensors

12

Gilmar S. Erzinger[1], Francine Schmoeller[1], Luiz H. Pinto[1], Luiz Américo[1], Ruth Hemmersbach[2], Jens Hauslage[2] and Donat-P. Häder[3]

[1]University of Joinville Region—UNIVILLE, Joinville, SC, Brazil, [2]Institute of Aerospace Medicine German Aerospace Center (DLR), Cologne, Germany, [3]Friedrich-Alexander University, Erlangen-Nürnberg, Germany

12.1 Introduction

The phenomenon of light emission by living organisms is called bioluminescence. It occurs in widely distributed organisms and in remarkably diverse species with the exception of terrestrial vertebrates (amphibians, birds, reptiles, and mammals) and higher plants [1]. Among the light-emitting species are bacteria, dinoflagellates, fungi, fish, insects, shrimps, and squids, which includes terrestrial, freshwater, and marine species in almost 50% of the different phyla in the animal and plant kingdoms [1].

In bioluminescence there is a chemical-physical transduction, that is, a transformation of chemical energy into light energy. Bioluminescence is generated by highly exothermic, enzymatically catalyzed chemical reactions in which the energy of chemical bonds of organic compounds is converted preferably into visible light. In these reactions, molecules generally called luciferins are oxidized by oxygen, producing electronically excited molecules that decay by emitting light. These reactions are catalyzed by enzymes called luciferases [2].

This system has been well studied in the crustacean *Cypridina* and both luciferin and luciferase have been isolated and characterized in this species. Several fish have a luciferin-luciferase system similar to that of crustaceans. An analogous case occurs in the protozoa *Noctiluca* and *Gonyaulax*, responsible for the luminescence of seawater when it is disturbed by mechanical agitation [3,4].

In some cases, bioluminescence is generated by the organism itself, and in others by bacteria, which cooperate in symbiosis with the host. One example of the latter is the deep-sea fish known as humpback anglerfish, humpback blackdevil, and Johnson's anglerfish (*Melanocetus johnsonii*) that can be found at depths up to 4500 m but generally remains above 1500 m [4].

In addition to the two systems mentioned above, there are others that are independent of the presence of luciferin or luciferase, which occur in the bacteria and fungi that emit light. Another system was found in the coelenterates, where bioluminescence does not depend on the presence of oxygen; the phenomenon occurs in the presence of calcium ions with specific proteins [5,6].

Bioassays. DOI: http://dx.doi.org/10.1016/B978-0-12-811861-0.00012-7

12.2 Historical aspects

Nealson and Hastings observed a peculiar characteristic luminescence in Gram-negative, symbiotic *Vibrio fischeri* bacteria (also referred to as *Photobacterium fischeri*) colonizing specialized luminous organs of certain fish and cephalopods such as squid (*Euprymna scolopes*) [7,8]. This symbiosis enables the squid to produce bioluminescence which plays a role in attraction of prey or camouflage. These researchers found that the bacterium *V. fischeri*, which colonizes the squid, is able to provide invisibility to their host. At night, when the squid feeds, luminescence from the light organs is directed downward and is regulated so that it simulates the intensity of moonlight and/or stars, thus avoiding attracting predators [9]. The marine bacterium occurs naturally in both the free-living planktonic state and as symbionts [10,11]. However, the luminescence is only observed when the bacteria colonize the organs of the hosts, and not when they are free-living. From an evolutionary point of view it seems to be consistent that the bacteria keep a strict control of bioluminescence since the mechanism by which light is produced requires a high amount of energy.

Moreover, studies carried out in liquid medium showed that light output is only activated in the presence of a large number of cells [12]. Initially, cultures of *P. fischeri* show minimal luminescence in the beginning of culture growth. This is due to a substance which can be dialyzed from the nutrient broth but which does not affect growth. It seems that the inhibitor may be binding to the luciferase. When the population reaches a critical size this inhibitor is removed and bioluminescence increases [13]. Subsequently, it was demonstrated that no luminescence was initiated by removal of the inhibitor, but by the accumulation of an activator molecule or self-inductor [2,8]. This molecule produced and secreted by the bacteria activated luminescence when a sufficiently high concentration was reached. Thus, the bacteria are able to realize the cell density by detecting the concentration of the autoinducer.

V. fischeri is a marine, Gram-negative, and facultative anaerobic bacterium which naturally emits a blue-green light under favorable environmental conditions and higher oxygen concentrations of >0.5 mg L^{-1} [14]. The bacterium belongs to the family Vibrionaceae, which contains species that engage in both cooperative and pathogenic interactions with host animals. *V. fischeri* is found throughout the world in temperate and subtropical regions, existing as either a free-living saprophyte, as a member of the microbial gut community in many marine mammals, or as a light organ symbiont in several species of squid and fish [15].

Once colonization has taken place, the symbiotic *V. fischeri* cells trigger a series of developmental events that transform the light organ from an instrument of symbiont acquisition to the functional adult light organ. Symbiosis-specific changes in light-organ morphology are signaled by the envelope components lipopolysaccharide and peptidoglycan of *V. fischeri* which work synergistically to induce widespread programmed cell death (apoptosis) in the ciliated epithelial fields on the surface, and subsequently, the complete regression of this "symbiont-harvesting apparatus" over a period of 4–5 days [16,17].

Previously, *V. fischeri* was named *Achromobacter fischeri*, but to highlight the phylogenetic distance from other species of vibrios it was cataloged as *P. fischeri* [18,19]. A more recent reclassification to *Aliivibrio* has been proposed, taking into account a 97.4% sequence similarity with the upper 16S rRNA among the species *V. fischeri, Vibrio logei, Vibrio salmonicida,* and *Vibrio wodanis*, compared to other species found in the family of Vibrionaceae [20]. However, few authors have adopted the new name, shown by a large number of current publications that still use the original name *V. fischeri*.

12.3 Metabolism of microbial luminescence

The luminescence activity involves electron transport systems associated with the production of light, which are directly affected by harmful environmental factors to the cellular metabolism. If a disturbance of the electron transport chain is manifested as a light reduction, probably a toxic substance is present in a bioavailable form capable of penetrating the cell [21]. The bioluminescence of bacteria derives from a reaction between molecular oxygen, a coenzyme (a mononucleotide flavin in the reduced state – $FMNH_2$) and a long-chain aldehyde, catalyzed by the enzyme luciferase; the products are FMN (in the oxidized state), water, and carboxylic acid (Fig. 12.1). During this reaction a photon is emitted. This reaction is highly specific for $FMNH_2$, whose autoxidation is avoided by remaining bound to the luciferase enzyme [22].

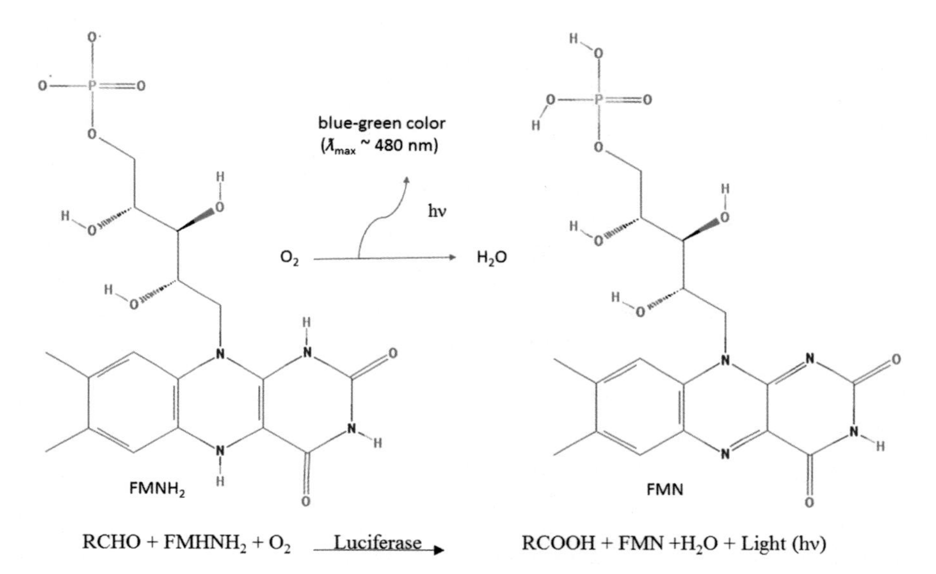

Figure 12.1 General outline of the reaction catalyzed by luciferase in *Vibrio fischeri* [19].

Light is continuously produced by living cells as the oxidized flavin (FMN) and the carboxylic acid formed in the reaction are recycled in parallel reactions [23]. The light emission is a process with high energy costs. Hastings and Nealson [24] estimated an expenditure of six adenosine triphosphate (ATP) molecules for each photon, assuming 100% reaction efficiency. This would explain why the luminescence phenomenon is expressed only when physiologically necessary [22]. Luciferase is an enzyme composed of two different subunits, an α chain (40 kDa) and a β chain (35 kDa). It does not require metals acting as cofactors or prosthetic groups [22]. The α and β subunits of luciferase are encoded by dislocation and luxB genes, which in *V. fischeri* are located adjacent to the lux operon, comprised of 18 other genes (luxC, D, and E) responsible for encoding enzymes involved in the complex carboxylic acid reductase for the synthesis of the aldehyde [25]. The luxCDABE genes cloned from some bacterial species (*Photobacterium*, *Vibrio*, and *Xenorhabdus*) exhibit similarity in some sequences, indicating evolutionary conservation [23,26]. The bioluminescent activity in bacteria is directly related to cell density, a phenomenon known as quorum sensing.

12.4 The phenomenon of quorum sensing

The molecular signals, also called autoinducers or pheromones, regulate specific genes that characterize the behavior of the population, in this manner providing the exploitation of the environment much more efficiently than would be possible for individual cells. Very low densities of pathogenic strains, for example, have a limited chance to overcome the defense system of a host (animal or plant) and successfully promote an attack on this organism. In this case, a delay in the expression of virulence factors can be a highly interesting strategy by the pathogen until a sufficient number of cells is reached, since the defense system of the host would be activated only when the situation is more favorable for the pathogen. Similarly, beneficial bacteria responsible for nitrogen fixation may employ this facility to optimize the formation of nodules in plant roots. Bacteria capable of producing antagonistic substances that act as biocontrol agents, killing the pathogenic microorganisms or inhibiting their proliferation, also benefit from the perception mechanism of quorum sensing (QS). In this case, the antagonistic substances are produced only when a critical population density of bacteria is reached by reducing the chances of onsetting the pathogen resistance in ways that make it difficult to control these populations.

QS is the communication mechanism by which bacteria regulate the expression of sets of genes in response to a specific cell density [27,28]. QS is an example of multicellular behavior in prokaryotes which regulates many physiological processes, including bioluminescence, antibiotic biosynthesis, plasmid transfer via conjugation and the production of virulence factors in pathogens of plants and animals. Once a specific quorum is reached the population is also capable of promoting multicellular differentiation, development, and fruiting body sporulation, as well as activation of secondary metabolism processes. The formation of biofilms has also been considered as a response to the QS mechanism. McClean et al. [29] detected the presence of molecular signals only on submerged rocks where there was the formation of biofilms. In extracts from rocks without biofilms these clues were absent.

The autoinducers are small molecules produced by bacterial cells. These molecules exit the cells, sometimes by simple diffusion, and accumulate in the environment in a quantity proportional to the cell number. Through the QS mechanism, the bacteria are able to detect the concentration of self-inducers, and thus realize the population size. From a certain threshold concentration of autoinducers, these signals serve as co-inducers and start regulating the transcription of target genes, and the transcription products ensure presumably some advantage for the bacterial cells in this particular condition. In vitro studies have shown that bacteria use a variety of flags, including nutritional status and density, to sense and respond to biotic factors [30,31].

12.5 Autoinducers in Gram-negative bacteria

The molecule produced by *V. fischeri* was isolated, characterized, and identified as N-(3-oxohexanoyl) homoserine lactone (3-oxo-C6-HSL) for the first time in 1981 by Eberhard et al. [32]. The analysis of genes involved in QS in *V. fischeri* was initiated by Engebrecht et al. [33] who showed that bacteria regulate the expression of genes encoding for bioluminescence in response to the density of the population, leading to a basic model. QS in this bacterium is currently a paradigm for other similar systems.

Research on the regulation of bioluminescence in *V. fischeri* led to the discovery of bacterial QS systems via N-acyl homoserine lactones (AHLs). From this discovery, the identification of the QS mechanism in different bacterial groups began by identification of their molecular chemical signals [34–36]. These extracellular signals are known generically as pheromones [28,37–39], self-inducers, or autoinductors [7,26,40].

One of the best-studied QS systems is the control of luminescence in *V. fischeri* [41]. The two genes involved are luxI and luxR. The proteins, LuxI and LuxR, control expression of the luciferase operon (luxICDABE) required for light production (Fig. 12.2). In particular, the first gene directs the synthesis of N-acyl homoserine lactone which is the key molecule of the system and diffuses in and out of the cell membrane and increases in concentration with increasing cell density; the second codes for a protein with two functional domains, the cytoplasmic receptor autoinducer and the DNA-binding transcriptional activator [33].

The QS, formerly known as "inductors", is a mechanism that triggers transcription of the lux genes only at high cell concentrations [23]. In its natural habitat, the luminescent bacteria can be found anywhere, being isolated from seawater samples collected at the surface and at depths of up to 1000 m [23]. However, under these conditions, the population of bacteria commonly found is very low, on the order of

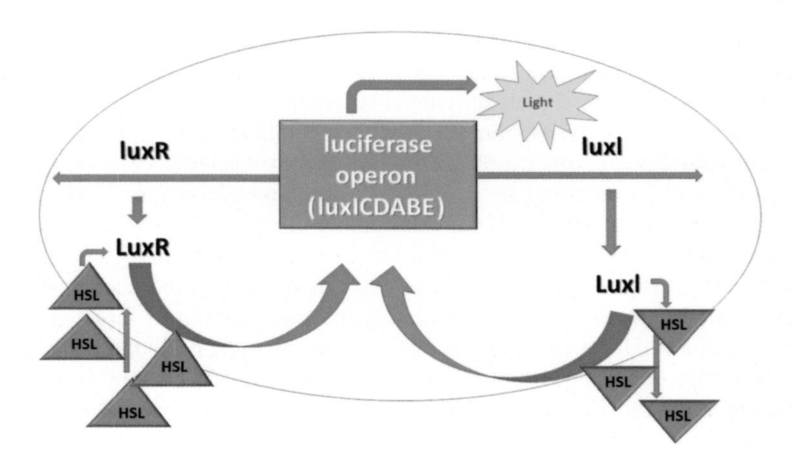

Figure 12.2 Schematic representation of the *Vibrio fischeri* quorum sensing system.

10^2 CFU mL^{-1} (colony-forming unit), resulting in the induction of luminescent activity [42]. Bioluminescence becomes evident when these bacteria occur in symbiotic association with higher organisms, when they reach very high concentrations (10^{10}–10^{11} CFU mL^{-1}), located in specific organs known as luminous organs [43]. This settlement ensures emission of light by the host and plays an interesting role in the attraction of prey or for camouflage. For example, at night, when the squid species *E. scolopes* feeds, luminescence from the light organs is directed to the seabed and regulated in such a way as to simulate the intensity of light of the moon and the stars thereby preventing the projection of a shadow detected by predators [44]. The symbiotic relationship between the squid and the bacteria is complemented by the expulsion of 90%–95% of the bacterial population each morning. The remaining 5%–10% of the bacteria multiply and repopulate the light organ over the day [45]. Thus, the expelled cells are widespread and can settle in other individuals.

The test with *V. fischeri* has a higher accuracy compared to bioassays with *Daphnia* sp. and fish, in addition to a satisfactory statistical significance achieved by responses produced from a large number of cells on the order of 10^6 cells per milliliter [22]. Since *V. fischeri* is a marine bacterium, the addition of sodium chloride (NaCl) to the test solution is required to reach a salt concentration of approximately 20 g L^{-1}. A 20% sucrose solution also promotes osmotic protection of cells [46,47]. Salt concentrations of less than 5 g L^{-1} may disrupt the cell membrane by the osmotic pressure difference [22]. The temperature and pH also affect the results of the bioassay. Therefore it is recommended to keep the temperature at 10–25°C and the pH at 6.0–8.5 [22,48]. The concentration of potassium (K^+) is related to the intracellular transcriptional activity of the lux genes [49]. The magnesium concentration found in sea water promotes the formation of flagella in *V. fischeri*, allowing their mobility and colonization of light organs of their hosts [50]. Therefore, the solutions used during preparation of *V. fischeri* for testing include

K^+ and Mg^{2+} in an appropriate composition [22]. Anomalies in the *V. fischeri* assay typically are related to the characteristics of environmental samples such as color, turbidity, salinity, and the presence of organic solvents [48]. Samples that exhibit a high concentration of nonsedimenting particulate material should be clarified by centrifugation or filtration. In the latter case, acetate or cellulose nitrate filters are not recommended, which can be toxic to the sample. For analysis of drinking water in which chlorine was used as a bactericide, it must be removed using a sodium thiosulfate solution, 1% (w/v), to avoid masking the toxicity from other substances [51].

12.6 Effect of inhibitors on bioluminescence emission of *V. fischeri*

The light inhibition in the test with *V. fischeri* may vary according to the nature of the toxic compound, as demonstrated in experiments with organic compounds and heavy metals. While organic compounds trigger a rapid and sustained response over time, heavy metals inhibit more slowly, depending on the applied concentration. Due to this behavior, the manufacturer of the Microtox test recommends multiple exposure intervals, such as 5, 15, and 30 min [51].

Thompson et al. observed the QS mechanism by *V. fischeri* for the first time in the late 1960s, when enhanced light emission was found only during the latter half of the growth curve in liquid culture [52]. In this process the bacteria produce a compound called "autoinducer" (AI) able to cross the cell wall and accumulate in the external environment. At low cell density the AI is in a low concentration, and rises until it reaches a critical concentration required for activation of bioluminescence genes [43]. AI produced by *V. fischeri* was identified as N-(3-oxohexanoyl) homoserine lactone (Fig. 12.3), in 1981 by Eberhard et al. [32,44].

According to the manual, the Microtox system for determining acute toxicity may generate different light levels. Different chemicals affect organisms in acute testing at different rates and produce different response curves. Typically there are three kinds of light level response curves for acute Microtox tests. The Microtox System Software states that it is necessary to make several light readings (It) at 5, 15, and sometimes 30 min for the test medium of the sample to obtain more accurately the determination of LD_{20} and LD_{50} [47].

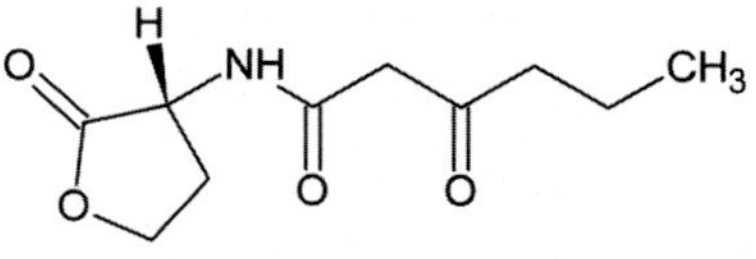

Figure 12.3 Molecular structure of the autoinducer N-(3-oxohexanoyl) homoserine lactone (3-oxo-C6-HSL) [43].

Phenol, for example, induces a very quick reaction. The light output falls drastically, rises, and then decreases again slightly over time (Fig. 12.4). For organic compounds a typical light response curve shows a kinetics at which the light output drops more gradually than for phenol and then levels off (Fig. 12.5).

For heavy metals the bioluminescence decay rate is essentially constant over a prolonged period of time and it is concentration dependent. The response curves indicate that for metals several light readings are recommended such as at 5, 15, and 30 min (Fig. 12.6) [51].

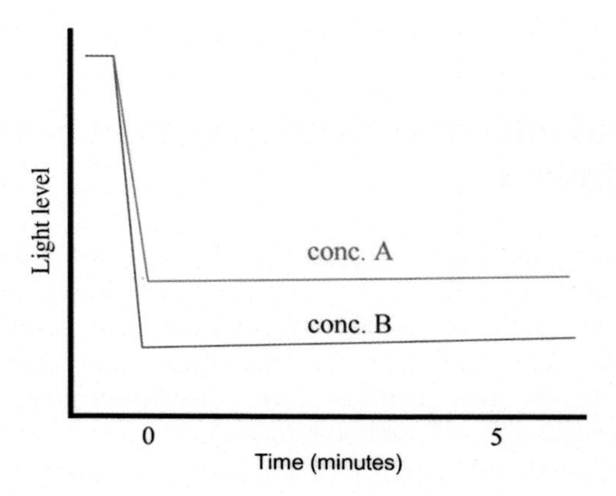

Figure 12.4 Typical time-response curves for phenol exposure at two concentrations in the Micotox system.
Source: Redrawn after Azur. Acute_Overview: AZUR environmental. Available from: http://www.coastalbio.com/images/Acute_Overview.pdf; 1998 [accessed 20.05.16].

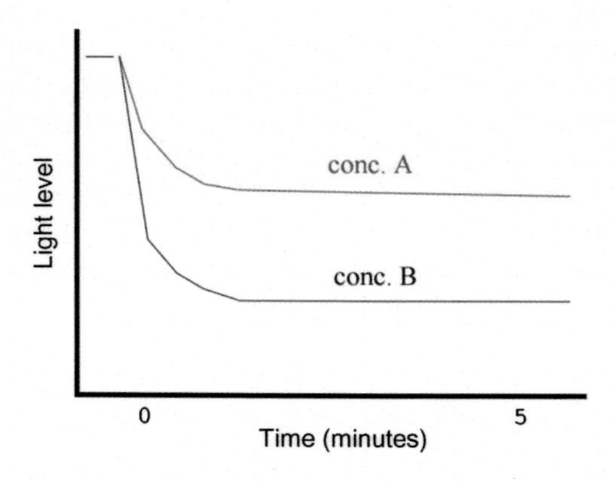

Figure 12.5 Typical time-response curves for exposure to organic compounds at two concentrations in the Micotox system.
Source: Redrawn after Azur. Acute_Overview: AZUR environmental. Available from: http://www.coastalbio.com/images/Acute_Overview.pdf; 1998 [accessed 20.05.16].

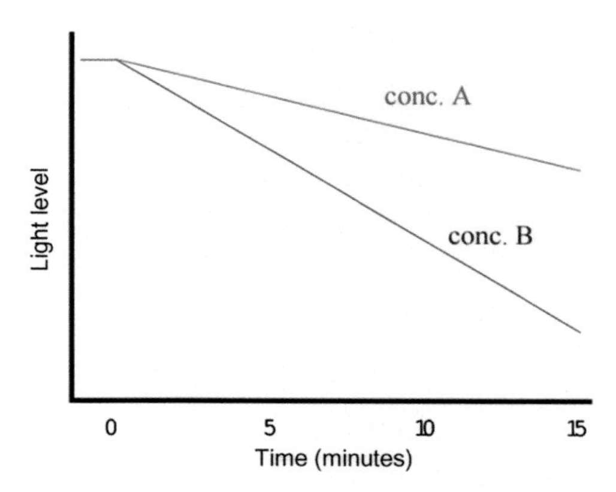

Figure 12.6 Typical time-response curves for exposure to heavy metals at two concentrations in the Micotox system.
Source: Redrawn after Azur. Acute_Overview: AZUR environmental. Available from: http://www.coastalbio.com/images/Acute_Overview.pdf; 1998 [accessed 20.05.16].

The toxic effects caused by metals in *V. fischeri* are assigned to the free ion form [53], where electrostatic interactions between the ions present in the reaction medium, and the metal diffusion through the membrane, contribute to a slow response [54]. Sensitivity of *V. fischeri* for different metals was observed by Fulladosa et al. [55], who found rapid inhibition caused by Ag^{+} and Hg^{2+} but low sensitivity to Cr^{6+}, Cd^{2+}, As^{3+}, and As^{5+}, due to detoxification reactions and exopolymer secretion as defense mechanisms. The inhibition of luminescence caused by the most toxic substances is irreversible. In phenolic inhibitors, the hydrophobic portion of the molecule disrupts the plasma membrane of the cell [56], or they are in direct association with the enzymes involved in the light production mechanism, such as luciferase and $NADH^{-}$dehydrogenase [57]. Both luciferase and the NADH dehydrogenase, involved in the regeneration of $FMNH_2$, contain hydrophobic domains susceptible to the action of phenols. According to Ismailov et al. [57], the reaction of the luciferase with its substrates aldehyde and $FMNH_2$ in *Vibrio harveyi* was inhibited by phenol and chlorinated derivatives, respectively, by competitive inhibition and noncompetitive mechanisms.

12.7 Commercial tests with *V. fischeri*

After 30 years testing with *V. fischeri* found wide applications, especially with standardization in 1993 by DIN standard 38412 part 34, and in 1998 by ISO 11348, Parts 1, 2, and 3 [14]. In 2006, ABNT published the NBR 15411: 2006 – "Aquatic Ecotoxicology—Determination of the inhibitory effect of water samples on the light emission of *V. fischeri* (Luminescent bacteria test)" [58], which, based on ISO 11348, recommends 30 min exposure. It is also divided in three parts: Part 1—Method using freshly grown bacteria; Part 2—Method using dehydrated bacteria;

Part 3—Method using freeze-dried bacteria. The main difference between the three parts is the type of preservation and the sensitivity of the bacteria. While Part 3 mentions that the freeze-dried bacteria respond with between 20%–80% inhibition of light emission after 30 min of contact with 2.2 mg L^{-1} Zn^{2+} (as $ZnSO_4.7H_2O$), parts 1 and 2 define the same inhibition conditions by applying an increased concentration of Zn^{2+} (25 mg L^{-1}), and therefore these formulations are less sensitive compared to the lyophilized bacteria. The observation is similar for another recommended reference substance, 3,5-dichlorophenol (6 mg L^{-1} for frozen and 3.5 mg L^{-1} for lyophilized bacteria), but not for potassium dichromate (4 mg L^{-1} of Cr^{6+} for frozen and 18.7 mg L^{-1} for lyophilized bacteria). Bertoletti [59] described that CETESB updated in 2001 the technical standard L5.227: "toxicity test with the luminescent bacterium *V. fischeri*: test method", which recommends a 15 min exposure time and the use of freeze-dried bacteria, for which an EC_{50} value ($ZnSO_4.7H_2O$) between 3 and 10 mg L^{-1} should be obtained (0.6 to 2.2 mg L^{-1} Zn^{2+}). Some systems and their luminescent biological reagents are available for the realization of bioluminescent assays, marketed generally as lyophilized formulations. Lyophilization (freeze drying) aims at the formation of a product structurally intact and ready for reconstitution that constitutes the method of preserving most microorganisms used in industry. This process begins with freezing the product at a controlled speed and in the presence of a cryoprotectant, to allow formation of tiny ice crystals. The free water is removed by sublimation, with the application of vacuum and temperature below a critical value in order to prevent melting of the frozen material and consequent destruction of the product. The remaining water molecules are removed during the second drying stage, in which the temperature is gradually increased. The product obtained has a very low moisture content, and can be stored and preserved for longer periods when sealed under vacuum or an inert atmosphere [60]. One of the first methods used, and certainly the best known, uses the luminescent marine bacterium *V. fischeri*, which is often referred to as Microtox assay [22]. Currently we can find on the market different luminometers by different manufacturers. The three most widely spread Luminometer systems are ToxAlert 10 (Merck), Microtox (Azur Environmental), and LUMIStox (HACH) for determining acute toxicity tests. The bioluminescence is produced by the marine bacterium, *V. fischeri* NRRL B 11177. In 1991, Kaiser and Palabrica already collected toxicity data obtained through bioluminescent assays to over 1300 chemicals [61].

Jennings et al. [62] conducted a study comparing three commercial systems that assess the chemical toxicity with the bioluminescent *Photobacterium* (*V. fischeri*) (Table 12.1). These data provide comparisons of toxicity which demonstrate that, when used under standard conditions, these bioluminescence-based toxicity assays produce very similar results.

Other products on the market are Microtox Acute Reagent and SOLO Microtox (Strategic Diagnostics Inc.), LUMISTox (LCK 480, Hach/Dr. Lange, Germany), ToxAlert (Merck, Germany), BioFix Lumi (Macherey-Nagel, Germany), ToxScreen I and II (CheckLight Bioluminescent Diagnostic Reagents, Israel), Bioluminex (ChromaDex, USA), and BioluxLyo (UBiotecha, Brazil). The Microtox test was developed in the early 1970s, in response to a need of the oil industry in California

Table 12.1 Mean percentage inhibition for reference concentrations of Zn^{++} after 5, 15, and 30 min contact time in ToxAlert 10, Microtox, and LUMIStox systems.

Assay system	Final [Zn^{++}] (mg L^{-1})	Percentage inhibition			
		5 min	15 min	30 min	*n*
ToxAlert 10	4.50	0.66 ± 0.19	31.29 ± 1.31	66.30 ± 0.96	76
Microtox	2.11	3.99 ± 0.42	26.72 ± 0.94	61.36 ± 1.55	75
LUMIStox	25.00	6.35 ± 0.94	36.94 ± 1.36	54.30 ± 1.35	79

The [Zn^{++}] are final concentrations recommended by each manufacturer to give approximately 50% inhibition.
Source: After Jennings VL, Rayner-Brandes MH, Bird DJ. Assessing chemical toxicity with the bioluminescent photobacterium (Vibrio fischeri): a comparison of three commercial systems. Water Res 2001;35(14):3448–3456.

to get a faster toxicity test which was practical in relation to the usual tests on fish and invertebrates. Since 2000, Strategic Diagnostics Incorporated (SDI), located in Newark, DE, USA, is responsible for marketing the entire line of Microtox products, including the reagent, the light-measuring device (luminometer Microtox Model 500 Analyzer), and accessories. Currently the Microtox luminometer is being marketed by Azur Environmental. Over the past decade, the number of publications using Microtox nearly doubled. In 2005, Johnson had already reported 1400 scientific publications using the bioluminescent method, which according to this author attests the worldwide acceptance of this test [48]. An example of the Microtox impact on the international scientific community was the publication of a "Toxicity Index", which shows the compilation of the results of toxicity after 5, 15, and 30 min for approximately 1300 compounds listed in the Chemical Abstracts Service (CAS) [61]. In the same line of Microtox products, SDI offers the DeltaTox and Mutatox systems. DeltaTox lends itself to the actual testing of environmental samples in the field and offers a portable luminometer, SOLO Microtox reagent, and kit solutions. The Mutatox reagent is based on applying a nonluminescent mutant strain of *V. fischeri* to detect genotoxic effects. The presence of genotoxic compounds causes mutations, signaled by the restoration of luminescence [63]. The ToxScreen II reagent has the advantage to use the *Photobacterium leiognathi* SB bacteria line, which, according to the manufacturer, is tolerant to a higher temperature range (between 18 and 35°C) in contrast to the normally employed *V. fischeri* with an optimum temperature at 26°C. Moreover, with ToxScreen II it is possible to preserve the bacterial suspension for seven days, allowing the customer to open new ampoules for each new test [64]. BioluxLyo is the first reactant using *V. fischeri* developed in Brazil. The formulations available are used to perform five to ten tests with a single ampoule, each test comprising nine dilutions of a sample and a negative control [65]. According to the manufacturer, Bioluminex represents a breakthrough in bioassays, as it combines in a single test separation of components of a sample by thin layer chromatography (TLC), followed by addition of a suspension of *V. fischeri*. The spots formed by chromatography, which appear as dark spots, correspond to compounds that alone would present toxicity [66]. A common question is the applicability of the tests with

V. fischeri which refers to the ecological relevance of using a marine bio-indicator to analyze freshwater samples. The addition of sodium chloride to the analyzed liquid matrix introduces a new variable in their chemical composition [18], which may cause precipitation of some compounds present in the sample.

An alternative to the use of marine organisms is the transfer of luminescence genes into bacterial strains isolated from various sources, such as water, soil and sludge wastewater treatment systems. However, numerous studies comparing *V. fischeri* and freshwater organisms such as microcrustaceans, fish, and algae, attest to the high correlation between these tests [61,67], and the increased sensitivity of the bacteria to detecting organic compounds and metals, as well as convenience and lower cost per analysis running. High sensitivity refers to the narrow light regulatory mechanism of production associated with the central pathways of cellular metabolism.

There are some significant differences in the methodology used for each system. The ToxAlert 10 system uses only one series of sample dilutions, while the Microtox and LUMIStox system use sample dilutions in duplicate, and the average value of two readings is used in subsequent calculations. In the Microtox system the first serial dilutions are measured before the second set is initiated. In the LUMIStox system, dilution series are executed in parallel together and duplicate dilutions are measured directly one after the other [62].

12.8 Methods for bioluminescence measurements

The methods employ bioluminescent organisms which naturally emit light, or bacteria isolated from nature and converted into luminescent strains by genetic manipulation [18]. The response may manifest itself by inhibition of light caused by toxic substances in the sample being analyzed or production of luminescence triggered by a particular compound, involving genetically modified organisms (GMOs). The bioluminescence phenomenon can be observed in some free-living marine bacteria or in symbiotic association with some higher organisms. Marine species belonging to the family Vibrionaceae are divided between the *Vibrio* and *Photobacterium* genera. In the first, the species *V. harveyi*, *Vibrio splendidus*, *V. fischeri*, and *V. logei* are known. Some strains of *Vibrio cholerae* species, pathogenic to humans, are also able to emit light but are usually found in freshwater. Among the three species of *Photobacterium* only *P. phosphoreum* and *P. leiognathi* are luminescent. Other luminescent species are found in terrestrial or freshwater environments, especially from the genus *Xenorhabdus* [19]. The *Vibrio* genus covers marine bacteria with straight or curved bacillus shape. They are distinct in relation to *Photobacterium* due to the presence of a polar flagellum (although some members have peritrichal flagella) covered by a sheath, and its inability to accumulate poly-β-hydroxybutyrate. Both genera require sodium (Na^+) in the sample to be analyzed, and therefore bacteria are abundant in the marine environment, found free in the seawater, in the intestinal tract, and on the animal body surface. Present saprophytic activity can decompose chitin and alginic acid [68,69]. Luminescent tests can be applied to investigate the existence of toxicity in samples of water, sediment, or soil; it is widely used for detecting toxicity in effluents [14]. The tests with luminescent bacteria are often used for a preliminary assessment ("screening") from a battery of tests, due to the greater speed of execution and lower maintenance costs [18]. Characterized as bioassays they require only amounts of samples and reagents on the order of microliters, which allows the simultaneous performance of various analyses, and need less space for the materials and instrumentation [70]. Currently, naturally luminescent species such as *V. fischeri*, *V. harveyi*, *P. leiognathi*, and *Pseudomononas fluorescens* are used [18].

An additional source of variation relates to the detailed procedure adopted in different bioassays and how bacteria are reconstituted before use. The components of the media used to sustain the bacteria for the duration of the test also varies widely. Consequently, the inhibition of light production may depend, to a greater or lesser extent, on which bioassay system is used [71,72].

In a toxicity test with *V. fischeri* the natural luminescence before and after an exposure to a wastewater sample or a chemical is measured. In the presence of toxic substances, the light intensity decreases due to an inhibition of metabolic processes of the cell [73]. This reduction caused by a sample should be compared to the effect of luminescence produced in a negative control (typically, 2% NaCl solution) and optionally by employing a positive control with reference substances such as heavy metals. In acute toxicity tests exposure intervals of 5, 15, and 30 min are used. The difference between the initial and final light intensities, corresponding to an inhibition value, is expressed as percentage caused by a particular concentration (in the case of known substances) or dilution (for water and wastewater) of the sample. The results can be expressed as a toxicity factor for bacteria (FTB) or effective concentration (EC) [14]. The FTB is numerically equal to the dilution factor (DF) and equivalent to the lowest dilution of a series, in which the inhibition of luminescence is less than 20%. This parameter is used for the classification of different wastes in almost all the international approved standards and also for the calculation of the permitted percentage of this effluent in the receiving water body, taking into account the flow rate and the effluent toxicity itself. The EC corresponding to the sample concentration at which a given inhibition value is observed is expressed as CEx, where x is the percentage. Commonly, the values CE_{20}, CE_{50}, and CE_{80} are determined and the EC_{50} is used to determine the effect of environmental toxicity [74].

12.9 Genetically modified microorganisms

Advances in molecular biology have enabled the cloning and expression of the lux operon in nonluminescent bacteria, finding wide applications in industry, medicine, and microbial ecology, and in particular, environmental biotechnology [22]. Parallel to *V. fischeri*, several genetically modified microorganisms (GMO) have been developed which are listed in Table 12.2. With few exceptions, the

Table 12.2 GMO luminescent bacteria used for detection of toxicity in wastewater [75]

Modified strain	Parental lineage	Source luxCLUBE	Response mode	Type of detected toxic substances
DPD 2511	*Escherichia coli*	*Vibrio fischeri*	Light output	Oxidative stress
DPD 2540	*E. coli*	*V. fischeri*	Light output	Dissolution cell membrane
DPD 2794	*E. coli*	*V. fischeri*	Light output	Modifying DNA
TV 1061	*E. coli*	*V. fischeri*	Light output	Protein modification
GC 2	*E. coli*	*Xenorhabdus luminescens*	No light emission	Toxic substances in general
Shk1	*Pseudomonas fluorecens*	*V. fischeri*	No light emission	Toxic substances in general
BS 530	*Pseudomonas pseudoalcaligenes*	*Photorhabdus luminescens*	No light emission	Toxic substances in general
BS 566	*Pseudomonas putida*	*P. luminescens*	No light emission	Toxic substances in general
BS 675	*P. putida*	*P. luminescens*	No light emission	Toxic substances in general
BS 678	*P. putida*	*P. luminescens*	No light emission	Toxic substances in general

luminescent GMOs were obtained by the insertion of the luxCDABE operon from *V. fischeri* [75].

Basically, most luminescent GMOs are constructed by the insertion of a promoter region in front of the operon luxCDABE (or part of the luxAB genes) devoid of their native promoter into a plasmid. This is then transferred into a strain that provides a desirable characteristic associated with the promoter of interest [76]. These systems represent a potential for the detection of pollutants in soil and water, because the degradation of most of these compounds promoted by microorganisms is mediated by an operon located in plasmids. The expression of genes related to this process, together with the observation of a response easily measured as the light emission, can be a way to determine the concentration and bioavailability of a particular group of toxic substances [25]. Furthermore, it is possible to classify toxic agents according to their mechanism of action (rupture of the cell membrane, structural modification of proteins or genetic material), whose onset is associated with the production of light

[75,77]. Some applications include the construction of a strain (*P. fluorescens* HK44) capable of detecting and quantifying salicylate and naphthalene in a bioavailable form by light emission [25]. A biosensor for the detection of substances such as benzene, toluene, ethylbenzene, xylene (BTEX), and fuel constituents, in aqueous samples was constructed by inserting the tod-luxCDABE sequence into the chromosome of *Pseudomonas putida* F1 line [76]. To detect heavy metals, biosensors were developed by the fusion of the resistance *mer* operon and the luxCDABE operon to detect mercury at concentrations as low as 0.1 ppb [78].

Despite the large amount of constructed assays, luciferase systems have some drawbacks which include dependence on substrate concentration, the physiological state of the bacteria, and the instability of the enzyme luciferase associated with suppression or activation factors present in the sample to be tested [22]. Another strategy used for the construction of biomarkers is the insertion of the GFP (green fluorescent protein) gene into the chromosome of the host cell for production of the GFP protein, not naturally found in terrestrial microorganisms, which offers great flexibility to environmental monitoring assays [77].

12.10 Bioluminescent dinoflagellates serve as bioassay to indicate mechanical stress

Dinoflagellates represent one of the fastest mechanosensitive reporter systems in nature to visualize the impact of shear and hydrodynamic forces on unicellular organisms in fluids. They react on a corresponding stimulus within a time range of 15–20 ms [79–81] by showing a fascinating bioluminescent response (Fig. 12.7). Shear stress and hydrodynamic gradients produce a deformation of the cell

Figure 12.7 Ocean bioluminescence—a fascinating spectacle produced by living organisms, Sea Sparkle at the Yacht Port of Zeebrugge in Belgium.
Source: Photo by Hans Hillewaert with permission.

membrane resulting in a detectable bioluminescence visualized by means of photon counting probes [3,80,82]. As a conclusion the emitted light of the dinoflagellates can be used to quantify the mechanical stress [80]. The sensitivity of dinoflagellates was tested in *Pyrocystis lunula* by atomic force microscopy and showed a threshold of 7.2 ± 3.4 μN after a cell deformation of 2.1 ± 0.65 μm on a deformed area of 1.4% of the cell surface [83].

The bioluminescence capacity is controlled by a circadian rhythm. One hour after the onset of darkness at night *Pyrocystis noctiluca* has the ability to flash for 6 h; after this period it switches to glowing for another 6 h [84]. In order to use the organisms in laboratory experiments, the light/dark cycle can be shifted by cultivating them in light during the night and keeping them in the dark during the day. The mechanism of light emission and bioluminescence as a form of chemiluminescence, where the reaction chemicals—luciferin for the substrate and luciferase for the enzyme—are produced by living organisms, is already described in Section 12.3.

Mechanical shearing of dinoflagellates induces intracellular signaling, which is so far only partly understood. A membrane-based shear receptor has been postulated in this cascade, which triggers an increase in cytosolic Ca^{2+}. The subsequent proton flux turns on the reaction between luciferin and luciferase and thereby the emission of light. Reaction scintillions as reaction caverns in the vacuole are described as source of the bioluminescence [85–87].

The ecological reason for bioluminescence is defense and hunting. Predators touch the prey when feeding on them and produce water streams due to their swimming activities, which in turn results in luminescence of the dinoflagellates, which might disturb the predator [88]. Dinoflagellates are prey of especially small crustaceans. The further biological advantage of bioluminescence is obviously the attraction of secondary predators [89,90]. It was demonstrated that both fish [90] and a cephalopod (*Sepia*) [91] prey at night more efficiently on crustaceans in the presence of luminescent dinoflagellates [89,90,92–94].

The bioluminescence of dinoflagellates is induced by a velocity gradient of fluids and can be utilized as a bioassay and optical indicator for hydrodynamic and shear stresses [95]. Experiments showed that the intensity of the emitted light is a function of the shear stress level and cell concentration [96]. To measure the bioluminescence resulting from shear forces a photomultiplier-based detector system is recommended. Changes in the bioluminescence produced by shear forces acting on the outer cell wall can be measured by changes in the relative photon count.

We used dinoflagellates as highly sensitive reporter organisms in order to contribute to the question of whether shear forces are induced in rotation platforms, which have been constructed to mimic micogravity conditions on ground, having a pivotal role in preparing space-based experiments. In the case of shear stress the organism reacts with a detectable bioluminescence emission. Clinostats and Random Positioning Machines (RPM) are commonly applied in gravitational cell biology to simulate micrgravity conditions.

To exclude potential nongravitational cellular responses to commonly used methodologies, *P. noctiluca* was exposed to varying operational modes. With cells in an RPM rotating around two axes with random velocities and directions, we observed significantly greater mechanical and shear stress compared with a clinostat (rotation around one axis perpendicular to gravity) experiments applying constant rotation around one axis. We conclude that in contrast to RPM, one axis clinorotation induces substantially less nongravitational responses through shear forces. Therefore, dinoflagellates provided a suitable bioassay for evaluation of mechanical stress.

12.11 Conclusions

The need to improve the efficiency and effectiveness of monitoring and evaluation of the quality of water increased interest in innovative methodologies. Within this concept biomonitoring is the systematic use of living organisms or their responses to environmental quality. Most tests for evaluating the biological effects of toxicity, i.e., the impacts of toxic chemicals in living cells require large laboratories that are based on manual procedures and require hours, days, or weeks to obtain results. Due to all these disadvantages/limitations, recent studies deal with the use of rapid, reproducible, and cost-effective bacterial assays for screening and evaluation. Among the bacterial bioassays, the inhibition test of *V. fischeri* luminescence is the most common. Based on the literature, the *V. fischeri* inhibition test is the most sensitive, cost effective, and easy to operate, and requires only 5−30 min to predict toxicity.

References

[1] Meighen EA. Molecular biology of bacterial bioluminescence. Microbiol Rev 1991;55 (1):123−42.
[2] Eberhard A. Inhibition and activation of bacterial luciferase synthesis. J Bacteriol 1972;109:1101−5.
[3] Cussatlegras A-S, Le Gal P. Bioluminescence of the dinoflagellate *Pyrocystis noctiluca* induced by laminar and turbulent Couette flow. J Exp Mar Biol Ecol 2004;310 (2):227−46.
[4] Orlov AM, et al. First record of humpback anglerfish (*Melanocetus johnsonii*) (Melanocetidae) in Antarctic waters. Polar Res 2015;34.
[5] Randall D, Burggren W, French K. Eckert animal physiology. 5th ed. New York: Macmillan; 2002.
[6] Nelson DL, Cox MM. Lehninger's principles of biochemistry. 4th ed. New York: W.H. Freeman; 2017 7th ed.: MacMillan.
[7] Nealson KH. Autoinduction of bacterial luciferase. Arch Microb 1977;112(1):73−9.
[8] Nealson KH, Platt T, Hastings JW. Cellular control of the synthesis and activity of the bacterial luminescent system. J Bact 1970;104(1):313−22.
[9] Visick KL, McFall-Ngai MJ. An exclusive contract: specificity in the *Vibrio fischeri-Euprymna scolopes* partnership. J Bact 2000;182(7):1779−87.
[10] Ruby RH. Delayed fluorescence from *Chlorella*: II. Effects of electron transport inhibitors DCMU and NH_2OH. Photochem Photobiol 1977;26:293−8.

[11] Ruby E, Nealson K. Symbiotic association of *Photobacterium fischeri* with the marine luminous fish *Monocentris japonica*: a model of symbiosis based on bacterial studies. Biol Bull 1976;151(3):574–86.

[12] Greenberg EP. Quorum sensing in Gram-negative bacteria. Washington, DC: ASM news 1997;63, 371–377.

[13] Kempner E, Hanson F. Aspects of light production by *Photobacterium fischeri*. J Bact 1968;95(3):975–9.

[14] Knie JLW, Lopes EWB. Testes ecotoxicológicos: métodos, técnicas e aplicações. In: FATMA/GTZ, 2004.

[15] Nealson KH, Hastings JW. The luminous bacteria. In: Balows A, Trüper HG, Dworkin M, Harder W, Schleifer K-H, editors. The prokaryotes, a handbook on the biology of bacteria: ecophysiology, isolation, identification, applications. Berlin Heidelberg New York: Springer; 1991. p. 625–39.

[16] Koropatnick TA, Engle JT, Apicella MA, Stabb EV, Goldman WE, McFall-Ngai MJ. Microbial factor-mediated development in a host-bacterial mutualism. Science 2004;306(5699):1186–8.

[17] Foster JS, Apicella M, McFall-Ngai M. *Vibrio fischeri* lipopolysaccharide induces developmental apoptosis, but not complete morphogenesis, of the *Euprymna scolopes* symbiotic light organ. Dev Biol 2000;226(2):242–54.

[18] Girotti S, Ferri EN, Fumo MG, Maiolini E. Monitoring of environmental pollutants by bioluminescent bacteria. Analyt Chim Acta 2008;608(1):2–29.

[19] Blum LJ. Bio- and chemi-luminescent sensors. Singapore: World Scientific; 1997.

[20] Urbanczyk H, Ast JC, Higgins MJ, Carson J, Dunlap PV. Reclassification of *Vibrio fischeri*, *Vibrio logei*, *Vibrio salmonicida* and *Vibrio wodanis* as *Aliivibrio fischeri* gen. nov., comb. nov., *Aliivibrio logei* comb. nov., *Aliivibrio salmonicida* comb. nov. and *Aliivibrio wodanis* comb. nov. Int J Syst Evol Microbiol 2007;57(12):2823–9.

[21] Steinberg CEW, Lorenz R, Spieser OH. Effects of atrazine on swimming behavior of zebrafish, *Brachydanio rerio*. Water Res 1995;29:981–5.

[22] Nunes-Halldorson VdS, Duran NL. Bioluminescent bacteria: lux genes as environmental biosensors. Braz J Microbiol 2003;34(2):91–6.

[23] Hastings JW. Bioluminescence. Cambridge, MA: Harvard University; 2004 [2 March 2016]. Available from: mips.stanford.edu/public/abstracts/hastings.pdf

[24] Hastings JW, Nealson KH. The symbiotic luminous bacteria. In: Starr MP, Stolp H, Trüper HG, Balows A, Schlegel HG, editors. The prokaryotes. A handbook on habitats, isolation, and identification of bacteria. Berlin: Springer-Verlag; 1981. p. 1322–45.

[25] Heitzer A, Webb OF, Thonnard JE, Sayler GS. Specific and quantitative assessment of naphthalene and salicylate bioavailability by using a bioluminescent catabolic reporter bacterium. Appl Env Microbiol 1992;58(6):1839–46.

[26] Meighen EA. Bacterial bioluminescence: organization, regulation, and application of the lux genes. FASEB J 1993;7:1016–22.

[27] Cha C, Gao P, Chen Y-C, Shaw PD, Farrand SK. Production of acyl-homoserine lactone quorum-sensing signals by gram-negative plant-associated bacteria. Mol Plant Microbe Interact 1998;11(11):1119–29.

[28] Hardman AM, Stewart GS, Williams P. Quorum sensing and the cell-cell communication dependent regulation of gene expression in pathogenic and non-pathogenic bacteria. Antonie Van Leeuwenhoek 1998;74(4):199–210.

[29] McClean KH, Winson MK, Fish L, Taylor A, Chhabra SR, Camara M, et al. Quorum sensing and *Chromobacterium violaceum*: exploitation of violacein production and inhibition for the detection of N-acylhomoserine lactones. Microbiology 1997;143(12):3703–11.

[30] Wirth R. Sex pheromones and gene transfer in *Enterococcus faecalis*. Res Microbiol 2000;151(6):493–6.
[31] Gray KM, Pearson JP, Downie JA, Boboye B, Greenberg EP. Cell-to-cell signaling in the symbiotic nitrogen-fixing bacterium *Rhizobium leguminosarum*: autoinduction of a stationary phase and rhizosphere-expressed genes. J Bact 1996;178(2):372–6.
[32] Eberhard A, Burlingame AL, Eberhard C, Kenyon GL, Nealson KH, Oppenheimer NJ. Structural identification of autoinducer of *Photobacterium fischeri* luciferase. Biochemistry 1981;20(9):2444–9.
[33] Engebrecht J, Nealson K, Silverman M. Bacterial bioluminescence: isolation and genetic analysis of functions from *Vibrio fischeri*. Cell 1983;32(3):773–81.
[34] Kaprelyants AS, Kell DB. Do bacteria need to communicate with each other for growth? Trends Microbiol 1996;4(6):237–42.
[35] Costerton JW, Lewandowski Z, Caldwell DE, Korber DR, Lappin-Scott HM. Microbial biofilms. Annu Rev Microbiol 1995;49(1):711–45.
[36] Kaiser D, Losick R. How and why bacteria talk to each other. Cell 1993;73(5):873–85.
[37] Harris SJ, Shih YL, Bentley SD, Salmond GP. The hexA gene of *Erwinia carotovora* encodes a LysR homologue and regulates motility and the expression of multiple virulence determinants. Mol Microbiol 1998;28(4):705–17.
[38] Moré MI, Finger LD, Stryker JL. Enzymatic synthesis of a quorum-sensing autoinducer through use of defined substrates. Science 1996;272(5268):1655.
[39] Stephens K. Pheromones among the procaryotes. CRC Crit Rev Microbiol 1986;13 (4):309–34.
[40] Fuqua C, Winans SC, Greenberg EP. Census and consensus in bacterial ecosystems: the LuxR-LuxI family of quorum-sensing transcriptional regulators. Annu Rev Microbiol 1996;50(1):727–51.
[41] Schaefer AL, Hanzelka BL, Eberhard A, Greenberg EP. Quorum sensing in *Vibrio fischeri*: probing autoinducer-LuxR interactions with autoinducer analogs. J Bact 1996;178(10):2897–901.
[42] Turovskiy Y, Kashtanov D, Paskhover B, Chikindas ML. Quorum sensing: fact, fiction, and everything in between. Adv Appl Microbiol 2007;62:191–234.
[43] Fuqua WC, Winans SC, Greenberg EP. Quorum sensing in bacteria: the LuxR-LuxI family of cell density-responsive transcriptional regulators. J Bact 1994;176(2):269.
[44] Rumjanek NG, Fonseca MD, Xavier G. Quorum sensing. Revista Biotecnologia Ciência & Desenvolvimento 2004;33:35.
[45] Visick KL, Ruby EG. *Vibrio fischeri* and its host: it takes two to tango. Curr Opin Microbiol 2006;9(6):632–8.
[46] Ankley GT, Peterson GS, Amato JR, Jenson JJ. Evaluation of sucrose as an alternative to sodium chloride in the Microtox® assay: comparison to fish and cladoceran tests with freshwater effluents. Environ Toxicol Chem 1990;9(10):1305–10.
[47] Hinwood A, McCormick M. The effect of ionic solutes on EC50 values measured using the Microtox test. Environ Toxicol 1987;2(4):449–61.
[48] Johnson BT. Microtox® acute toxicity test. Small-scale freshwater toxicity investigations. Netherlands: Springer; 2005. p. 69–105.
[49] Watanabe H, Inaba H, Hastings JW. Effects of aldehyde and internal ions on bioluminescence expression of *Photobacterium phosphoreum*. Arch Microb 1991;156(1):1–4.
[50] O'Shea TM, DeLoney-Marino CR, Shibata S, Aizawa S-I, Wolfe AJ, Visick KL. Magnesium promotes flagellation of *Vibrio fischeri*. J Bact 2005;187(6):2058–65.
[51] Azur. Acute_Overview: AZUR environmental. Available from: http://www.coastalbio.com/images/Acute_Overview.pdf; 1998 [accessed 20.05.16].

[52] Hastings J, Greenberg EP. Quorum sensing: the explanation of a curious phenomenon reveals a common characteristic of bacteria. J Bact 1999;181(9):2667–8.

[53] Deheyn DD, Bencheikh-Latmani R, Latz MI. Chemical speciation and toxicity of metals assessed by three bioluminescence-based assays using marine organisms. Environ Toxicol 2004;19(3):161–78.

[54] Petala M, Tsiridis V, Kyriazis S, Samaras P, Kungolos A, Sakellaropoulos G, editors. Evaluation of toxic response of heavy metals and organic pollutants using Microtox acute toxicity test. In: Proceedings of the International Conference on Environmental Science and Technology; 2005.

[55] Fulladosa E, Desjardin V, Murat J-C, Gourdon R, Villaescusa I. Cr (VI) reduction into Cr (III) as a mechanism to explain the low sensitivity of *Vibrio fischeri* bioassay to detect chromium pollution. Chemosphere 2006;65(4):644–50.

[56] Heipieper H-J, Keweloh H, Rehm H-J. Influence of phenols on growth and membrane permeability of free and immobilized *Escherichia coli*. Appl Env Microbiol 1991;57 (4):1213–17.

[57] Ismailov AD, Pogosyan SI, Mitrofanova TI, Egorov NS, Netrusov AI. Bacterial bioluminescence inhibition by chlorophenols. Appl Biochem Microbiol 2000;36(4):404–8.

[58] ABNT. Ecotoxicologia aquática. Rio de Janeiro: Associação Brasileira De Normas Técnicas; 2006 Contract No.: NBR 15.411-3:2006.

[59] Bertoletti E. Controle ecotoxicológico de efluentes líquidos no Estado de São Paulo. In: CETESB, Série Manuais; 2008.

[60] Bjerketorp J, Håkansson S, Belkin S, Jansson JK. Advances in preservation methods: keeping biosensor microorganisms alive and active. Curr Opin Biotechnol 2006;17 (1):43–9.

[61] Kaiser KLE, Palabrica VS. *Photobacterium phosphoreum* toxicity data index. Water Poll Res J Can 1991;26:361–431.

[62] Jennings VL, Rayner-Brandes MH, Bird DJ. Assessing chemical toxicity with the bioluminescent photobacterium (*Vibrio fischeri*): a comparison of three commercial systems. Water Res 2001;35(14):3448–56.

[63] Kroon A.G.M., Mullem A.V. MutatoxTM test. Aquasense [20 September 2016]. Available from: http://www.microlan.nl/cms/spaw2/uploads/files/applicaties/Mutatox_test_Aqua Sense.pdf phpMyAdmin90a1592554ecf76ebe7c76e2d531280f = ?

[64] Checklight. ToxScreen-II test 2016 [20 September 2016]. Available from: http://www.checklight.biz.

[65] UBIOTECH. Frequently asked questions from users of the product Biolux®Lyo10 [20 September 2016]. Available from: www.ubiotech.com.br.

[66] Chromadex. Bioluminex 2016 [20 September 2016]. Available from: http://www.chromadex.com/Literature/Catalog/Bioluminex.pdf.

[67] Umbuzeiro GdA, Rodrigues PF. O teste de toxicidade com bactérias luminescentes e o controle da poluição das águas. Mundo Saúde (Impr) 2004;28(4):444–9.

[68] Brock TD, Madigan MT. Biology of microorganisms. 6th ed. Upper Saddle River, NJ: Prenctice-Hall Inc; 1991.

[69] Stanier RY. Microbial pathogenesis. In: Stanier RY, Ingraham JI, Wheelis ML, Painter RR, editors. General microbiology. UK: Macmillan Education; 1986. p. 621–34.

[70] Johnson BT, Long ER. Rapid toxicity assessment of sediments from estuarine ecosystems: a new tandem in vitro testing approach. Environ Toxicol Chem 1998;17(6):1099–106.

[71] Kahru A, Borchardt B. Toxicity of 39 MEIC chemicals to bioluminescent photobacteria (the Biotox test): correlation with other test systems. ATLA-Altern Lab Anim 1994;22:147–60.

[72] Guzzella L. Comparison of test procedures for sediment toxicity evaluation with *Vibrio fischeri* bacteria. Chemosphere 1998;37(14):2895–909.

[73] Parvez S, Venkataraman C, Mukherji S. A review on advantages of implementing luminescence inhibition test (*Vibrio fischeri*) for acute toxicity prediction of chemicals. Environ Int 2006;32(2):265–8.

[74] USEPA. Freshwater ecosystems: United States environmental protection agency [12 December 2015]. Available from: http://www.epa.gov/bioindicators/aquatic/freshwater.html.

[75] Ren S. Assessing wastewater toxicity to activated sludge: recent research and developments. Environ Int 2004;30(8):1151–64.

[76] Applegate B, Kehrmeyer S, Sayler G. A chromosomally based tod-luxCDABE whole-cell reporter for benzene, toluene, ethybenzene, and xylene (BTEX) sensing. Appl Env Microbiol 1998;64(7):2730–5.

[77] Lei Y, Chen W, Ashok M. Microbial biosensors. Analyt Chim Acta 2006;568 (1):200–10.

[78] Selifonova O, Burlage R, Barkay T. Bioluminescent sensors for detection of bioavailable Hg (II) in the environment. Appl Env Microbiol 1993;59(9):3083–90.

[79] Widder EA, Case JF. Two flash forms in the bioluminescent dinoflagellate, *Pyrocystis fusiformis*. J Comparat Physiol 1981;143(1):43–52.

[80] Latz MI, Bovard M, VanDelinder V, Segre E, Rohr J, Groisman A. Bioluminescent response of individual dinoflagellate cells to hydrodynamic stress measured with millisecond resolution in a microfluidic device. J Exp Biol 2008;211(17):2865–75.

[81] Eckert R. Bioelectric control of bioluminescence in the dinoflagellate *Noctiluca*: II. Asynchronous flash initiation by a propagated triggering potential. Science 1965;147:1142–5.

[82] Latz MI, Nauen JC, Rohr J. Bioluminescence response of four species of dinoflagellates to fully developed pipe flow. J Plankton Res 2004;26(12):1529–46.

[83] Tesson B, Latz MI. Mechanosensitivity of a rapid bioluminescence reporter system assessed by atomic force microscopy. Biophys J 2015;108(6):1341–51.

[84] Hardeland R., Hoppenrath M. Bioluminescence in Dinoflagellates [24 February 2017]. Available from: http://tolweb.org/notes/?note_id = 5621.

[85] von Dassow P, Latz MI. The role of Ca^{2+} in stimulated bioluminescence of the dinoflagellate *Lingulodinium polyedrum*. J Exp Biol 2002;205:2971–86.

[86] Jin K, Klima JC, Deane G, Dale Stokes M, Latz MI. Pharmacological investigation of the bioluminescence signaling pathway of the dinoflagellate *Lingulodinium polyedrum*: evidence for the role of stretch-activated ion channels. J Phycol 2013;49(4):733–45.

[87] Mallipattu S, Haidekker M, Von Dassow P, Latz M, Frangos J. Evidence for shear-induced increase in membrane fluidity in the dinoflagellate *Lingulodinium polyedrum*. J Comp Physiol A 2002;188(5):409–16.

[88] Maldonado EM, Latz MI. Shear-stress dependence of dinoflagellate bioluminescence. Biol Bull 2007;212(3):242–9.

[89] Abrahams MV, Townsend LD. Bioluminescence in dinoflagellates: a test of the burgular alarm hypothesis. Ecology 1993;74(1):258–60.

[90] Mesinger A, Case J. Dinoflagellate luminescence increases susceptibility of zooplankton to teleost predation. Mar Biol 1992;112(2):207–10.

[91] Fleisher KJ, Case JF. Cephalopod predation facilitated by dinoflagellate luminescence. Biol Bull 1995;189(3):263–71.

[92] White HH. Effects of dinoflagellate bioluminescence on the ingestion rates of herbivorous zooplankton. J Exp Mar Biol Ecol 1979;36(3):217–24.

[93] Buskey EJ, Swift E. Behavioral responses of oceanic zooplankton to simulated bioluminescence. Biol Bull 1985;168:263–75.
[94] Esaias WE, Curl HC. Effect of dinoflagellate bioluminescence on copepod ingestion rates. Limnol Oceanogr 1972;17(6):901–6.
[95] Latz MI, Rohr J. Luminescent response of the red tide dinoflagellate *Lingulodinium polyedrum* to laminar and turbulent flow. Limnol Oceanogr 1999;44(6):1423–35.
[96] Deane G, Stokes MD. A quantitative model for flow-induced bioluminescence in dinoflagellates. J Theor Biol 2005;237(2):147–69.

Image processing for bioassays

Marcus Jansen, Stefan Paulus, Kevin Nagel and Tino Dornbusch
LemnaTec GmbH, Aachen, Germany

13.1 Introduction

Bioassays are essential tools in many disciplines comprising plant science, ecotoxicology, and agronomy. They deliver essential information on developmental processes, gene functions, or effects of external factors. Thereby they enable tackling grand challenges in ensuring human nutrition, coping with climatic uncertainties, as well as preventing or mitigating environmental damage [1–5]. It is well known that the human population in many parts of the world is still growing and the demand for affordable and nutritious food strongly increases. Agricultural production needs to augment crop yield substantially to keep pace with this development [6–8]. Particularly gaining welfare in developing countries drives the demand for animal-derived products that potentiate the request for productive agriculture to provide plants for feeding livestock.

Plant breeders and producers of fertilizer and plant protectants face the challenge to supply highly efficient products that preserve resources and avoid unwanted secondary effects. Crop yield, plant growth performance, stress resilience, and resource use are primary targets in plant improvement. For fertilizers, cultivation adjuvants, and protectants specificity of action, resource efficiency, and absence of toxicity are highly desirable. Achieving these targets demands dedicated tools to screen for novel candidates, test their functions and effects as well as their interaction with the whole system of agricultural production. A multitude of molecular and biochemical methods that assist achieving such goals is well established and continuously improved [9]. Complementary to molecular measurements, assessments of phenotypic properties are highly informative for evaluating the targeted function. Noninvasive sensor-based measurements deliver comprehensive sets of phenotypic traits and properties, their spatial distribution as well as their temporal changes. Thus, such methods are powerful tools to assess both internal functions as well as external effects at multiple levels. A central element of such noninvasive tools is to acquire and analyze images of biological objects [10]. Technically, a basic measurement platform consists of cameras or related sensors together with dedicated illumination and suitable imaging environment. The imaging environment prevents external light and provides homogeneous backgrounds for image acquisition. Central elements are software packages that capture sensor data and translate the measured values into parameters that deliver information of biological relevance. Enormous technical progress was achieved in the recent two decades. Beginning with analyzing images of quite simple objects such as duckweed fronds or *Arabidopsis rosettes*, phenotyping technology was developed for more sophisticated

Bioassays. DOI: http://dx.doi.org/10.1016/B978-0-12-811861-0.00013-9

applications later. These comprise 3D measurements in visible or nonvisible wavelength ranges, or analyses of fluorescence light.

13.2 Sensor-equipped phenotyping platforms

For use in scientific experiments, breeding, product development, environmental monitoring, or precision agriculture, sensors are mounted on dedicated platforms that allow proper sensor operation targeted to capture phenotypic properties of the plants or other biological objects. Depending on the use case, platforms are designed for laboratory, greenhouse, or field measurements. The setups range from manually operated instruments to fully automated multisensor platforms. LemnaTec laboratory Scanalyzers are either manually-operated cabinets with a single sensor type, or they have a mechanical automation that moves sensors to larger amounts of preloaded samples, e.g., leaf discs in multiwell plates (Fig. 13.1A, B). The latter types allow high-throughput measurements with multiple sensors. Such laboratory Scanalyzers are dedicated to seedlings, rosette-forming plants (e.g., *Arabidopsis*), or biological samples on Petri dishes or multiwell plates.

Scanalyzers for larger growth rooms and greenhouses allow top- and side-view measurements of potted plants and therefore are referred to as 3D-Scanalyzers (Fig. 13.1C). Such platforms usually comprise automated transport of pots on conveyer systems (Fig. 13.1D) that enable automated measurements of up to several hundred pots. Combined with weighing and water supply stations, the irrigation of the pots is controlled by the automated system, too, and thereby enables applying

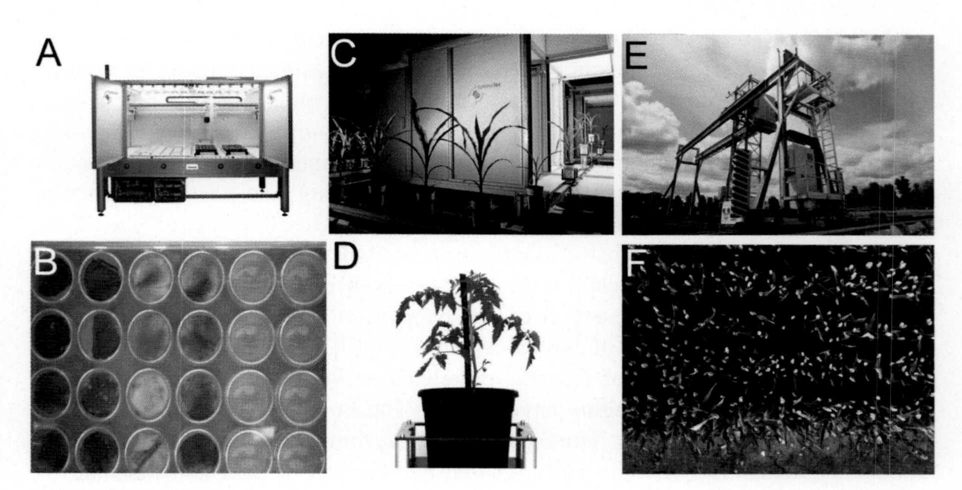

Figure 13.1 LemnaTec Scanalyzers together with typical sample images. Laboratory HTS Scanalyzer (A) is used e.g., for leaf discs in well-plates (B). 3D Scanalyzers (C) operate in greenhouses and measure potted plants (D). Field Scanalyzers (E) perform canopy measurements (F).

various water supply scenarios, e.g., drought experiments. Field Scanalyzers (Fig. 13.1E) are positioning setups that carry a group of downwards-viewing sensors across plot areas, both for field experiments and for indoor cultivation with plants growing in the ground or large immobile containers.

13.3 Sensors for noninvasive plant measurements

The overall phenome of plants describing the entity of the outer appearance consists of various properties, such as size, morphology, or colors that are displayed by the organs of the plant. To capture different aspects of the plants' phenome, geometric and radiometric information is measured with various noninvasive sensors. Currently a broad range of sensors, comprising cameras operating in RGB, IR, and NIR ranges, hyperspectral cameras, fluorescence and luminescence cameras, as well as laser scanners are used in phenotyping. To measure and analyze phenotypic properties, LemnaTec provides platforms for the coincidental use of different sensors together with a joint storage system and a collective analysis software platform.

As an example, the new Field Scanalyzer platform (Fig. 13.1E) carries a multisensor box including sensors that cover the electromagnetic spectrum between 380–1700 nm, such as RGB cameras, hyperspectral cameras (VNIR & ExVNIR), and a thermal camera. Moreover, there are 3D laser-scanners and environmental sensors, which measure e.g., light conditions, air humidity, or air temperature. Such technology records a comprehensive set of images and related data types of the plants (e.g., an RGB photograph of a wheat canopy, Fig. 13.1F), comprising a broad range of radiometric information. Other Scanalyzer platforms with one or multiple sensors have been designed and are available for greenhouse and laboratory uses.

13.3.1 RGB cameras

Plants generally have a low reflectance in the visible part of the electromagnetic spectrum, because chlorophylls, carotenoids, and other pigments absorb large parts of the radiation. There is a characteristic peak of reflectance around 550 nm, due to low absorbance of the pigments in this wavelength region [11]. This reflectance makes plants look green and enables the detection of plant tissue in RGB images. The characteristic green color is the most important factor that allows for the separation of plants from the background in image analysis procedures. Therefore, colors with good contrast towards green, e.g., white or black, are chosen as background material when setting up phenotyping platforms. Time-resolved image analysis measuring dimensions of visible plant tissue enables the monitoring of the growth of plants, and changes in growth are very sensitive indicators of plant stress, e.g., water limitation. Reduced growth rates frequently occur already at mild stress levels, before other factors such as stomatal conductance or photosynthesis are affected [11–13]. Therefore, growth monitoring is an essential and very powerful

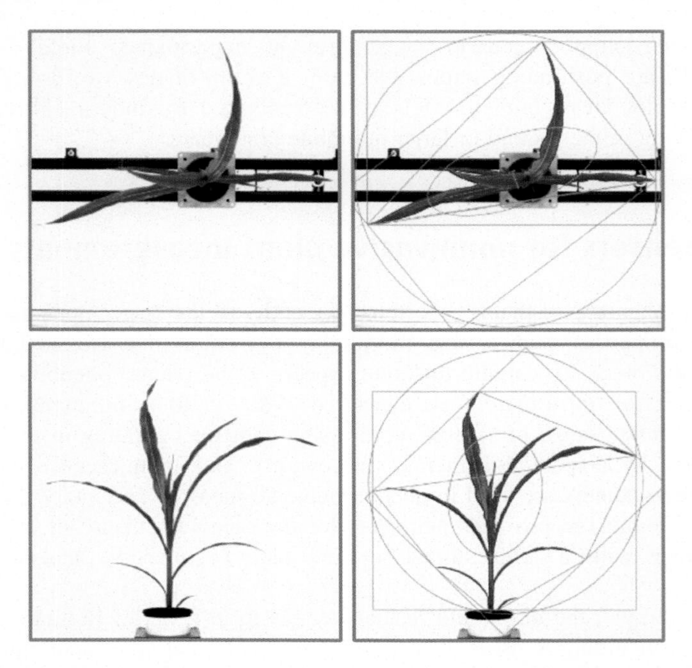

Figure 13.2 Top- and side-view images of a maize plant in a 3D Scanalyzer imaging unit. *Red lines* denote image processing results that comprise separating the plant from the background, measuring the visible surfaces, and analyzing geometric parameters.

tool in plant phenotyping. Beyond analyzing size and growth derived thereof, RGB images, particularly when taken from different view angles, give access to dimensions and morphological data on the plants and on surface color information (Fig. 13.2).

Morphological factors give additional insight into the phenotypic state of plants compared to what size and growth data would do alone. For instance, in a study by Jansen et al. [14], two populations of *Arabidopsis thaliana* plants grown at different light and irrigation levels had the same average size but markedly differed in compactness at the end of the measurement period. In this case, the response to environmental factors was much better resolved when combining size and shape information derived from the image processing. The third factor that derives from RGB image processing is the color information that can carry valuable information on environmental responses, because deviations from the original leaf color frequently indicate stress, diseases, or senescence of the tissue. In turn, sophisticated image processing tools are required to correctly allocate the observable color distribution to their respective biological causes. This in turn opens the field for broad future developments for phenotyping of stress and disease responses. RGB image processing not only applies for shoots and leaves of plants, but also for example for roots [15], mites [16], or yeast colonies [17].

13.3.2 Infrared cameras

Whereas RGB cameras record electromagnetic radiation in the spectral range between 380 and 780 nm, infrared (IR) cameras capture thermal radiation in the range of 7500–13,000 nm and thereby capture the intensity of IR radiation that is emitted from the object surface. Such IR radiation correlates to temperature, e.g., that of the plant tissue. Temperatures in plant tissues largely depend on external thermal energy. Whereas examples for plants heating up their tissue actively are very rare, active cooling by transpiration is common in leaves. Transpiration for leaf cooling is essential because plants have to expose their leaves to the incoming sunlight in order to collect energy for photosynthesis [18]. Transpiration works through opening stomata, which allows water evaporating from the leaves and coincidently entering CO_2 that is required for photosynthesis. Transpiration cools down leaf tissue temperatures, usually below ambient air temperatures. However, in water-limited conditions, full-rate transpiration quickly depletes water resources, and therefore water limitations triggers closing the stomata [18,19]. In such cases, temperatures of leaf tissue increase and plants must handle the trade-off between water loss, risking tissue dry-out, and temperature increase, risking tissue heat-damage. Moreover, other stress factors, such as pathogen attacks, change stomatal regimes and thus alter transpiration and thereby change heat emission [20]. Overall there are various factors in plant physiology that have an impact on surface IR emission, thus recording such is meaningful for many research purposes. When using IR measurements for recording information on such processes and status, it must be considered that all these measurements act indirectly. The use of IR information for assessing plant physiological properties is an active field of research and method development, and it is to be expected that various applications will arise in the future. When sensing plants with IR cameras, changes in environmental temperatures prior to or during the measurement may critically change the heat emission and thereby influence the result. For instance, when moving a camera above the sample and thereby creating shade on the plant, it is likely that IR emission responds. Even more prominently, when moving a plant out of a growth environment into a measuring platform for IR recording, changes in the surrounding temperature are highly probable and therefore IR emission of the plant will differ from the situation in the growth environment. Researchers need to consider such technical issues and consider them for experimental design and data interpretation.

13.3.3 Near-infrared cameras

As all biological processes require sufficient availability of water, monitoring water content and availability is a useful and essential tool in plant biology. For many purposes, e.g., when monitoring growth processes, it is important to assess water content noninvasively rather than harvesting and drying the material to weigh water content. Water content in plant tissue causes reflectance of electromagnetic radiation in near- (700–1300 nm) and mid- (1300–2500 nm) infrared wavelength ranges [11,21,22]. The transition from visible light to near-infrared (NIR) radiation is characterized by a sharp increase of reflectance from leaf surfaces, a phenomenon called red edge. Near-infrared (NIR, also called short-wave infra-red, SWIR) radiation is largely reflected by plant leaves in canopies due to scattering in the mesophyll of individual leaves and through layers of overlapping leaves. Such scattering effects decrease in wavelength ranges beyond 1300 nm due to absorption by the water that is present inside the leaves. At 1450 nm, 1930 nm, and 2500 nm there are characteristic water absorption bands [11,23]. The water present in the leaves and causing the absorption is the target of NIR imaging, as the ratio of current water content to water content of fully turgescent leaves determines the water status of the leaf. A fully turgescent leaf has no water deficit, and can be assumed to have 100% of the maximum possible water content. Water deficit, e.g., caused by low soil water availability or by high air temperatures, eventually combined with wind, decreases water content in leaves. Leaves counteract water loss by closing stomata and slowing down water-consuming processes such as photosynthesis, but this cannot fully prevent losing water from leaf tissue. Leaf water loss changes reflectance in the NIR and mid-infrared (MIR), and changes become particularly obvious at the water absorption bands [23]. When using NIR reflectance in plant phenotyping, specific reflective properties of the plant material must be considered. NIR reflectance is inversely proportional to water content of the measured materials.

To map NIR signals spatially to plants, NIR-sensitive cameras can be applied to generate NIR images. In such images, regions with high water content and consequently low NIR reflectance appear dark, whereas dryer regions have higher NIR reflectance and appear bright. Using image analysis techniques, such reflectance patterns can be translated into color-coded maps describing relative water content of plant tissue.

13.3.4 Fluorescence recording cameras

Different from other camera applications, fluorescence light emission is recorded instead of the reflectance of incident radiation. Nevertheless, such incident radiation is required as excitation for the subsequent fluorescence emission. Therefore, precise knowledge of the light source that does the excitation is required. Such may be natural light when measuring outside, but in experimental setups excitations light frequently originates from artificial light sources. Recording cameras usually are equipped with filters that are specific to the targeted type of fluorescence. The cameras are very sensitive to light because fluorescence light is usually very weak. High light sensitivity in turn decreases the resolution of the camera. Recording fluorescence can have different targets. First it aims at static autofluorescence that can be emitted from tissue constituents such as pigments or from fluorescent biomarkers that are applied in experiments. Second it aims at dynamic recording of fluorescence transients of the chlorophyll.

Autofluorescence assays that record fluorescence of plant tissue constituents can serve for monitoring stress and senescence, for instance as response to salinity as an adverse environmental factor [24]. In this study, monitoring the fluorescence signal served as phenotypic indicator for senescence. In molecular biology, the use of green fluorescent protein (GFP) and other fluorescent biomarkers is a widespread tool for studying gene expression and protein localization. Recording related

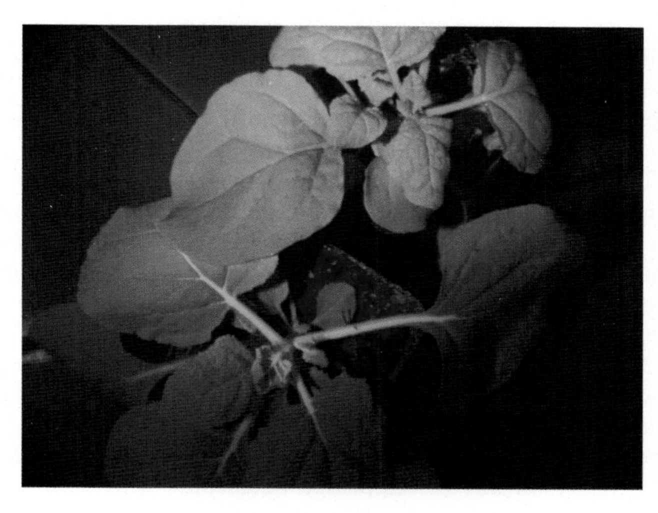

Figure 13.3 *Nicotiana benthamiana* plants without (upper plant) and with (lower plant) constitutive GFP expression.

fluorescence (Fig. 13.3) together with image processing allows detailed characterization of biomarker distribution within plant tissues.

Chlorophyll fluorescence analysis is a common tool to infer the state of the photosynthetic apparatus (cf. Chapter 9: Photosynthesis assessed by chlorophyll fluorescence, this volume). Fluorescence analysis addresses the status and functions of the chlorophylls of the light-harvesting complexes. It must be pointed out that measuring chlorophyll fluorescence is not a direct measurement of the whole photosynthetic process, which comprises the chlorophyll-related light reactions and the Calvin cycle (so-called "dark reaction"). Carbon fixation from aerial CO_2 is part of the Calvin cycle reactions. Early studies revealed that there is a relationship of visible fluorescence and photosynthesis [25]. Due to the interrelationship of the different types of energy dissipation in chlorophyll, measuring fluorescence emission can serve as tool for calculating photosynthetic capacity and activity [26,27]. However, fluorescence emission is a minor function of chlorophyll whereas electrochemical energy transfer is the main and intended function that leads to photosynthesis. Multiple scientific studies were conducted for interpreting fluorescence signals and drawing conclusions on photosynthetic functions as well as on the status of the photosystems. Based on such studies, chlorophyll fluorescence analysis became a widespread tool in plant physiology and ecophysiology [26–28]. Thus, chlorophyll fluorescence measurements are state-of-the-art procedures in current plant phenotyping assays [29]. When measuring chlorophyll fluorescence, multiple factors influence the result and need consideration for data interpretation. Besides the genetic properties of the plant, in particular diurnal rhythms—as well as environmental factors and their fluctuations—change photosynthetic processes and thereby lead to variations in measurable fluorescence [30]. Measuring chlorophyll fluorescence requires fluorescence light sensors that are combined with illumination units. Such setups allow not only recording the current state of fluorescence but also inducing changes in fluorescence by changing illumination conditions. Recording fluorescence induction curves enables calculating parameters related to chlorophyll functions. Sensors can either be spot sensors, usually combined with leaf clips, or specific cameras that record fluorescence with spatial resolution and display fluorescence maps across the plant surface. For comprehensive biological analyses, it is desirable to have data on photosynthetic parameters at whole leaf level, or even better for all visible leaves of a plant. Therefore, imaging measurements are favorable compared to spot-measurements in terms of spatial resolution. Chlorophyll fluorescence images allow pixel-wise evaluation of data and thereby enable assessing fluorescence differences between leaves and even within leaves [30–33]. There are two main measuring principles in use, the direct fluorescence method and the modulated fluorescence method, both having specific advantages and drawbacks [28,30]. Phenotyping platforms can harbor measuring heads of both operation types, depending on the scientific and technical demands, for instance the GROWSCREEN Fluoro platform containing a pulse amplitude modulated (PAM) fluorometer [34], or the Field Scanalyzer carrying a camera for direct fluorescence measurements [35]. Chlorophyll fluorescence measurements serve as tools for estimating photosynthetic parameters in many fields of plant sciences and breeding. They are used for assessing plant productivity, effects of management, fertilizer application, agrochemical application, or environmental factors. Changes in chlorophyll fluorescence are considered as an indicator of stress and many plant diseases.

In a water deficit study, mild water deficit evoked strong changes in the effective quantum yield (Fig. 13.4C) that relates to operating photosynthesis, indicating that

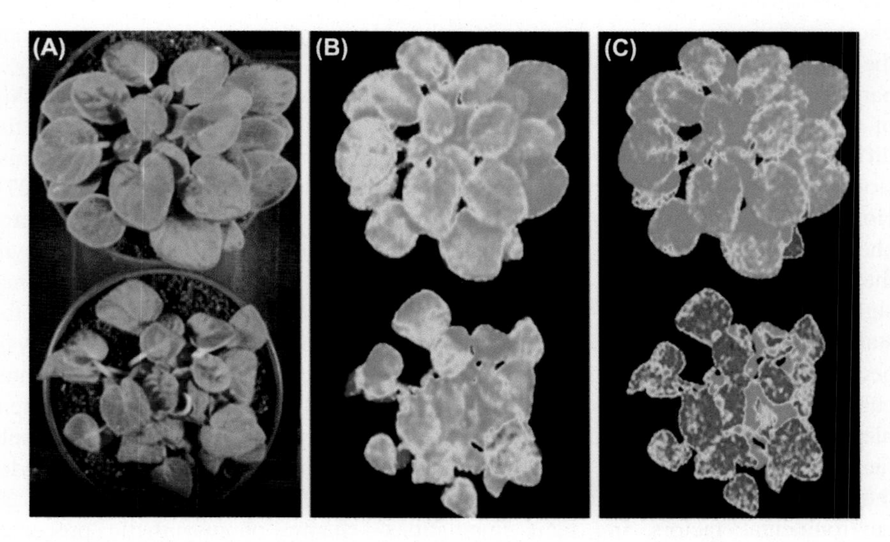

Figure 13.4 Well-watered (upper row) and water deficient (lower row) plants assessed for maximum quantum yield of PSII (Fv/Fm; B) and effective quantum yield of PSII (Fv′/Fm′; C); RGB-photographs as comparison (A).

photosynthesis was impaired at low water availability. The maximum quantum yield, however, was nearly equal in both plants (Fig. 13.4B), indicating that there was no substantial damage to photosystem II. Testing mutants, transgenic lines, or novel cultivars frequently includes chlorophyll fluorescence measures to detect changes in photosynthetic capacity. Moreover, chlorophyll fluorescence measurements support data on quality and storage processes of vegetables, flowers, and fruits [30].

13.3.5 Hyperspectral cameras

Depending on their developmental stage and their interaction with the environment, plants produce and accumulate a broad variety of substances, which can be localized within cells, in cell walls, intercellular space, in the cuticle, or as epicuticular compounds. Such biochemical constituents comprise pigments, water, proteins, and other metabolites that influence spectral reflectance in the visible light range but also in nonvisible wavelength bands, particularly NIR radiation. The influence on reflectance depends on the chemical properties of the metabolites and their spatial distribution on or below the plant surfaces. Metabolites can absorb or reflect incident radiation and thereby contribute to the overall spectral absorbance and reflectance of the plant surface. Therefore, spectral properties of plants carry information on the physiological state of the plant and serve as indicators for plant performance. In addition, orientation of plant organs and their physical surface structures contribute to spectral reflectance. Taken together, complex sets of factors make up the spectral reflectance measurable at the plant surfaces, with single compounds attributing with overlapping peaks. Such overlaps require sophisticated data analysis procedures to attribute the measured signals to their respective biochemical origins.

Various approaches of optical spectroscopy, and in particular using hyperspectral imaging, have been established for characterizing vegetation with sensor technology [14,36,37]. This includes remote sensing with satellites and aerial vehicles, as well as proximal sensing with equipment carried closely to the plants. By application of multi- and hyperspectral sensors, spectra of plant surface reflectance are collected (Fig. 13.5) and analyzed for calculating characteristic parameters such as vegetation indices [38]. Approaches using the complete spectra were implemented to analyze differences between different diseases, e.g., to distinguish between sugar beet rust and powdery mildew [39], resulting in so called metro maps showing the dynamics of disease development using automated data mining routines [40].

The advantage of hyperspectral image analysis over the analysis of standard RGB images is the possibility to monitor and quantify changes in the reflectance of plants and single organs when the presence of e.g., diseases cannot be seen in the visible spectrum.

The following text focuses on hyperspectral imaging from short distances, in many cases under controlled illumination in contrast to unmanned aerial vehicle (UAV), plane, or satellite imaging. Nevertheless, many examples in literature derived from remote sensing applications and validation for proximal sensing need further work. Unlike hyperspectral spectrometers which acquire data from a limited measuring spot, hyperspectral cameras produce a 2D projection of the measured object and measure spectra pixel-wise. This delivers a data-cube consisting of a

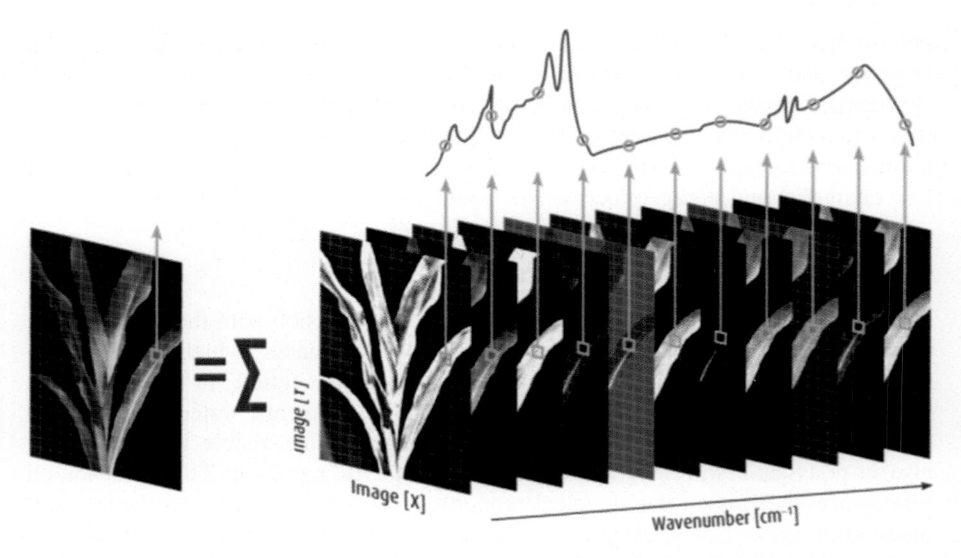

Figure 13.5 A hyperspectral image is a data-cube consisting of different images for every observed wavelength.

Figure 13.6 Hyperspectral data: reflectance of a plant at different wavelengths.

stack of images of the object's reflectance throughout a range of wavelengths (Fig. 13.6). Such data cubes generally are large, and targeted data evaluation is necessary for adequate use of the data for phenotyping purposes.

Spectral reflectance data frequently serve for calculating vegetation indices. These indices, e.g., the biomass indicator NDVI (normalized difference vegetation index), are calculated from a few wavelengths of the measured spectra. Typically, the intensities of two or three spectral bands are applied against each other to characterize plant properties such as physiological or structural characteristics [10], pigment content [41,42], or responses to biotic [36] and abiotic stress. The choice of wavelengths is determined either by spectral properties of the measured target or by empirical experience and correlation of measurements to other observations such as biochemical components or biomass data.

13.3.6 Laser scanners

During the last years, technology has risen to image the complete morphological structure of plants [43,44]. Access to the 3D shape enables the derivation of parameters that are hidden when using 2D images. Monitoring of growth of individual leaves and their movements, e.g., during processes controlled by the circadian clock, is only possible if full access to the geometry is granted [45].

For the data acquisition, there are different techniques like structure-from-motion approaches that use various RGB images from different points of view [46], time-of-flight cameras [47], structured light [48], or laser scanners [49,50] that use active laser triangulation or a phase shift method to measure 3D data. During the past years, research has focused on laser scanning due to the high resolution and the fast acquisition process. Point to point distances of 0.25 mm and similar point accuracy are possible and enable high precision plant imaging in 3D on laboratory, greenhouse, and field scales. Approaches for high-throughput measurements have been developed, as well as methods for the automated classification of 3D point clouds to separately analyze the plants organs like leaf, stem, or ears [51]. Technologies and methods established for analyzing 2D projections cannot be simply transferred to 3D images. Nevertheless, during the last years, science has taken a big step forward in creating automated pipelines for 3D plant parameterization [40], similar to what has been done before in 2D space.

As these first steps have been undertaken on the laboratory or greenhouse scale, the new challenge is to bring this knowledge to the field. Plant phenotyping in 3D enables the generation of new parameters like the calculation of height maps, or by imaging at different points in time the visualization of growth maps (Fig. 13.7).

A growth map requires the subtraction of two point clouds of one sample generated at different subsequent times. Therefore, a high positioning accuracy of the sensor movement is essential.

For the classification of a 3D point cloud, a geometry-driven feature calculation is needed. Further information like color, fluorescence, IR, or hyperspectral reflectance data can be combined with the 3D data with sophisticated sensor fusion techniques. As the native output of a laser scanner is XYZ data, geometry-based features are ideal. These can be used as feature input for a classification using machine learning routines like support-vector-machines, boosted trees, or random forest approaches [52]. First results of a machine-learning based approach for the classification of leaves on a field point were generated with the Field Scanalyzer in a wheat experiment (Fig. 13.8C, F).

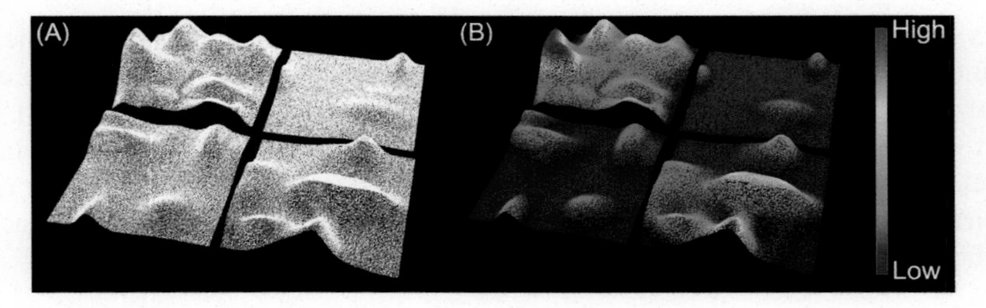

Figure 13.7 Coarse synthetic point cloud as an example for field data visualized with white points using shading for a better characterization (A). A height map can be extracted by encoding the height with color (B). This depicts the first step for the generation of a growth map where growth is encoded by color.

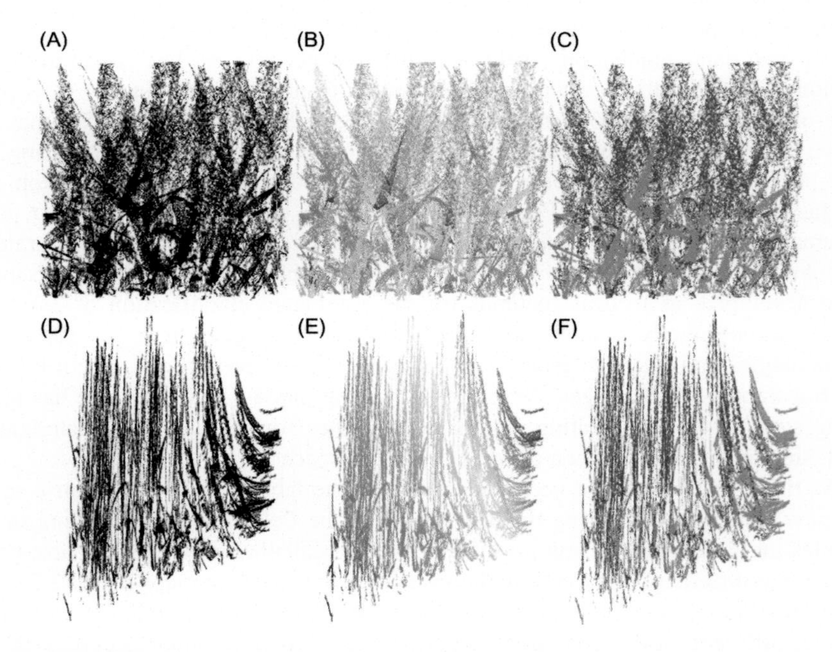

Figure 13.8 The XYZ point cloud can be visualized in Euclidian space with a view from above (A, B, C) or as a side view (D, E, F). (A) and (D) show the pure XYZ visualization, (B) and (E) show a coloring per the distance to the sensor. (C) and (F) show first classification results for the detection of leaf points within the point cloud using Surface Feature Histograms and a Random Forest classifier.

A highly accurate classification of the point cloud is essential for inclusion within an automated parameterization pipeline comprising point cloud acquisition, outlier elimination, classification of different organs, and finally parameterization by using e.g., primitives like triangles (leaves) or cylinders (stem).

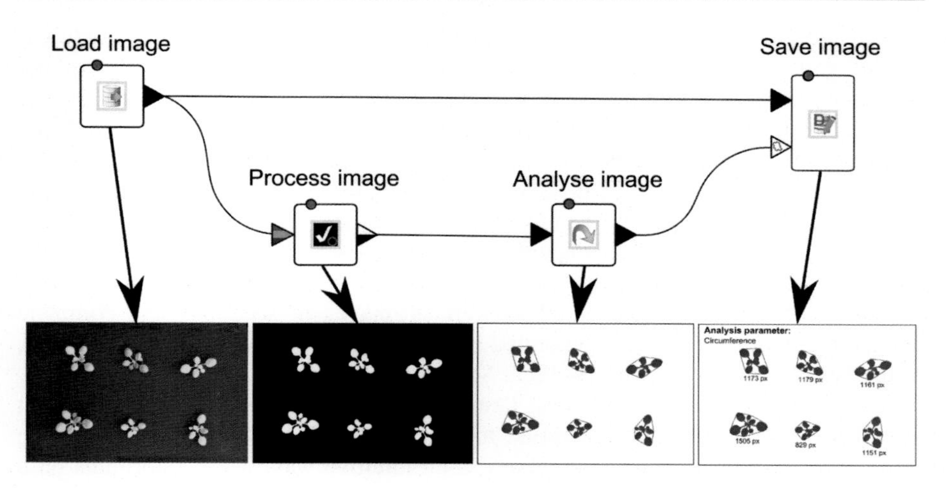

Figure 13.9 Example of a basic data processing pipeline for *Arabidopsis* image analysis built with the LemnaGrid software platform.

13.4 Environmental sensors

Phenotypes strongly depend on environmental factors and the influence of the environment on the phenotypic appearance of an individual plant is probably as high as the influence of the plant genome [4]. Therefore, information on factors such as temperature, light, air humidity or air flow in the growth environment and—if this is different—in the measuring environment are valuable and essential for interpreting phenotypic data. Similarly, information on soil properties and water availability amend phenotypic datasets. Dedicated sensors help to record and monitor different environmental data and storing such data together with the coincident phenotypic parameters of the plants exposed to these environments enables interpreting them in a manner of plant-environment interactions Fig. 13.9.

13.5 Software for sensor data capture and analysis

LemnaTec Scanalyzer systems cover complete phenotyping workflows that comprise plant or sensor positioning, data acquisition, data storage, data access, data processing, data analysis, and data visualization steps. The flexible workflow can comprise using external analysis tools and export of processed data into common statistical software packages, particularly into R software [53]. The essential part of the phenotyping systems is the software package and within the software package the image processing and analysis tools have highest importance in delivering phenotypic data out of the measured signals. The image analysis software LemnaGrid provides tools for the processing of RGB, IR, NIR, fluorescence, and hyperspectral images. Foreground/background segmentation, separation by color, and quantification of geometrical properties like diameter, bounding box, and perimeter or the convex hull, are typical processing steps for RGB images of plants. For hyperspectral images the calculation of a broad range of vegetation indices is possible. More sophisticated problems can be solved using state-of-the-art machine learning methods for the prediction of measured nitrogen levels using hyperspectral cameras on field scale. An advantage of the LemnaTec image processing software is its applicability on images almost independent of the source sensor. Furthermore, it works independently of the scale of the images. Images from laboratory experiments with microtiter plates can be processed as well as images of seedlings and complete plants. The latter are usually generated by the LemnaTec Greenhouse Scanalyzer, or the new multisensory platform Field Scanalyzer.

The image data analysis platform LemnaGrid is a computer vision software that provides a library of image operation tools, called grid devices. Users can create image processing and analysis pipelines in an intuitive way, using algorithms on a graphical programming interface. Users can establish re-usable and customizable workflows and they need not learn a programming language. Therefore, operators of the phenotyping platforms can run their analysis without involving data and information scientists. The workflow establishment functions with dragging and dropping grid devices from a library panel to the grid designer window, and then connecting the devices in a logical manner. Analysis pipelines can be re-used for subsequent experiments and shared among collaborators, thus inter-experimental comparability is promoted.

13.5.1 Image processing for shoot phenotypes

A basic workflow consists of four main devices: an image processing module that is a database reader for loading images, together with an image processing and an image analysis module that converts the image into objects and characterizes these. Finally, results and processed images are written to the database.

In a more complex application case, digitized images of agar-germinated *Arabidopsis* seedlings in a multiwell plate are loaded and processed in a pipeline made out of predefined functions. Using color converters and thresholds, plants are separated from the background, and pixel-areas recognized as plants are converted to objects. These objects are analyzed to describe their properties such as surface area, geometrical properties, and colors. Applying the procedure to an image acquired with a HTS Scanalyzer, the software recognizes the plants separately from the background and determines sizes and geometrical factors for each individual plant (Fig. 13.10). Rectangles drawn around the plants indicate that the software has measured length and width of the visible plant and such results are delivered as numerical values in output tables.

The same image processing principles can be applied to other plant species, too. Pot-grown crop plants such as soybeans are frequently imaged from top and side views in 3D-Scanalyzers that are mounted in greenhouses (Fig. 13.11).

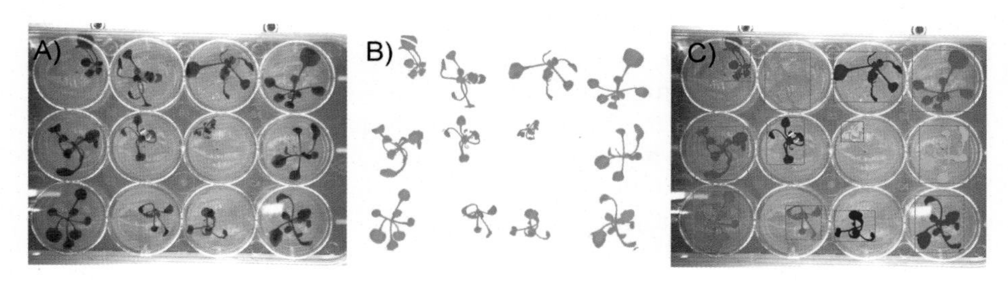

Figure 13.10 *Arabidopsis* seedling analysis with a LemnaGrid image processing pipeline. From the original image (A) the background is removed (B) and individual plants are further characterized (C).

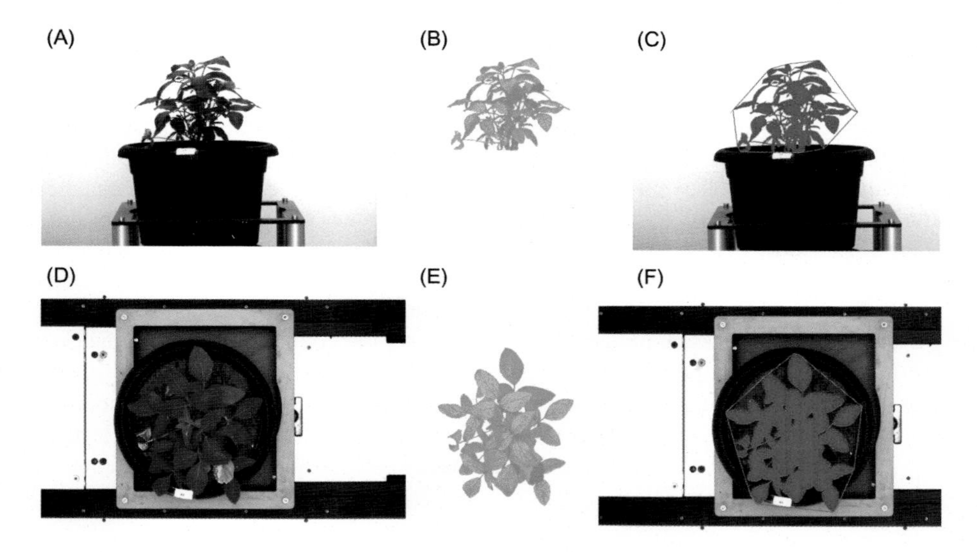

Figure 13.11 Soybean plant top- (D–F) and side- (A–C) view analysis with a LemnaGrid image processing pipeline. From the original image (A, D) the background is removed (B, E) and the plant is further characterized for color distribution (*blue* to *green* color-coding in B, E) and for geometry (lines around the plant in C, F).

Taken together, LemnaGrid enables an automated evaluation, but still leaves the possibility that users refine results in complex analysis cases.

13.5.2 Image processing for root phenotypes

Compared to shoots, roots are more challenging, both in terms of data acquisition and data processing. Root data acquisition relies on making roots accessible to the sensors, and comprises either excavating the root system, i.e., a single-point destructive measurement, or cultivating the plants in specialized containers that allow access to the roots. However, such cultivation is largely artificial compared to normal growth conditions. Nevertheless, once image- and sensor data have been acquired, LemnaTec software provides tools for processing the data towards biologically interpretable parameters. Such parameters comprise length and width of the whole root system as well as of the roots as such. Thereby, discriminating root types is enabled and allows characterizing them individually. Branching patterns can be captured with the image processing tools, too.

13.6 Applications in phenotyping and ecotoxicology

13.6.1 Ecotoxicology

In ecotoxicology studies, simple model organisms are used to evaluate potential harmful effects of compounds on living organisms. Model species serve as representative samples for larger classes, e.g., duckweeds fronds are taken to represent higher plants. Test substances are added to the organisms following standardized protocols, and organisms are monitored for developmental changes afterwards, which comprise size and shape changes as well as discoloration (Fig. 13.12 left).

LemnaTec was founded with the scope to provide an imaging-based technology for duckweed inhibition tests, which are frequently used to assess plant toxicity for various substance classes. Image processing allows the counting of fronds, measuring their size and shape, and analyzing their color (Fig. 13.12A–F). Many examples of LemnaTec duckweed biotests were published in recent years, comprising examples of technical application in water management [54,55], nanoparticle testing [56], and chemical catalysts [57]. The Duckweed test proved to be valuable in

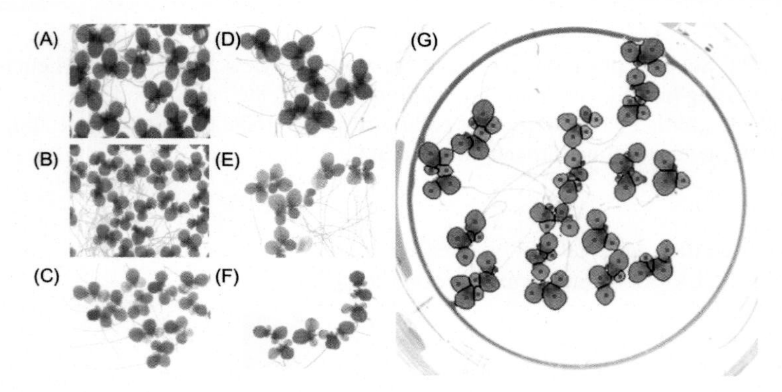

Figure 13.12 Typical phenotypic changes of healthy fronds (A) in comparison to stressed fronds (B–F; B, reduced area; C, chlorosis; D, reduced number; E, old necrosis; F, young necrosis), and duckweed fronds after analysis with LemnaGrid—*red dots* denote frond counts, single detected fronds are encircled by dark lines (G).

the food and agriculture sector, too, for instance when testing food additives [58], herbicide toxicology [59,60], or characterizing active ingredients from natural sources [61,62]. Similarly, medical applications deserve duckweed screens when testing potential toxic effects in human and veterinary drugs [63–66], or for homeopathic drugs [67,68]. In many cases testing procedures follow standardized guidelines, e.g., Organisation for Economic Co-operation and Development (OECD) or ISO protocols.

Based on this first camera-based biotest, further applications were developed in which image processing tools delivered data on phenotypic properties of organisms. Such applications in ecotoxicology comprise invertebrate feeding and motility assays. In such tests, compounds are screened for influences on either target- or nontarget organisms with respect to potential use in agriculture. Saran and colleagues used termite feeding assays monitored with a camera mounted on a LemnaTec HTS Scanalyzer to analyze the effectiveness of an insecticide [69]. In particular, they measured the feeding cessation of treated termites by monitoring the size changes of filter paper discs that were eaten by the termites.

13.6.2 Phenotyping in plant science

In basic research, genetic and physiological questions are frequently approached using model species such as *A. thaliana*. Although not fully representative for all plants, *Arabidopsis* studies are valuable tools to evaluate gene functions and signaling pathways. Results gained with model studies deliver highly valuable support for analyzing functions in other species, e.g., in crop plants. A central question when analyzing gene functions and physiological pathways is whether certain changes in the genome and the physiology turn into phenotypic alterations of the organism. Moreover, temporal occurrence of such changes and external factors promoting or suppressing such changes provide essential data on the physiological functions. Analyzing the resistance mechanisms of plants to aphid infestations, Avila and colleagues investigated the effect of α-dioxygenase (α-DOX) activity in the defense against aphids, as shown with expression studies on aphid-challenged tomato plants. Knocking down α-DOX in *A. thaliana* increased the susceptibility to aphids, pointing at a functional role of α-DOX and the connected oxylipin-pathway in plant-aphid-interactions. As manipulating such pathways may have unwanted side effects, the authors tested whether the α-DOX mutants were phenotypically changed compared to the corresponding ecotype Col-0. Thereby images captured with an HTS Scanalyzer and processed with LemnaGrid proved that there were no significant phenotypic differences between mutants and Col-0, as they had the same size ranges and shape parameter distributions [70].

Basic research on leaf growth and leaf angle changes in response to light-dark cycles was analyzed by Dornbusch and colleagues. Using an HTS Scanalyzer equipped with an RGB camera and a laser scanner they monitored leaf elevation angles and sizes of *A. thaliana* plants several times throughout the day. Monitoring leaf length, elongation rates, elevation angles, and movement rates in response to constant light or dark as well as to long- and short-day conditions provided detailed insight into diel cycles of leaf development. In combination with studies on mutants affected in phytochrome signaling the paper provided insight in the regulatory processes of leaf development and light responses [45].

13.6.3 *Plant phenotyping in crop research*

Crop improvement is one of the most urgent tasks in current plant breeding, as an enormous increase in demand for plant-derived products will rise in the near future due to the growing human population and the depletion of fossil resources. In addition, environmental and aerial pollution has a negative impact on soil, water, and climate, and thereby creates increasing stress for growing crops. Therefore, harnessing crops against stress factors is a second essential task for plant breeding. A third challenge is exploiting novel types of crops for producing raw materials, biochemicals, bio-fuels, or novel types of food and feed. All such efforts demand phenotypic validation of genetic improvement and breeding progress, meaning that crop researchers and breeders need to measure phenotypic performance of candidates and novel genotypes. Many studies in applied research and early breeding stages include tests with potted indoor-grown plants. In such cultivation situations, phenotypic assessments can be run in sensor-to-plant or plant-to-sensor mode.

The pant-to-sensor mode is commonly present in LemnaTec 3D Scanalyzers that are installed in several phenotyping greenhouses. The platforms consist of conveyor systems on which carriers with pots can move around. With this technique, plants are brought to measuring boxes as well as to weighing and watering units in a regular manner, allowing day-by-day acquisition of phenotypic data and monitoring pot water status combined with defined irrigation. Whereas all 3D Scanalyzers comprise a measurement box with top and side view RGB cameras, systems may comprise one or more additional boxes with different types of sensors. Hairmansis and colleagues used the RGB and fluorescence cameras in the 3D Scanalyzer at the Plant Accelerator in Adelaide for assessing growth and senescence of rice plants in response to saline soil. Soil salinity is one of the major threats to crop cultivation and analyzing mechanisms of salinity tolerance are highly desired for crop breeding. Rice cultivation in particular is limited by saline soils and therefore the researchers combined phenotypic and molecular data on the response of rice cultivars to different levels of salinity in a time-course analysis. Phenotypic changes in response to salinity comprised growth depression and early senescence. Cultivar-specific occurrence of the stress symptoms in combination with ion-tolerance mechanisms of the leaf and shoot tissues provided clues for the mechanisms of ionic and osmotic components of salt stress responses. These in turn set the basis for future breeding towards stress-resilient rice lines [24]. Similarly, the Plant Accelerator's phenotyping platform was used to assess salinity stress in wheat cultivars. The researchers linked phenotypic responses to transcriptomic changes of the plants under salt stress. Dynamic changes of phenotypes, expressed in growth rate alterations, conveyed valuable information on stress behavior [71].

13.6.4 Field phenotyping

Except for horticulture where production partly takes place in greenhouses, crop production generally includes field cultivation of plants. From this perspective crop researchers and breeders need field experiments in order to acquire realistic and reliable data on the performance of the plants. It is known that growing plants under controlled conditions is quite artificial compared to natural or agricultural growth conditions [72]. In particular, as comparative studies show in many cases, results gained in indoor cultivation do not fully correlate with results from field cultivation [73].

Traditional methods in field phenotyping either depend on harvesting the plants of interest, taking representative samples, or on manual measurements of selected plants in a certain plot area. However, in such cases either plants can be measured only once or data are available only for a minor subset of all plants. When cultivating plants in plots, central plants are likely to be not accessible for manual measurements without damaging the outer plants. Drawbacks of traditional methods drive demand for noninvasive and automated field phenotyping platforms.

Challenges in field phenotyping are different from and more complex than those occurring with indoor cultivated plants. In the field, plants are fixed in the soil, so only sensor-to-plant machines can be used. Moreover, in most cases plants grow in dense canopies so that no side-view sensors are applicable and the field of view contains no separately standing single plants. Measurements therefore relate to plots or parts of plots rather than to individual plants. Bringing sensors to the plants can happen in different ways, and related with different modes of operation, speed, resolution, and positioning accuracy. Sensor towers serve for positioning a set of sensors at elevated positions next to plot areas and sensors usually rotate on top of the tower to enlarge the field of view. The height of the tower determines the view angle and thereby the size of the measured area. Sensors acquire data neither horizontally nor vertically but at sloping angles that change from one plant position to the next. Field robots and autonomously moving vehicles can move sensors in horizontal or vertical positions across plants, providing advantage for data acquisition. Usually they have limited view angels, but due to their motility they can scan large plot areas and they are flexible in operation at different locations. Similarly, airborne sensor systems are very mobile and flexible in application. Airborne systems such as drones or blimps can reach inner parts of dense vegetation covers more easily than field robots that usually drive next to the vegetation and carry sensors on an extension arm. Usually, the payload of airborne systems is very low, and therefore only a very limited number of sensors can be mounted at once, whereas payload of field robots is much higher, allowing for the carrying of more sensors. Positioning of field robots and airborne systems depend on GPS signals that limit the accuracy of positioning the sensors at defined positions in a field area. When measuring time-lapse experiments, limited accuracy impairs the revisiting of a specific plant in subsequent measurements. The LemnaTec Field Scanalyzer is an example for a multisensor platform with limited mobility but high positioning accuracy (Fig. 13.13).

Figure 13.13 Field Scanalyzer at Rothamsted Research.

As it operates on rails, it can move only within the rail area; however positions are approached with accuracy in the millimeter range, so repeated measurements on single plants are possible. Due to the high payload of the gantry construction, various sensors for measuring plant properties and environmental factors are mounted for simultaneous operation. In the first months after the establishment of the first Field Scanalyzer at Rothamsted Research (Great Britain) in spring 2015, pilot tests revealed versatile use cases such as counting wheat ears with RGB image processing, calculating canopy surface models from laser-scanner generated point clouds, monitoring canopy surface heat emission with IR cameras, or recording quantum yield of photosystem II [35,74]. There is large field of applications for the calculation of vegetation indices using data from the hyperspectral cameras mounted on the Field Scanalyzer.

13.7 Conclusions and outlook

Various sensor types deliver a broad range of phenotypic data on plants and other organisms. Whereas images in the visible light range still are most frequently used and carry already a large proportion of the available information, other sensors are gaining importance. The latter increasingly give insight into physiological aspects of the phenomes and thereby enable closer links to biochemical and molecular data. Prospects of image based bioassays comprise physiological insights that are based on spectrally resolved images and on sensor fusion applications. Mapping spectral reflectance data together with 3D spatial information will strongly improve the recording and analyzing of phenotypic data. Next generation image processing will allow the detection target properties of the organisms from data recordings with strong background noise. This will be important because recordings from outdoor cultivations will strongly gain importance. In contrast to recordings from controlled environments, presence of background noise occurs regularly in records taken from fields or even nature. Applications beyond research and controlled testing environments, those in breeding, ecological studies, and precision agriculture will mainly comprise outdoor-generated measurements and will therefore require sophisticated data processing tools.

References

[1] Araus JL, Cairns JE. Field high-throughput phenotyping: the new crop breeding frontier. Trends Plant Sci 2014;19(1):52–61.
[2] Chaudhary E, Sharma P. Duckweed as a test organism for ecotoxicological assessment of wastewater. Int J Sci Res 2014;3:2073–5.
[3] Dhondt S, Wuyts N, Inzé D. Cell to whole-plant phenotyping: the best is yet to come. Trends Plant Sci 2013;18(8):428–39.
[4] Fiorani F, Schurr U. Future scenarios for plant phenotyping. Annu Rev Plant Biol 2013;64:267–91.
[5] Hassan SH, Van Ginkel SW, Hussein MA, Abskharon R, Oh S-E. Toxicity assessment using different bioassays and microbial biosensors. Environ Int 2016;92:106–18.
[6] Beddington J. Food security: contributions from science to a new and greener revolution. Philos Trans R Soc Lond B Biol Sci 2010;365(1537):61–71.
[7] Tilman D, Cassman KG, Matson PA, Naylor R, Polasky S. Agricultural sustainability and intensive production practices. Nature 2002;418(6898):671–7.
[8] Tilman D, Balzer C, Hill J, Befort BL. Global food demand and the sustainable intensification of agriculture. Proc Natl Acad Sci USA 2011;108(50):20260–4.
[9] Rajasundaram D, Selbig J. More effort—more results: recent advances in integrative 'omics' data analysis. Curr Opin Plant Biol 2016;30:57–61.
[10] Rascher U, Blossfeld S, Fiorani F, Jahnke S, Jansen M, Kuhn AJ, et al. Non-invasive approaches for phenotyping of enhanced performance traits in bean. Funct Plant Biol 2011;38(12):968–83.
[11] Berger B, de Regt B, Tester M. High-throughput phenotyping of plant shoots. In: Normanly J, editor. High-throughput phenotyping in plants. Totowa, NJ: Humana Press; 2012. p. 9–20.
[12] Boyer J. Differing sensitivity of photosynthesis to low leaf water potentials in corn and soybean. Plant Physiol 1970;46(2):236–9.
[13] Saab IN, Sharp RE. Non-hydraulic signals from maize roots in drying soil: inhibition of leaf elongation but not stomatal conductance. Planta 1989;179(4):466–74.
[14] Jansen M, Pinto F, Nagel KA, van Dusschoten D, Fiorani F, Rascher U, et al. Non-invasive phenotyping methodologies enable the accurate characterization of growth and performance of shoots and roots. In: Tuberosa R, Graner A, Frison E, editors. Genomics of plant genetic resources. Dordrecht, The Netherlands: Springer; 2014. p. 173–206.
[15] Nagel G, Hegemann P, Schwärzel M. Algen, Froscheier und Putzfimmel bei Fliegen. Neues Werkzeug zur Licht-gesteuerten Manipulation von Zellen und Tieren. BIOforum 2007;30(6):46–7.

[16] Prullage JB, Tran HV, Timmons P, Harriman J, Chester ST, Powell K. The combined mode of action of fipronil and amitraz on the motility of *Rhipicephalus sanguineus*. Vet Parasitol 2011;179(4):302–10.
[17] Reekmans R, De Smet K, Chen C, Van Hummelen P, Contreras R. Old yellow enzyme interferes with Bax-induced NADPH loss and lipid peroxidation in yeast. FEMS Yeast Res 2005;5(8):711–25.
[18] Jones H, Rotenberg E. Energy, radiation and temperature regulation in plants. In: Fullerlove G, editor. Encyclopedia of life sciences. Chichester: John Wiley & Sons, Ltd, Nature Publishing Group; 2001.
[19] Jones HG. Stomatal control of photosynthesis and transpiration. J Exp Bot 1998;49:387–98.
[20] Chaerle L, Hagenbeek D, De Bruyne E, Valcke R, Van Der Straeten D. Thermal and chlorophyll-fluorescence imaging distinguish plant-pathogen interactions at an early stage. Plant Cell Physiol 2004;45(7):887–96.
[21] Hunt ER, Rock BN. Detection of changes in leaf water content using near-and middle-infrared reflectances. Remote Sens Environ 1989;30(1):43–54.
[22] Penuelas J, Filella I, Biel C, Serrano L, Save R. The reflectance at the 950–970 nm region as an indicator of plant water status. Int J Remote Sens 1993;14 (10):1887–905.
[23] Knipling EB. Physical and physiological basis for the reflectance of visible and near-infrared radiation from vegetation. Remote Sens Environ 1970;1(3):155–9.
[24] Hairmansis A, Berger B, Tester M, Roy SJ. Image-based phenotyping for non-destructive screening of different salinity tolerance traits in rice. Rice 2014;7(1):1.
[25] Hirsch A. Neue Versuche zur Kohlensäureassimilation. Naturwiss 1931;19(48):964.
[26] Baker NR. Chlorophyll fluorescence: a probe of photosynthesis in vivo. Annu Rev Plant Biol 2008;59:89–113.
[27] Maxwell K, Johnson GN. Chlorophyll fluorescence – a practical guide. J Exp Bot 2000;51:659–68.
[28] Schreiber U. Pulse-amplitude-modulation (PAM) fluorometry and saturation pulse method: an overview. In: Chlorophyll a fluorescence. Springer; 2004. p. 279–319.
[29] Humplík JF, Lazár D, Husičková A, Spíchal L. Automated phenotyping of plant shoots using imaging methods for analysis of plant stress responses–a review. Plant Methods 2015;11(1):1.
[30] Strasser RJ, Srivastava A, Tsimilli-Michael M. The fluorescence transient as a tool to characterize and screen photosynthetic samples. In: Probing photosynthesis: mechanisms, regulation and adaptation. 2000. p. 445–483.
[31] Govindjee Nedbal G. Seeing is believing. Photosynthetica 2000;38(4):481–2.
[32] Lichtenthaler H, Langsdorf G, Lenk S, Buschmann C. Chlorophyll fluorescence imaging of photosynthetic activity with the flash-lamp fluorescence imaging system. Photosynthetica 2005;43(3):355–69.
[33] Nedbal L, Whitmarsh J. Chlorophyll fluorescence imaging of leaves and fruits. In: Chlorophyll a Fluorescence. Springer; 2004. p. 389–407.
[34] Jansen M, Gilmer F, Biskup B, Nagel KA, Rascher U, Fischbach A, et al. Simultaneous phenotyping of leaf growth and chlorophyll fluorescence via GROWSCREEN FLUORO allows detection of stress tolerance in *Arabidopsis thaliana* and other rosette plants. Funct Plant Biol 2009;36(11):902–14.
[35] Dornbusch T, Hawkesford M, Jansen M, Nagel K, Niehaus B, Paulus S, et al. Digital field phenotyping by LemnaTec. Aachen: LemnaTec; 2015.

[36] Bauriegel E, Herppich WB. Hyperspectral and chlorophyll fluorescence imaging for early detection of plant diseases, with special reference to *Fusarium* spec. infections on wheat. Agriculture 2014;4(1):32–57.

[37] Govender M, Chetty K, Bulcock H. A review of hyperspectral remote sensing and its application in vegetation and water resource studies. Water SA 2007;33(2):145–52.

[38] Jackson RD, Huete AR. Interpreting vegetation indices. Prev Vet Med 1991;11 (3):185–200.

[39] Mahlein A-K, Steiner U, Hillnhütter C, Dehne H-W, Oerke E-C. Hyperspectral imaging for small-scale analysis of symptoms caused by different sugar beet diseases. Plant Methods 2012;8(1):1.

[40] Wahabzada M, Paulus S, Kersting K, Mahlein A-K. Automated interpretation of 3D laserscanned point clouds for plant organ segmentation. BMC Bioinformat 2015;16 (1):248.

[41] Blackburn GA. Quantifying chlorophylls and caroteniods at leaf and canopy scales: an evaluation of some hyperspectral approaches. Remote Sens Environ 1998;66 (3):273–85.

[42] Gamon J, Surfus J. Assessing leaf pigment content and activity with a reflectometer. New Phytol 1999;143(1):105–17.

[43] Omasa K, Hosoi F, Konishi A. 3D lidar imaging for detecting and understanding plant responses and canopy structure. J Exp Bot 2007;58(4):881–98.

[44] Paulus S, Behmann J, Mahlein A-K, Plümer L, Kuhlmann H. Low-cost 3D systems: suitable tools for plant phenotyping. Sensors 2014;14(2):3001–18.

[45] Dornbusch T, Michaud O, Xenarios I, Fankhauser C. Differentially phased leaf growth and movements in *Arabidopsis* depend on coordinated circadian and light regulation. Plant Cell 2014;26(10):3911–21.

[46] Rose JC, Paulus S, Kuhlmann H. Accuracy analysis of a multi-view stereo approach for phenotyping of tomato plants at the organ level. Sensors 2015;15(5):9651–65.

[47] Kraft M, Salomão de Freitas N, Munack A. Test of a 3D time of flight camera for shape measurements of plants. In: CIGR Workshop on Image Analysis in Agriculture. 2010. p. 108–115.

[48] Nguyen TT, Slaughter DC, Max N, Maloof JN, Sinha N. Structured light-based 3D reconstruction system for plants. Sensors 2015;15(8):18587–612.

[49] Paulus S, Dupuis J, Riedel S, Kuhlmann H. Automated analysis of barley organs using 3D laser scanning: An approach for high throughput phenotyping. Sensors 2014;14 (7):12670–86.

[50] Paulus S, Schumann H, Kuhlmann H, Léon J. High-precision laser scanning system for capturing 3D plant architecture and analysing growth of cereal plants. Biosyst Eng 2014;121:1–11.

[51] Paproki A, Sirault X, Berry S, Furbank R, Fripp J. A novel mesh processing based technique for 3D plant analysis. BMC Plant Biol 2012;12(1):63.

[52] Paulus S, Dupuis J, Mahlein A-K, Kuhlmann H. Surface feature based classification of plant organs from 3D laserscanned point clouds for plant phenotyping. BMC Bioinformat 2013;14(1):1.

[53] R Development Core Team. R: A language and environment for statistical computing. R Foundation for Statistical Computing, Vienna, Austria; 2008. ISBN 3-900051-07-0, http://www.R-project.org.

[54] Mendonça E, Picado A, Silva L, Anselmo A. Ecotoxicological evaluation of cork-boiling wastewaters. Ecotoxicol Environ Saf 2007;66(3):384–90.

[55] Mendonça E, Picado A, Paixão SM, Silva L, Barbosa M, Cunha MA. Ecotoxicological evaluation of wastewater in a municipal WWTP in Lisbon area (Portugal). Desalination Water Treat 2013;51(19-21):4162–70.

[56] Picado A, Paixão SM, Moita L, Silva L, Diniz MS, Lourenço J, et al. A multi-integrated approach on toxicity effects of engineered TiO_2 nanoparticles. Front Environ Sci Eng 2015;9(5):793–803.

[57] Stolte S, Bui H, Steudte S, Korinth V, Arning J, Białk-Bielińska A, et al. Preliminary toxicity and ecotoxicity assessment of methyltrioxorhenium and its derivatives. Green Chem 2015;17(2):1136–44.

[58] Stolte S, Steudte S, Schebb NH, Willenberg I, Stepnowski P. Ecotoxicity of artificial sweeteners and stevioside. Environ Int 2013;60:123–7.

[59] Grossmann K, Hutzler J, Tresch S, Christiansen N, Looser R, Ehrhardt T. On the mode of action of the herbicides cinmethylin and 5-benzyloxymethyl-1, 2-isoxazolines: putative inhibitors of plant tyrosine aminotransferase. Pest Manage Sci 2012;68(3):482–92.

[60] Tresch S, Niggeweg R, Grossmann K. The herbicide flamprop-M-methyl has a new antimicrotubule mechanism of action. Pest Manage Sci 2008;64(11):1195–203.

[61] Diers JA, Bowling JJ, Duke SO, Wahyuono S, Kelly M, Hamann MT. Zebra mussel antifouling activity of the marine natural product aaptamine and analogs. Mar Biotechnol 2006;8(4):366–72.

[62] Meepagala KM, Sturtz G, Wedge DE, Schrader KK, Duke SO. Phytotoxic and antifungal compounds from two Apiaceae species, *Lomatium californicum* and *Ligusticum hultenii*, rich sources of Z-ligustilide and apiol, respectively. J Chem Ecol 2005;31 (7):1567–78.

[63] Białk-Bielińska A, Stolte S, Arning J, Uebers U, Böschen A, Stepnowski P, et al. Ecotoxicity evaluation of selected sulfonamides. Chemosphere 2011;85(6):928–33.

[64] Kołodziejska M, Maszkowska J, Białk-Bielińska A, Steudte S, Kumirska J, Stepnowski P, et al. Aquatic toxicity of four veterinary drugs commonly applied in fish farming and animal husbandry. Chemosphere 2013;92(9):1253–9.

[65] Maszkowska J, Stolte S, Kumirska J, Łukaszewicz P, Mioduszewska K, Puckowski A, et al. Beta-blockers in the environment: Part II. Ecotoxicity study. Sci Total Environ 2014;493:1122–6.

[66] Wagil M, Białk-Bielińska A, Puckowski A, Wychodnik K, Maszkowska J, Mulkiewicz E, et al. Toxicity of anthelmintic drugs (fenbendazole and flubendazole) to aquatic organisms. Environ Sci Pollut Res 2015;22(4):2566–73.

[67] Jäger T, Scherr C, Simon M, Heusser P, Baumgartner S. Development of a test system for homeopathic preparations using impaired duckweed (*Lemna gibba* L.). J Altern Complement Med 2011;17(4):315–23.

[68] Scherr C, Simon M, Spranger J, Baumgartner S. Duckweed (*Lemna gibba* L.) as a test organism for homeopathic potencies. J Altern Complement Med 2007;13(9):931–7.

[69] Saran RK, Ziegler M, Kudlie S, Harrison D, Leva DM, Scherer C, et al. Behavioral effects and tunneling responses of eastern subterranean termites (Isoptera: Rhinotermitidae) exposed to chlorantraniliprole-treated soils. J Econ Entomol 2014;107 (5):1878–89.

[70] Avila CA, Arevalo-Soliz LM, Lorence A, Goggin FL. Expression of α-DIOXYGENASE 1 in tomato and *Arabidopsis* contributes to plant defenses against aphids. Mol Plant-Microbe Interact 2013;26(8):977–86.

[71] Takahashi F, Tilbrook J, Trittermann C, Berger B, Roy SJ, Seki M, et al. Comparison of leaf sheath transcriptome profiles with physiological traits of bread wheat cultivars under salinity stress. PLoS One 2015;10(8):e0133322.

[72] Passioura JB. Viewpoint: the perils of pot experiments. Funct Plant Biol 2006;33 (12):1075–9.
[73] Parent B, Shahinnia F, Maphosa L, Berger B, Rabie H, Chalmers K, et al. Combining field performance with controlled environment plant imaging to identify the genetic control of growth and transpiration underlying yield response to water-deficit stress in wheat. J Exp Bot 2015;66(18):5481–92.
[74] Virlet N, Sabermanesh K, Sadeghi-Tehran P, Hawkesford MJ. Field Scanalyzer: an automated robotic field phenotyping platform for detailed crop monitoring. Funct Plant Biol 2016;44(1):143–53.

Express detection of water pollutants by photoelectric recording from algal cell suspensions

14

Elena G. Govorunova and Oleg A. Sineshchekov
University of Texas Health Science Center at Houston, Houston, TX, United States

14.1 Introduction

Microorganisms such as unicellular photosynthetic algae are very convenient in bioassays for testing environmental pollution because of the low cost and ease of their cultivation and handling. Moreover, algae were found to be more sensitive to toxic substances in water and soil elutriates as compared to other test organisms [1]. Several physiological parameters can be used as a readout: growth rate [2–5], photosynthetic activity [6–10], pigmentation [11,12], glutathione metabolism [13,14], the life cycle [15], and/or cell shape and ultrastructure [16]. However, changes in these parameters accumulate only after a relatively long exposure of test microorganisms to toxic substances. In contrast, changes in specific behavioral responses such as phototaxis, i.e., orientation of the swimming direction with respect to the direction of the light stimulus [17–19], or gravitaxis, orientation in the Earth's gravitational field [20], are observed minutes or even seconds after the exposure of the organism to toxic agents and therefore provide the basis for express toxicity tests [21]. Moreover, these responses are more sensitive to environmental pollutants than motility itself (assessed as percentage of motile cells or the mean swimming rate) [7,22,23]. However, faithful measurements of specific behavioral responses requires tracking of individual cells (real-time motion analysis) such as that implemented in the ECOTOX system, and relies on considerable computational power [24,25].

Probing only initial stages of the sensory cascade instead of measuring the final behavioral response improves the time resolution and sensitivity of the measuring system. In chlorophyte (green) flagellates such as the model organism *Chlamydomonas reinhardtii*, the earliest detectable event in the signal transduction chain of phototaxis is generation of a cascade of photoelectrical responses in the cell membrane [26]. These responses appear on the millisecond time scale, can be easily quantified by a simple population assay and are extremely sensitive to environmental pollutants, which makes their measurement ideal for express bioassays to test toxic substances in water [27].

Bioassays. DOI: http://dx.doi.org/10.1016/B978-0-12-811861-0.00014-0

14.2 Channelrhodopsin-mediated photosensory transduction in green flagellate algae

Phototactic orientation in chlorophyte flagellate algae is mediated by a specialized photoreceptor system that is based on two type 1 (microbial-type) rhodopsins, initially called *Chlamydomonas* sensory rhodopsins A and B [28], but today better known under the names channelrhodopsins (ChRs) 1 and 2 [29,30]. Their peak absorption wavelengths are 510 and 470 nm, respectively. ChR molecules are confined to a small area of the plasma membrane overlaying the eyespot which in chlorophyte algae consists of one to several layers of carotenoid globules subtended by thylakoid membranes within the chloroplast [31]. The eyespot is located asymmetrically within the cell and acts as a shading/reflection device during the cell's helical swimming [32]. When the axis of the cell's rotation coincides with the direction of light, the illumination of the photoreceptor membrane does not change during the rotation cycle and the swimming continues in the original direction. However, when the swimming path deviates from the direction of light, the photoreceptor illumination becomes cyclically modulated by the eyespot. This is perceived by the cell as the signal to change the swimming direction until the photoreceptor illumination during the rotation cycle becomes constant again.

Photoexcitation of ChRs results in generation of inward photoreceptor currents (PC) across the plasma membrane, leading to its depolarization [26]. Two current components can be distinguished: the early PC that appears <30 μs after an excitation flash and saturates at high light intensity, and the late PC that develops after a light-dependent delay and saturates at low light intensity [33,34]. Although both ChRs are responsible for PC generation, photoexcitation of ChR1 primarily contributes to the early PC, whereas that of ChR2 mostly initiates the late PC [28]. The late PC reflects opening of unidentified secondary Ca^{2+} channels coupled to photoactivation of ChRs by an as yet unknown mechanism, which leads to >1000-fold amplification of the sensory signal [35].

If membrane depolarization by PC exceeds a certain threshold, voltage-gated Ca^{2+} channels in the flagellar membrane open, which leads to a massive influx of Ca^{2+} into the intraflagellar space (called the regenerative response [26] or flagellar current (FC) [36]). An increase in the intraflagellar Ca^{2+} concentration results in an abrupt change in the axoneme beating pattern from breast stroke to undulation [37] causing the photophobic, or photoshock behavior [38]. The *CAV2* gene encoding flagellar voltage-gated Ca^{2+} channels has recently been cloned [39]. In contrast to the photoshock response, photoorientation requires asymmetric changes in the beating of the two cell's flagella [40,41], in which different sensitivity of the two axonemes to Ca^{2+} is thought to play a role [42]. Mechanisms by which subthreshold membrane depolarization by PC leads to alteration of the intraflagellar Ca^{2+} concentration during phototaxis remain elusive.

Practically all so far tested species of green algae (*C. reinhardtii*, *Haematococcus pluvialis, Spermatozopsis similis, Hafniomonas reticulata*, *Polytomella magna*, *Volvox carteri*, and *Platymonas (Tetraselmis) subcordiformis*) exhibit very similar photoelectric cascades reflecting contribution of at least two spectrally shifted ChRs [43–45]. Indeed, at least two ChR transcripts have been identified in a number of chlorophyte species by an ongoing algal transcriptome sequencing project [46]. Photocurrents similar to those in chlorophyte flagellates were also detected in the phylogenetically distant alga *Cryptomonas* [47]. Cryptophyte genomes harbor many opsin genes some of which encode proteins acting as light-gated channels conducting anions [48] or cations [49], which are thought to mediate PC in cryptophyte cells.

In contrast, phototactic flagellate algae from other phyla such as *Ochromonas danica* (Stramenopiles) or the classical object of photobehavior research *Euglena gracilis* (Euglenozoa) do not generate PC. In *Euglena*, a flavoprotein, photoactivated adenylyl cyclase has been identified as the receptor for step-up photophobic response leading to photoavoidance [50]. The chemical nature of photoreceptors in *Ochromonas* and many other photomotile flagellates outside Chlorophyta and Euglenozoa lineages remains currently unknown.

14.3 Principles of photoelectric measurements in cell suspensions

The first demonstration of PC in chlorophyte flagellates has been achieved in individual cells by means of the suction pipette technique [26], which can only be used in microorganisms with elastic cell walls. In contrast, application of a technically simpler population assay for photoelectric recording is not limited by the cell wall structure [51]. Another advantage of this assay is its instantaneous averaging of the responses of millions of cells in a population, which makes the results more reproducible than measurements in individual cells, and yields a higher signal-to-noise ratio. Also important is that the population assay provides a possibility of recording from freely swimming flagellates under fully physiological conditions.

Three modifications of this assay have been developed (Fig. 14.1). In the unilateral (UL) mode (Fig. 14.1, top) a short flash is delivered along the line connecting the electrodes immersed in a suspension of non-oriented algal cells. The eyespot acts as a directional antenna so that the number of photons captured by the photoreceptor of an individual cell is determined by the angle of the light incidence on the eyespot surface. Upon photoexcitation, the cells oriented with their eyespots towards the light source generate maximal PC, whereas those with the eyespots away from the light source generate minimal current, so that the difference signal is picked up by the electrodes. The sign of this difference signal is considered positive by convention. The amplitude of the signal depends on the amplitude of microscopic photocurrents generated by individual cells and the degree of shading/reflection of the excitation light by the eyespot and the chloroplast (i.e., on the directional sensitivity of the photoreceptor apparatus).

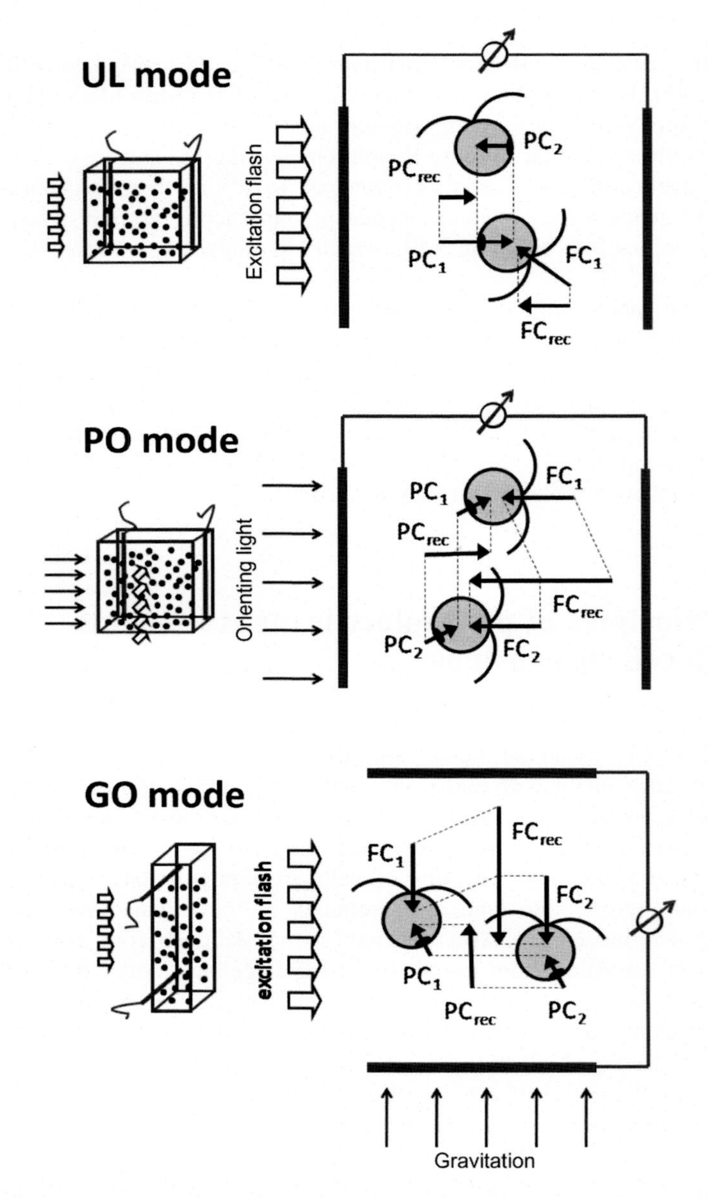

Figure 14.1 Three modifications of the population assay for photoelectric recording. Left: a macroscopic view of the measuring cuvettes; right: schematic presentation of the principles of measurements. PC_1, PC_2, FC_1 and FC_2, photocurrents generated by individual cells in a suspension. PC_{rec} and FC_{rec}, macroscopic resultant currents recorded by the electrodes. *Source*: Modified with permission from Govorunova EG, Altschuler IM, Häder DP, Sineshchekov OA. A novel express bioassay for detecting toxic substances in water by recording rhodopsin-mediated photoelectric responses in Chlamydomonas cell suspensions. Photochem Photobiol 2000;72:320–326.

In retinal-reconstituted carotenoid-deficient *C. reinhardtii* mutants that also lack normal photosynthetic pigments, the sign of the net PC recorded in the UL mode is opposite to that in the wild-type cells [52]. It is explained by focusing of the incident light on the opposite side of the cell ("lens effect") by almost transparent cell bodies. Although FC is an all-or-nothing event, its probability is higher in the cells oriented with their eyespots towards the excitation flash, which again results in an incomplete compensation by the signal from cells in the opposite orientation. The sign of FC measured in the UL mode depends on the angle between the eyespot and the flagella. In most *C. reinhardtii* strains it is slightly more than 90 degrees, so that the sign of FC recorded in the UL mode is negative. In *H. pluvialis*, an alga with a different eyespot position within the cell, both PC and FC recorded in the UL mode have the same direction [51].

The second modification (the phototactically pre-oriented (PO) mode) of the population assay takes advantage of pre-orientation of the cells with phototactically active continuous light (Fig. 14.1, center). In this case the two electrodes are placed one after another along the direction of the pre-orienting light, and the excitation flash is delivered at the 90 degrees angle. Under these conditions, only PC from photooriented cells contribute to the recorded signal, whereas the difference signal from non-oriented cells is not detected. The signal amplitude is proportional to the cosine of the angle between the direction of the current and the line connecting the electrodes. In other words, it depends not only on the amplitude of microscopic photocurrents generated by each cell, but also on the degree of orientation of the cells in a suspension. If the former is constant (which can be independently tested by measurements in the UL mode), this mode of measurements can be used for an instant estimation of the degree of photoorientation. The sign of this signal is determined by the sign of phototaxis and the position of the eyespot during helical swimming. In most *C. reinhardtii* strains the eyespot position is such that a positive signal corresponds to negative phototaxis. *C. reinhardtii* swims in helices of quite narrow cone angles with the flagella-bearing pole of the cell looking forward. Thus, a projection of the FC on the line connecting the electrodes is greater than that of the PC that is almost perpendicular to this line. Therefore, the PO mode of measurements is better suited for recording the FC than the UL mode.

Finally, the third modification of the population assay (the gravitactically pre-oriented (GO) mode) is similar to the second, but uses gravitaxis instead of phototaxis for pre-orientation of the cells (Fig. 14.1, bottom). Green flagellates demonstrate strong negative gravitaxis (upward swimming), the mechanism of which is still debated [53,54]. When the electrodes are placed one above the other, only photocurrents from cells oriented along the vertical line are recorded. Provided that intrinsic photosensitivity of the cells is unchanged, this method can serve for instantaneous estimation of the degree of graviorientation. Again, the sign of this signal is determined by the sign of gravitaxis and the position of the eyespot, and in most *C. reinhardtii* strains a positive signal corresponds to negative gravitaxis.

A limitation of the population method for photoelectric measurements is that it requires cells to be immersed in low ionic strength medium. Therefore, it is best suited for freshwater algae. Nevertheless, it has also successfully been used for photocurrent recording from the marine species *P. subcordiformis*, by transferring the cells to low ionic strength medium shortly before the experiment [45]. Correct interpretation of the results obtained by the population assay requires clear understanding of its principles and sources of possible artifacts, such as those due to the photoeffect on the electrodes that occurs if they are not completely shielded from the excitation light. Also, it has to be taken into account that besides ChR-mediated photocurrents the electrodes register photoinduced charge displacement in photosynthetic membranes [51]. This signal can be distinguished from PC by its faster kinetics and an action spectrum that corresponds to that of photosynthesis.

14.4 Detection of heavy metal ions in water

Heavy metals are a group of elements with a high atomic weight and density above 3.5 g cm^{-3}. Human activities have led to their wide distribution in the environment, raising concerns over their potential detrimental effects on human health. As, Cd, Hg, and Pb are included in the World Health Organisation's list of 10 chemicals of major public concern (http://www.who.int/ipcs/assessment/public_health/chemicals_phc/en/). Although B, Co, Cu, Cr, Fe, Mb, Mn, Mo, Ni, Se, and Zn in low quantities are essential for biological metabolism (http://www.who.int/nutrition/publications/micronutrients/9241561734/en/), all heavy metals are toxic at concentrations in the micromolar range. Heavy metals are one of the major contaminants in aquatic ecosystems and may cause structural and functional damage to the cells that lead to carcinogenesis or apoptosis [55].

In flagellate algae heavy metals interfere with physiological and biochemical processes including photosynthesis, respiration, protein synthesis, and chlorophyll synthesis [56], induce ultrastructural changes [12], and impair motility, phototaxis, and gravitaxis [57]. Changes in any of these parameters can in principle be used to perform toxicity testing, but they are noticeable on different time scales. For instance, morphological and pigmentation changes were observed in *C. reinhardtii* only after 48 h exposure to 10 μM Cu^{2+} [12], and transient changes in gene transcription, after 46 h exposure to 8 μM Cd^{2+} [58]. Changes in *C. reinhardtii* motility appeared after 24 h [59], but showed high variance and were found unreliable for the purpose of water pollution detection [59].

For measurements of the influence of heavy metal cations on ChR-mediated photoelectric responses *C. reinhardtii* vegetative cells were converted to gametes that are more sensitive to environmental factors. Homogenous gamete populations can be easily obtained by nitrogen starvation, which makes gametes an ideal choice for standardized assays. Control (not poisoned) *C. reinhardtii* gamete suspensions generate large PC in the UL mode of measurements and display robust negative phototaxis and negative gravitaxis which gives rise to positive net PC in the PO and GO modes (Fig. 14.2, *dashed lines*). The inhibition of photocurrents was observed already 1 min after the addition of heavy metal ions (Fig. 14.3A) and was highly

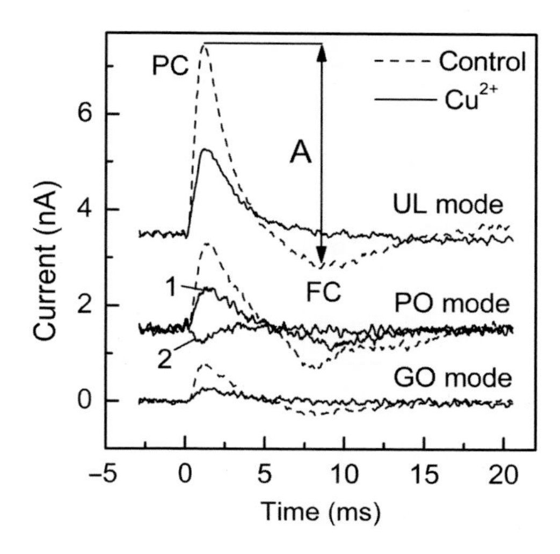

Figure 14.2 Inhibition of photoinduced electrical responses in *Chlamydomonas reinhardtii* by Cu^{2+} ions. *PC*, photoreceptor current; *FC*, flagellar current. A, PC–FC peak-to-peak amplitude. The responses were measured 3 min after the addition of $CuSO_4$ (*solid lines*) or blank medium (*dashed lines*) to identical cell samples. Final concentrations of $CuSO_4$ were: UL mode, 10^{-5} M; PO mode: (1) 10^{-6} M, (2) 10^{-5} M; GO mode, 3×10^{-5} M.
Source: Modified with permission from Govorunova EG, Altschuler IM, Häder DP, Sineshchekov OA. A novel express bioassay for detecting toxic substances in water by recording rhodopsin-mediated photoelectric responses in Chlamydomonas cell suspensions. Photochem Photobiol 2000;72:320–326.

reproducible (Fig. 14.3B). The currents measured in the PO mode were more sensitive to all tested heavy metal ions in the sequence $Cu^{2+} < Pb^{2+} < Zn^{2+} < Cd^{2+}$ than those measured in the UL and GO modes (Table 14.1).

Both PC and FC were inhibited by heavy metal cations. The inhibition of PC likely reflects direct action of heavy metal cations on photoreceptor molecules rather than structural damage of the photoreceptor apparatus that is known to occur after much longer exposure to these toxic agents [60]. This conclusion is also corroborated by the observation that all heavy metal ions that were found to suppress photocurrents in *C. reinhardtii* are also known to inhibit ion channels in animal tissues [61–64]. Also, the absorption maximum and photocycle kinetics of bacteriorhodopsin—a protein that belongs to the same superfamily as ChRs—were altered by the addition of heavy metal cations [65]. Finally, the kinetics and current-voltage relationships of channel currents generated by ChRs heterologously expressed in cultured animal cells are influenced by heavy metal cations [66,67].

When measurements were performed in the PO mode, at low Cu^{2+} concentrations (< 3 μM) the amplitude of PC decreased (curve 1 in Fig. 14.2) as when measured in the UL mode, but upon an increase in the Cu^{2+} concentration the sign of PC changed to negative (curve 3 in Fig. 14.2), which reflected a change in the

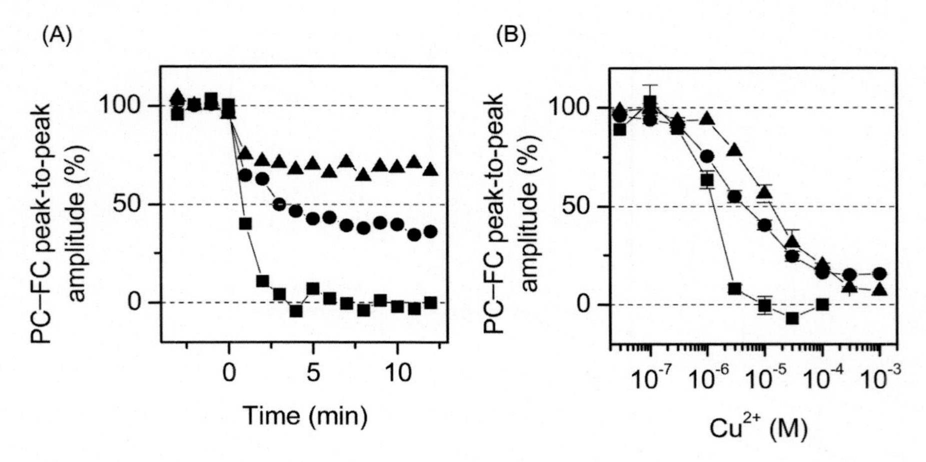

Figure 14.3 (A) The time course of the inhibitory effect of Cu^{2+} ions on the PC–FC peak-to-peak amplitude measured in the UL mode (*circles*), PO mode (*squares*), and GO mode (*triangles*). $CuSO_4$ at the final concentration 4.5×10^{-6} M was added at the time zero. (B) The dependence of the PC–FC peak-to-peak amplitude on the concentration of Cu^{2+} ions measured in the UL mode (*circles*), PO mode (*squares*) and GO mode (*triangles*). The effect was measured 3 min after the addition of $CuS0_4$ to *Chlamydomonas reinhardtii* cells. Error bars show square deviation of the mean ($n = 4$ samples).
Source: Modified with permission from Govorunova EG, Altschuler IM, Häder DP, Sineshchekov OA. A novel express bioassay for detecting toxic substances in water by recording rhodopsin-mediated photoelectric responses in Chlamydomonas cell suspensions. Photochem Photobiol 2000;72:320–326.

Table 14.1 EC_{50} (concentrations for 50% inhibition of the response) determined with three modes of photoelectric recording in *Chlamydomonas reinhardtii* cell suspensions

Toxic substance	UL mode	PO mode	GO mode
Cu^{2+}	5.0×10^{-6} M	1.4×10^{-6} M	1.3×10^{-5} M
Zn^{2+}	3.1×10^{-5} M	3.8×10^{-6} M	4.5×10^{-5} M
Cd^{2+}	2.0×10^{-5} M	8.0×10^{-6} M	2.0×10^{-5} M
Pb^{2+}	7.1×10^{-6} M	2.2×10^{-6} M	5.4×10^{-6} M
Formaldehyde	13.3×10^{-3} M	4.7×10^{-3} M	3.0×10^{-3} M[a]

[a]Concentration for 50% stimulation of the response.
Source: Modified with permission from Govorunova EG, Altschuler IM, Häder DP, Sineshchekov OA. A novel express bioassay for detecting toxic substances in water by recording rhodopsin-mediated photoelectric responses in Chlamydomonas cell suspensions. Photochem Photobiol 2000;72:320–326.

phototaxis sign from negative to positive. The sign of phototaxis (i.e., whether the cells swim toward or away from the light source) in chlorophyte flagellates is regulated by many factors, among which are photosynthesis and other processes of energy conversion [34,68]. Therefore the observed switch of phototaxis upon heavy metal poisoning likely results from inhibition of energy metabolism.

FC measured in the UL mode was inhibited by all tested heavy metal cations to an even greater degree than PC (see Fig. 14.4, top panel for Cu^{2+} effect). These results are consistent with the earlier observation that the photophobic response in *C. reinhardtii* that is initiated by FC was completely inhibited by Cu^{2+} at the concentrations that only partially impaired phototaxis [69]. A stronger inhibition of FC than PC by Cd^{2+} has been directly shown by measurements in individual *C. reinhardtii* cells by the suction pipette technique [70,71]. One possible explanation is that the voltage-gated CAV2 channels are more sensitive to heavy metal cations than ChRs and putative secondary Ca^{2+} channels responsible for the late PC component. Also, a small heavy metal-induced decrease in PC amplitude to the level when it causes only below-threshold membrane depolarization would lead to a dramatic decrease in the probability of FC generation, which can contribute to the observed higher sensitivity of FC than PC to heavy metal poisoning. In the PO mode both PC and FC were inhibited by Cu^{2+} to the same degree (Fig. 14.4, center panel). This can be explained by a larger projection of FC in phototactically pre-oriented cells on the measuring direction in the PO mode as compared to that in non-oriented cells in the UL mode, which compensates a stronger inhibition of microscopic FC by heavy metal ions. At Cu^{2+} concentrations $>100\ \mu M$ when PC tested in the UL mode was not yet completely inhibited, no response could be recorded in the PO mode indicating that downstream elements of the signal transduction chain in phototaxis are more sensitive to Cu^{2+} poisoning then ChRs.

In the UL mode Cd^{2+} was more toxic than Zn^{2+}, which reflects the higher sensitivity of CAV2 channels and possibly other elements of the photosensory cascade downstream from ChRs to Cd^{2+}. Zn^{2+} is the least toxic ion also to photocurrents recorded in the GO mode, whereas Pb^{2+} is the most toxic. The currents measured in the GO mode were less sensitive to Cu^{2+} and Zn^{2+}, as compared to those recorded in the UL mode (Table 14.1). One possible explanation is stimulation of gravitactic orientation by these ions, which compensates for inhibition of photocurrents detected in the UL mode, or a switch from positive (downward swimming) to negative (upward swimming) gravitaxis. As mentioned above, the positive sign of PC measured in the GO mode corresponds to negative net gravitaxis, but it does not necessarily mean that all individual cells swim upward, simply that negative gravitaxis is predominant in a cell population. A switch from positive to negative gravitaxis upon the addition of Cu^{2+}, Hg^{2+}, and, to a lesser extent, Cd^{2+} was also detected in young *Euglena* cultures by motion analysis [72]. *Chlamydomonas* photocurrents recorded in the GO mode first gradually decreased upon an increase in Pb^{2+} concentration from 1 to $10\ \mu M$ as those in the UL mode and then started increasing until they exceeded those in untreated cells at a concentration of $\sim 100\ \mu M$, after which the currents decreased again and finally dropped to zero

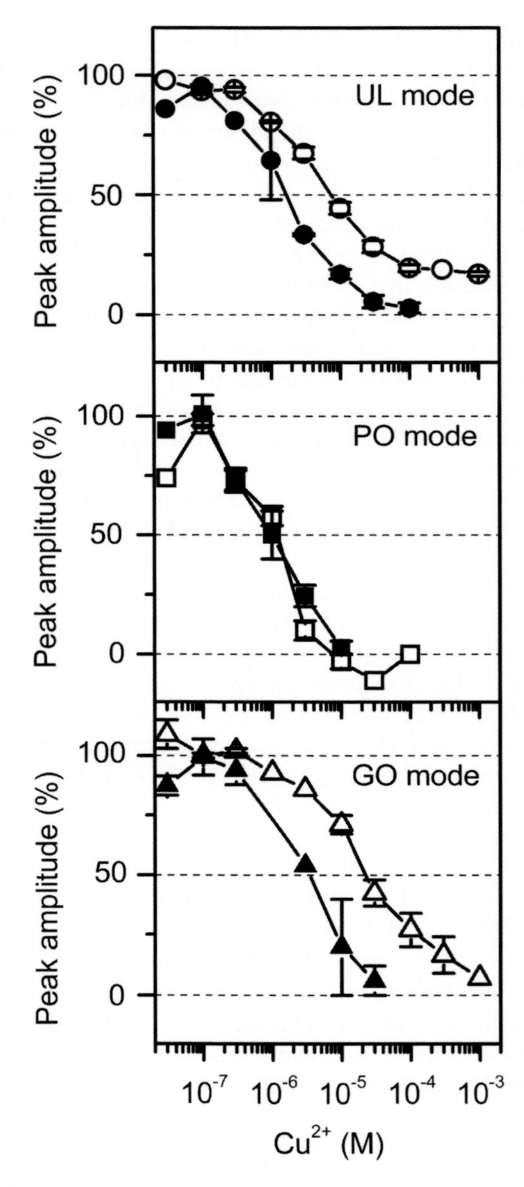

Figure 14.4 The dependence of the PC (*open symbols*) and FC (*closed symbols*) peak amplitudes on the concentration of Cu^{2+} ions. The effect was measured 3 min after the addition of $CuS0_4$ to *Chlamydomonas reinhardtii* cells. Error bars show square deviation of the mean ($n = 4$ samples).
Source: Modified with permission from Govorunova EG, Altschuler IM, Häder DP, Sineshchekov OA. A novel express bioassay for detecting toxic substances in water by recording rhodopsin-mediated photoelectric responses in Chlamydomonas cell suspensions. Photochem Photobiol 2000;72:320–326.

at ~1 mM. This complex behavior likely reflects specific influence of Pb^{2+} on *C. reinhardtii* gravitaxis, because parallel changes in photocurrents were not observed in the UL mode.

The mechanism of gravitaxis in *Chlamydomonas* remains poorly understood. On the one hand, cells with the flagella immobilized by chemical agents or mutations settle in suspension with their anterior, flagella-bearing end pointing upward [53,73], which supports the notion that gravitaxis in this organism occurs due to purely physical reasons, i.e., the longitudinal density gradient or the asymmetry of cell body that causes shape orientation. On the other hand, the existence of gravitaxis mutants, physical characteristics of which are not different from those of the wild type, argues for a physiological gravity-sensing mechanism [73]. Specific influence of Pb^{2+} on gravitactic orientation in *C. reinhardtii* detected by PC measurements in the GO mode may reflect inhibition of elements of the putative gravisensory system by this ion. On the other hand, the effect of Pb^{2+} on the rotational diffusivity that contributes to shape orientation, i.e., purely physical mechanism of gravitaxis, also cannot be excluded.

14.5 Detection of organic pollutants in water

Formaldehyde is one of ubiquitous organic environmental contaminants recently classified as carcinogen, also known to cause adverse reproductive and developmental effects in humans upon prolonged and chronic exposure [74]. It is a colorless irritant gas highly soluble in water. Formaldehyde readily diffuses across the cell membrane and causes single-strand breaks and cross-linking of DNA, and inhibition of RNA biosynthesis [75]. Among approximately 6,000 open reading frames tested by DNA microarray analysis in yeast, ~1200 were affected by formaldehyde [76].

The addition of formaldehyde caused inhibition of *C. reinhardtii* photocurrents measured in the UL and PO modes as did that of heavy metal ions (Fig. 14.5). Already after 1 min a 50% decrease in the current amplitude was measured, and the inhibition reached saturation after 7 min (Fig. 14.6A). In the PO mode a dose-dependent reversal of the PC sign from positive to negative was observed (Fig. 14.5, curve 2), indicating that the addition of formaldehyde also induced a switch from negative to positive phototaxis in *C. reinhardtii* gametes, as did the addition of heavy metal cations. Similar effects of different toxic agents on the sign of phototaxis confirm its regulation by general cell metabolism impaired by the addition of toxic compounds.

When PC was measured in the GO mode, the addition of low formaldehyde concentrations ($<1\%$) induced a dramatic transient dose-dependent increase in the signal amplitude (Fig. 14.6, *triangles*). It could only be explained by an increase in the precision of negative gravitactic orientation, because PC recorded in the UL mode decreased in the presence of formaldehyde (Fig. 14.6, *circles*). Formaldehyde

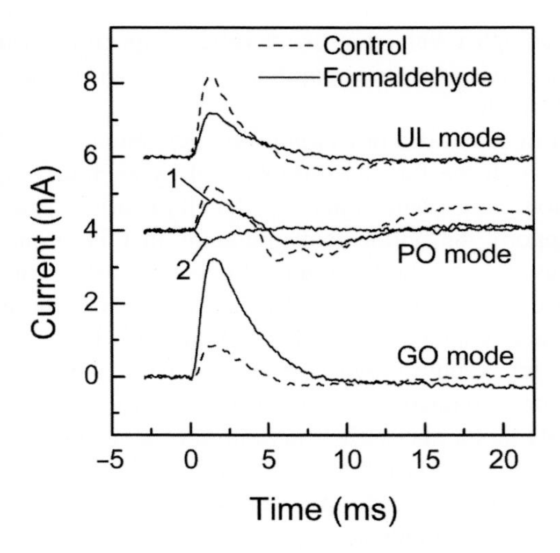

Figure 14.5 The influence of formaldehyde on the photoinduced electrical responses in *Chlamydomonas reinhardtii*. Traces recorded in the UL and GO modes were obtained by averaging 4 successive individual responses measured from 5 to 8 min after the addition of formaldehyde or medium to a cell suspension at the final concentration of 0.03%. Traces recorded in the PO mode were obtained by averaging four successive individual responses measured after: (1) 5 to 8 min; (2) 10 to 13 min after the addition of 0.01% formaldehyde to a cell suspension.
Source: Modified with permission from Govorunova EG, Altschuler IM, Häder DP, Sineshchekov OA. A novel express bioassay for detecting toxic substances in water by recording rhodopsin-mediated photoelectric responses in Chlamydomonas cell suspensions. Photochem Photobiol 2000;72:320–326.

treatment also reduced the average swimming velocity, which could explain the observed transient increase in the degree of gravitactic orientation. An inverse relationship between the precision of gravitactic orientation and swimming velocity was found in a motion analysis study of the influence of light on gravitaxis in *C. reinhardtii* [77]. However, a similar inhibition of motility by ethanol did not induce an increase in the degree of gravitactic orientation, which means that the effect of formaldehyde measured in the GO mode was at least partially due to the influence of this agent on other unknown factors that contribute to gravitaxis in *C. reinhardtii*, rather than a simple inhibition of cell motility.

In contrast to heavy metal ions that showed different toxicity towards PC and FC, formaldehyde inhibited them both to the same degree (Fig. 14.7), which suggests a non-specific mechanism of its action rather than, e.g., a block of specific ion channels. Formaldehyde easily binds to proteins and causes their cross-linking and aggregation [78,79]. Proteins have been identified as the major target of formaldehyde toxicity in yeast [76]. It is plausible that inhibition of photocurrents in *C. reinhardtii* by this agent is also caused by its general adverse effects on cellular proteins.

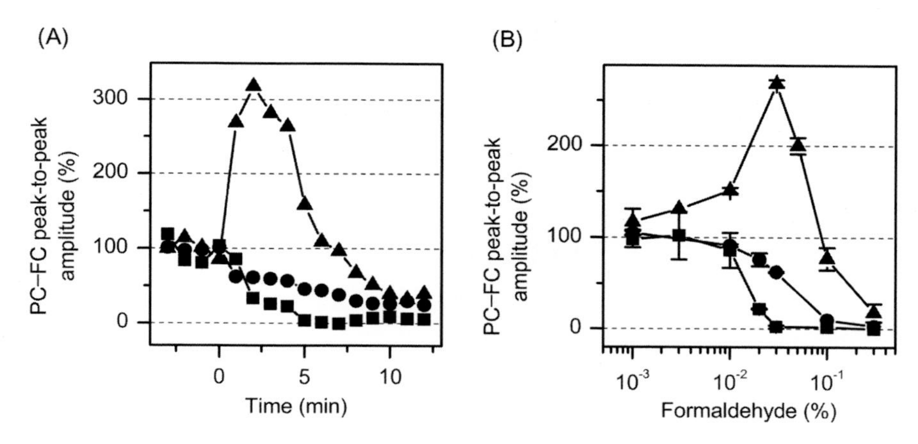

Figure 14.6 (A) The time course of the influence of formaldehyde on the PC–FC peak-to-peak amplitude measured in the UL mode (*circles*), PO mode (*squares*), and GO mode (*triangles*). Formaldehyde was added at time zero at the following final concentrations: the UL mode, 0.03%; the PO mode, 0.01%; the GO mode, 0.03%. (B) Dependence of the PC–FC peak-to-peak amplitude on formaldehyde concentration measured in the UL mode (*circles*), PO mode (*squares*), and GO mode (*triangles*). The effect was measured 1 min after the addition of formaldehyde to *Chlamydomonas reinhardtii* cells. Error bars show square deviation of the mean ($n = 4$ samples).
Source: Modified with permission from Govorunova EG, Altschuler IM, Häder DP, Sineshchekov OA. A novel express bioassay for detecting toxic substances in water by recording rhodopsin-mediated photoelectric responses in Chlamydomonas cell suspensions. Photochem Photobiol 2000;72:320–326.

14.6 Comparison with ECOTOX

The sensitivity of photoelectric responses in *C. reinhardtii* to heavy metal ions was at least several times higher than that of gravitaxis in *E. gracilis* monitored with a computer-controlled video imaging and motion analysis system ECOTOX (Fig. 14.8), and at least an order of magnitude higher than measurements of motility in *E. gracilis* [7,72,80] and *Haematococcus lacustris* [7].

The sensitivity of photoelectric responses in *C. reinhardtii* to formaldehyde was lower, even in the most sensitive PO mode, than that of *Euglena* gravitaxis (Fig. 14.8). The difference in the sensitivities of the test organisms to different pollutants have been well documented in the literature [81,82]. As all tested species of chlorophyte flagellates possess a two-rhodopsin photoreceptor system and generate photocurrents upon photoexcitation [43], it is possible that some of them are more sensitive to organic pollutants (e.g., due to a different structure of the cell wall that is less protective against pollution) and will be therefore better suited for their testing than the model alga *C. reinhardtii.*

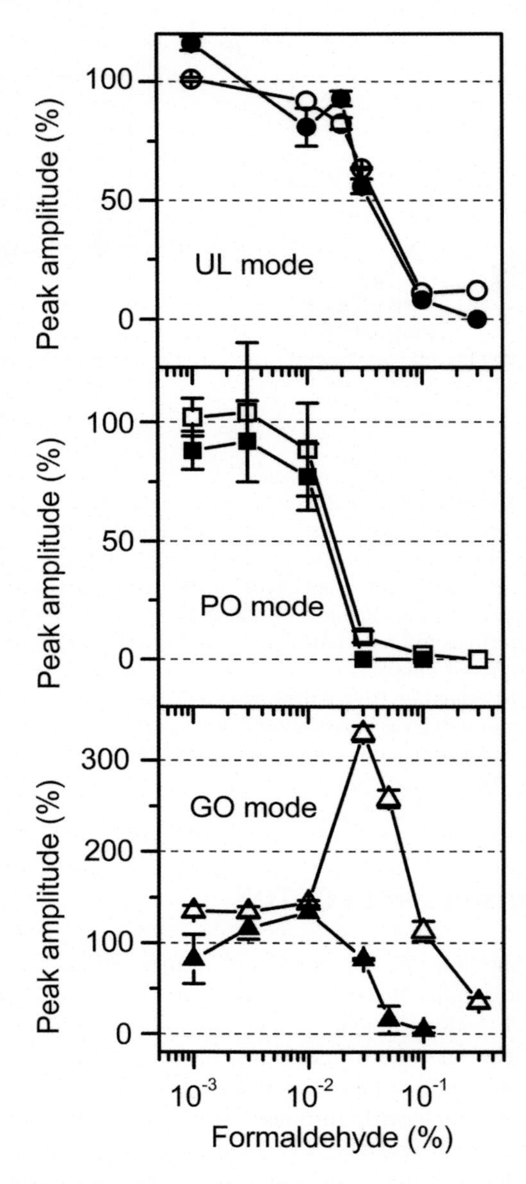

Figure 14.7 The dependence of the PC (*open symbols*) and FC (*closed symbols*) peak amplitudes on the concentration of formaldehyde. The effect was measured 1 min after the addition of formaldehyde to *Chlamydomonas reinhardtii* cells. Error bars show square deviation of the mean ($n = 4$ samples).

Source: Modified with permission from Govorunova EG, Altschuler IM, Häder DP, Sineshchekov OA. A novel express bioassay for detecting toxic substances in water by recording rhodopsin-mediated photoelectric responses in Chlamydomonas cell suspensions. Photochem Photobiol 2000;72:320–326.

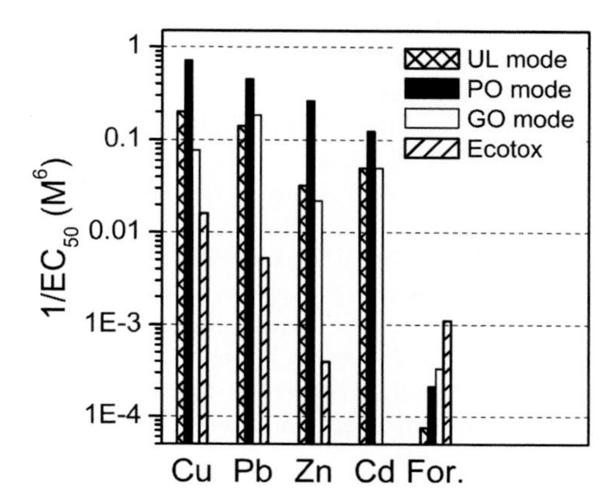

Figure 14.8 Comparative sensitivities of the three modifications of the photoelectric bioassay using *Chlamydomonas reinhardtii* [27] and the ECOTOX system using *Euglena gracilis* [24] to the heavy metal ions and formaldehyde.
Source: Modified with permission from Govorunova EG, Altschuler IM, Häder DP, Sineshchekov OA. A novel express bioassay for detecting toxic substances in water by recording rhodopsin-mediated photoelectric responses in Chlamydomonas cell suspensions. Photochem Photobiol 2000;72:320–326.

14.7 Conclusions

ChR-mediated photoelectric responses recorded in *C. reinhardtii* cell suspensions are highly sensitive to environmental pollutants, such as heavy metal ions and toxic organic molecules. Changes in photocurrents were already observed one to several minutes after the addition of the toxic substances to cell samples. Therefore, measurements of ChR-mediated photocurrents can be recommended to assay water toxicity. The test procedure can be easily standardized and completed in a short time and will provide highly reproducible results. Technical simplicity and low costs are other important advantages to consider, especially when the testing of a large number of water samples is required. Further improvement of the assay can possibly be achieved by switching to other freshwater green flagellate species, some of which might be more sensitive to specific pollutants, such as formaldehyde, for which *C. reinhardtii* was less sensitive than *E. gracilis* that does not generate PC. However, extension of this assay to testing saltwater samples is more problematic, because photoelectric measurements in cells suspensions are not feasible at high ionic strength.

References

[1] Thomas JM, Skalski JR, Cline JF, McShane MC, Miller WE, Peterson SA, et al. Characterization of chemical waste site contamination and determination of its extent using bioassays. Environ Toxicol Chem 1986;5:487–501.

[2] Fennikoh KB, Hirshfield HI, Kneip TJ. Cadmium toxicity in planktonic organisms of a fresh water food web. Environ Res 1978;15:357–67.

[3] Lin KC, Lee YL, Chen CY. Metal toxicity to *Chlorella pyrenoidosa* assessed by a short-term continuous test. J Hazard Mater 2007 Apr 2;142(1-2):236–41. PubMed PMID: 16971040. Epub 2006/09/15. eng.

[4] Rial D, Murado MA, Menduina A, Fucinos P, Gonzalez P, Miron J, et al. Effects of spill-treating agents on growth kinetics of marine microalgae. J Hazard Mater 2013;263 (Pt 2):374–81. PubMed PMID: 23911058. Epub 2013/08/06. eng.

[5] Nagai T, Taya K, Annoh H, Ishihara S. Application of a fluorometric microplate algal toxicity assay for riverine periphytic algal species. Ecotoxicol Environ Saf 2013 Aug;94:37–44. PubMed PMID: 23706602. Epub 2013/05/28. eng.

[6] Irmer U, Wachholz I, Schaefer H, Lorch DW. Influence of lead on *Chlamydomonas reinhardtii* Volvocales Chlorophyta: accumulation, toxicity and ultrastructural changes. Environ Exp Bot 1986;26:97–106.

[7] Braune W, Häder D-P, Hagen C. Copper toxicity in the green alga *Haematococcus lacustris*: flagellates become blind by copper (II) ions. Cytobios 1994;77:29–39.

[8] van der Heever JA, Grobbelaar JU. The use of oxygen evolution to assess the short-term effects of toxicants on algal photosynthetic rates. Water SA 1997;23:233–7.

[9] Ding G, Wouterse M, Baerselman R, Peijnenburg WJ. Toxicity of polyfluorinated and perfluorinated compounds to lettuce (*Lactuca sativa*) and green algae (*Pseudokirchneriella subcapitata*). Arch Environ Contam Toxicol 2012;62(1):49–55. PubMed PMID: 21626016. Epub 2011/06/01. eng.

[10] Umar L, Alexander FA, Wiest J. Application of algae-biosensor for environmental monitoring. Conf Proc IEEE Eng Med Biol Soc 2015;2015:7099–102. PubMed PMID: 26737928. Epub 2016/01/07. eng.

[11] Huang GL, Mao Y, Headley JV, Sun HW. Temporal changes in the toxicity of pentachlorophenol to *Chlorella pyrenidosa* algae. J Environ Sci Health B 2003;38(5):551–9. PubMed PMID: 12929714. Epub 2003/08/22. eng'.

[12] Rodriguez MC, Barsanti L, Passarelli V, Evangelista V, Conforti V, Gualtieri P. Effects of chromium on photosynthetic and photoreceptive apparatus of the alga *Chlamydomonas reinhardtii*. Environ Res 2007;105(2):234–9. PubMed PMID: 17346694. Epub 2007/03/10. eng.

[13] Niederer C, Behra R, Harder A, Schwarzenbach RP, Escher BI. Mechanistic approaches for evaluating the toxicity of reactive organochlorines and epoxides in green algae. Environ Toxicol Chem 2004;23(3):697–704. PubMed PMID: 15285364. Epub 2004/08/03. eng.

[14] Jamers A, Blust R, De Coen W, Griffin JL, Jones OA. Copper toxicity in the microalga *Chlamydomonas reinhardtii*: an integrated approach. Biometals 2013;26(5):731–40. PubMed PMID: 23775669. Epub 2013/06/19. eng.

[15] Xylander M, Braune W. Influence of nickel on the green alga *Haematococcus lacustris* Rostafinski in phases of its life cycle. J Plant Physiol 1994;144(1):86–93.

[16] Bauer DE, Conforti V, Ruiz L, Gomez N. An in situ test to explore the responses of *Scenedesmus acutus* and *Lepocinclis acus* as indicators of the changes in water quality in lowland streams. Ecotoxicol Environ Saf 2012;77:71–8. PubMed PMID: 22088329. Epub 2011/11/18. eng.

[17] Nultsch W, Häder D-P. Photomovement in motile microorganisms. II. Photochem Photobiol 1988;47:837–69.

[18] Witman GB. *Chlamydomonas* phototaxis. Trends Cell Biol 1993;3:403–8.

[19] Jekely G. Evolution of phototaxis. Phil Trans R Soc Lond B Biol Sci 2009;364 (1531):2795−808. PubMed PMID: 19720645. Pubmed Central PMCID: 2781859. Epub 2009/09/02. eng.
[20] Häder D-P, Lebert M, Richter P, Ntefidou M. Gravitaxis and graviperception in flagellates. Adv Space Res 2003;31:2181−6.
[21] Pettersson M, Ekelund NG. Effects of the herbicides roundup and avans on *Euglena gracilis*. Arch Environ Contam Toxicol 2006;50(2):175−81. PubMed PMID: 16317487. Epub 2005/12/01. eng.
[22] Häder D-P, Lebert M, Tahedl H, Richter P. The Erlanger flagellate test (EFT): Photosynthetic flagellates in biological dosimeters. J Photochem Photobiol B Biol 1997;40:23−8.
[23] Shitanda I, Takada K, Sakai Y, Tatsuma T. Amperometric biosensing systems based on motility and gravitaxis of flagellate algae for aquatic risk assessment. Anal Chem 2005;77:6715−18.
[24] Tahedl H, Häder D-P. Fast examination of water quality using the automatic biotest ECOTOX based on the movement behavior of a freshwater flagellate. Water Res 1999;33:426−32.
[25] Streb C, Richter P, Ntefidou M, Lebert M, Häder D-P. ECOTOX - biomonitoring based on real time movement analysis of unicellular organisms. J Gravit Physiol 2002;9(1): P345−6. PubMed PMID: 15002607. Epub 2004/03/09. eng.
[26] Litvin FF, Sineshchekov OA, Sineshchekov VA. Photoreceptor electric potential in the phototaxis of the alga *Haematococcus pluvialis*. Nature 1978;271:476−8.
[27] Govorunova EG, Altschuler IM, Häder DP, Sineshchekov OA. A novel express bioassay for detecting toxic substances in water by recording rhodopsin-mediated photoelectric responses in *Chlamydomonas* cell suspensions. Photochem Photobiol 2000;72:320−6.
[28] Sineshchekov OA, Jung K-H, Spudich JL. Two rhodopsins mediate phototaxis to low- and high-intensity light in *Chlamydomonas reinhardtii*. Proc Natl Acad Sci USA 2002;99:8689−94.
[29] Nagel G, Ollig D, Fuhrmann M, Kateriya S, Musti AM, Bamberg E, et al. Channelrhodopsin-1: a light-gated proton channel in green algae. Science 2002;296 (5577):2395−8. PubMed PMID: 12089443. Epub 2002/06/29. eng.
[30] Nagel G, Szellas T, Huhn W, Kateriya S, Adeishvili N, Berthold P, et al. Channelrhodopsin-2, a directly light-gated cation-selective membrane channel. Proc Natl Acad Sci USA 2003;100(24):13940−5. PubMed PMID: 14615590. Pubmed Central PMCID: 283525. Epub 2003/11/15. eng.
[31] Sager R, Palade GE. Structure and development of the chloroplast in *Chlamydomonas*. I. The normal green cell. J Biophys Biochem Cytol 1957;481−94.
[32] Foster K-W, Smyth RD. Light antennas in phototactic algae. Microbiol Rev 1980;44:572−630.
[33] Sineshchekov OA, Litvin FF, Keszthelyi L. Two components of photoreceptor potential of the flagellated green alga *Haematococcus pluvialis*. Biophys J 1990;57:33−9.
[34] Sineshchekov OA, Govorunova EG. Rhodopsin-mediated photosensing in green flagellated algae. Trends Plant Sci 1999;4:58−63.
[35] Sineshchekov OA, Govorunova EG, Spudich JL. Photosensory functions of channelrhodopsins in native algal cells. Photochem Photobiol 2009;85(2):556−63. PubMed PMID: 19222796. Epub 2009/02/19. eng.
[36] Harz H, Hegemann P. Rhodopsin-regulated calcium currents in *Chlamydomonas*. Nature 1991;351:489−91.

[37] Bessen M, Fay RB, Witman GB. Calcium control of isolated flagella axonemes of *Chlamydomonas*. J Cell Biol 1980;86:446–55.

[38] Schmidt JA, Eckert R. Calcium couples flagella reversal to photostimulation in *Chlamydomonas reinhardtii*. Nature 1976;262:713–15.

[39] Fujiu K, Nakayama Y, Yanagisawa A, Sokabe M, Yoshimura K. *Chlamydomonas* CAV2 encodes a voltage- dependent calcium channel required for the flagellar waveform conversion. Curr Biol 2009;19(2):133–9. PubMed PMID: 19167228. Epub 2009/01/27. eng.

[40] Sineshchekov OA. Photoreception in unicellular flagellates: bioelectric phenomena in phototaxis. In: Douglas RD, editor. Light in biology and medicine. II. New York: Plenum Press; 1991. p. 523–32.

[41] Rüffer U, Nultsch W. Flagellar photoresponses of *Chlamydomonas* cells held on micropipettes: II. Change in flagellar beat pattern. Cell Motil Cytoskeleton 1991;18:269–78.

[42] Kamiya R, Witman GB. Submicromolar levels of calcium control the balance of beating between the two flagella in demembranated models of *Chlamydomonas*. J Cell Biol 1984;98:97–107.

[43] Sineshchekov OA, Spudich JL. Sensory rhodopsin signaling in green flagellate algae. In: Briggs WR, Spudich JL, editors. Handbook of Photosensory Receptors. Weinheim: Wiley-VCH; 2005. p. 25–42.

[44] Braun FJ, Hegemann P. Two light-activated conductances in the eye of the green alga *Volvox carteri*. Biophys J 1999;76:1668–78.

[45] Govorunova EG, Sineshchekov OA, Li H, Janz R, Spudich JL. Characterization of a highly efficient blue-shifted channelrhodopsin from the marine alga *Platymonas subcordiformis*. J Biol Chem 2013;288(41):29911–22. PubMed PMID: 23995841. Pubmed Central PMCID: 3795289. Epub 2013/09/03. eng.

[46] Klapoetke NC, Murata Y, Kim SS, Pulver SR, Birdsey-Benson A, Cho YK, et al. Independent optical excitation of distinct neural populations. Nat Methods 2014;11 (3):338–46. PubMed PMID: 24509633. Pubmed Central PMCID: 3943671. Epub 2014/02/11. eng.

[47] Sineshchekov OA, Govorunova EG, Jung K-H, Zauner S, Maier U-G, Spudich JL. Rhodopsin-mediated photoreception in cryptophyte flagellates. Biophys J 2005;89:4310–19.

[48] Govorunova EG, Sineshchekov OA, Liu X, Janz R, Spudich JL. Natural light-gated anion channels: a family of microbial rhodopsins for advanced optogenetics. Science 2015;349(6248):647–50.

[49] Govorunova EG, Sineshchekov OA, Spudich JL. Structurally distinct cation channelrhodopsins from cryptophyte algae. Biophys J 2016;110(11):2302–4. PubMed PMID: 27233115. Pubmed Central PMCID: 4906376. Epub 2016/05/29. eng.

[50] Iseki N, Matsunaga S, Murakami A, Ohno K, Shiga K, Yoshida K, et al. A blue-light-activated adenylyl cyclase mediates photoavoidance in *Euglena gracilis*. Nature 2002;415:1047–51.

[51] Sineshchekov OA, Govorunova EG, Der A, Keszthelyi L, Nultsch W. Photoelectric responses in phototactic flagellated algae measured in cell suspension. J Photochem Photobiol B Biol 1992;13:119–34.

[52] Sineshchekov OA, Govorunova EG, Der A, Keszthelyi L, Nultsch W. Photoinduced electric currents in carotenoid-deficient *Chlamydomonas* mutants reconstituted with retinal and its analogs. Biophys J 1994;66:2073–84.

[53] Kam V, Moseyko N, Nemson J, Feldman LJ. Gravitaxis in *Chlamydomonas reinhardtii*: characterization using video microscopy and computer analysis. Int J Plant Sci 1999;160:1093–8.

[54] Roberts AM. Mechanisms of gravitaxis in *Chlamydomonas*. Biol Bull 2006;210 (2):78–80. PubMed PMID: 16641513. Epub 2006/04/28. eng.

[55] Beyersmann D, Hartwig A. Carcinogenic metal compounds: recent insight into molecular and cellular mechanisms. Arch Toxicol 2008;82(8):493–512. PubMed PMID: 18496671. Epub 2008/05/23. eng.

[56] Rai LC, Gau JP, Kumar HD. Phycology and heavy metal pollution. Biol Rev 1981;56:99–151.

[57] Bean B, Harris A. Selective inhibition of flagellar activity in *Chlamydomonas reinhardtii* by nickel. J Protozool 1979;26:235–40.

[58] Jamers A, Blust R, De Coen W, Griffin JL, Jones OA. An omics based assessment of cadmium toxicity in the green alga *Chlamydomonas reinhardtii*. Aquat Toxicol 2013;126:355–64. PubMed PMID: 23063003. Epub 2012/10/16. eng.

[59] Danilov RA, Ekelund NG. Effects of Cu^{2+}, Ni^{2+}, Pb^{2+}, Zn^{2+} and pentachlorophenol on photosynthesis and motility in *Chlamydomonas reinhardtii* in short-term exposure experiments. BMC Ecol 2001;1:1.

[60] Visviki I, Rachlin JW. Acute and chronic exposure of *Dunaliella salina* and *Chlamydomonas bullosa* to copper and cadmium: effects on ultrastructure. Arch Environ Contam Toxicol 1994;26:154–62.

[61] Staruschenko A, Dorofeeva NA, Bolshakov KV, Stockand JD. Subunit-dependent cadmium and nickel inhibition of acid-sensing ion channels. Dev Neurobiol 2007;67 (1):97–107. PubMed PMID: 17443775. Epub 2007/04/20. eng.

[62] Neal AP, Guilarte TR. Molecular neurobiology of lead (Pb^{2+}): effects on synaptic function. Mol Neurobiol 2010;42(3):151–60. PubMed PMID: 21042954. Pubmed Central PMCID: 3076195. Epub 2010/11/03. eng.

[63] Chen J, Myerburg MM, Passero CJ, Winarski KL, Sheng S. External Cu^{2+} inhibits human epithelial Na + channels by binding at a subunit interface of extracellular domains. J Biol Chem 2011;286(31):27436–46. PubMed PMID: 21659509. Pubmed Central PMCID: 3149337. Epub 2011/06/11. eng.

[64] Yang W, Manna PT, Zou J, Luo J, Beech DJ, Sivaprasadarao A, et al. Zinc inactivates melastatin transient receptor potential 2 channels via the outer pore. J Biol Chem 2011;286(27):23789–98. PubMed PMID: 21602277. Pubmed Central PMCID: 3129160. Epub 2011/05/24. eng.

[65] Ariki M, Lanyi JK. Characterization of metal ion-binding sites in bacteriorhodopsin. J Biol Chem 1986;261(18):8167–74. PubMed PMID: 3722147. Epub 1986/06/25. eng.

[66] Tanimoto S, Sugiyama Y, Takahashi T, Ishizuka T, Yawo H. Involvement of glutamate 97 in ion influx through photo-activated channelrhodopsin-2. Neurosci Res 2013;75 (1):13–22. PubMed PMID: 22664343. Epub 2012/06/06. Eng.

[67] Watanabe S, Ishizuka T, Hososhima S, Zamani A, Hoque MR, Yawo H. The regulatory mechanism of ion permeation through a channelrhodopsin derived from *Mesostigma viride* (MvChR1). Photochem Photobiol Sci 2016;15(3):365–74. PubMed PMID: 26853505. Epub 2016/02/09. eng.

[68] Takahashi T, Watanabe M. Photosynthesis modulates the sign of phototaxis of wild-type *Chlamydomonas reinhardtii*. FEBS Lett 1993;336:516–20.

[69] Bean B, Yussen P. Photoresponses of *Chlamydomonas*: differential inhibitions of phototactic and photophobic responses by low concentrations of Cu^{2+}. J Cell Biol 1979;83:351a.

[70] Holland E-M, Braun F-J, Nonnengaesser C, Harz H, Hegemann P. The nature of rhodopsin-triggered photocurrents in *Chlamydomonas*. I. Kinetics and influence of divalent ions. Biophys J 1996;70:924–31.

[71] Govorunova EG, Sineshchekov OA, Hegemann P. Desensitization and dark recovery of the photoreceptor current in *Chlamydomonas reinhardtii*. Plant Physiol 1997;115:633–42.

[72] Stallwitz E, Häder D-P. Effects of heavy metals on motility and gravitactic orientation of the flagellate, *Euglena gracilis*. Eur J Protistol 1994;30:18–24.

[73] Yoshimura K, Matsuo Y, Kamiya R. Gravitaxis in *Chlamydomonas reinhardtii* studied with novel mutants. Plant Cell Physiol 2003;44:1112–18.

[74] Duong A, Steinmaus C, McHale CM, Vaughan CP, Zhang L. Reproductive and developmental toxicity of formaldehyde: a systematic review. Mutat Res 2011;728 (3):118–38. PubMed PMID: 21787879. Pubmed Central PMCID: 3203331. Epub 2011/07/27. eng.

[75] Chang CC, Gershwin ME. Perspectives on formaldehyde toxicity: separating fact from fantasy. Regul Toxicol Pharmacol 1992;16(2):150–60. PubMed PMID: 1438995. Epub 1992/10/01. eng.

[76] Yasokawa D, Murata S, Iwahashi Y, Kitagawa E, Nakagawa R, Hashido T, et al. Toxicity of methanol and formaldehyde towards *Saccharomyces cerevisiae* as assessed by DNA microarray analysis. Appl Biochem Biotechnol 2010;160(6):1685–98. PubMed PMID: 19499198. Epub 2009/06/06. eng.

[77] Sineshchekov OA, Lebert M, Häder D-P. Effects of light on gravitaxis and velocity in *Chlamydomonas reinardtii*. J Plant Physiol 2000;157:247–54.

[78] Toews J, Rogalski JC, Kast J. Accessibility governs the relative reactivity of basic residues in formaldehyde-induced protein modifications. Anal Chim Acta 2010;676(1-2):60–7. PubMed PMID: 20800743. Epub 2010/08/31. eng.

[79] Jiang W, Schwendeman SP. Formaldehyde-mediated aggregation of protein antigens: comparison of untreated and formalinized model antigens. Biotechnol Bioeng 2000;70 (5):507–17. PubMed PMID: 11042547. Epub 2000/10/24. eng.

[80] Stallwitz E, Häder D-P. Motility and phototactic orientation of the flagellate *Euglena gracilis* impaired by heavy metal ions. J Photochem Photobiol B Biol 1993;18:67–74.

[81] Wong PTS, Burnison G, Chau YK. Cadmium toxicity to freshwater algae. Bull Environ Contam Toxicol 1979;23:487–90.

[82] Magdaleno A, Velez CG, Wenzel MT, Tell G. Effects of cadmium, copper and zinc on growth of four isolated algae from a highly polluted Argentina river. Bull Environ Contam Toxicol 2014;92(2):202–7. PubMed PMID: 24297640. Epub 2013/12/04. eng.

Fish

Peter-Diedrich Hansen
Technical University of Berlin, Berlin, Germany

15.1 Introduction

This chapter describes a suite of standardized and validated fish bioassays according to EN ISO and OECD. The problem of acute toxicity testing and replacement of the whole animal test with fish cell lines will be discussed. After extended research activities in testing fish cell lines for the replacement of the "acute fish test" in the wastewater levy act (AbwAG), it became evident that fish tissue cultures as supplement or replacement for the fish bioassay are much less sensitive than whole animal testing. Consequently, the "acute fish test" with the whole organism was again the only accepted fish test until the more sensitive test—with the determination of the non-acute-poisonous effects of wastewater on fish eggs by dilution limits—was ready for standardization and validation. The standardization and validation of this "fish egg assay" was conducted by a working subgroup of the newly-formed working group at DIN: "Sub-animal testing (group T)". From now on the acute fish assay was replaced by the fish egg bioassay. Whole animal testing with fish remained only in the highly advanced "real time" fish bioassays with fish monitors for drinking water, surface water, and effluents. The onsite fish monitors are meanwhile well established and help to understand the complexity of environmental signaling.

Some other sub-animal fish related chemical tests are developed with fish eggs and fish embryos followed by ISO and OECD, i.e., the Early Life Stage Test (ELST). Especially in the marine field, the sub-lethal effects of emerging new contaminants on fish eggs and larvae are considered for in situ testing and are well accepted for effects monitoring according to the marine conventions for protecting the sea. Examples are given in this chapter. Beside the fish egg test, other sub-animal testing with endpoints like neurotoxicity, genotoxicity, endocrine effects, and immunotoxicity will soon become a standard according to EN ISO. There are already standards like the Cholinesterase Inhibition Test after DIN for a biomarker of neurotoxicity, or the DNA-unwinding test of genotoxicity, or vitellogenin in fish for endocrine effects, or phagocytosis for immuntoxicity. Fish is one of the key organisms for bioassays and biomarkers (biochemical responses) in the function of a valid indicator value for effects directed monitoring in aquatic ecosystems.

The available variety of standards of fish bioassays after ISO, EN ISO, and OECD contributes very well to the Water Framework Directive (WFD) concerning the "good ecological status" and the "good chemical status" and the Regulation of Chemicals by REACH (Registration, Evaluation, Authorization, and Restriction of Chemicals).

Bioassays. DOI: http://dx.doi.org/10.1016/B978-0-12-811861-0.00015-2

15.2 Acute fish toxicity testing

Almost without exception, present-day fish tests incorporate (1) a series of test-aquaria, each with a different but more or less constant concentration of the toxicant; (2) a group of similar fish in length and weight and of course species—usually ten in each aquarium with a volume of 10 L; (3) observation on fish mortality during exposure pf 48 h or prolonged observation of 96 h; and (4) final results expressed as the concentration tolerated by the median or "average" fish.

The endpoint mortality is defined by the immobility of the fish (instability of swimming performance). Chemicals may cause effects concerning physiology or biochemical responses (biomarker) but also effluents. According to the §7a of the German Federal Water Act (WHG) and the wastewater levy act (AbwAG) the fish species *Leuciscus idus melanotus* (golden orfe) is exposed to diluted wastewater [1]: 1 volume wastewater + n volume dilution water (n = 1, 2, 3, 4, 5, 7).

The test fish golden orfe are limited to a certain size (5–8 cm and $K = 0.8-1.1\ g\ cm^{-3}$) to guarantee a constant range of sensitivity. As the golden orfe is an European fish that reproduces only once a year (April to June) and grows up to a size of 30–40 cm, the breeder has to ensure the desired size 5–8 cm by keeping the fishes in a cold environment at low food supply. Care has to be taken to guarantee healthy test fish after a prolonged time under these conditions [1]. The inter-laboratory reproducibility of the fish test has been evaluated in test series together with the regional authorities and the stakeholders. Only minor differences of + 1 G level were found [1], variations which are inevitable if the real value is exactly between two *G* levels. The acute fish test, which has been carried out since the beginning of the 1980s, can be regarded as an established and reliable procedure. Meanwhile the new terms and definitions for progressing testing are published by the International Standard Organisation (ISO) and always updated to today [2].

The acute fish toxicity test was always under discussion because of the wastewater fee calculated by the dilution factors G_F or LID (G_F = Toxicity Fish; LID = lowest ineffective dilution). Because of the ethical review committee and animal welfare it was not acceptable to gain money through the death of fish (DIN 38412 part 20) after the wastewater levy act (AbwAG). According to the "3R" concept it was necessary to refine, reduce, and replace animal testing. Fish testing in general was under discussion by fish experts in the new DIN working group "fish test" [3]. Critical deliberations and investigations were made on the acute fish test [4–6] after DIN and OECD but also on a long-term test [7] according to the German Chemical Act. At the same time the DIN working group "fish test" became involved by testing and validation of the results in the refinements of the acute fish test DIN 38412 part 20/31 [5]. The exposed numbers of test fishes were reduced from 10 to 5 and finally to 3. But still the acute fish test was under discussion and for replacement by the fish cell line test [8].

15.3 Replacement of the fish bioassay by the fish cell line test

The state of the art concerning tissue cultures as a supplementary method to the fish assay with the golden orfe (*L. idus melanotus*) pursuant to the wastewater levy act (AbwAG) was investigated by a research project and by the members of the DIN working group "fish test" [3]. A standard operating procedure (S.O. P.) was introduced which is based on Ahne's studies on the use of fish tissue cultures for toxicity tests in order to replace the fish assay [8]: "Determination of toxicity in wastewater with the cell line R_1 of the rainbow trout". The validation and standardization of this assay was then conducted by the newly-formed working group at DIN "Sub-animal Testing". The principle of the test method was to measure the influence of toxic compounds of the wastewater on the vitality of fish cells. In this method (measure values), criteria reflecting the vital functions of the cells were:

1. Preservation of the attachment of the cells to the surface of the culture vessel.
2. Capability to accumulate the vital dye neutral red in intact lysosomes of the fish cells.

In Ahne's studies [8] primary cell lines were investigated, particularly primary cell line D_1 from the golden orfe, but later changing to the primary cell line of the rainbow trout R_1 [9,10]. In parallel to the validation and standardization of this assay by the DIN group "Sub-animal Testing" the outcome and first results by the project members of the research project on fish cell-lines for the toxicological evaluation of wastewater samples [11] showed clearly that there was only one fish cell line acceptable, the permanent fish cell line RTG-2 (*Oncorhynchus mykiss*, formerly *Salmo gairdneri*) established from the gonadal cells of a rainbow trout [11] available on request by the DMSZ (German Collection of Microorganism and Cell cultures: RTG2/ACC352). The research project was supported and guided by the ZEBET (Zentralinstitut für Alternativmethoden zu Tierversuchen) of the former Bundesgesundheitsamt and the European Centre for the Validation of Alternative Methods (ECVAM) [12]. The ZEBET has changed its name to BfR (ZEBET-BfR, Bundesinstitut für Risikobewertung) (Table 15.1).

Table 15.1 Example of the calculation of the Dilution factor G_F (F for fish) in the DIN 38412 part 31 or internationally LID (lowest ineffective dilution). The lowest number of G or LID is per definition the Dilution step G_F/LID where fish survive

Wastewater	Dilution water	Dilution step G/LID	Number of dead fish
1	4	5	2
1	5	6	0
1	7	8	0

The validation and standardization activity by the working group at DIN "Sub-animal Testing" was always updated by the newest results of the extended research project.

The native wastewater matrix was not a sufficient medium for testing toxicity by a fish cell line. There were a lot of interferences with undissolved substances and/or colored wastewater samples, which can falsify the test results and/or reduce reproducibility. Additionally, the microtiter plates have to be suitable for cell cultures and, if necessary, their suitability has to be examined. For the test procedure, tissue culture grade microtiter plates (96 wells, flat bottom) are employed. Only cells of a current passage (flasks with a monolayer of approximately 90% confluence) are used. The lots of FCS (fetal calf serum) have to be tested for their suitability to support the growth of the fibroblast-like gonadal RTG-2 cells.

An inverted microscope is necessary for checking the RTG-2 stock cultures and for controlling trypsinization. A counting chamber or cell counter has to be used to determine the number of cells after trypsinization. A microtiter plate reader is necessary for the measurements of the values. Extinction is measured at 540 nm with a reference wavelength at 690 nm. Linearity of the measurement has to be checked. The final relevant G_Z-value is the lowest dilution level G of the test batch in which the mean OD (optical density) is reduced by less than 50% compared with the mean OD of the negative control.

After the validation of the test results of an international ring exercise (DIN 38415-part5) with the RTG-2 cell line (ring exercise with at minimum 9 external laboratories) and several extended comparing studies by industry and governmental authorities concerning the acute "fish assay" [5] and the new fish RTG-2 cell line assay showed clearly that the fish RTG-2 cell line assay was significantly less sensitive. The decision was made to delete the RTG-2 fish cell line (G_Z-values) from wastewater testing and the acute "fish assay" [5] was not replaced at this time.

15.4 "Real time" fish monitors for wastewater effluents, at river sites and fish monitors for protecting of drinking water supply

Fish bioassays with whole animals still became popular for the "real time" fish monitoring and "early warning" of pollutants in the outlets of wastewater, at river sites, or concerning protection of drinking water supply. For drinking water there are two fish monitors available [13,14]. The fish warning test after Geller and Mäckle is working with the Nile pike (*Gnathonemus petersi*) in a flow-through test chamber with electrodes to receive the current pulses given by the fish [13]. The measurement values are the frequency of the current pulses. The test water has to be degassed before it flows through the test chamber with the exposed fish. The second commercially available fish monitor for drinking water is the bbe Fish Biomonitor ToxProtect64", this monitor is designed as an early warning system [14].

The robust fish (i.e., *Leucaspius delineatus*) pass through 80 light barriers sending their signals by a transducer and data processor. The monitor is very sensitive to detect contaminants as well as cyanobacteria and biotoxins spills or accidents like terrorist attacks.

In the well known and world wide established fish monitor after Besch & Juhnke and Scharf with the fish are swimming against a strong water stream and moving in case of weakness by pollutants against a photo barrier in the back of the arena chamber [15].

Another example for a fish monitor is the automatic WRc MK III-Fish Monitor developed by Water Research Center (WRc, Medmenham Laboratory) [16]. This fish monitor was tested over a period of one and a half years along 685 km of the River Rhine [17,18] and detected more than ten disasters and alarms (increasing toxicities). The WRc-Monitor was investigating the exposure of 8 adult fish (rainbow trout, 25–30 cm length) in flow-through channels [18,19]. The fish produce small oscillating voltages in the water by changing the ventilation frequency picked up by electrodes. The individual signals of the eight exposed fish are digitized for data transfer every 20 s [19]. The latest developments in "real time fish monitoring" are demonstrated in Fig. 15.1: the bbe Fish Toximeter.

The bbe Fish Toximeter is a containment with a very low flow-through, the fishes (i.e., golden orfe) swim in front of a light screen, and their swimming route is registered by a video camera (Fig. 15.1B). The toxicity detection is based on the video analysis of the fish behavior with regard to (1) average swimming speed; (2) speed distribution; (3) average distance of the individual fishes; (4) distribution of the fish in the containment; and (5) destination and circles of the moving fish.

The parameters are collected automatically by the video camera through video image analysis. The Toximeter is very similar to the so-called "BehavioQuant" [20,21] but without freeze branding of the fish (zebrafish). There is no "fish training" on swimming against a strong water current (Besch & Juhnke and Scharf fish monitor) [15].

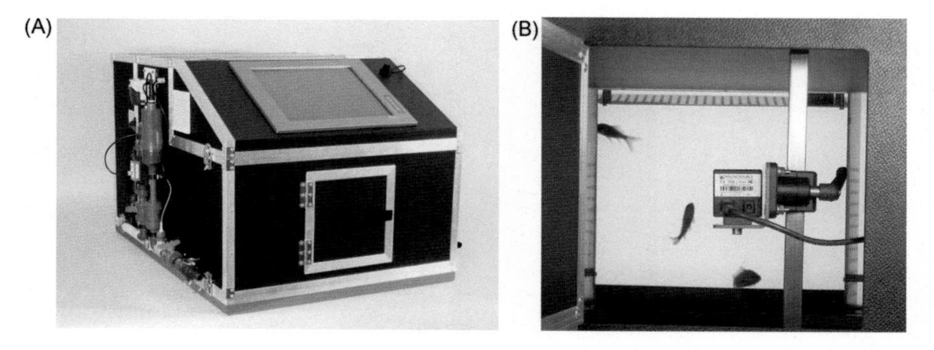

Figure 15.1 (A) bbe Fish Toximeter with a low flow-through. (B) Fish tank from inside with the video camera for video image analysis, and fish (*Leuciscus idus melanotus*).

15.5 Static, semi-static and dynamic fish bioassays

The working group "fish test" [3] concentrated their activities on the OECD long-term fish test (14-day study) [7] according to the German Chemical Act with the zebrafish (*Danio rerio*). Beside the long-term fish test [7] additional fish bioassays were investigated concerning growth [22–25] with juvenile fish. Experience has been gained from growth tests with zebrafisch (*D. rerio*) [22,26] and medaka (*Oryzias latipes*) [25,27]. In Fig. 15.2 the fish growth of a mouthbreeder

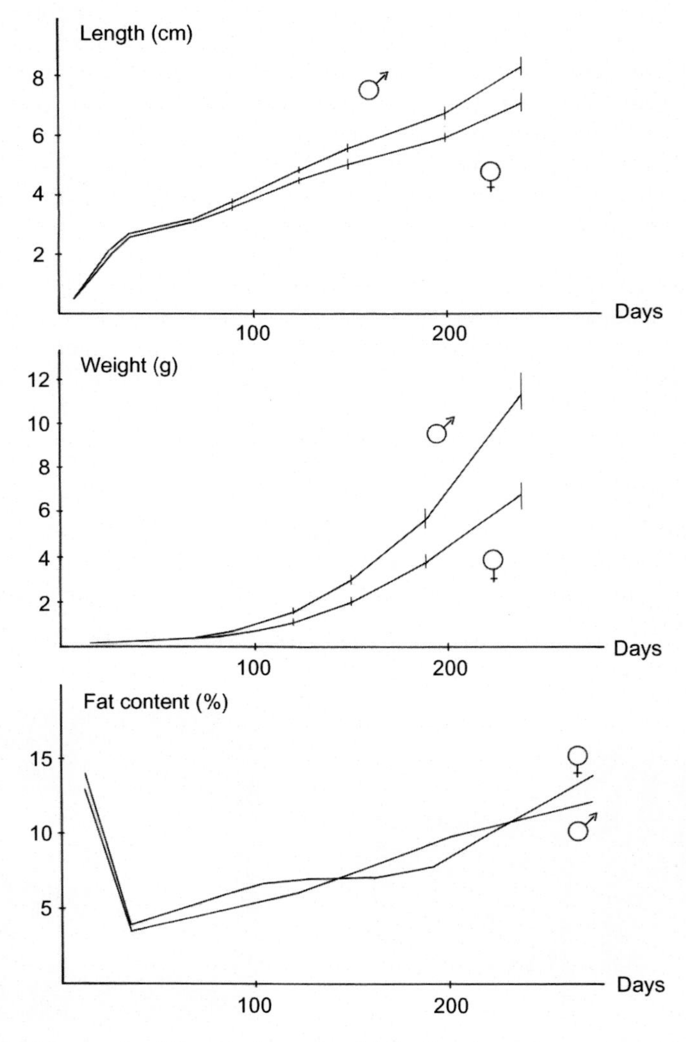

Figure 15.2 Fish growth of the mouthbreeder cichlid *Pseudotropheus zebra* with length, weight, and fat content distinction by male and female.

(*Pseudotropheus zebra*) is demonstrated. These species were investigated in chemical testing and endocrine effects (vitellogenin expression) because of their clear male or female distinction.

With increasing sophisticated fish bioassays the experimental design had to be changed from static to semi-static and finally to a dynamic flow-through system. With long-term fish assays and growth of juvenile fish the physiological conditions become more complex in terms of feeding the fish and keeping them under physiologically correct conditions concerning water quality control parameters like temperature, pH value, oxygen measurements, temporary and permanent hardness of the water, light cycle, etc. in order to keep all stress parameters as low as possible. Even semi-static testing with growing or developing fish larvae is very time consuming. The exchange of the water has to be done at least every day and consequently the flow-through design of a dynamic fish bioassay is the more relevant test design, avoid adding to the stress of the fish by changing the abiotic parameters sequentially [28]. A good example of the advantage of a dynamic fish bioassay is the bioaccumulation in fish [29,30]. To measure the BCF (Bioaccumulation Factor) there are two routes, the aqueous and the dietary route [30]. The OECD Guideline 305 [30] offers for both routes the semi-static fish test and the flow-through fish test.

The BCF is per definition the ratio of the test substance concentration in the test fish (C_f) to the concentration in the test water (C_w) at steady-state [30]. In Fig. 15.3 the results of exposure and bioaccumulation are demonstrated for different fish species and dynamic as well as static testing.

In Fig. 15.4 an experimental design is shown for the bioaccumulation test [30] of volatile and bio-accumulating substances without aeration. The sufficient oxygen concentration is supplied for the exposed fish by a layer of a gas mixture (31% O_2 and 69% N_2).

The OECD guideline 305 [30] supplies fish bioassays with methods for external exposure (water matrix) and internal exposure (dietary route). Beside the bioaccumulation there are many biochemical responses well known by effects monitoring in the field. These biochemical response effects are used in a broad sense by the term biomarker [31]. The International Programme on Chemical Safety (IPCS) describes biomarkers and risk assessment and gives definitions on biomarkers [31]. The evaluation of risks concerning bioaccumulation of chemicals is regulated in ANNEX XIII of REACH [32]: PBT: persistent and bioaccumulative chemicals, BCF > 2000; vPvB: very persistent and very bioaccumulative chemicals, BCF > 5000. There is as well a link to the WFD [32] concerning risk assessment (acceptable and non-acceptable risk) [33]. The use of fish-biomarkers (biochemical responses) in multiarrays (Fig. 15.5) for effects monitoring [34,35] is complementary to chemical analysis since they can alert for presence of newly emerging contaminants. Bringing fish-biomarkers and effects related bioassays into the WFD [32,33], the effect-related approaches and biomarkers are relevant in the context of the so-called QN (Quality Norms) of environmental relevant substances [33] and the "good chemical status" [32] of waterways.

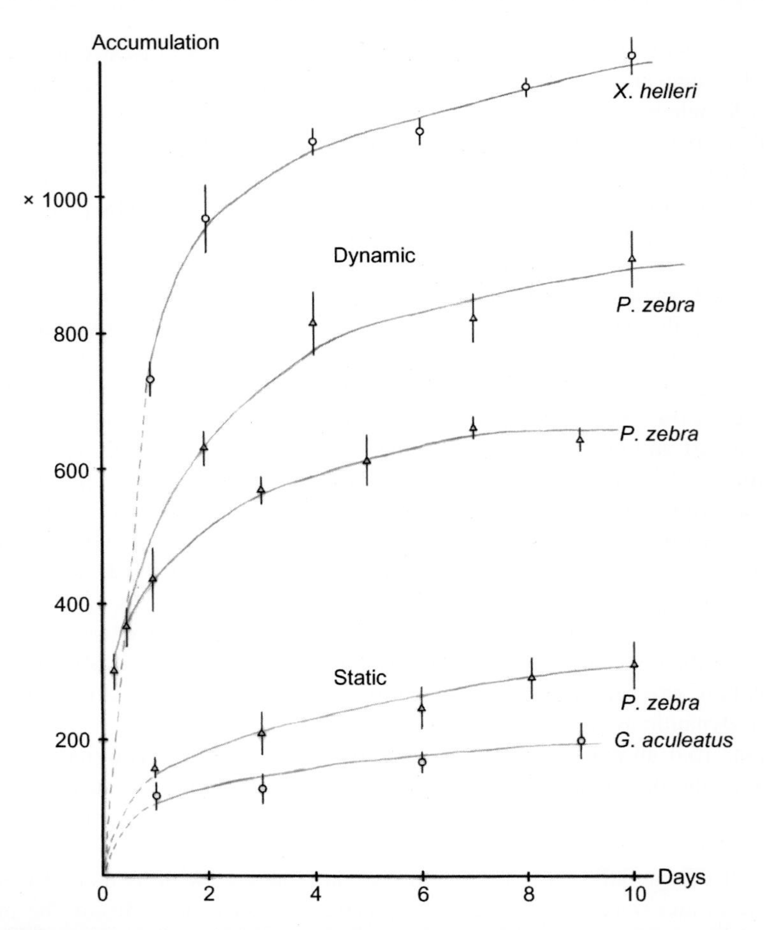

Figure 15.3 Accumulation of Lindane (-BHC) by three fish species (Green swordtail *Xiphophorus helleri*, Malawi zebra Cichlid *Pseutropheus zebra*, Stickleback *Gasterosteus aculeatus*).

15.6 Fish biomarker bioassays

It is very important to transfer the monitored biochemical responses (biomarker) in fish under the WFD [32] into an operational effect-related standard. The fish biomarkers serve finally as the basis for environmental protection against hazardous substances [32,36].

In Fig. 15.5 the common fish biomarkers are visualized. The standardization for the biochemical and physiological measurements on fish starts with the sampling of fish, and the handling and preservation of samples [37]. An analysis of the exploration of fish biomarkers and sediment assays [34,38] along coastal transects generates important information for coastal zone management and fisheries. The

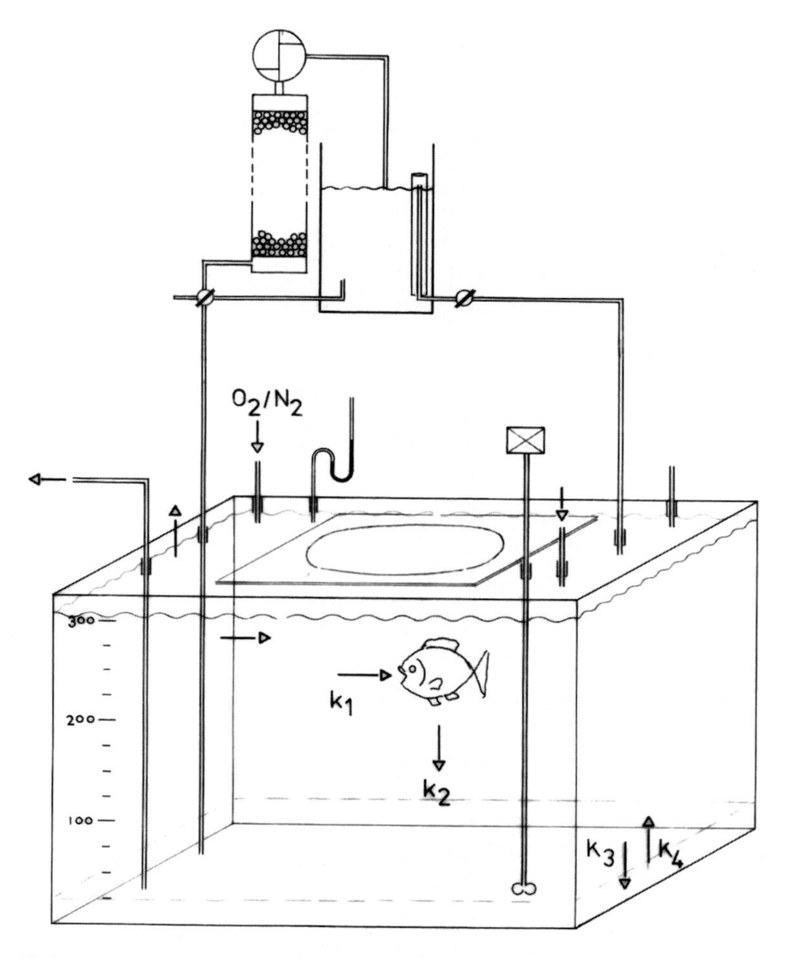

Figure 15.4 Bioaccumulation test for volatile chemicals [28,29]. Test water under a layer of a gas mixture (31% O_2 and 69% N_2) to guarantee sufficient oxygen content without aeration for the exposed fish. Uptake and decontamination of the exposed fish, and uptake and decontamination of the sediment.

neurotoxic contamination in fish (Fig. 15.5) is measured by cholinesterase (ChE) inhibition. There are two ways to get the sample from the fish brains or from muscle tissues from the back of the fish [39]. Much more difficult is the measurement of biomarkers concerning immunotoxic effects [40,41]. Very well investigated are biomarkers concerning biotransformation in fish liver measurement of phase I by the induction of the Mixed-Functional Oxigenase (MFO) and the substrate Ethoxiresorufin (EROD); the measurable product is resorufin [42]. The determination of ethoxyresorufin-*O*-deethylase (EROD) became an ISO Standard in 2007 [42]. The basis and fundamental science of the EROD fish assay was done by Burk and Mayer [43]; later the EROD activity measurements in fish were optimized by

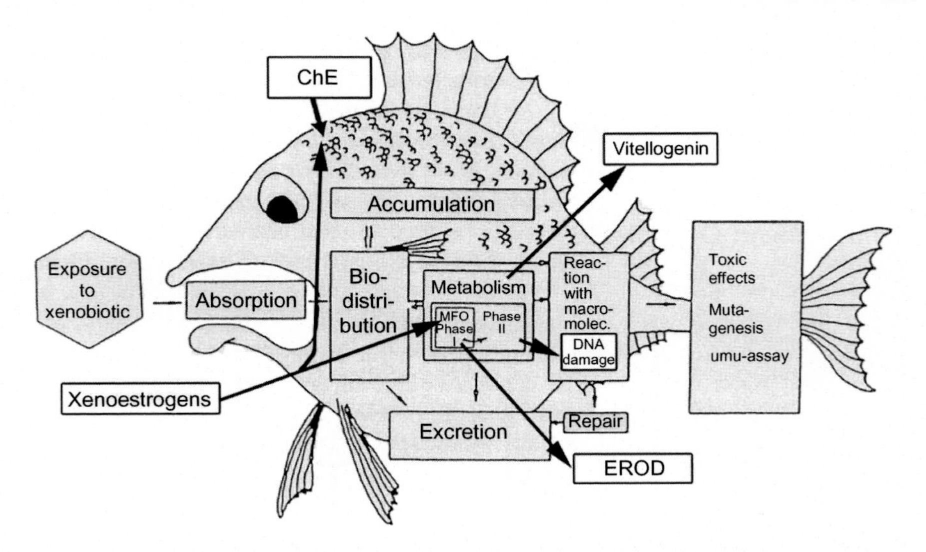

Figure 15.5 Fish bioassays and biochemical responses (biomarker) concerning neurotoxicity (ChE), biotransformation (MFO = Mixed-Functional Oxygenase: EROD = Ethoxiresorufin), endocrine effects (Vitellogenine), and genotoxicity (DNA unwinding).

fluorescence plate-readers [44,45]. Because of the widely used fish biomarker EROD and the need for comparable data an inter-laboratory comparison of measurements of Ethoxyresorufin *O*-deethylase activity in the dab (*Limanda limanda*) was organized [46] by ICES (International Council for Exploration of the Sea). Meanwhile the biomarker EROD is a standardized method according to ISO [42]. The fish species are monitored concerning the induction of EROD in extended effects monitoring programs in inland [47] and coastal waters [48]. New chemicals and drugs are tested i.e., dissolved in fish oil and peritoneal injection. During the internal exposure (incubation approx. 7 days) the individual fish are kept in a flow-through containment. After the incubation of the fish [37] the liver is dissected [42] and the induction of the EROD is measured by fluorescence spectrometry [42]. As a matter of routine, several biomarkers like cholinesterase inhibition, EROD, endocrine effects, and genotoxicity are investigated in parallel to get more complex information on the effect-related health status of the fish [49]. For newly emerging contaminants incubation experiments in a suite of flow-through tanks will be set up to keep the fish in a comfortable environment. For the EROD induction assays the fish are individually injected peritoneal by a syringe with the contaminant in a solvent (fish oil) for internal exposure. The other method, not so comfortable for the fish, is to blow the solved contaminant (fish oil) sealed in a capsule orally via tubing into the gut. For testing the endocrine effects in fish there is an ISO standard available for measuring values of the biomarker vitellogenin [50]. Vitellogenin (Vtg) is a large phospholipoglycoprotein produced as the yolk protein precursor in the liver of oviparous vertebrates, such as fish. Vtg protein is secreted from hepatocytes through the secretory pathway, enters the circulation and is taken up by the growing oocyte. Normally the Vtg is only present in female fish but many studies have shown that it is also present in males [51–53] because of estrogenic and endocrine disrupting compounds (EDC). The vitellogenin induction in fish as an indicator of exposure to environmental estrogens became of interest because the increasing preponderance of female fish populations has become a matter of concern not only for scientists but also for environmental authorities. For example, in the extensive waterways of Berlin, Germany, 70% of the fish population became female. Two questions were investigated: [1] is the biomarker vitellogenin (vitellogenin synthesis assay) an appropriate tool for determining endocrine effects; and [2] do chemical pollutants in effluents, such as oestradiol, phthalates, alky-phenols, and alkyl-ethoxylates, influence sex differentiation [51]. An extended sampling [37,50] in the Berlin waterways by the Berlin Fishery Authorities gave evidence that there is a feminization of different fish species in the populations of fish [52]. The international interest was stimulated by reports from the United Kingdom (UK) of female characteristics in fish taken from waters that receive effluents from sewage treatment plants. The exposure experiments with male fish in the containments of the so-called onsite "WaBoLu–Aquatox" monitoring system (see Chapter 20: Applications for the real environment; this volume) in mixtures of definite amounts of effluent and dilution water in the flow-through system showed clearly a dilution correlated significant induction of vitellogenin in male fish [51]. The dilution steps (10%, 20%, 30% and 40% effluent) are relevant with regard to the effluent loading of the Berlin waterways depending on the season of the year. Overall the results with the on-site exposed fish (*O. mykiss*) showed clearly that the LOEC for the endocrine effects of 17β-oestradiol is very close to the environmental concentration in the effluents and consequently in the waterways. In other studies with different fish species like the Japanese medaka (*O. latipes*), fathead minnow (*Pimephales promelas*), zebrafish (*D. rerio*), and the three-spined stickleback (*Gasterosteus aculeatus*), the biomarker vitellogenin and fish sexual development tests [54] was confirmed in inter-laboratory validation exercises from 2006–2010. An important amendment is a short-term screening test (21-day fish assay) for estrogenic and androgenic activity and aromatase inhibition [55]. Most of the studies for androgenic substances are done with fathead minnow (*P. promelas*) [56] and medaka (*O. latipes*) [57].

15.7 Bioassays and fish reproduction: fertilization, embryo-larvae, early life stage, lifescycle ("reciprocal" bioassays)

Parameters such as growth [22,25] and reproduction are often considered as a better predictor of effects that may ultimately arise on a population and ecosystem level. The so-called "chronic bioasssays" [2,7] do not necessarily cover a large part of a life cycle of the fish, but they may also focus more or less on that part of a fish life cycle that is thought to be very sensitive to stress. This could be for instance fertilization tests such as the "in situ" embryo larvae tests of fish [58]. For describing the reproductive development in fish it is very important for the scientific and governmental dialog to use a standardized terminology [59].

Several advanced bioassays are able to detect the complex toxic potential of multiple unknown contaminants at low concentrations [58]. For an integrated assessment it is necessary to use a complementary combination of several test methods [60–64].

The reproduction test includes the gametogenesis in the parental fish and the fertilization. The embryo-larvae test incorporates the embryonal development, hatching rate, and yolk-sac larvae stages. The ELST Toxicity Test covers additionally the transitional step from endogenic to exogenic feeding and the postlarval to juvenile stages.

The life cycle test comprises the sexual maturity, the running ripe (stage Vi), and the reproduction (F_2). In Fig. 15.6 the updated and harmonized terminology [59] of the test systems in fish reproduction is shown.

The sublethal effects of pollutants and bio-accumulating contaminants on fish eggs and larvae are described in [23,58,65]. In field studies running ripe flounders *Platichthys flesus* were caught with a bottom trawl at different locations in the Baltic Sea. Eggs and milt of running ripe parental fish were mixed and then sea water was added; the eggs (1000–3000 per incubating jar) and the spermatic fluid were incubated for 10–15 min to ensure proper fertilization [66]. The pelagic eggs were then carefully washed with sea water and transferred to 500 mL incubation jars containing 300 mL Baltic Sea water. Incubation was carried out at 6.8 + 0.9°C and a salinity of $27.5^0/_{00}$. The sea water used for the experiments had been pumped from a depth of 16 m in the western Baltic Sea and served as "standard incubating medium" during the experiment. Dead and unfertilized eggs were recorded and

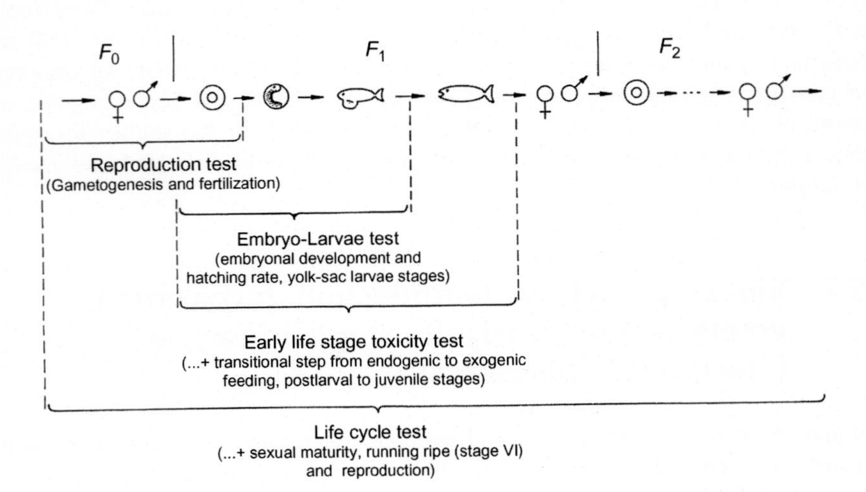

Figure 15.6 Fish bioassays, reproduction toxicology, and standardized terminology [59].

removed once a day when the incubating water was changed. Each incubating trial was carried out with three replicates. Viable hatch of straight and healthy appearing larvae (Fig. 15.6) were determined at the end of each experiment. The fertilization, the incubating of the eggs, the embryonal development, and hatching of the larvae were done onboard a research vessel.

The methodological approach [58] chosen in this study seems to be very promising by combining in situ exposure of adults during the sensitive phase of gonad development with experimental control of embryo performance during development [58]. It is very important to standardize these experiments carefully [2,23,37] and for interpretation of the data an advanced statistical analysis is needed [23]. The experiments with the pelagic flounder eggs showed clearly that it is quite difficult to include the whole lifecycle in a semi-static incubation, but meanwhile the technology is already there to improve this assay with a flow-through system. The whole lifecycle in the described method [58,65] is reduced to the valid endpoint of the larval stage IIα and IIβ. To include the whole lifecycle, experiments were done with spring spawning herring (*Clupea harengus*) [66]. This fish species is a substrate spawner with demersal (sticky) eggs. The experimental design is quite different to fish with pelagic eggs [58,65]. Running ripe Baltic herring were caught during late April with a drift net located near the coast of Travemünde (Germany). The net was operated overnight and lifted during early morning hours (5:00 a.m.). Live females were transferred to a flow-through containment onboard the fisherman's boat. Live transport was 2.5 h to the laboratory. The eggs were stripped from the females in single rows, transferred to glass plates, stored for 15 min at 8°C in sea water of $20^{0}/_{00}$ salinity, and fertilized with a sperm suspension derived from 10 males. After allowing 15 min contact time, the glass plates and the attached eggs were washed several times in clean sea water and transferred to 500 mL incubation jars containing 300 mL Baltic Sea water. Three glass plates carrying between 605–1615 eggs were prepared from each female. Therefore, the results for interpretation were the means of three replicates [58,65].

Incubation was conducted at 6.5–7.5°C and a salinity of $20^{0}/_{00}$. The incubation jars were continuously aerated. Water exchange was undertaken every 48 h until hatching. During the hatching period water was exchanged daily, and the total number of hatched larvae was counted and viable hatch (straight and healthy larvae) determined. Since all eggs were incubated under standardized conditions, the quality of the egg material (i.e., the gonad burden of the parental fish with various trace contaminants) was considered as a major source of possible variation in hatching success [66].

Such fish assays ("reciprocal bioassays") are very time consuming and costly but they are at a very high level for political decision making, commercial potential for fisheries regulations and international conventions for conservation measures in legal aspects.

There is a need to control marginal conditions for reproducible fish bioassays: standard procedures needed to hold up in court cases related to violating the WFD on habitat and fish recruitment. Malformation in North Sea pelagic embryos [67] of dab, whiting, cod, flounder and plaice in the southern North Sea have been successfully monitored by quantitative investigations of developmental defects in eggs immediately after capture. Incidences were higher in near-coastal waters known to receive high pollution loads.

15.7.1 "Reciprocal fish bioassay" with running ripe contaminated fish, fertilization, egg incubation and influence of solar ultraviolet-B on pelagic fish embryo developments

In experiments, embryos of dab (*L. limanda*) taken from sites located along a gradient of chemical stress in the Southern Atlantic were exposed to solar ultraviolet-B radiation and allowed to recover for a defined period for repair [68,69]. The hypothesis was that there might be an interaction between exposure to chemical contaminants and the ability of embryos to recover by repair from UV-induced DNA breaks and generation of thymine dimers [69]. To measure possibly UV-mediated DNA damage in fish-egg embryos both procedures the DNA unwinding assay and the assay for the determination of thymine dimers were applied. The DNA Unwinding Assay (DNA alkaline denaturation assay) is monitoring the genotoxicity in the eukaryotic tissues of the fish-egg embryos (Fig. 15.7).

15.7.1.1 Permeability of the fish-eggs and multixenobiotic resistance-mediating transport (MXRtr)

To get more information concerning standardization of the fish eggs and relevant sublethal effects for the fish-eggs embryos beside the DNA damages the uptake of fluorescine, the accumulation of Rhodamin B (Rh B) combined with the specific ionic blocker, Verapamil (Ver) was used to estimate the export activity of the MXRtr in the fish eggs. The developmental stages of fish-eggs embryos and the uptake of fluoresceine were investigated as a parameter to quantify the permeability by the load of fluorescein in the fish-egg embryos (green fluorescence, 450–490 nm). The penetration of fluorescein was measured by fluorescent microscopy and microfluorometry. The red fluorescence of accumulated Rh B in the eggs was measured as well by contact microfluorometry (excitation 510–560 nm, dichroic mirror 575 nm, barrier filter >590 nm) with a 16 × objective and mirror (25 μm in diameter). The fish-egg embryos were incubated with 0.5 μM Rh B and 10 μM of Ver for 10 min. The mean difference between Rh B fluorescence with and without inhibitor Ver is proportional to the amount of Rh B eliminated by the MXRtr system per hour. There was a high deviation of 5%–40% in the early stages especially in stage I a, a lower deviation of the stages I b, II α, II β, II γ and III α was found in the range of 30% (± 10%). The mean permeability of the stage III β was approximately 20% (± 5%). The permeability of the later stages III γ, IV α, and IV β was increasing up to 35% (± 5%). The permeability of the later stages was nearly comparable with the controls. In the exposure experiments with SONSI 2 (sunshine simulator) and SONSI 7 in the early stages of I $\alpha\gamma$ and Ib γ of the fish-egg embryos, the permeability was increased by 14% against the controls.

Figure 15.7 "Reciprocal-fish-bioassays" with running ripe contaminated fish, egg incubation, and influence of solar ultraviolet-B on pelagic fish embryos. Fish egg exposure to UV-B under controlled conditions for developing fish embryos. The exposure-cuvette was made out of silica glass and for the exposure equipped with 250 eggs. The calibration of the instrumental setup for the measurements and simulation of solar radiation—SONSI (UV-B, 290 nm and 4645-6568 Ws m^{-2})—was achieved with a 1000 W quartz halogen light source in the laboratory of the Alfred-Wegener-Institut (AWI) before and after each cruise [69].

15.7.1.2 Eco-genotoxicity: influence of solar ultraviolet-B on pelagic fish embryos: DNA fragmentation and thymine dimers

Running ripe dab were caught from polluted gradients. The eggs were stripped from the chemically stressed sites and exposed to a SONSI over three cruises with comparison to the solar reference spectrum measured on board the RV Walther

Herwig III at 54.5°N, 7.0°E. The eco-genotoxic impact on stressed fish-eggs embryos exposed to UV radiation was assessed by DNA unwinding [69]. Thymine dimers cause no lesions in the DNA molecule like DNA strand breaks (SBs). The formation of thymine dimers is a result of the breaking of H-bridging covalent binding sites between, e.g., thymine and adenine and replacement by a covalent binding cyclobutane thymine dimer between two thymine molecules, which are located side--by-side at the DNA strand. The cyclobutane-thymine dimers are very stable. The repair is facilitated by special enzymes like DNA-polymerases and topoisomerases that perform the genetic code repair.

The natural unwinding of DNA's double helix starts at either strand end or at strand nicks and are generated by topoisomerases. Additional unwinding points occur at apurinic and apyrimidinic (AP) sites, SBs, and repair patches. Unwinding can also start at reactive oxygen species (ROS)-induced lesions in the DNA double helix or in the genetic code. Those additional unwinding points combine to enhance DNA's normal unwinding rate under alkaline conditions [69]. DNA damage in the form of thymine dimers differs distinctly from DNA SBs.

The experiments were carried out with defined and realistic irradiation qualities and quantities. DNA damage was measured by DNA unwinding and thymine dimers measured by an ELISA. Results showed a time- and dosage-dependent influence of UV-B radiation on the sublethal parameter genotoxicity (DNA damages) [69]. The repair of the DNA SBs, as well as the thymine dimers, were investigated by the post-replication repair in embryo cells [69].

The degree of unwinding is calculated as F (double-stranded remaining fraction) and the negative logarithm (-log F) which is proportional to the average number of unwinding points. Increased $-\log F$ values in samples over control values indicate a genotoxic impact. The DNA unwinding results were calculated as P-values ($P = -\log F_x/\text{-}\log F_0$) and only results >1 are discussed as effects.

The enzymes DNA ligases and DNA polymerases are responsible for the repair of DNA SBs. These enzymes are monitoring the DNA molecule to check the H-bridging covalent binding sites and the nucleoside gaps to replace these gaps by the corresponding nucleotide.

The enzymes DNA ligases and DNA polymerases are activated by UV radiation. In general there are lesions on the DNA double helix or in the genetic code because of the ability of the ROS including some oxygen-centered free radicals to modulate genes and protein expression and alter DNA polymerases.

UV-B and the UV-C radiation generate lesions in DNA such as cyclobutyl pyrimidine dimers and 6-4 pyrimidine dimers which block the DNA replication and transcription. Because of the importance of these premutagenic lesions and their lethal consequences and carcinogenic potential if unrepaired, an immunochemical method (ELISA) was developed to measure the thymine dimers.

The three individual experiments with fish-egg embryos of the developmental stages II δ – III gives comparable responses. At the first, third, and fourth day there was an increase in the thymine dimers and only on the fourth and fifth day was there some increase in the DNA unwinding. The dosis was at the fifth day with 14.330 Ws m^{-2} quite high but also exposures on days with a low dosis (third day of the experiment) with 5.836 Ws m^{-2} resulted in an increase in the thymine dimers [69].

The experimental design of experiments with an instrumental set-up for simulation of solar radiation (SONSI) exposure, with always the same fish-egg quality and embryo developmental stages and 5 times radiations by SONSI, gives a clear picture by the results: with the increasing mortality from the second to the third day of the experiment the DNA unwinding data were increasing as well and one day in advance of the maximum peak of the DNA damages also the thymine dimers gave a significant high peak signal. On the fourth day there was again an increase of the effects but no effects on the fifth day (repair). In the final phase of the experiments the results may demonstrate the effect of the UV irradiation by increasing DNA repair through the increase in the transcriptional processes in the fish-egg embryos. The developmental stage of the fish-egg embryos at the beginning of the experiment was Ib β and at the end of the experiment III β-IIIγ in the controls. Under SONSI the development of the fish-egg embryos was slightly delayed to IIβ-IIIα.

In summary there are sub-lethal effects (damages and repair) at the molecular level which can be demonstrated by DNA unwinding and the formed thymine dimers [69]. But it has to be concluded that under the present general weather conditions in spring and at the present levels of environmental ozone, allowing for a reduction to 180 DU, the embryonic development of North Sea spring spawning is not endangered by UV-B radiation [68,69].

Probably still newly emerging contaminants are involved in the reduction of the reproductive capacity of fish species, as indicated [48,66,67,69]. In order to discover other causative contaminants in situ in the "low dose" range, the "reciprocal-fish-bioassays" will have to be continued on a broader scale, using a larger number of experimental fish and a wider range of residues.

References

[1] Hansen P-D. The potential and limitations of new technical approaches to ecotoxicology monitoring. In: Richardson M, editor. Environmental toxicity assessment. London: Taylor & Francis; 1995. p. 13–28.

[2] ISO 6107. Water quality vocabulary – New terms and definitions in: ISO/TC 147/SC5/ N 179.

[3] Bresch H, Gode P, Hamburger B, Hansen P-D, Juhnke I, Krebs F, et al. Deliberations and investigations of the DIN working group "fish test" on a long-term test according to the German Chemicals Act. Z Wasser Abwasser Forsch 1986;19:47–51.

[4] OECD 203. Fish acute toxicity test. OECD guideline for the testing of chemicals, Section 2: effects of biotic systems. http://dx.doi.org/10.1787/9789264069961-en; 2009.

[5] DIN 38412 part 31 (formerly part 20). Fish Test (L31)—Golden Orfe (*Leuciscus idus melanotus*), German standard methods for the examination of water, wastewater and sludge; bioassays (group L), determining the tolerance of fish to the toxicity of wastewater by use of a dilution series (L31). Berlin: Beuth Verlag; 1989.

[6] Stuhlfauth T. Ecotoxicological monitoring of effluents. In: Richardson M, editor. Environmental toxicology assessment. London: Taylor & Francis; 1995. p. 187–98.

[7] OECD 204 (deleted 2014). Fish, prolonged toxicity test: 14-day study. OECD guideline for the testing of chemicals, Section 2: effects of biotic systems. http://dx.doi.org/10.1787/9789264069985-en; 1984.
[8] Ahne W. Untersuchungen über die Verwendung von Fischzellkulturen für Toxizitätsbestimmungen zur Einschränkung und Ersatz des Fischtests. Zbl Bakt Hyg I Abt Orig B 1985;180:480–504.
[9] Babich H, Borenfreund E. Cultured fish cells for the ecotoxicity testing of aquatic pollutants. Environ Toxicol 1987;2(2):119–33.
[10] Hansen P-D, Schwanz-Pfitzner J, Tillmanns GM. Ein Fischzellkulturtest as Ergänzungs- oder Ersatzmethode zum Fischtest. Bundesgesundheitsblatt 1989;32 (8):343–6.
[11] Schulz M, Lewald B, Kohlpoth M, Rusche B, Lorenz KHJ, Unruh E, et al. Fischzellinien in der toxikologischen Bewertung von Abwasserproben. ALTEX 1995;12(4):188–95.
[12] ECVAM Status Report 2003 – ECVAM's response to the changing political environment for alternatives: consequences of the European Union Chemicals and Cosmetics Policies. ATLA 31, p. 473–481
[13] Geller Mäckle H. Kontinuierlicher Biotest zur Toxizitätsüberwachung von Trinkwasser. DVGW-Schriftenreihe 1977;14:173–81.
[14] Moldaenke C, Baganz D, Staaks G. Monitoring of toxins in drinking water by the ToxProtec64 fish monitor. TECHNEAU report D.3.4.12, p. 1–8.
[15] Scharf B. A fish test alarm device for the continual recording of acute toxic substances in water. Arch Hydrobiol 1979;85(2):250–6.
[16] Evans GP, Johnson D, Withell C. Development of the WRC Mk III fish monitor: description of the system and its response to some commonly encountered pollutants, WRc Publication, TR 233.
[17] Löbbel H-J, Stein P. Messung der Kiemendeckelbewegung mit dem WRc-Fischmonitor im on-line Betrieb am Niederrhein. In: Steinhäuser KG, Hansen P-D, editors. Biologische testverfahren. Schriftenreihe des vereins für wasser-, boden- und lufthygiene. Stuttgart/New York: Gustav Fischer Verlag; 1992. p. 323–32.
[18] Irmer U. Continuous biotests for water monitoring of the river rhine—summary, recommendations, description of test methods, uba Texte 58/94, ISSN 0722-186X; 1994, p. 12–14, ANNEX XXI–XXII.
[19] Stein P, Hansen P-D, Löbbel J. Erprobung des WRc–Fischmonitors zur Störfallüberwachung am Rhein. In: Pluta H-J, Knie J, Leschber R, editors. Biomonitore in der Gewässerüberwachung. Stuttgart, Jena, New York: Gustav Fischer Verlag; 1994. p. 75–85.
[20] Blühbaum-Gronau E, Spieser OH, Krebs F. Bewertungkriterien für einen Verhaltensfischtest zur kontinuierlichen Gewässerüberwachung. In: Steinhäuser KG, Hansen P-D, editors. Biologische testverfahren. Schriftenreihe des vereins für wasser-, boden- und lufthygiene. 89. Stuttgart/New York: Gustav Fischer Verlag; 1992. p. 323–32.
[21] Baillieul M, Scheunders P. On-line determination of the velocity of simultaneously moving organisms by image analysis for the detection of sublethal toxicity. Water Res 1998;32(4):1027–34.
[22] OECD 215. Fish, juvenile growth test. OECD guideline for the testing of chemicals, Section 2: effects of biotic systems. http://dx.doi.org/10.1787/9789264070202-en; 2000.
[23] ISO 20281. Water quality – guidance document on statistical interpretation of eco-toxicological data – ANNEX 4: analysis of a "fish growth" data set (according to CECD GL 204/215-ISO 10229). Berlin: Beuth Verlag; 2006.

[24] Nagel R, Bresch H, Caspers N, Hansen P, Markert M, Munk R, et al. Effect of 3, 4-dichloroaniline on the early life stages of the zebrafish (*Brachydanio retio*): results of a comparative laboratory study. Ecotoxicol Environ Saf 1991;21(2):157–64.
[25] Holcombe G, Benoit D, Hammermeister D, Leonard E, Johnson R. Acute and long-term effects of nine chemicals on the Japanese medaka (*Oryzias latipes*). Arch Env Contamin Toxicol 1995;28(3):287–97.
[26] Crossland N. A method to evaluate effects of toxic chemicals on fish growth. Chemosphere 1985;14(11-12):1855–70.
[27] Benoit DA, Holcombe GW, Spehar RL. Guidelines for conducting early life toxicity test with Japanese medaka (Oryzias latipes). Duluth, MN: Environmental Research Laboratory; 1991.
[28] Mount DI, Brungs WA. A simplified dosing apparatus for fish toxicology studies. Water Res 1967;1(1):21IN123–2IN229.
[29] Van der Oost R, Beyer J, Vermeulen NP. Fish bioaccumulation and biomarkers in environmental risk assessment: a review. Environ Toxicol Pharmacol 2003;13(2):57–149.
[30] OECD 305. Bioaccumulation in fish: aqueous and dietary exposure (bioaccumulation: Semi-static fish test 305 B; Test for the degree of bioconcentration in fish 305 C; flow-through fish test 305 E. OECD Guideline for the Testing of Chemicals, Section 3: Degradation and Accumulation. http://dx.doi.org/10.1787/9789264070462-en; 2012.
[31] IPCS – International Programme on Chemical Safety. Environmental Health Criteria 155. Biomarkers and risk assessment: concepts and principles. Geneva: World Health Organisation; 1993, p. 11–58.
[32] EU. Directive 2000/60 EC of the European Parliament and of the Council establishing a framework for the community action in the field of water policy or short the EU Water Framework Directive (WFD). Official Journal of the European Union; 2000, L327, p. 1–69.
[33] Hansen P-D. Risk assessment of emerging contaminants in aquatic systems. TrAC Trends Anal Chem 2007;26(11):1095–9.
[34] Hansen P-D, Blasco J, DelValls T, Poulsen V, Van den Heuvel-Greve M. Biological analysis (bioassays, biomarkers, biosensors). In: Barceló D, Petrovic M, editors. Sustainable management of sediment resources, vol 2, sediment quality and impact assessment of pollutants. Amsterdam, London, New York: Elsevier; 2007. p. 131–57.
[35] Hansen P-D. Biomarkers. In: Markert BA, Breure AM, Zechmeister HG, editors. Bioindicators & biomonitors, principles, conceps and applications. Amsterdam, Boston, London, Oxford, Paris, San Diego, San Francisco, Singapore, Sydney, Tokyo: Elsevier; 2003. p. 203–20.
[36] REACH-Regulation (EC) No. 1907/2006 of the European Parliament and of the Council concerning the Registration, Evaluation, Authorisation and Restriction of Chemicals (REACH). Official Journal of the European Union; 2006, L396, p. 1–852.
[37] ISO 23893-1. Water quality – biochemical and physiological measurements on fish – Part 1: sampling of fish, handling and preservation of samples, TC 147, SC5. Biological methods, Reference ISO 23893-1-2007 (E). Berlin: Beuth Verlag; 2007, p. 1–18.
[38] Schipper CA, Lahr J, van den Brink PJ, George SG, Hansen P-D, Silva de Assis HC, et al. A retrospective analysis to explore the applicability of fish biomarkers and sediment bioassays along contaminated salinity transsects. ICES J Mar Sci 2009;66 (10):2089–105.
[39] Sturm A, De Assis HDS, Hansen P-D. Cholinesterases of marine teleost fish: enzymological characterization and potential use in the monitoring of neurotoxic contamination. Mar Environ Res 1999;47(4):389–98.

[40] Stave J, Roberson B, Hetrick F. Chemiluminescence of phagocytic cells isolated from the pronephros of striped bass. Dev Comparat Immunol 1983;7(2):269–76.

[41] Skouras A, Broeg K, Dizer H, Westernhagen H, Hansen P-D, Steinhagen D. The use of innate immune responses as biomarkers in a programme of integrated biological effects monitoring on flounder (*Platichthys flesus*) from the southern North Sea. Helgoland Mar Res 2003;57(3):190.

[42] ISO 23893-2. Water quality – biochemical and physiological measurements on fish – Part 2: determination of ethoxyresorufin-O-deethylase (EROD), TC 147, SC5. Biological methods, Reference ISO 23893-2-2007 (E). Berlin: Beuth Verlag; 2007, p. 1–14.

[43] Burke A, Robinson LF. The Southern Ocean's role in carbon exchange during the last deglaciation. Science 2012;335(6068):557–61 PubMed PMID: WOS:000299769200036. English.

[44] Eggens ML, Galgani F. Ethoxyresorufin-O-deethylase (EROD) activity in flatfish: fast determination with a fluorescence plate-reader. Mar Environ Res 1992;33(3):213–21.

[45] Hansen P-D. Enzymatische Verfahren zur Erfassung der Biotransformation (Entgiftungsaktivität) in der Fischleber - ein Beitrag zum biologischen Effektsmonitoring. In: FWid GDCh, editor. Biochemische methoden zur schadstofferfassung im wasser - Möglichkeiten und grenzen. Weinheim: VCH Verlagsgesellschaft; 1993. p. 54–62.

[46] Stagg R, Addison R. An inter-laboratory comparison of measurements of ethoxyresorufin O-de-ethylase activity in dab (*Limanda limanda*) liver. Mar Environ Res 1995;40 (1):93–108.

[47] Pluta H-J. Investigations on biotransformation (mixed function oxydase activities) in fish liver. In: Braunbeck T, Hanke W, Segner H, editors. Fish ecotoxicology and ecophysiology. Weinheim: VCH Verlag; 1993. p. 14–28.

[48] von Westernhagen H, Krüner G, Broeg K. Ethoxyresorufin O-deethylase (EROD) activity in the liver of dab (*Limanda limanda* L.) and flounder (*Platichthys flesus* L.) from the German Bight. EROD expression and tissue contamination. Helgoland Mar Res 1999;53(3):244–9.

[49] Herbert A, Hansen P-D. Mixed-functional oxydase (MFO) annd DNA damage in fish 1995;81(1):113–18.

[50] ISO 2393-3. Water quality—biochemical and physiological measurements on fish—Part 3: determination of Vitellogenin, TC 147, SC5 Biological methods. Reference ISO 23893-3-2007 (E). Berlin: Beuth Verlag; 2007, p. 1–13.

[51] Hansen P-D, Dizer H, Hock B, Marx A, Sherry J, McMaster M, et al. Vitellogenin–a biomarker for endocrine disruptors. TrAC Trends Anal Chem 1998;17(7):448–51.

[52] Huschek G, Hansen P-D. Ecotoxicological classification of the Berlin river system using bioassays in respect to the European Water Framework Directive. Environ Monit Assess 2006;121(1):15–31.

[53] Sherry J, Gamble A, Hodson P, Salomon K, Hock B, Marx A, et al. Vitellogenin induction in fish as an indicator of exposure to environmental estrogens. In: Rao SS, editor. Impact assessment of hazardous aquatic contaminants. 6. Boca Raton, London, New York, Washington: Lewis Publishers; 1999. p. 125–60.

[54] OECD 234. Fish sexual development test. OECD guideline for the testing of chemicals, Section 2: effects of biotic systems. http://dx.doi.org/10.1787/9789264122369-en; 2011.

[55] OECD 230. 21-Day fish assay. A short-term screening for oestrogenic and androgenic activity, and aromatase inhibition. OECD guideline for the testing of chemicals, Section 2: effects of biotic systems. http://dx.doi.org/10.1787/9789264076228-en; 2009.

[56] Ankley GT, Jensen KM, Makynen EA, Kahl MD, Korte JJ, Hornung MW, et al. Effects of the androgenic growth promoter 17-β-trenbolone on fecundity and reproductive endocrinology of the fathead minnow. Environ Toxicol Chem 2003;22 (6):1350–60.

[57] Seki M, Yokota H, Matsubara H, Maeda M, Tadokoro H, Kobayashi K. Fish full life-cycle testing for androgen methyltestosterone on medaka (*Oryzias latipes*). Environ Toxicol Chem 2004;23(3):774–81.

[58] von Westernhagen H, Rosenthal H, Dethlefsen V, Ernst W, Harms U. Bioaccumulating substances and reproductive success in Baltic flounder Platichthys flesus. Aquat Toxicol 1981;1(2):85–99.

[59] Brown-Peterson NJ, Wyanski DM, Saborido-Rey F, Macewicz BJ, Lowerre-Barbieri SK. A standardized terminology for describing reproductive development in fishes. Mar Coast Fish 2011;3(1):52–70.

[60] DIN 38415-T6. German standard methods for the examination of water, wastewater and sludge; Subanimal testing (group T), Part 6: toxicity to fish; Determination of the non-acute-poisonous effects of wastewater to fish eggs by dilution limits. Berlin: Beuth Verlag; 2000.

[61] EN ISO 15088. Water quality—determination of the acute toxicity of wastewater to zebrafish eggs (*Danio rerio*) (ISO 15088:2007). Berlin: Beuth Verlag; 2008.

[62] ISO 15088. Water quality—determination of the acute toxicity of wastewater to zebrafish eggs (*Danio rerio*) (EN ISO 2008, DIN EN ISO 2009). Berlin: Beuth Verlag; 2009

[63] OECD 212. Fish, short-term toxicity test on embryo and sac-fry stages. OECD guideline for the testing of chemicals, Section 2: effects of biotic systems. http://dx.doi.org/10.1787/9789264070141-en; 1998.

[64] OECD 210. Fish, early life stage toxicity test. Paris: OECD Publishing; 1992.

[65] von Westernhagen H. Sublethel effects of pollutants on fish eggs and larvae. Fish Physiol 1988;XIA(4):253–346.

[66] Hansen P-D, von Westernhagen H, Rosenthal H. Chlorinated hydrocarbons and hatching success in Baltic herring spring spawners. Mar Environ Res 1985;15(1):59–76.

[67] Dethlefsen V, Von Westernhagen H, Cameron P. Malformations in North Sea pelagic fish embryos during the period 1984–1995. ICES J Mar Sci 1996;53(6):1024–35.

[68] Dethlefsen V, von Westernhagen H, Tüg H, Hansen PD, Dizer H. Influence of solar ultraviolet-B on pelagic fish embryos: osmolality, mortality and viable hatch. Helgoland Mar Res 2001;55(1):45–55.

[69] Hansen P-D, Wittekindt E, Sherry J, Unruh E, Dizer H, Tüg H, et al. Genotoxicity in the environment (eco-genotoxicity). In: Barcelo D, Hansen P-D, editors. Biosensors for Environmental Monitoring of Aquatic Systems, Handbook of Environmental Chemistry. Vol. 5 Part J. Berlin, Heidelberg: Springer Publishers; 2009. p. 203–26.

Bioassays for solar UV radiation

16

Donat-P. Häder[1] *and Gerda Horneck*[2]
[1]Friedrich-Alexander University, Erlangen-Nürnberg, Germany,
[2]German Aerospace Center (DLR), Cologne, Germany

16.1 Introduction

Solar radiation is a stress factor for the biota. All exposed organisms are damaged and have to repair injury inflicted especially by short-wavelength solar UV radiation. UV can be divided into UV-C (<280 nm); however this radiation is effectively absorbed by ozone and oxygen in the stratosphere. Also part of UV-B (280−315 nm) is absorbed by stratospheric ozone, while most of UV-A (315−400 nm) is transmitted to the surface of the Earth [1]. Since the 1980s stratospheric ozone has been found to be catalytically destroyed by anthropogenic production and release of chlorofluorocarbons and other trace gases [2,3]. Due to the implementation of the Montreal Protocol and its later amendments, production and emission has been effectively phased out, but it will take until about 2065 for stratospheric ozone levels to return to pre-1980 values [1,3]. In addition, continuing climate change will affect the development of ozone depletion and the exposure of organisms to enhanced solar UV radiation [4]. The most obvious effects of stratospheric ozone depletion is the Antarctic ozone hole [5], but over the Arctic frequent ozone holes also have been reported [6]. Smaller ozone losses have been reported for Northern and Southern mid latitudes while the originally high equatorial UV levels have not changed considerably.

Even though solar UV-B radiation has only limited penetration into the water column, aquatic ecosystems are affected by solar UV because primary producers such as phytoplankton and macroalgae have to position themselves close to the water surface in order to harvest sufficient solar visible radiation to drive their photosynthesis [7]. In addition, temperature increases due to global climate change enhance stratification in the water column and reduce the depth of the mixing layer where phytoplankton dwell and thus expose them to stronger solar UV-B levels [7,8]. However, many aquatic primary producers have developed strategies to mitigate the deleterious effects of solar short-wavelength radiation on the DNA and other vital biomolecules including synthesis of screening pigments such as mycosporine-like amino acids (MAAs) and scytonemin [9,10], vertical migration [11,12], and repair mechanisms for damaged DNA [13,14] and the photosynthetic apparatus [15].

Depending or their geographic position, terrestrial ecosystems face increased solar UV radiation [16]. Exposure to short-wavelength radiation alters leaf morphology and root structure as well as photosynthetic capacity and growth [17−19]. However, as aquatic primary producers, terrestrial plants have also developed

Bioassays. DOI: http://dx.doi.org/10.1016/B978-0-12-811861-0.00016-4

mitigating strategies against solar UV-B including synthesis of UV-absorbing substances [20,21], quenching of reactive oxygen species [22], and repair of DNA damage [23].

Solar UV also affects human health. It has been linked with the occurrence of skin cancer such as melanoma after UV exposure [24,25]. Several eye problems such as pterygium and ocular melanoma are attributed to excessive exposure to solar UV [25,26]. On the other hand, solar UV plays an important role in the synthesis of vitamin D which has been shown to play a key role in musculoskeletal development, myopia, and asthma, to name only a few [27–29]. Solar UV-B mainly damages the DNA by the induction of cyclobutane pyrimidine dimers (CPD) [30,31]. In contrast to algae, plants, and most animals, primates lack the DNA repair enzyme CPD photolyase and have to rely on other repair mechanisms such as excision repair [32].

The irradiance of solar and artificial radiation can be measured using spectroradiometers or filter radiometers for specific wavelength bands [33,34]. The Eldonet network is a global network including more than 40 stations in 24 countries on five continents [35,36]. These instruments either use an integrating sphere or a teflon diffuser to collect light from a hemisphere above a horizontal surface [33,37]. The radiation is weighted with the cosine of the zenith angle; thus horizontally impinging radiation has a value of 0 regardless of its irradiance, while vertically impinging radiation is multiplied by a factor of 1.

This physically defined way of measuring radiation is far from being represented by the irradiation hitting a human body. E.g., while standing the face and the eyes are more or less perpendicular to a horizontal surface. Also the irradiances measured over a day do not imply how long a person was exposed to solar radiation being outdoors. Therefore there is a need for personal dosimeters which measure the accumulated solar radiation a person is exposed to over a defined period of time [38,39]. There are two different types of dosimeters, one being based on a chemical reaction induced by solar UV radiation [40,41] and the other represented by a biological probe which is affected by UV [42]. These dosimeters, such as the DLR biofilm dosimeter [43], have to be small to be worn like a button on the outer clothes, and the accumulated dose of solar UV radiation has to be easily retrieved from the device.

16.2 Criteria for reliable biological UV dosimeters

When constructing a biological UV dosimeter a number of criteria need to be met [44,45]. First of all the observed endpoint must be indicative of the radiation stress, e.g., it must be sensitive and specific in the UV-B and UV-A wavelength range. The spectral dependency should be as close as possible with that of the expected UV-inflicted damage in humans or other organisms and thus be indicative of the biological effect. This can be verified by comparing the output of the biological dosimeter with that of other bioassays as well as electronic spectroradiometers and other instruments. The error should be estimated. A number of controls need to be implemented such as dark control and UV-free light exposure. The output of the bioassay must be reproducible, which can be achieved by standardized construction and procedures. In addition, the output should be independent of environmental factors such as temperature and humidity.

The output must follow the reciprocity law over a wide dynamic range of irradiances and the angular response should follow the cosine law. The bioassay needs to be robust and have a long shelf life. The biological sample must be protected against physical and chemical damage. In order to be useful in routine measurements it must be robust, easy to handle, allow comfortable wearing, and be safe to the environment and cost effective [45].

16.3 The DLR biofilm dosimeter

In order to produce a device to routinely quantify exposure to solar UV radiation for humans the DLR biofilm dosimeter has been developed [46]. The basic idea is to use dried spores of *Bacillus subtilis* immobilized on a transparent polyester sheet in the form of a button which can be worn on the cloths when a person is exposed to solar radiation (Fig. 16.1) [47]. After exposure the biofilm is incubated in a growth medium (Difco nutrient broth) in which the spores germinate and are allowed to undergo several cell divisions. Then the growth is stopped by washing the biofilm in distilled water and drying it at 80° C. Proteins synthesized by the growing bacteria are stained and quantified by photometry. Their density is a measure to determine the damaging effect of the solar UV exposure. The dried spores have a long shelf life and biofilms have been found not to lose their sensitivity over a period of 9 months at a humidity <70% [46]. The sensitivity is almost independent of the temperature in the range of −20−70°C and the humidity of 37%−80%,

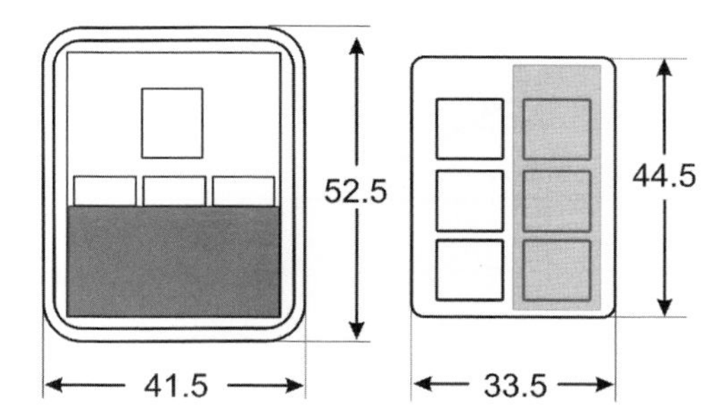

Figure 16.1 The DLR biofilm dosimeter. Housing made of plastics (left) with the biofilm (right). Three of the measurement areas are covered with a Mylar foil which cuts off UV-B. The lower half is covered so that it serves as a dark control for calibration.
Source: Redrawn after Rettberg P, Sief R, Horneck G. The DLR-Biofilm as personal UV dosimeter. In: Baumstark-Khan C, Kozubek S, Horneck G, editors. Fundamentals for the assessment of risks from environmental radiation. Dordrecht: Kluwer; 1999. p. 367−370 [47].

The angular sensitivity of the biofilm follows a cosine function with the highest deviation of −18% at an incidence angle of 45 degrees.

In a first attempt, spores of the wild-type strain of *B. subtilis* (Marburg, Inst. of Microbiology, Frankfurt, Germany) were used. This strain is capable of repairing UV-induced lesions of the DNA. Since the spores are in a dormant stage the UV lesions accumulate during exposure and are not repaired in the dried state. Thus the response obeys the reciprocity law. In a second approach the strain TKJ 6321 (uvrA10 sspl polA151 hisH101 metB101) was used which has a defect in the DNA excision repair mechanism [48]. Both strains show a similar sensitivity in the range of 280−330 nm [46].

The biological activity is being determined from a calibration curve which is produced using simulated solar radiation (Sol 2) (Fig. 16.2). It shows the biological activity measured as optical density of the stained proteins in percent vs the irradiation time which corresponds to a known fluence (UV dosis). The uncertainty has been calculated to be less than 5%.

In order to produce reliable data from a measurement device for the effect of solar UV radiation, its weighting function has to be as close as possible to the action spectrum of the biological damage [49−51]. There are many action spectra available, e.g., for plant damage, skin cancer, DNA damage, erythema, or malignant melanoma [49,52−55]. Fig. 16.3 shows the action spectrum of the DLR biofilm in comparison to the standard CIE (Commission International de l'Eclairage) action spectrum for erythema induction as well as that of an electronic dosimeter [58].

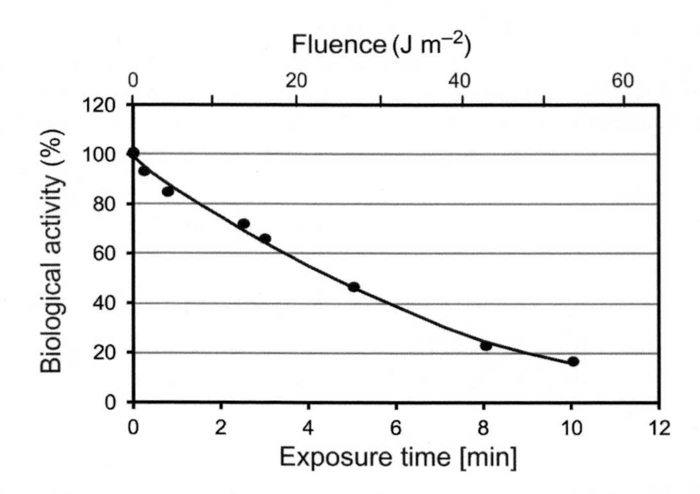

Figure 16.2 Typical calibration curve showing the UV-induced decrease in biological activity of the wild type *Bacillus subtilis* spores exposed to simulated solar radiation (Sol 2) at room temperature and about 50% r.h.
Source: Redrawn after Quintern L, Horneck G, Eschweiler U, Bücker H. A biofilm used as ultraviolet-dosimeter. Photochem Photobiol 1992;55(3):389−395 [46].

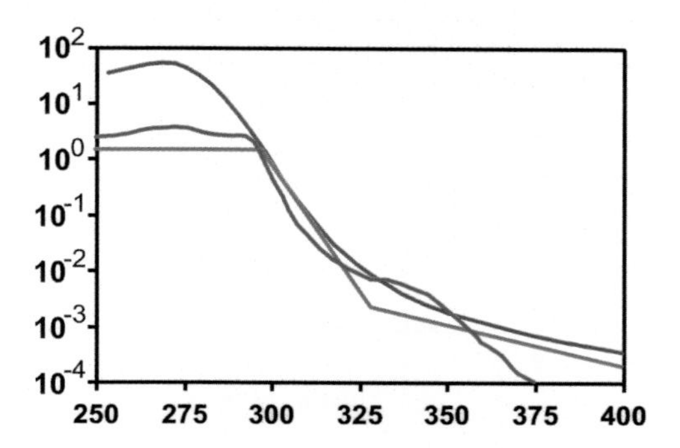

Figure 16.3 Comparison of the action spectrum of the DLR biofilm dosimeter (*red line*) with the sensitivity curve of the X2000 datalogger (*green line*) and the standard CIE spectrum for erythema induction (*blue line*).
Source: Redrawn after Cockell C, Horneck G, Rettberg P, Arendt J, Scherer K, Facius R, et al. Human exposure to ultraviolet radiation at the Antipodes-a comparison between an Antarctic (67°S) and Arctic (75°N) location. Polar Biol 2002;25(7):492–499 [43]; Cockell CS, Scherer K, Horneck G, Rettberg P, Facius R, Gugg-Helminger A, et al. Exposure of Arctic field scientists to ultraviolet radiation evaluated using personal dosimeters. Photochem Photobiol 2001;74:570–578 [56]; Rettberg P, Horneck G. Biologically weighted measurement of UV radiation in space and on Earth with the biofilm technique. Adv Space Res 2000;26(12):2005–2014 [57].

16.4 Other biological monitoring systems

Quite a number of different biological systems have been employed in bioassays to indicate UV stress, including several bacteria such as various strains of *Escherichia coli* and *B. subtilis* as well as flagellates such as *Euglena gracilis* and dinoflagellates [51]. In addition, biomolecules have been employed such as uracil, in which dimer formation is observed, and DNA in which UV radiation induces dimers as well [45,59]. The radiation-resistant cyanobacterium *Chroococcidiopsis* was exposed in a dried monolayer in the Atacama Desert, Yungay, Chile [60]. In the absence of liquid water the enzymatic repair of UV-induced damage is inhibited and consequently the organisms were killed within one day of solar UV exposure. However covering the organisms by a 1 mm layer of gypsum and mineral grains prevented UV-inflicted damage during 8 days' exposure. These findings have implications on the evolution of cyanobacteria during the early development of life on Earth with 100 times higher UV radiation than today [61,62] as well as for the UV exposure and the habitability of Mars [63,64]. Other bioassays used leaf epidermis from *Brassica oleracea* and *Sinapis alba*, seedlings of *Lepidium*, *Petroselinum*, *Anethum*, or *Daucus*, cultivars of *Helianthus annuus* and *Zea mays*, the microcrustacean *Daphnia*, human skin cells, as well as whole ecosystems [51]. Measured endpoints were dimer formation, flavonoid synthesis, growth rate, mortality, erythema, and pyrimidine dimer formation as well as growth and photosynthesis or species composition in the case of the ecosystem test.

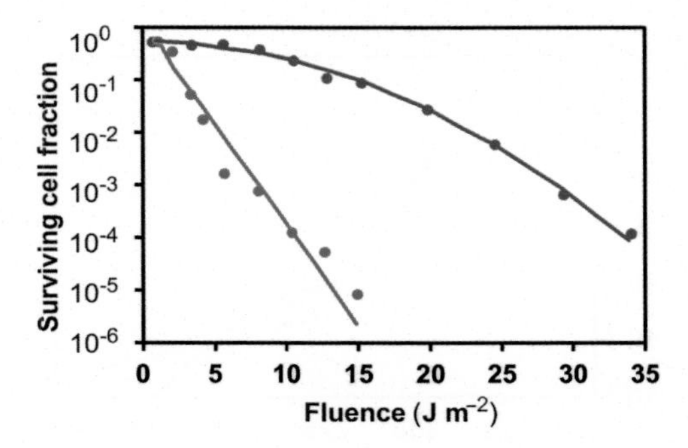

Figure 16.4 Survival of UV-C exposed Chinese hamster ovary cells comparing a UV repair-proficient strain (AA8) (*red curve*) with a UV repair-deficient strain (UV5) (*blue curve*). *Source*: Redrawn after Baumstark-Khan C, Hellweg CE, Scherer K, Horneck G. Mammalian cells as biomonitors of UV-exposure. Anal Chim Acta 1999;387(3):281–287 [66].

Chinese hamster ovary cells with different repair capacities have also been used to quantify the expression of the green fluorescent protein indicating UV-C-induced cell damage [65]. The RoDos biological UV dosimeter is based on these ovary cells growing in a monolayer on a UV transparent foil [66]. Two cell types have been used: the repair-proficient AA8 and the UV5 strain defective in nucleotide excision repair. UV exposure was done with UV-C (254 nm) and simulated solar radiation (Fig. 16.4). The repair-deficient strain turned out to be more than 4 orders of magnitude more sensitive to UV-C exposure than the repair-proficient strain based on survival and relative cell growth. After UV exposure the cells were allowed to grow for three days at 37°C and then fixed and stained. The cell growth was quantified by image analysis as relative absorbance. By sing this technique and employing cut-off filters an action spectrum could be generated.

Another bioassay to measure UV exposure is based on the deactivation of the phage T7 [67–71]. The phage is highly sensitive to solar or artificial UV because it is not capable of repair of UV-inflicted damage. After UV exposure its effectiveness to lyse bacteria growing in a lawn on an agar medium is determined. An action spectrum for the UV inhibition has been constructed in the range of 240–514 nm, which resembles the UV sensitivity of DNA. From the measurements the minimum erythemal dose (MED) exposure can be calculated [67]. This bioassay has been subjected to a comparison with the *B. subtilis* biofilm and an electronic spectroradiometer showing high congruence.

16.5 Monitoring of environmental UV radiation

The DLR biofilm dosimeters were used to monitor the UV exposure of several groups in Germany. Fig. 16.5 shows the data obtained with 12 school children wearing one set of dosimeters during three weekends (6 days) and wearing another

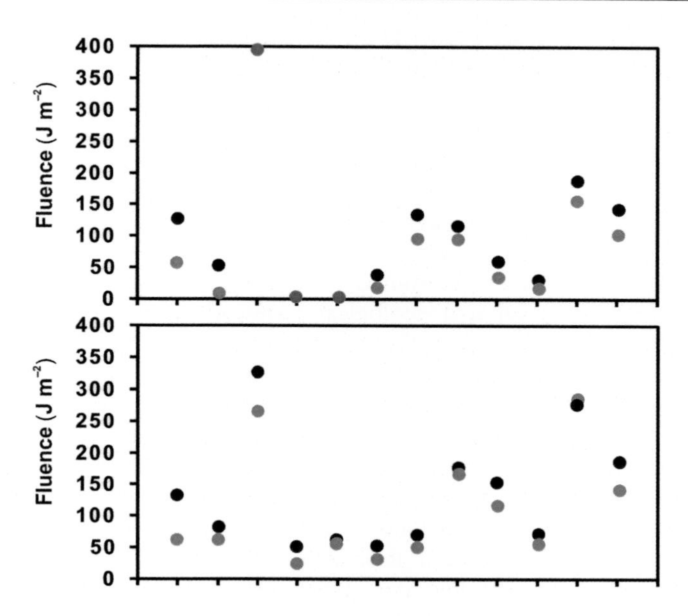

Figure 16.5 UV doses 12 school children (abscissa) were exposed to in May/June 1997. Black symbols UV-A and UV-B, *blue symbols* UV-A, *red symbols* less than 10 J m^{-2} and green symbols more than 400 J m^{-2} (all values biofilm weighted). The top panel represents the data obtained during three weekends (6 days) and the bottom panel the data from two working weeks (10 days).
Source: Redrawn after Rettberg P, Sief R, Horneck G. The DLR-Biofilm as personal UV dosimeter. In: Baumstark-Khan C, Kozubek S, Horneck G, editors. Fundamentals for the assessment of risks from environmental radiation. Dordrecht: Kluwer; 1999. p. 367–370 [47].

one during two working weeks (10 days) [47]. Each child wrote a protocol about the outdoor activity. Measurements with Robertson-Berger meters and a spectroradiometer were performed in parallel. The data show a large variability, which reflects the personal outdoor behavior.

A similar experiment was carried out with elementary school children for four seasons at five sites in Japan, including Sapporo, Tsukuba, Tokyo, Miyazaki, and Naha between 26° N and 43° N [72]. The 1-week long measurements showed that boys collected larger doses than girls. 8.1% of the children were exposed to more than 1 MED per day and 1.8% to more than 2 MED; UV exposure was highest in spring and summer. The biologically effective doses of solar UV radiation at four sites in Japan and Europe have been compared using spore dosimetry and spectral photometry [73]. Shapes of the effectiveness spectra were very similar at the four sites while the dose rates for spore inactivation showed some variations.

The long-term dosimetry of solar UV radiation was determined in Antarctica using the repair-deficient strain TKL 6321 of *B. subtilis* in order to assess the increased solar UV-B radiation under the ozone hole [74]. The dried spores were exposed under quartz discs in a box open to solar radiation at the German Antarctic

Georg von Neumayer Station (70° 37′S). As expected, the biological inactivation was highest under extremely depleted ozone concentrations and was similar to the inactivation values during the Antarctic summer. When the solar radiation was filtered through a WG 335 filter, which blocks all of the UV-B and removes short-wavelength UV-A, the survival of the spores was several orders of magnitude higher, indicating that UV-B is the most damaging part of solar radiation, despite the fact that the irradiances in that wavelength range are much lower than in the UV-A range.

For comparison, the DLR biofilm dosimeter was evaluated in the Arctic by scientists during biological and geological fieldwork working at the Haughton impact structure on Devon Island (75°N) and the Canadian High Arctic during a 24 h daylight period [56]. Under clear skies the total daily erythemally-weighted exposure was up to 5.8 standard erythemal doses (SED) as measured with an electronic dosimeter (X2000). Under overcast skies this value dropped by a mean of 54%. From this study it was estimated that a scientist would accumulate about 80 SED on his face during the month of June when working mostly outdoors in contrast to medical personnel or computer scientists who received about half of this dose spending most of their time in their tents. Pointing the electronic dosimeter in different directions revealed that the erythemal weighted irradiance was about 5×10^{-2} W m^{-2} at 13.00 h on 24 July under clear skies when positioned horizontally, while the reading was 6.26×10^{-2} W m^{-2} when oriented directly toward the sun, and 2×10^{-2} W m^{-2} when oriented vertically at 90 degrees from the sun. Biologically weighted measurements of UV radiation were carried out comparing UV measurement in Antarctica, several locations in Europe, and extraterrestrial radiation in space [57].

In order to compare human exposure to solar UV radiation, measurements were carried out at two polar locations, one being at the British Rothera Station in the Antarctic (67°S) and the other in the Canadian High Arctic (75°N) [43]. Human exposure depends on a number of environmental and occupational factors including time spent outdoors, orientation to the Sun, clothing, and sun protection as well as direct vs. diffuse UV fluxes, zenith angle, latitude, albedo, and aerosols in the atmosphere. Fig. 16.6 shows the erythemally-weighted solar irradiance in dependence of the cosine of the zenith angle for both sites measured during two clear days. Measurements were performed during 1–15 December 2000 at Rothera Station and during 1–31 July 2000 on Devon Island in the Canadian High Arctic. Obviously the UV exposure is about 4.3 times higher in the Antarctic location than in the Arctic even in the absence of ozone depletion (324 ± 21 DU), which may be due to widespread snow and ice covers and lower atmospheric pollution. In fact, melting of snow covers between 5 and 15 cm thickness increased the inactivation of *B. subtilis* spores by up to ten times [75]. This value is two times higher than that produced by 50% ozone depletion. In contrast, snow and ice cover of a few millimeters reduced the biologically weighted irradiances by up to 55%.

The diurnal profile of erythemal doses was recorded in El Arenosillo, Spain during a measurement campaign using the biofilm in parallel to a Yankee UVB-1

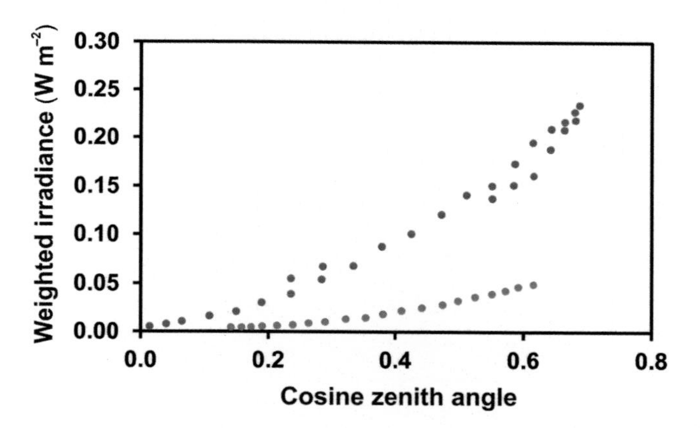

Figure 16.6 Comparison of erythemally-weighted irradiances in the Arctic (*blue symbols*) and Antarctic (*red symbols*) in dependence of the cosine zenith angle.
Source: Redrawn after Cockell C, Horneck G, Rettberg P, Arendt J, Scherer K, Facius R, et al. Human exposure to ultraviolet radiation at the Antipodes-a comparison between an Antarctic (67°S) and Arctic (75°N) location. Polar Biol 2002;25(7):492–499 [56].

dosimeter (Fig. 16.7). Both measurements agree fairly well. At noon the biological effective irradiance corresponded to 3–4 MED/h. These measurements were part of a project which included five locations in Germany (Westerland, Cologne, Erlangen, Schauinsland, Neuherberg) and three sites in Spain (El Arenosillo, Maspalomas in Gran Canaria and Torrejón de Ardoz in Madrid) [76].

Spore dosimetry was also used to determine the daily irradiance of and personal exposure to solar UV radiation in four locations in the Northern Hemisphere, including Naha and Sapporo (Japan), Thessaloniki (Greece), and Abisko (Northern Sweden) [77]. The data show a wide distribution of irradiances that strongly depend on the solar zenith angle affected by latitude.

16.6 Applications in space

Solar radiation is filtered by the atmosphere before reaching the Earth surface. As indicated above, the UV-C component is completely removed and most of UV-B is attenuated. Therefore much higher solar UV irradiance is measured in space than on the Earth surface [78]. The Lithopanspermia experiments have been devised to determine if various organisms ranging from bacteria and lichens to fungal spores could survive the harsh extraterrestrial conditions partially covered with artificial meteorite material proving the possibility that early life forms could have survived long space travel and arrived on Earth protected inside meteorites [79].The DLR biofilm was exposed during the Spacelab

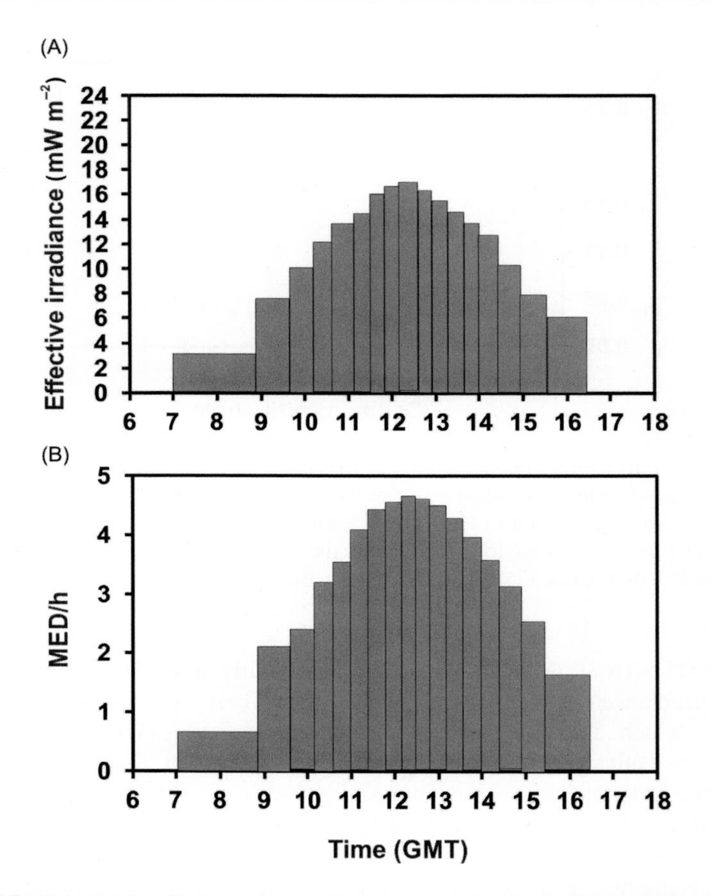

Figure 16.7 Biologically effective solar irradiance measured with the DLR biofilm dosimeters (top panel) in comparison with the measured MED/h using the Yankee UVB-1 Biometer (bottom panel) on 5 May 1997 in El Arenosillo at Huelva, Spain (27°N) under clear skies. *Source*: Redrawn after De la Torre R, Horneck G, Rettberg P, Luccini E, Vilaplana J, Gil M. Monitoring of biologically effective UV irradiance at El Arenosillo (INTA), Andalucía, in Spain. Adv Space Res 2000;26(12):2015–2019 [76].

mission D-2, in the experiment RD-UVRAD, for defined time intervals, to extraterrestrial solar radiation. This was filtered through a system to simulate different ozone column thicknesses. [80]. Using the EXPOSE facility on the International Space Station (ISS) the extraterrestrial UV radiation has been measured using the DLR spore dosimeter [50,81]. In addition, a number of other biological samples have been exposed to solar radiation [82,83]. The T7 phage was used in parallel to the uracil biological dosimeter in a vacuum-tight case [84]. In the same facility solar radiation was measured with the R3D dosimeter which had been developed by scientists from Erlangen, Germany in conjunction with colleagues from the Bulgarian academy of science [85,86].

References

[1] Steinbrecht W, Claude H, Schönenborn F, McDermid IS, Leblanc T, Godin-Beekman S, et al. Ozone and temperature trends in the upper stratosphere at five stations of the Network for the Detection of Atmospheric Composition Change. Int J Rem Sens 2009; 30(15-16) 3875–3886. Pubmed Central PMCID: 080930.

[2] Li F, Stolarski RS, Newman PA. Stratospheric ozone in the post-CFC era. Atmos Chem Phys 2009;9:2207–13. Pubmed Central PMCID: 4/12/2008

[3] Hoffmann L, Hoppe C, Müller R, Dutton G, Gille J, Griessbach S, et al. Stratospheric lifetime ratio of CFC-11 and CFC-12 from satellite and model climatologies. Atmos Chem Phys 2014;14(22):12479–97.

[4] UNEP EEAP. Environmental effects of ozone depletion and its interactions with climate change: progress report, 2015. Photochem Photobiol Sci 2016;15(2):141–74. PubMed PMID: WOS:000370421800001.

[5] Salby M, Titova EA, Deschamps L. Changes of the Antarctic ozone hole: controlling mechanisms, seasonal predictability, and evolution. J Geophys Res 2012;117(D10): D10111.

[6] Manney GL, Santee ML, Rex M, Livesey NJ, Pitts MC, Veefkind P, et al. Unprecedented Arctic ozone loss in 2011. Nature 2011;469–75.

[7] Häder D-P, Williamson CE, Wängberg S-Å, Rautio M, Rose KC, Gao K, et al. Effects of UV radiation on aquatic ecosystems and interactions with other environmental factors. Photochem Photobiol Sci 2015;14(1):108–26.

[8] Gao KS, Xu JT, Gao G, Li YH, Hutchins DA, Huang BQ, et al. Rising CO_2 and increased light exposure synergistically reduce marine primary productivity. Nat Clim Change 2012;2(7):519–23. PubMed PMID: WOS:000306249500016.

[9] Singh SP, Sinha RP, Klisch M, Häder D-P. Mycosporine-like amino acids (MAAs) profile of a rice-field cyanobacterium *Anabaena doliolum* as influenced by PAR and UVR. Planta 2008;229:225–33.

[10] Singh SP, Ha S-Y, Sinha RP, Häder D-P. Photoheterotrophic growth unprecedentedly increases the biosynthesis of mycosporine-like amino acid shinorine in the cyanobacterium Anabaena sp., isolated from hot springs of Rajgir (India). Acta Physiologiae Plantarum 2014;36:389–97. 2013/10/30. English.

[11] Mohovic B, Gianesella SMF, Laurion I, Roy S. Ultraviolet B-photoprotection efficiency of mesocosm-enclosed natural phytoplankton communities from different latitudes: Rimouski (Canada) and Ubatuba (Brazil). Photochem Photobiol 2006;82 (4):952–61.

[12] Richter PR, Häder D-P, Gonçalves RJ, Marcoval MA, Villafañe VE, Helbling EW. Vertical migration and motility responses in three marine phytoplankton species exposed to solar radiation. Photochem Photobiol 2007;83(4):810–17. PubMed PMID: 34739.

[13] Richa Sinha RP, Häder D-P. Physiological aspects of UV-excitation of DNA. In: Barbatti B, Borin AC, Ullrich AC, editors. Photoinduced phenomena in nucleic acids. topics in current chemistry. Berlin Heidelberg: Springer; 2014. p. 1–46.

[14] Li C, Ma L, Mou S, Wang Y, Zheng Z, Liu F, et al. Cyclobutane pyrimidine dimers photolyase from extremophilic microalga: remarkable UVB resistance and efficient DNA damage repair. Mutat Res Fund Mol Mech Mut 2015;773:37–42.

[15] Sicora C, M t Z, Vass I. The interaction of visible and UV-B light during photodamage and repair of Photosystem II. Photosynth Res 2003;75:127–37.

[16] Verdaguer D, Llorens L, Bernal M, Badosa J. Photomorphogenic effects of UVB and UVA radiation on leaves of six Mediterranean sclerophyllous woody species subjected to two different watering regimes at the seedling stage. Environ Exp Bot 2012;79:66–75. PubMed PMID: WOS:000302592900009. English.

[17] Furness NH, Jolliffe PA, Upadhyaya MK. Ultraviolet-B radiation and plant competition: experimental approaches and underlying mechanisms. Photochem Photobiol 2005;81(5):1026–37.

[18] Feng H, An L, Chen T, Quiang W, Xu S, Zhang M, et al. The effect of enhanced ultraviolet-B radiation on growth, photosynthesis and stable carbon isotope composition (ë ^{13}C) of two soybean cultivars (*Glycine max*) under field conditions. Environ Exp Bot 2003;49:1–8.

[19] Abdullaev A, Djumaev B, Karimov KK. Influence of UV-radiation on the photosynthesis and photosynthetic carbon metabolism in high mountainous plants. BMC Plant Biol 2005;5(Suppl. 1):S1.

[20] Flint SD, Searles PS, Caldwell MM. Field testing of biological spectral weighting functions for induction of UV-absorbing compounds in higher plants. Photochem Photobiol 2004;79(5):399–403.

[21] Hilal M, Rodríguez-Montelongo L, Rosa M, Gallardo M, Gonz lez JA, Interdonato R, et al. Solar and supplemental UV-B radiation effects in lemon peel UV-B-absorbing compound content - seasonal variations. Photochem Photobiol 2008;84:1480–6.

[22] Hideg Ü, Kálai T, Kós PB, Asada K, Hideg K. Singlet oxygen in plants—its significance and possible detection with double (fluorescent and spin) indicator reagents. Photochem Photobiol 2006;82(5):1211–18.

[23] Kaiser G, Kleiner O, Beisswenger C, Batschauer A. Increased DNA repair in *Arabidopsis* plants overexpressing CPD photolyase. Planta 2009;230(3):505–15. PubMed PMID: ISI:000268393600006.

[24] Lemus-Deschamps L, Makin JK. Fifty years of changes in UV Index and implications for skin cancer in Australia. Int J Biometeorol 2012;56(4):727–35. PubMed PMID: 21870202. Epub 2011/08/27. eng.

[25] Rivolta C, Royer-Bertrand B, Rimoldi D, Schalenbourg A, Zografos L, Leyvraz S, et al. UV light signature in conjunctival melanoma; not only skin should be protected from solar radiation. J Hum Genet 2016;61(4):361–2. PubMed PMID: 26657935.

[26] Vojnikovic B, Njiric S, Coklo M, Toth I, Spanjol J, Marinovic M. Sunlight and incidence of pterygium on Croatian Island Rab--epidemiological study. Coll Antropol 2007;31(Suppl 1):61–2. PubMed PMID: 17469753.

[27] Hollams E, Teob S, Kusel M, Holt B, Holt K, Inouye M, et al. Vitamin D over the first decade and susceptibility to childhood allergy and asthma. J Aller Clin Immunol 2016;. Available from: http://dx.doi.org/10.1016/j.jaci.2016.07.032.

[28] Tideman JW, Polling JR, Voortman T, Jaddoe VW, Uitterlinden AG, Hofman A, et al. Low serum vitamin D is associated with axial length and risk of myopia in young children. Eur J Epidemiol 2016;31(5):491–9. PubMed PMID: 26955828. Pubmed Central PMCID: 4901111.

[29] Wu F, Wills K, Laslett LL, Oldenburg B, Seibel MJ, Jones G, et al. Cut-points for associations between vitamin D status and multiple musculoskeletal outcomes in middle-aged women. Osteoporos Int 2016. Available from: http://dx.doi.org/10.1007/s00198-016-3754-9.

[30] Richa Sinha RP, Häder D-P. Physiological aspects of UV-excitation of DNA. In: Barbatto M, Borin AC, Ullrich S, editors. Topics in current chemistry: photoinduced

phenomena in nucleic acids II: DNA fragments and phenomenological aspects, vol. 355. Berlin Heidelberg: Springer; 2015. p. 203–48.
[31] Fitch ME, Nakajima S, Yasui A, Ford JM. *In vivo* recruitment of XPC to UV-induced cyclobutane pyrimidine dimers by the *DDB2* gene product. J Biol Chem 2003;278 (47):46906–10.
[32] Murray HC, Maltby VE, Smith DW, Bowden NA. Nucleotide excision repair deficiency in melanoma in response to UVA. Exp Hematol Oncol 2015;5:6. PubMed PMID: 26913219. Pubmed Central PMCID: 4765239.
[33] Seckmeyer G, Bais A, Bernhard G, Blumthaler M, Drüke S, Kiedron P, et al. Instruments to measure solar ultraviolet radiation WMO/TD-No. 1538, part 4: array spectroradiometers (vol. no. 191). Geneva: World Meterological Organization; 2010.
[34] Riechelmann S, Schrempf M, Seckmeyer G. Simultaneous measurement of spectral sky radiance by a non-scanning multidirectional spectroradiometer (MUDIS). Meas Sci Technol 2013;24(12):125501.
[35] Häder D-P, Lebert M. ELDONET - European light dosimeter network. In: Ghetti F, Checcucci G, Bornman JF, editors. Environmental UV radiation: impact on ecosystems and human health and predictive models. IV. Earth and environmental sciences - Vol. 57. The Netherlands: Springer; 2006. p. 95–108.
[36] Häder D-P, Lebert M, Schuster M, del Ciampo L, Helbling EW, McKenzie R. ELDONET–A decade of monitoring solar radiation on five continents. Photochem Photobiol 2007;83:1348–57.
[37] Häder D-P, Lebert M, Marangoni R, Colombetti G. ELDONET–European Light Dosimeter Network hardware and software. J Photochem Photobiol B Biol 1999;52:51–8.
[38] Heydenreich J, Wulf HC. Miniature personal electronic UVR dosimeter with erythema response and time-stamped readings in a wristwatch. Photochem Photobiol 2005;81 (5):1138–44.
[39] Seckmeyer G, Klingebiel M, Riechelmann S, Lohse I, McKenzie RL, Ben Liley J, et al. A critical assessment of two types of personal UV dosimeters. Photochem Photobiol 2012;88(1):215–22.
[40] Ohko Y, Hashimoto K, Fujishima A. Kinetics of photocatalytic reactions under extremely low-intensity UV illumination on titanium dioxide thin films. J Phys Chem A 1997;101(43):8057–62.
[41] Schneider L, Bloch W, Kopp K, Hainzl A, Rettberg P, Wlaschek M, et al. 8-Isoprostane is a dose-related biomarker for photo-oxidative ultraviolet (UV) B damage in vivo: a pilot study with personal UV dosimetry. Br J Dermatol 2006;154 (6):1147–54.
[42] Horneck G, Rettberg P, Facius R. Biological UV dosimetry as tool for assessing the risks from an increased environmental UV-B radiation. In: Baumstark-Khan C, Kozubek S, Horneck G, editors. Fundamentals for the assessment of risks from environmental radiation. Netherlands: Kluwer Academic Publishers; 1999. p. 451–6.
[43] Cockell C, Horneck G, Rettberg P, Arendt J, Scherer K, Facius R, et al. Human exposure to ultraviolet radiation at the Antipodes-a comparison between an Antarctic (67°S) and Arctic (75°N) location. Polar Biol 2002;25(7):492–9.
[44] Baumstark-Khan C, Kozubek S, Horneck G. Fundamentals for the assessment of risks from environmental radiation. Berlin: Springer Science & Business Media; 2013.
[45] Rettberg P, Horneck G. Intrinsic and extrinsic biomarkers for the assessment of risks from environmental UV radiation. J Epidemiol 1999;9(6sup):78–83.

[46] Quintern L, Horneck G, Eschweiler U, Bücker H. A biofilm used as ultraviolet-dosimeter. Photochem Photobiol 1992;55(3):389–95.
[47] Rettberg P, Sief R, Horneck G. The DLR-Biofilm as personal UV dosimeter. In: Baumstark-Khan C, Kozubek S, Horneck G, editors. Fundamentals for the assessment of risks from environmental radiation. Dordrecht: Kluwer; 1999. p. 367–70.
[48] Tanooka H, Munakata N, Kitahara S. Mutation induction with UV-and X-radiations in spores and vegetative cells of *Bacillus subtilis*. Mutat Res Fund Mol Mech Mut 1978;49(2):179–86.
[49] Horneck G. Quantification of the biological effectiveness of environmental UV radiation. J Photochem Photobiol B Biol 1995;31:43–9.
[50] Horneck G. Biological monitoring of radiation exposure. Adv Space Res 1998;22 (12):1631–41.
[51] Horneck G. Biological UV dosimetry. In: Häder D-P, editor. The effects f ozone depletion on aquatic ecosystems. Austin, TX: R.G.Landes Company; 1997. p. 119–42.
[52] Ballare CL, Caldwell MM, Flint SD, Robinson A, Bornman JF. Effects of solar ultraviolet radiation on terrestrial ecosystems. Patterns, mechanisms, and interactions with climate change. Photochem Photobiol Sci 2011;10(2):226–41. PubMed PMID: WOS:000286835400003.
[53] Kazantzidis A, Bais AF, Zempila MM, Kazadzis S, den Outer PN, Koskela T, et al. Calculations of the human Vitamin D exposure from UV spectral measurements at three European stations. Photochem Photobiol Sci 2009;45(8):45–51.
[54] Lindberg C, Horneck G, Bücker H. UV action spectrum for photoproduct formation in DNA of *B. subtilis* spores. Radiat Biol 1991;59:573.
[55] Quaite FE, Sutherland BM, Sutherland JC. Action spectrum for DNA damage in alfalfa lowers predicted impact of ozone depletion. Nature 1992;358:576–8.
[56] Cockell CS, Scherer K, Horneck G, Rettberg P, Facius R, Gugg-Helminger A, et al. Exposure of Arctic field scientists to ultraviolet radiation evaluated using personal dosimeters. Photochem Photobiol 2001;74:570–8.
[57] Rettberg P, Horneck G. Biologically weighted measurement of UV radiation in space and on Earth with the biofilm technique. Adv Space Res 2000;26(12):2005–14.
[58] CIE. Erythema reference action spectrum and standard erythema dose. Vienna, Austria: CIE Standrad Bureau, Commission Internationale de l'Eclairage; 1998.
[59] Horneck G. Biological dosimetry of solar UV radiation. Trends Photochem Photobiol 1997;4:67–78.
[60] Cockell CS, McKay CP, Warren-Rhodes K, Horneck G. Ultraviolet radiation-induced limitation to epilithic microbial growth in arid deserts - Dosimetric experiments in the hyperarid core of the Atacama Desert. J Photochem Photobiol B Biol 2008;90 (2):79–87. PubMed PMID: 35198.
[61] Rettberg P. The influence of space parameters like solar ultraviolet radiation on the survival of microorganisms. NATO Sci Ser 2005;366:126.
[62] Cockell CS, Horneck G. The history of the UV radiation climate of the earth-theoretical and space-based observations. Photochem Photobiol 2001;73:447–51.
[63] Kolb C, Abart R, Bérces A, Garry J, Hansen AA, Hohenau W, et al. An ultraviolet simulator for the incident Martian surface radiation and its applications. Int J Astrobiol 2005;4(3/4):241.
[64] Cockell CS, Schuerger AC, Billi D, Friedmann EI, Panitz C. Effects of a simulated martian UV flux on the cyanobacterium, *Chroococcidiopsis* sp. 029. Astrobiol 2005;5 (2):127–40.

[65] Baumstark-Khan C, Palm M, Wehner J, Okabe M, Ikawa M, Horneck G. Green fluorescent protein (GFP) as a marker for cell viability after UV irradiation. J Fluoresc 1999;9 (1):37–43.
[66] Baumstark-Khan C, Hellweg CE, Scherer K, Horneck G. Mammalian cells as biomonitors of UV-exposure. Anal Chim Acta 1999;387(3):281–7.
[67] Rontó G, Gáspár S, Bérces A. Phages T7 in biological UV dose measurement. J Photochem Photobiol B Biol 1992;12:285–94.
[68] Rontó G, Gáspár S, Gugolya Z. Ultraviolet dosimetry in outdoor measurements based on bacteriophage T7 as a biosensor. Photochem Photobiol 1993;59:209–14.
[69] Gaspar S, Berces A, Ronto G, Grof P. Biological effectiveness of environmental radiation in aquatic systems, measurements by T7-phage sensor. J Photochem Photobiol B Biol 1996;32:183–7.
[70] Fekete A, Vink AA, Gáspár S, Berces A, Modos K, Ronto G, et al. Assessment of the effects of various UV sources on inactivation and photoproduct induction in phage T7 dosimeter. Photochem Photobiol 1998;68:527–31.
[71] Fekete A, Vink AA, Gáspár S, Modos K, Berces A, Rontó G, et al. Influence of phage proteins on formation of specific UV DNA photoproducts in phage T7. Photochem Photobiol 1999;69:545–52.
[72] Ono M, Munakata N, Watanabe S. UV exposure of elementary school children in five Japanese cities. Photochem Photobiol 2005;81(2):437–45.
[73] Munakata N, Kazadzis S, Bais AF, Hieda K, Rontó G, Rettberg P, et al. Comparisons of spore dosimetry and spectral photometry of solar-UV radiation at four sites in Japan and Europe. Photochem Photobiol 2000;72:739–45.
[74] Puskeppeleit M, Quintern LE, el Naggar S, Schott JU, Eschweiler U, Horneck G, et al. Long-term dosimetry of solar UV radiation in Antarctica with spores of *Bacillus subtilis*. Appl Env Microbiol 1992;58:2355–9.
[75] Cockell CS, Rettberg P, Horneck G, Wynn-Williams DD, Scherer K, Gugg-Helminger A. Influence of ice and snow covers on the UV exposure of terrestrial microbial communities: dosimetric studies. J Photochem Photobiol B Biol 2002;68(1):23–32.
[76] De la Torre R, Horneck G, Rettberg P, Luccini E, Vilaplana J, Gil M. Monitoring of biologically effective UV irradiance at El Arenosillo (INTA), Andalucía, in Spain. Adv Space Res 2000;26(12):2015–19.
[77] Munakata N, Makita K, Bolsée D, Gillotay D, Horneck G. Spore dosimetry of solar UV radiation: applications to monitoring of daily irradiance and personal exposure. Adv Space Res 2000;26(12):1995–2003.
[78] Rettberg P, Scherer K, Horneck G. Biological responses to solar UV radiation in space and on Earth. In: Ehrenfrend P, Angerer O, Battrick B, editors. Exo-/astrobiology, proceedings of the first European workshop (ES SP-496) 2001. p. 247–250.
[79] de la Torre R, Sancho LG, Horneck G, de los Ríos A, Wierzchos J, Olsson-Francis K, et al. Survival of lichens and bacteria exposed to outer space conditions—Results of the Lithopanspermia experiments. Icarus 2010;208:735–48.
[80] Horneck G, Rettberg P, Rabbow E, Strauch W, Seckmeyer G, Facius R, et al. Biological dosimetry of solar radiation for different simulated ozone column thicknesses. J Photochem Photobiol B Biol 1996;32:189–96.
[81] Horneck G, Wynn-Williams DD, Mancinelli RL, Cadet J, Munakata N, Ronto G, et al. Biological experiments on the expose facility of the International Space Station. In: Proceedings of the 2nd European Symposium on the Utilisation of the International Space Station, ESTEC, Noordwijk, The Netherlands, 16–18 November 1998 (ESA SP-433) 1999. p. 459–468.

[82] Horneck G, Zell M. Special collection on EXPOSE-E mission. Astrobiol 2012;12 (5):373–528.

[83] Horneck G, Panitz C, Zell M. Special issue: EXPOSE-R. Int J Astrobiol 2015;14 (3):3–142.

[84] Bérces A, Egyeki M, Fekete A, Horneck G, Kovács G, Panitz C, et al. The PUR experiment on the EXPOSE-R facility: biological dosimetry of solar extraterrestrial UV radiation. Int J Astrobiol 2015;14(01):47–53.

[85] Häder D-P, Dachev T. Measurement of solar and cosmic radiation during spaceflight. Surv Geophys 2003;24:229–46.

[86] Häder D-P, Richter P, Schuster M, Dachev T, Tomov B, Georgiev P, et al. R3D-B2 - Measurement of ionizing and solar radiation in open space in the BIOPAN 5 facility outside the FOTON M2 satellite. Adv Space Res 2009;43(8):1200–11.

A comparison of commonly used and commercially available bioassays for aquatic ecosystems

17

Azizullah Azizullah[1] *and Donat-P. Häder*[2]
[1]Kohat University of Science and Technology (KUST), Kohat, Pakistan,
[2]Friedrich-Alexander University, Erlangen-Nürnberg, Germany

17.1 Introduction

As mentioned in the introductory chapter of this volumen, chemical analysis of drinking water, wastewater, and natural aqueous habitats is of limited value because of the immense number of natural and human-made chemicals, the enormous time required, and the high price tag for routine monitoring [1]. One alternative is the employment of bioassays in the assessment of toxicity and detection of potential pollutants in aquatic ecosystems, groundwater, wastewater, and water for human consumption [2].

One disadvantage of the use of bioassays is that they do not identify the chemical nature of the pollutant in the water, but they indicate the presence of a toxic substance [3], so that a targeted chemical analysis can commence once a signal is received from the bioassay [4]. Early bioassays used fish and other animals, plants, and algae as bioindicator organisms, which had several disadvantages [5–8]. Use of vertebrate animals is problematic because of ethical considerations, and plants usually require a long time to develop a visible and quantifiable response.

When developing a bioassay a number of points have to be taken into account. First of all a suitable organism, biomolecule or biological process needs to be selected [9]. An organism should be easy to cultivate and maintained even in labs without sophisticated facilities, such as cell lines, bacteria, fungi (yeasts), or multicellular organisms such as plants and invertebrate and vertebrate animals [10]. When choosing vertebrates ethical considerations need to be kept in mind, since nowadays in many countries it is difficult to obtain the required legal documents to conduct experiments with vertebrates [11]. Next, one or several endpoints need to be defined for the assessment [12]. Often it is a good idea to choose several endpoints in order to have a broader repertoire of responses affected by diverse groups of toxic substances. Mortality, growth, and reproduction are common endpoints [13,14]. However, these responses usually take a long time on the order of days to weeks to show the effect of a toxin; most of the time it is desirable to obtain a result if a sample is toxic or not in a very short time in order to initiate preventive measures [15]. For online monitoring it is indispensable to carry out a real time evaluation [16,17].

Bioassays. DOI: http://dx.doi.org/10.1016/B978-0-12-811861-0.00017-6

The bioassay should be reliable and the responses reproducible. For this purpose positive and negative controls need to be conducted. Positive controls use a known toxic concentration such as a defined concentration of a heavy metal. When applied this should result in a predictable response with low variability [18]. In the negative control no toxin is applied and no effect should be visible in the selected endpoints [19]. E.g., the acute hepatopancreatic necrosis disease, caused by the bacterium *Vibrio parahaemolyticus*, which results in substantial mortality, has been studied in penaeid shrimp in Southeast Asia and Latin America [20]. While infected shrimp showed a high mortality, the negative controls (not infected) showed no mortality. The results obtained with the bioassay need to be reproducible on a day-to-day basis [21].This can best be proven in interlaboratory comparisons of the same or similar bioassays based on the same organism and endpoints [22,23]. The result of a test needs to be independent of chemical and physical environmental factors such as temperature, pH, and other nontoxic ingredients in water samples such as nutrients. The measured values should follow the Bunsen-Roscoe reciprocity law over a wide dynamic range of concentrations of the toxic substance [24] which states that when doubling the concentration the response will double. However, in practice most bioassays and reactions follow a sigmoidal dependence between concentration and response [25]. At low concentrations there is no visible response. Increasing concentrations result in first slow and then faster rising reactions, and at even higher concentrations the response levels off and reaches a saturation which may be 100%. At values around 50% inhibition the reaction tends to be linear and for this reason EC_{50} values are calculated which indicate the toxin concentration which causes a 50% inhibition of the endpoint response [26,27]. Another important criterion is sensitivity. The bioassay must give a measurable result at low concentrations of the toxin which is on the order or below of the concentration which results in damage to human health or the environment [28,29].

Other considerations for a practical bioassay are robustness of the instrument and a long shelf life in the case of biomaterials such as extracted DNA, lyophilized bacteria, or plant seeds [30]. It should be easy to handle and usable without lengthy training for the personnel which uses it in routine measurements [12,26]. It needs to allow high throughput and require small samples [31]. For field applications it needs to be portable [32]. And finally a bioassay needs to be cost effective both in terms of the initial investment and the running costs during continuous usage [33]. This is especially important for applications in developing countries where funding of expensive instrumentation and running costs is limited [34].

17.2 An overview of commonly used commercial bioassays

There are a large number of commercially available bioassays on the market. Some have been designed for a specialized task. E.g., a single plant-based bioassay has been developed to detect the western corn rootworm (*Diabrotica virgifera virgifera*

LeConte) which is a serious economic pest of corn (*Zea mays*) in the United States [35]. Another bioassay was designed to assess jasmonate-dependent defenses triggered by pathogens, herbivorous insects, and beneficial rhizobacteria [36]. A suite of bioassays has been developed to compare the effects of silver, titanium dioxide, and silica dioxide nanoparticles with the dental disinfectant chlorohexidine on *Streptococcus* mutants [37].

In contrast, more general bioassays have been produced for toxicity profiling of marine surface sediments [38]. Vasquez et al. have reviewed the currently available bioassays for evaluation of the biological potency of pharmaceuticals in treated wastewater [39]. A number of common bioassays are not restricted to specific groups of toxicants or pollutants but rather respond to heavy metals, persistent organic pollutants, organic toxic substances, pesticides, and other toxicants, however with different sensitivity to the different groups of chemicals [40,41].

A number of widely distributed bioassay systems are based on the inhibition of bioluminescence produced by bacteria or eukaryotic cells [42]. These instruments are capable of detecting heavy metals as well as natural and anthropogenic organic chemicals in a wide range of ecosystems such as surface water, groundwater, wastewater, soil, sediment, and marine habitats. The mode of action is based on the ability of the organism to produce light using a reaction mostly involving a luciferin catalyzed by the enzyme luciferase (cf. Chapter 12: Bioluminescence systems in environmental biosensors, this volume). In the presence of toxic substances this bioluminescence decreases in intensity which can be quantified to determine the toxicity. One disadvantage of several bioassays in this category is that they use marine bacteria; therefore NaCl has to be added to each sample. One way around this problem is to transfer the lux genes into freshwater bacteria.

In contrast, the bioassay ECOTOX uses eukaryotic flagellates (cf. Chapter 10: Ecotox, this volume). *Euglena gracilis* is a photosynthetic unicellular organism which is motile and orients itself with respect to light and gravity [43–45]. There are several versions of the ECOTOX instrument available: stationary instruments for lab use and portable ones for field applications. All use several endpoints such as motility, swimming velocity, phototactic and gravitactic orientation as well as size and form parameters [26,46,47]. The organism is very sensitive to toxic components in its aqueous environment and responds by changing its swimming velocity, losing its orientation with respect to light and gravity, or by changing its cell shape (rounding off). Therefore the instruments have been successfully used to monitor ecosystems [48], municipal and industrial wastewaters [49–51], the efficiency of wastewater treatment plants [52], heavy metal pollution [53,54] and the toxicity of detergents [55], pesticides [56], and fertilizers [57]. One advantage of the ECOTOX system is its online analysis of potentially toxic samples using real time image analysis of the moving flagellates, which permits constant monitoring of aqueous ecosystems and early warning of excessive pollutant concentrations.

The microcrustacean *Daphnia* and related organisms are recognized as sensitive bioindicators to monitor pollution and toxicity in aquatic ecosystems (cf. Chapter 11: Daphniatox, this volume). In some bioassays using these organisms growth, reproduction, and mortality is being monitored which may take several days to obtain useful results [58,59]. The *Daphnia* test has been standardized by several national and international agencies and certified in standard methods such as the German DIN 38412 L-30 and the OECD test 202 [60]. This allows interlaboratory comparisons and internationally accepted test results. More recent instruments use computer-aided online image analysis to monitor swimming velocity, percentage of motile organisms as well as several form and size factors such as length and roundness in real time [12,61]. These instruments not only allow fast analysis of toxic samples within minutes but also to detect minute changes in organism form and length, which are invisible to the naked eye. Instruments utilizing *Daphnia* and related organisms have been employed to monitor heavy metal toxicity [12], organophosphorus and carbamate pesticides [62], sublethal toxicity of insecticides [59] and nematocides [63], toxicity in the sludge of industrial water treatment plants [64], and eco-genotoxicity testing [65] to name only a few applications.

Ulvatox is based on the marine Chlorophyte *Ulva pertusa* [66] found in the Mediterranean Sea as well as in the Pacific and Indian Oceans, but other *Ulva* species might as well be used [67] (cf. Chapter 7: Toxicity testing using the marine macroalga *Ulva pertusa*: method development and application, this volume). Under defined growth conditions of light, salinity, and temperature marginal cells of the thallus release zoospores which is accompanied with a color change since the cells appear white after the zoospores have left in contrast to the more centrally positioned green thallus cells [68]. This color change can be evaluated visually or by computer-assisted image analysis [69]. The release of zoospores is retarded or inhibited by the presence of toxins which is used as the endpoint for this Ulvatox bioassay. The test has a high sensitivity and has been used to investigate a wide range of metals such as silver, arsenic, cadmium, cobalt, chromium, copper, iron, mercury, manganese, nickel, lead, and zinc which are often found in industrial wastewaters [70]. The assay meets the required standards of toxicity norms for several metals in Korea as indicated by their EC_{50} values and showing that algae are not less sensitive than animals to toxicants [71,72]. In fact, the Ulvatox instrument has been shown to be more sensitive to commonly used test substances than in tests with bioluminescent bacteria, *Daphnia*, *Lemna* as well as fish and sea urchins [68,73]. The bioassay has been applied to monitor industrial, livestock, and leather manufactory wastes as well as urban and rural sewage [68,70]. Further tests were carried out with metals, formalin, diesel fuel, and tributyltin oxide [74,75], as well as thiazolininediones [76]. Also the effectiveness of a three-step wastewater treatment plant has been assessed using the spore release in *Ulva* [77]. In addition to the automatic instrument an Ulva kit has been developed based on the same principle of inhibition of spore release from the thallus.

The bioassay Lemnatox uses the miniature aquatic angiosperm *Lemna minor* and related species commonly known as duckweed widespread in freshwater ponds and lakes [78]. Also the related genera *Spirodela polyrhiza* and *Wolffia arrhiza* have been used in bioassays [79,80]. A number of endpoints have been defined to determine toxicity, such as growth and increase in number of fronds, as well as survival, which has been summarized by Wang [81]. More recently image processing has been employed to count fronds and to measure their size and shape, which is also used to determine the color and thus the pigmentation of the fronds [82] (cf. Chapter 1: Introduction, this volume). The duckweed test has been employed in water management [83,84], nanoparticle testing [85], and monitoring of chemical catalysts [86]. It was employed to test for additives in food [87] as well as to analyze herbicide toxicology [88,89] or to characterize active ingredients from natural sources [90,91]. It has also been used to test for toxicity in human and veterinary drugs [92–95], or for homeopathic drugs [96,97]. The test has been standardized in OECD and ISO protocols [98,99]. Recently the ecotoxicity of nanoparticles magnetite has been determined with the duckweed test as well as with *Daphnia magna* [98,99]. Another comparison was made with duckweed and the *Vibrio fischeri* bioassay to assess toxicity of phenylurea herbicides [100]. One advantage of the test is the high sensitivity of the aquatic angiosperm; on the other hand monitoring growth and change in pigmentation requires a long time on the order of days.

The choice for a bioassay is generally based on its sensitivity, cost, and response time. The most commonly used bioassays are compared here in terms of these parameters.

17.3 Comparison of common bioassays

17.3.1 Sensitivity

The sensitivity of a bioassay or the toxicity level of a substance is usually quantified by different terminologies of effective concentrations. The most common ones are no observed effect concentration (NOEC) and half- (or median-) maximal effective concentration (EC_{50}), median lethal concentration (LC_{50}), and median lethal dose (LD_{50}). NOEC is the maximum concentration of a chemical with no significant effect on the population of the used bioassay. EC_{50} is the concentration of a chemical that causes 50% of the maximal effect. LC_{50} or LD_{50} is the concentration or dose (generally used for liquid and solid, respectively) of a chemical that will kill 50% of the population of the bioassay organisms. These values are generally derived from effect concentration (EC) curves. The bioassay organism is treated with a wide range of doses of the chemical under test and the observed effect (usually percent effect in comparison to the control) on a given parameter of the test organism is plotted against the concentration/dose of the chemical. The desired endpoint (NOEC or EC_{50} etc.) is calculated from the regression curve. A higher NOEC or $EC_{50}/LD_{50}/LC_{50}$ value obtained for a chemical reveals lesser toxicity and vice versa. In comparison of two or more bioassays with the same chemical, the one giving the lowest NOEC or EC_{50} is considered as the more sensitive bioassay. Although a number of terminologies are used in ecotoxicology for this purpose as discussed above in this chapter and elsewhere in this volume, hereafter in this chapter we use only the EC_{50} values for discussion on comparative sensitivities of different bioassays. A large number of bioassays with a number of different endpoints are used in ecotoxicology. However, covering all bioassays with all the endpoints is beyond the scope of this chapter. Therefore, only the few most common bioassays with their most common endpoints are compared.

The sensitivity of different bioassays can vary greatly for the same chemical, and a bioassay most sensitive for one chemical may not always be the most sensitive bioassay for other chemicals. It is not easy to identify a specific bioassay as the most sensitive one for monitoring water pollution. So many variations and conflicting reports exist on the sensitivity of the commonly used bioassays. A few examples of EC_{50} values for some heavy metals obtained with some common bioassays are shown in Table 17.1 which reflect large variations for different bioassays even for the same bioassay extracted from different studies.

The EC_{50} value of a substance with a particular bioassay may vary with exposure time. Therefore, many workers determine EC_{50} for the testing substances at different (increasing) incubation times. In many cases the adverse effect of the tested substance increases with increasing exposure time and a decrease in the EC_{50} value is observed, but in some cases the opposite effect can be seen due to the adaptation of the bioassay organism to the chemical under test.

The bioluminescence test using *V. fischeri* (Microtox) is one of the most commonly used bioassays and is claimed by many studies as the most sensitive bioassay. For example, toxicity analysis of industrial and urban wastewaters with a complex multi-trophic test battery consisting of six different tests (Algaltoxkit FTM with *Selenastrum capricornutum*, Daphtoxkit FTM with *D. magna*, Protoxkit FTM with *Tetrahymena thermophile*, Charatox, ThamnotoxkitTM, and Microtox), found that Microtox along with Charatox and ThamnotoxkitTM showed the highest sensitivity as compared to Algaltoxkit FTM, Daphtoxkit FTM, and Protoxkit FTM [124]. The assessment of water-extractable ecotoxic potential of contaminated soils, using five different bioassay methods also revealed the bioluminescence test of *V. fischeri* as the most sensitive test [125]. However many critiques, for example by Isidori et al. [126], are of the opinion that most of the toxicity data for chemicals (e.g., pharmaceuticals) obtained with *V. fischeri* tests after short exposure times (generally 5–30 min) underestimate the toxic potential of some chemicals, because many compounds interfere only slightly with the biosynthetic pathways of bioluminescence. Furthermore, there are a number of studies that proved *V. fischeri* bioluminescence test as the least sensitive. For example, a study on comparative sensitivity of three bioassays, *D. magna* (mobility), *V. fischeri* (bioluminescence), and *Pseudomonas putida* (growth inhibition), to four heavy metals including Cd, Mn, Pb, and Zn, revealed that *D. magna* was the most sensitive to Cd, Zn, and Mn, while *P. putida* was the most sensitive species to Pb [112]. A comparative study of Cd toxicity using a battery of standard test species [127] also revealed lower sensitivity of the bacterial test (IC_{50} 2400 mg L^{-1}) than *D. magna* (24 h IC_{50} 0.64 mg L^{-1}). Many other studies reported *D. magna* to be more sensitive than *V. fischeri* tests. For example, the 15 min EC_{50} obtained for Cd with *V. fischeri* was 36.08 mg L^{-1}, which was much higher than the 24 h EC_{50} for *D. magna* (0.29 mg L^{-1}) [128]. Even the growth inhibition test of bacteria, *P. putida*, was found to be more sensitive than the *V. fischeri* luminescence test to Cd, Pb, and Mn [112]. Of the four metals tested, only for Zn, the bioluminescence inhibition test

Table 17.1 **EC_{50} values of common bioassays for a few heavy metals**

	Ecotox	*Lemna*	Bacteria luminescence (*Vibrio fischeri*)	*Daphnia*	Fish (*Labeo rohita*)
Ag (mg L^{-1})	0.54 (3 min) [46] 0.6 (3 min) [104]	0.081 (7 days) [98] 160 (48 h) [105]	1.856 (30 min) [101]	0.0091 (48 h) [102]	30 (96 h) [103]
As (mg L^{-1})	141 (3 min) [46]	8.18 (5 min) [106]	35 (5 min) [107]	24 (48 h) [102]	21.1 (96 h) [108]
Cd (mg L^{-1})	202 (3 min) [52] 0.86 (24 h) [52]	0.323 (7 days) [98] 0.64 (96 h) [111]	3.31 (30 min) [109] 52.51 (30 min) [112]	0.0036 (48 h) [102] 0.17 (48 h) [112] 0.22 (24 h) [112] 0.31 (48 h) [113]	153.24 (96 h) [110]
Co (mg L^{-1})	4.3 (3 min) [46]	0.557 (7 days) [98]	135–177 (5 min) [107]	0.71 (48 h) [102]	106 [96] [114]
Cu (mg L^{-1})	4 (3 min) [46] 8 (3 min) [104] 19 (3 min) [52] 23 (24 h) [52]	0.33 (7 days) [115] 0.45 (96 h) [111]	0.33 (30 min) [109] 1.2 – 20 (5 min) [116] 34.4 (30 min) [104]	0.0013 (48 h) [102] 0.82 (48 h) [105]	72.72 (96 h) [110]
Cr (mg L^{-1})	27.8 (3 min) [46]	11.1 (7 days) [98] 240.4 (7 days) [117]	70 – 100 (5 min) [107] 7.83 (30 min) [116]	0.013 (48 h) [102] 9.07 (48 h) [113] 79.2 (48 h) [117]	64.22 (96 h) [114]
Ni (mg L^{-1})	7.9 (3 min) [46] 10 (24 h) [52] 235 (3 min) [52]	0.37 (7 days) [98] 1.90 (96 h) [111]	25.2 – 410 (5 min) [118]	0.65 (48 h) [102]	22.01 (96 h) [119] 92.3 (96 h) [103]
Fe (mg L^{-1})	–	186.8 (7 days) [117]	0.12 (30 min) [109]	2.3 (48 h) [102] 80.8 (48 h) [117]	49.75 (96 h) [119]

(*Continued*)

Table 17.1 (Continued)

	Ecotox	*Lemna*	Bacteria luminescence (*Vibrio fischeri*)	*Daphnia*	Fish (*Labeo rohita*)
Hg (mg L^{-1})	1 (3 min) [104] 38 (3 min) [52] 35 (24 h) [52]	0.48 (7 days) [120] 0.64 (96 h) [120] 0.683 (7 days) [98] 195 (48 h) [105]	0.35 (30 min) [104] [104] 0.67 (30 min) [109]	0.00065 (48 h) [102] 0.251 (48 h)[13]	0.018 (96 h) [121]
Pb (mg L^{-1})	40.1 (3 min) [46]	6.8 (96 h) [120] 5.5 (7 days) [120]	0.85 (30 min) [122] 1.32 (30 min) [109] 35.97 (30 min) [112]	0.29 (48 h) [102] 5.1 (48 h) [113] 74.73 (48 h) [112] 208.14 (24 h) [112]	22.11 (96 h) [119] 36.72 (96 h) [114] 34.2 (96 h) [123]
Zn (mg L^{-1})	164 (3 min) [46]	0.909 (7 days) [98] 5.5 (96 h) [111] 131 (7 days) [117]	4.64 (30 min) [112] 2–49 (30 min) [107]	0.41 (48 h) [112] 0.72 (48 h) [102] 1.8 (24 h) [112] 59.4 (48 h) [117]	26.23 (96 h) [119] 85.44 (96 h) [110] 165.3 (96 h) [103]

Time shown in parenthesis is exposure time. More than one value for the same metal with the same bioassay is extracted from different studies. References are indicated for each value.

with *V. fischeri* was more sensitive [112]. Ecotoxicological evaluation of metal plating effluents having high concentrations of Zn, Cr, Fe, and Cu with the Microtox and *Daphnia* assay concluded that Microtox was less sensitive than acute and chronic tests with *Daphnia* [129]. Another similar example can be found in a study for wastewater quality control in a wastewater treatment plant which revealed that the *V. fischeri* test had the poorest sensitivity to detect the adverse effects of various pollutants present in the samples, but *D. magna* and *S. capricornutum* were quite sensitive in many cases [130]. The conventional tests like algal growth inhibition and *D. magna* immobilization tests were recommended to be suitable for assessing toxicity of effluents discharges, but the rapid chemiluminescence and bioluminescence (*V. fischeri*) tests were not recommended for this purpose [131]. The bioluminescence test of bacteria is usually preferred due to its short response time, though its sensitivity and ecological relevancy has been questioned by many workers. Another short response time bioassay, Ecotox, which uses the freshwater flagellate *Euglena* as a test organism, needs 5–10 min for one measurement. Therefore, Ecotox can be a better alternative where assessment in a short time is required. Ahmed [52] compared the sensitivity of Ecotox with other common bioassays like *Daphnia* and fish for wastewater samples from the Bayerisches Landesamt für Umwelt, LFU, in Munich and concluded that orientation endpoints of Ecotox were far more sensitive than these other common bioassays for short-term assessment of wastewater. The sensitivity of Ecotox for pollutants like metals was found to be comparable to or even higher than other commonly used biotests such as *D. magna*, bioluminescence bioassay (Microtox), fish bioassays, genetoxicity tests (mutagenesis), and algal growth test [52].

As stated above, it may not always be true that a particular bioassay is always the most sensitive for all types of pollutants. The sensitivity can vary greatly with the nature of chemicals. For instance, aquatic toxicity assessment of four commonly used veterinary drugs, including doramectin (DOR), metronidazole (MET), florfenicol (FLO), and oxytetracycline (OXT), were tested for their aquatic toxicity using bacteria (*V. fischeri*), duckweed (*L. minor*), green algae (*Scenedesmus vacuolatus*), and crustaceans (*D. magna*). To OXT and FLO, duckweed (EC_{50} 3.26 and 2.96 mg L^{-1}, respectively) was the most sensitive followed by green algae (EC_{50} 40.4 and 18.0 mg L^{-1}) and bacteria (EC_{50} 108 and 29.4 mg L^{-1}) but the crustacean test (EC_{50} 114 and 337 mg L^{-1}) was the least sensitive, however, for DOR the crustacean test was the most sensitive with an EC_{50} of 6.37×10^{-5} mg L^{-1}. None of the used bioassay was sensitive to the tested concentration range of MET [93]. Another detailed study on the toxicity of thirteen different food supplements containing magnesium (Mg), chromium (Cr), iron (Fe), and zinc (Zn) using algal (*Scenedesmus subspicatus* and *Raphidocelis subcapitata*), Daphnia (*D. magna*) and duckweed (*L. minor*) bioassays revealed similar variation in sensitivity for heavy metals. EC_{50} values for a particular metal differed significantly among all the bioassay organisms used. The EC_{50} and order of sensitivity was for *S. subspicatus* Fe <Zn <Mg <Cr (EC_{50} 46.9, 59.8, 73.0, 88.1 mg L^{-1}, respectively), for *R. subcapitata* Fe <Zn <Mg <Cr (EC_{50} 44.9, 52.6, 62.2, 76.8 mg L^{-1}, respectively), for *D. magna* Zn <Cr <Fe < Mg (EC_{50} 59.4, 79.2, 80.8, 82.0 mg L^{-1}, respectively), and for *L. minor* Zn <Fe <Mg <Cr (EC_{50} 131.0, 186.8, 192.5, 240.4 mg L^{-1}, respectively) [117]. A study on comparative sensitivity of three bioassays, *D. magna* (immobility), *V. fischeri* (bioluminescence), and *P. putida* (growth inhibition), to four heavy metals (Cd, Mn, Pb, and Zn) revealed sensitivity ranking as: *D. magna* Cd > Zn > Mn > Pb; *V. fischeri* Zn > Pb > Cd > Mn, and *P. putida* Pb > Cd > Zn > Mn [112]. The sensitivity of different microbial based tests, including *Pseudomonas fluorescens*, *Saccharomyces cerevisiae* and the Microtox to heavy metal toxicity showed that the Microtox test was the most sensitive assay for detecting toxicity of zinc, copper, and mercury but not for cadmium, chromium, and nickel [132]. This variation in response of different bioassays to different chemicals and vice versa can be explained with the mode of action of a chemical, and the metabolic pathways of the bioassay organism. For example, the low sensitivity of *V. fischeri* to the veterinary drug oxytetracycline (OXT) was attributed to its mode of action on protein synthesis, which is not of importance during short-term bioluminescence testing [133].

The bioassays compared above mostly belong to different groups of organisms that can be an argument for their different sensitivities. But even closely related species were found to have large differences in their sensitivity to aquatic pollutants. For example, testing the toxicities of three surfactants using three different species of algae concluded that the most sensitive alga for one surfactant was not the most sensitive one for other surfactants [134]. A common surfactant (sodium lauryl sulfate) when tested with thirteen different species of algae and the EC_{100} values varied from 31 to 4000 mg L^{-1} [135]. An old literature review on surfactant toxicities to different algae [136] summarized that the effective concentrations of cationic, anionic, and nonionic surfactants for algae ranged from 0.003–10 mg L^{-1}, 0.003–4000 mg L^{-1} and 0.003–17,784 mg L^{-1}, respectively. Similar variations in response to detergents were reported using other closely related aquatic organisms. For example, the 48 h EC_{50} of sodium silicate for cf. *Ceriodaphnia dubia* was 1.47 mg L^{-1} [137] but the same chemical gave an EC_{50} of 494 mg L^{-1} when tested with *D. magna* [138]. Such conflicting observations are very common in literature, and thus it may not be easy to conclude which bioassay is more sensitive to aquatic pollutants.

The time factor, i.e., the duration of exposure, is an essential factor in determining the sensitivity of a bioassay to a pollutant. Different exposure times ranging from a few minutes to several days are commonly used depending on the type of bioassay and the end point used. For example, with bioluminescence test of bacteria usually a 5–30 min exposure duration is used. However, some critiques strongly argue that such extremely short-term tests may not be sufficient to show the effect on the tested organism and thus provides imprecise risk assessment and therefore recommend long-term bioassays to overcome this limitation. The sensitivity of the same bioassay generally varies greatly with a change in exposure time. For example, EC_{50} values of 26 different detergents for *D. magna* dropped from 4.0–1615 mg L^{-1} after 24 h exposure to 3.2–673 mg L^{-1} after 48 h [139]. Testing the toxicity of DAP with Ecotox at different exposure time, Azizullah et al. [57] found that in short-term experiments (3 min exposure) DAP affected motility in

Euglena at concentrations above 1.8 g L^{-1} but affected this parameter at a concentration above 0.63 g L^{-1} in longer incubation experiments (24 h or above). Similarly, the EC_{50} values for DAP obtained after 24 or 72 h exposure were much lower than that at 0 h. The 24 h EC_{50} and NOEC of Ni and Cd for motility parameter of *Euglena* were at least 20 times lower than those after immediate exposure (3 min) [52]. But further variation in the EC_{50} values on exposure longer than 24 h (up to 7 days) was negligible. Similarly, Herkovits [140] while testing Ni toxicity using amphibian embryos in a 7 days experiment found a decrease in effective concentration from 24 to 96 h of exposure but a further increase in exposure time to one week showed very minor changes in the effective concentrations. Comparing the results obtained by short- and long-term toxicity assessment of different water toxin samples with *Euglena* Ahmed concluded that 24 h tests were more sensitive for most of the tested pollutants [52]. Further incubation up to 7 days did not change the measured EC_{50} values much. Although some toxins could show most significant changes in EC_{50} after 1 h of treatment (such as Ni and 2,4-DCP) other toxins like Cd and PCP showed clear changes only after at least 24 h of incubation [52].

It may not always be true that an increased incubation time will result in increased sensitivity of a bioassay. For example, testing the toxicity of urea with Ecotox revealed that the adverse effect of urea did not increase with increasing exposure time but rather a decrease in the inhibitory effect was observed [57]. Similarly, a comparison of short- and long-term effects of two pesticides (carbufuran and malathion) on *Euglena* revealed that the effect was less pronounced in long-term exposure tests [56]. After 24–72 h exposure, 50 mg L^{-1} of carbofuran caused a 24%–32% inhibition of motility in *Euglena* but the same concentration inhibited motility only by 2% in the 7-days experiment. Similarly, this pesticide significantly affected swimming velocity, cell shape, and photosynthetic efficiency after 24–72 h exposure but did not affect these parameters during 7-days exposure. The authors concluded that after a long time exposure the cells of *Euglena* might have initiated some adaptation mechanisms which allow them to tolerate pesticide toxicity and hence showed little sensitivity. Some other studies suggested the same and found lower toxicity of heavy metals like aluminum and lead to *Euglena* in long-term exposure than in short-term exposure [141,142].

A change in exposure duration can even change the order of toxicity of chemicals. For example, Ahmed tested the toxicity of four toxic metals including Cd, Cu, Hg, and Ni using Ecotox with *Euglena* for short- (3 min) and long-term (24 h) exposure [52]. In the 3-min exposure test, the toxicity order was Cu $>$ Hg $>$ Ni $>$ Cd while after 24 h the toxicity order changed to Cd $>$ Ni $>$ Cu $>$ Hg. Another study found that the order of toxicity of Cd, Cu, and Hg to motility in *Euglena* was Hg $>$ Cd $>$ Cu [142]. These variations in toxicity magnitude for the same organism reflect that certain factors like growth conditions and age can influence the sensitivity of a bioassay to toxin. For example, the naupliar stages of *Acartia tonsa* was 28 times more sensitive than the adult stages to cypemethrin after 96 h exposure [143]. Similarly, other factors such as water temperature, water pH, and water hardness and form of a chemical (e.g., speciation of a metal) greatly influence the toxicity of a pollutant and hence sensitivity of a bioassay [140]. E.g., an increase in pH [6–9] or suspended sediment concentration (0–200 $mg\ L^{-1}$) significantly increased the toxicity of Roundup (a herbicide) to *C. dubia*, but there was no significant effects of change in temperature [144]. In contrast, Folmar et al. [145] observed that rainbow trouts and bluegills showed a higher sensitivity to Roundup with increasing temperature from 7 to 17°C and 17 to 27°C, respectively. In the case of metals speciation can affect their toxicity. For example, Cr(VI) was more toxic than Cr(III) and Cu(II) was more toxic than Cu(I) when tested with Microtox [146].

Due to large variations in the response of organisms to different chemicals, it is nowadays recommended that for toxicity assessment sets of sensitive species from different groups should be used to accurately assess toxic hazards of a chemical or pollutant of environmental concern. Or at least a bioassay with multiple endpoints should be used as different parameters of the same organisms may respond differently to the same pollutant, and different parameters of the same organism may respond differently to the same pollutant at different exposure times. Ecotox is an example of such bioassays which uses multiple parameters (motility, cell speed, cell shape, and a number of orientation parameters) of the test organism *Euglena* as endpoints. For wastewater quality assessment, its gravitactic orientation was recommended as the most sensitive endpoint [147]. An advantage of such bioassays is that they can more precisely predict the risk of a pollutant by assessing more than one parameter. For example, in many cases the motility was not affected so much by wastewater, but the orientation parameters detected wastewater pollution very sensitively in short-term tests [147].

17.3.2 Response time

Response time (here meaning the exposure and measurement time needed for an accurate and sensitive response) of a bioassay is one of the important factors in choosing a bioassay. As discussed above in this chapter as well as in other chapters of this volume, response times vary from a few minutes to several days or even sometimes to months depending upon the choice of bioassay and type of risk assessment. The shortest response time can definitely be the best choice for an ecotoxicologist. The majority of the common bioassays used in ecotoxicological assessment of water quality or water pollutants generally needs a time of 24 h to seven days, with the exception of bacterial bioluminescence and Ecotox assays that need a few minutes for measurement. For the bacterial bioluminescence assay generally 5–30 min exposure time is needed while Ecotox needs 6–10 min for a complete measurement of a sample. Based on exposure time, these two bioassays can be the best options if assessment is needed in a really short time. However, as discussed earlier, some critiques object that such short-term tests may sometimes underestimate the toxicity of a substance and such objections have mostly been faced by the bioluminescence bioassay. Studies published with Ecotox so far have mostly confirmed motility and especially gravitactic orientation to be reliable and very sensitive endpoints for such short-term toxicity assessment.

17.3.3 Costs

A comparison of different bioassays based on costs does not seem possible. The authors of almost every study claimed the bioassay they used as cost effective but none have provided the estimation of actual costs for any of the bioassays they used. So far only one recent study can be found that calculated the per sample costs for six bioassays including fish toxicity test (*Lebistes reticulatus*), Microtox (*V. fischeri*), Thamnotox (*Thamnocephalus platyurus*), Daphtox (*D. magna*), *L. minor*, and *Lepidium sativum* test [148]. The per sample costs for these tests were 145, 30, 25.4, 38.5, 10, and 9 €, respectively. It is not so easy to estimate the cost of a bioassay, as it is dependent on a number of factors that may vary greatly from area to area, and may vary according to the requirements of a specific test sample. According to Persoone and Van de Vel [149], the cost of a bioassay test can be mainly dependent on personnel, durable equipment, and consumables costs. Most of the bioassays are dependent on continuous culturing (i.e., maintenance) of live stocks of the test species, with the inherent technical requirements and costs of year-round culturing and maintenance [149]. According to this study, which was based on analysis of data from about 40 ecotoxicological laboratories from various European countries, most of the costs of bioassays were related to the maintenance of stock cultures and their daily culturing expenses of the test organisms, rather than to conducting the actual test. For some aquatic bioassays "culture/maintenance free" microbiotests have been developed where dormant stages ("cryptobiotic eggs") of the test species are stored for long periods of time and "hatched" at the time of performance of the assays [150]. Such microbiotests were first developed in the early 1990s at the Laboratory for Environmental Toxicology and Aquatic Ecology at Ghent University in Belgium. These assays were given the generic name "Toxkits" [150]. This technique, nowadays used for, e.g., bacteria and *Daphnia* tests, can minimize the culture maintenance charges for a bioassay. Proper studies are needed to estimate and compare the costs of different bioassays.

References

[1] Petrie B, Barden R, Kasprzyk-Hordern B. A review on emerging contaminants in wastewaters and the environment: current knowledge, understudied areas and recommendations for future monitoring. Water Res 2015;72:3–27.

[2] Brack W, Ait-Aissa S, Burgess RM, Busch W, Creusot N, Di Paolo C, et al. Effect-directed analysis supporting monitoring of aquatic environments—An in-depth overview. Sci Total Environ 2016;544:1073–118.

[3] Wernersson A-S, Carere M, Maggi C, Tusil P, Soldan P, James A, et al. The European technical report on aquatic effect-based monitoring tools under the water framework directive. Env Sci Eur 2015;27(1):1–11.

[4] Hu X, Shi W, Yu N, Jiang X, Wang S, Giesy JP, et al. Bioassay-directed identification of organic toxicants in water and sediment of Tai Lake, China. Water Res 2015;73:231–41.

[5] Skidmore J. Resistance to zinc sulphate of the zebrafish (*Brachydanio rerio* Hamilton-Buchanan) at different phases of its life history. Ann Appl Biol 1965;56(1):47–53.

[6] Adelman IR, Smith Jr LL. Fathead minnows (*Pimephales promelas*) and goldfish (*Carassius auratus*) as standard fish in bioassays and their reaction to potential reference toxicants. J Fish Board Can 1976;33(2):209–14.

[7] Troelstra S, Wagenaar R, Smant W, Peters B. Interpretation of bioassays in the study of interactions between soil organisms and plants: involvement of nutrient factors. New Phytol 2001;150(3):697–706.

[8] Rojíčková R, Dvořáková D, Maršálek B. The use of miniaturized algal bioassays in comparison to the standard flask assay. Environ Toxicol 1998;13(3):235–41.

[9] Farré M, Barceló D. Toxicity testing of wastewater and sewage sludge by biosensors, bioassays and chemical analysis. TrAC Trends Anal Chem 2003;22(5):299–310.

[10] Wieczerzak M, Namieśnik J, Kudłak B. Bioassays as one of the Green Chemistry tools for assessing environmental quality: a review. Environ Int 2016;94:341–61.

[11] van der Meulen-Frank M, Prins J-B, Waarts B-L, Hofstra W. Vertebrate animals used for experimental and other scientific purposes: principles and practice for legislation and protection. In: Glaudemans AWJM, Medema J, van Zanten AK, Dierckx RAJO, Ahaus CTBK, editors. Quality in nuclear medicine. Berlin, Heidelberg: Springer; 2017. p. 91–105.

[12] Häder D-P, Erzinger GS. Daphniatox–Online monitoring of aquatic pollution and toxic substances. Chemosphere 2017;167:228–35.

[13] Rodriguez-Ruiz A, Asensio V, Zaldibar B, Soto M, Marigómez I. Toxicity assessment through multiple endpoint bioassays in soils posing environmental risk according to regulatory screening values. Environ Sci Pollut Res 2014;21(16):9689–708.

[14] Neuparth T, Martins C, Carmen B, Costa MH, Martins I, Costa PM, et al. Hypocholesterolaemic pharmaceutical simvastatin disrupts reproduction and population growth of the amphipod *Gammarus locusta* at the ng/L range. Aquat Toxicol 2014;155:337–47.

[15] Tan L, Schirmer K. Cell culture-based biosensing techniques for detecting toxicity in water. Curr Opin Biotechnol 2017;45:59–68.

[16] Azizullah A, Richter P, Häder D-P. Responses of morphological, physiological, and biochemical parameters in *Euglena gracilis* to 7-days exposure to two commonly used fertilizers DAP and urea. J Appl Phycol 2012;24(1):21–33.

[17] Azizullah A, Murad W, Adnan M, Ullah W, Häder D-P. Gravitactic orientation of *Euglena gracilis* - a sensitive endpoint for ecotoxicological assessment of water pollutants. Front Env Sci 2013;1:4.

[18] Allinson G, Zhang P, Bui A, Allinson M, Rose G, Marshall S, et al. Pesticide and trace metal occurrence and aquatic benchmark exceedances in surface waters and sediments of urban wetlands and retention ponds in Melbourne, Australia. Environ Sci Pollut Res 2015;22(13):10214–26.

[19] Graser G, Walters FS. A standardized lepidopteran bioassay to investigate the bioactivity of insecticidal proteins produced in transgenic crops. In: Faye L, Gomord V, editors. Recombinant proteins from plants: methods and protocols. Berlin, Heidelberg: Springer; 2016. p. 259–70.

[20] Han JE, Tang KF, Piamsomboon P, Pantoja CR. Evaluation of a reliable non-invasive molecular test for the diagnosis of the causative agent of acute hepatopancreatic necrosis disease of shrimp. Aquac Rep 2017;5:58–61.

[21] Diana T, Kanitz M, Lehmann M, Li Y, Olivo PD, Kahaly GJ. Standardization of a bioassay for thyrotropin receptor stimulating autoantibodies. Thyroid 2015;25(2):169–75.

[22] Xia T, Hamilton Jr RF, Bonner JC, Crandall ED, Elder A, Fazlollahi F, et al. Interlaboratory evaluation of in vitro cytotoxicity and inflammatory responses to engineered nanomaterials: the NIEHS Nano GO Consortium. Environ Health Perspect 2013;121(6):683.
[23] Mehinto AC, Jia A, Snyder SA, Jayasinghe BS, Denslow ND, Crago J, et al. Interlaboratory comparison of in vitro bioassays for screening of endocrine active chemicals in recycled water. Water Res 2015;83:303–9.
[24] Li S, Wallis LK, Ma H, Diamond SA. Phototoxicity of TiO_2 nanoparticles to a freshwater benthic amphipod: are benthic systems at risk?. Sci Total Environ 2014;466:800–8.
[25] Lee DW, Jung JE, Yang YM, Kim JG, Yi HK, Jeon JG. The antibacterial activity of chlorhexidine digluconate against *Streptococcus mutans* biofilms follows sigmoidal patterns. Eur J Oral Sci 2016;124(5):440–6.
[26] Häder D-P, Erzinger GS. Ecotox – Monitoring of pollution and toxic substances in aquatic ecosystems. J Ecol Environ Sci 2015;3(2):22–7.
[27] Stypuła-Trębas S, Minta M, Radko L, Żmudzki J. Oestrogenic and (anti) androgenic activity of zearalenone and its metabolites in two in vitro yeast bioassays. World Mycotoxin J 2016;9(2):247–55.
[28] Medvedeva SE, Александр М, Родичева ЭК, Александр М. Bioluminescent bioassays based on luminous bacteria. Biology 2015;2(4):418–52.
[29] de Paiva Magalhães D, da Costa Marques MR, Baptista DF, Buss DF. Selecting a sensitive battery of bioassays to detect toxic effects of metals in effluents. Ecotoxicol Environ Saf 2014;110:73–81.
[30] Mascarin GM, Jackson MA, Behle RW, Kobori NN, Júnior ÍD. Improved shelf life of dried *Beauveria bassiana* blastospores using convective drying and active packaging processes. Appl Microbiol Biotechnol 2016;100(19):8359–70.
[31] Froment J, Thomas KV, Tollefsen KE. Automated high-throughput in vitro screening of the acetylcholine esterase inhibiting potential of environmental samples, mixtures and single compounds. Ecotoxicol Environ Saf 2016;130:74–80.
[32] Kokkali V, van Delft W. Overview of commercially available bioassays for assessing chemical toxicity in aqueous samples. TrAC Trends Anal Chem 2014;61:133–55.
[33] Malev O, Klobučar R, Tišler T, Drobne D, Trebše P, editors. Application of cost effective biological tools for monitoring of chemical poisoning. In: NATO Advanced Research Workshop in Nanotechnology to Aid Chemical and Biological Defence; 2014.
[34] Chouler J, Di Lorenzo M. Water quality monitoring in developing countries; can microbial fuel cells be the answer? Biosensors 2015;5(3):450–70.
[35] Gassmann A, Shrestha R, Jakka S, Dunbar M, Clifton E, Paolino A, et al. Evidence of resistance to Cry34/35Ab1 corn by western corn rootworm: root injury in the field and larval survival in plant-based bioassays. J Econ Entomol 2016;109:1872–80.
[36] Van Wees SC, Van Pelt JA, Bakker PA, Pieterse CM. Bioassays for assessing jasmonate-dependent defenses triggered by pathogens, herbivorous insects, or beneficial rhizobacteria. In: Goossens A, Pauwels L, editors. Jasmonate signaling: methods and protocols. Berlin, Heidelberg: Springer; 2013. p. 35–49.
[37] Besinis A, De Peralta T, Handy RD. The antibacterial effects of silver, titanium dioxide and silica dioxide nanoparticles compared to the dental disinfectant chlorhexidine on *Streptococcus mutans* using a suite of bioassays. Nanotoxicology 2014;8 (1):1–16.
[38] Vethaak AD, Hamers T, Martínez-Gómez C, Kamstra JH, de Weert J, Leonards PE, et al. Toxicity profiling of marine surface sediments: a case study using rapid screening

bioassays of exhaustive total extracts, elutriates and passive sampler extracts. Mar Environ Res 2017;124:81–91.

[39] Vasquez MI, Michael I, Kümmerer K, Fatta-Kassinos D. Bioassays currently available for evaluating the biological potency of pharmaceuticals in treated wastewater. In: Fatta-Kassinos D, Dionysiou D, Kümmerer K, editors. Wastewater reuse and current challenges. Berlin: Springer; 2015. p. 49–80.

[40] Arienzo M, Albanese S, Lima A, Cannatelli C, Aliberti F, Cicotti F, et al. Assessment of the concentrations of polycyclic aromatic hydrocarbons and organochlorine pesticides in soils from the Sarno River basin, Italy, and ecotoxicological survey by *Daphnia magna*. Environ Monit Assess 2015;187(2):52.

[41] de Castro-Català N, Kuzmanovic M, Roig N, Sierra J, Ginebreda A, Barceló D, et al. Ecotoxicity of sediments in rivers: Invertebrate community, toxicity bioassays and the toxic unit approach as complementary assessment tools. Sci Total Environ 2015;540:2097–396.

[42] Xu T, Close D, Smartt A, Ripp S, Sayler G. Detection of organic compounds with whole-cell bioluminescent bioassays. In: Thouand G, Marks R, editors. Bioluminescence: fundamentals and applications in biotechnology-Volume 1. Berlin: Springer; 2014. p. 111–51.

[43] Levandowsky M. Sensory responses in *Euglena*. ASM News 2001;67:299–303.

[44] Lebert M. Phototaxis of *Euglena gracilis* - flavins and pterins. In: Häder D-P, Lebert M, editors. Photomovement. Comprehensive series in photosciences. 1. Amsterdam, London, New York, Oxford, Paris, Shannon, Tokyo: Elsevier; 2001. p. 297–341.

[45] Häder D-P, Faddoul J, Lebert M, Richter P, Schuster M, Richter R, et al. Investigation of gravitaxis and phototaxis in *Euglena gracilis*. In: Sinha R, Sharma NK, Rai AK, editors. Advances in life sciences. New Delhi: IK International Publishing House; 2010. p. 117–31.

[46] Tahedl H, Häder D-P. Automated biomonitoring using real time movement analysis of *Euglena gracilis*. Ecotoxicol Environ Saf 2001;48(2):161–9.

[47] Azizullah A. Ecotoxicological assessment of anthropogenically produced common pollutants of aquatic environments [PhD thesis]. Erlangen, Germany: Friedrich-Alexander University; 2011.

[48] Streb C, Richter P, Sakashita T, Häder D-P. The use of bioassays for studying toxicology in ecosystems. Curr Top Plant Biol 2002;3:131–42.

[49] Danilov RA, Ekelund NGA. Influence of waste water from the paper industry and UV-B radiation on the photosynthetic efficiency of *Euglena gracilis*. J Appl Phycol 1999;11:157–63.

[50] Ahmed H, Häder D-P. Monitoring of waste water samples using the ECOTOX biosystem and the flagellate alga *Euglena gracilis*. J Water Air Soil Pollut 2011;216(1-4):547–60.

[51] Azizullah A, Richter P, Häder D-P. Ecotoxicological evaluation of wastewater samples from Gadoon Amazai Industrial Estate (GAIE), Swabi, Pakistan. Int J Environm Sci 2011;1(5):959–76.

[52] Ahmed H. Biomonitoring of aquatic ecosystems [PhD thesis]. Erlangen-Nürnberg: Friedrich-Alexander-Universität; 2010.

[53] Ahmed H, Häder D-P. A fast algal bioassay for assessment of copper toxicity in water using *Euglena gracilis*. J Appl Phycol 2010;22(6):785–92.

[54] Ahmed H, Häder D-P. Rapid ecotoxicological bioassay of nickel and cadmium using motility and photosynthetic parameters of *Euglena gracilis*. Environ Exp Bot 2010;69:68–75.

[55] Azizullah A, Richter P, Häder D-P. Toxicity assessment of a common laundry detergent using the freshwater flagellate *Euglena gracilis*. Chemosphere 2011;84(10):1392–400.
[56] Azizullah A, Richter P, Häder D-P. Comparative toxicity of the pesticides carbofuran and malathion to the freshwater flagellate *Euglena gracilis*. Ecotoxicology 2011;20(6):1442–54.
[57] Azizullah A, Nasir A, Richter P, Lebert M, Häder D-P. Evaluation of the adverse effects of two commonly used fertilizers, DAP and urea, on motility and orientation of the green flagellate *Euglena gracilis*. Environ Exp Bot 2011;74:140–50.
[58] Struewing KA, Lazorchak JM, Weaver PC, Johnson BR, Funk DH, Buchwalter DB. Part 2: Sensitivity comparisons of the mayfly *Centroptilum triangulifer* to *Ceriodaphnia dubia* and *Daphnia magna* using standard reference toxicants; NaCl, KCl and $CuSO_4$. Chemosphere 2015;139:597–603.
[59] Zein MA, McElmurry SP, Kashian DR, Savolainen PT, Pitts DK. Optical bioassay for measuring sublethal toxicity of insecticides in *Daphnia pulex*. Environ Toxicol Chem 2014;33(1):144–51.
[60] GSM. Examination of water, waste water and sludge, bio-assays (group L), Determining tolerance of *Daphnia* to toxicity of waste water by way of a dilution series. German Standard Methods. DIN 1989.
[61] Netto I. Assessing the usefulness of the automated monitoring systems Ecotox and DaphniaTox in an integrated early-warning system for drinking water [Master thesis]. Toronto Canada: Ryerson Univeristy; 2010.
[62] Pérez S, Rial D, Beiras R. Acute toxicity of selected organic pollutants to saltwater (mysid *Siriella armata*) and freshwater (cladoceran *Daphnia magna*) ecotoxicological models. Ecotoxicology 2015;24(6):1229–38.
[63] Gophen M. Bioassay indication of nemagon toxicity on *Daphnia magna* (Straus 1820). Int J Bioassays 2016;5(01):4739–41.
[64] Ahmadi M, Teymouri P, Ghalebi M, Jaafarzadeh N, Alavi N, Askari A, et al. Sludge characterization of an industrial water treatment plant, Iran. Desalin Water Treat 2014;52(28-30):5306–16.
[65] Pellegri V, Gorbi G, Buschini A. Comet Assay on *Daphnia magna* in eco-genotoxicity testing. Aquat Toxicol 2014;155:261–8.
[66] Verlaque M, Belsher T, Deslous-Paoli J-M. Morphology and reproduction of Asiatic *Ulva pertusa* (Ulvales, Chlorophyta) in Thau Lagoon (France, Mediterranean Sea). Cryptogamie Algologie 2002;23(4):301–10.
[67] Karthikeyan S, Balasubramanian R, Iyer C. Evaluation of the marine algae *Ulva fasciata* and *Sargassum* sp. for the biosorption of Cu (II) from aqueous solutions. Bioresour Technol 2007;98(2):452–5.
[68] Han T, Choi G-W. A novel marine algal toxicity bioassay based on sporulation inhibition in the green macroalga *Ulva pertusa* (Chlorophyta). Aquat Toxicol 2005;75(3):202–12.
[69] Han Y-S, Brown MT, Park GS, Han T. Evaluating aquatic toxicity by visual inspection of thallus color in the green macroalga *Ulva*: testing a novel bioassay. Environ Sci Technol 2007;41(10):3667–71.
[70] Han T, Kong J-A, Brown MT. Aquatic toxicity tests of Ulva *pertusa* Kjellman (Ulvales, Chlorophyta) using spore germination and gametophyte growth. Eur J Phycol 2009;44(3):357–63.
[71] Eklund BT, Kautsky L. Review on toxicity testing with marine macroalgae and the need for method standardization––exemplified with copper and phenol. Mar Pollut Bull 2003;46(2):171–81.

[72] Klaine SJ, Lewis MA. Algal and plant toxicity testing. In: Hoffman DJ, (Ed.) Handbook of ecotoxicology. Boca Raton, FL: Lewis Publishers, CRC Press, Inc; 1995. p. 163−84.

[73] Hooten RL, Carr RS. Development and application of a marine sediment pore-water toxicity test using *Ulva fasciata* zoospores. Environ Toxicol Chem 1998;17(5):932−40.

[74] Rai LC, Gaur JP, Kumar HD. Phycology and heavy-metal pollution. Biol Rev 1981;56:99−151.

[75] Oh J-j, Choi E-M, Han Y-S, Yoon J-H, Park A, Jin K, et al. Influence of salinity on metal toxicity to *Ulva pertusa*. Toxicol Environ Health Sci 2012;4(1):9−13.

[76] Kim E, Kim S-H, Kim H-C, Lee SG, Lee SJ, Jeong SW. Growth inhibition of aquatic plant caused by silver and titanium oxide nanoparticles. Toxicol Environ Health Sci 2011;3(1):1−6.

[77] Kim Y-J, Han Y-S, Kim E, Jung J, Kim S-H, Yoo S-J, et al. Application of the *Ulva pertusa* bioassay for a toxicity identification evaluation and reduction of effluent from a wastewater treatment plant. Front Env Sci 2015;3:2.

[78] Frédéric M, Samir L, Louise M, Abdelkrim A. Comprehensive modeling of mat density effect on duckweed (*Lemna minor*) growth under controlled eutrophication. Water Res 2006;40(15):2901−10.

[79] Ritchie RJ, Mekjinda N. Arsenic toxicity in the water weed *Wolffia arrhiza* measured using Pulse Amplitude Modulation Fluorometry (PAM) measurements of photosynthesis. Ecotoxicol Environ Saf 2016;132:178−85.

[80] Movafeghi A, Khataee A, Abedi M, Tarrahi R, Dadpour M, Vafaei F. Effects of TiO_2 nanoparticles on the aquatic plant *Spirodela polyrrhiza*: evaluation of growth parameters, pigment contents and antioxidant enzyme activities. J Environ Sci 2017; https://doi.org/10.1016/j.jes.2016.12.020, online 10 February 2017, in press.

[81] Wang W. Literature review on duckweed toxicity testing. Environ Res 1990;52:7−22.

[82] Cleuvers M, Ratte H-T. Phytotoxicity of coloured substances: is *Lemna* Duckweed an alternative to the algal growth inhibition test?. Chemosphere 2002;49(1):9−15.

[83] Mendonça E, Picado A, Silva L, Anselmo A. Ecotoxicological evaluation of cork-boiling wastewaters. Ecotoxicol Environ Saf 2007;66(3):384−90.

[84] Mendonça E, Picado A, Paixão SM, Silva L, Barbosa M, Cunha MA. Ecotoxicological evaluation of wastewater in a municipal WWTP in Lisbon area (Portugal). Desalin Water Treat 2013;51(19-21):4162−70.

[85] Picado A, Paixão SM, Moita L, Silva L, Diniz MS, Lourenço J, et al. A multi-integrated approach on toxicity effects of engineered TiO_2 nanoparticles. Front Environ Sci Eng 2015;9(5):793−803.

[86] Stolte S, Bui H, Steudte S, Korinth V, Arning J, Białk-Bielińska A, et al. Preliminary toxicity and ecotoxicity assessment of methyltrioxorhenium and its derivatives. Green Chem 2015;17(2):1136−44.

[87] Stolte S, Steudte S, Schebb NH, Willenberg I, Stepnowski P. Ecotoxicity of artificial sweeteners and stevioside. Environ Int 2013;60:123−7.

[88] Grossmann K, Hutzler J, Tresch S, Christiansen N, Looser R, Ehrhardt T. On the mode of action of the herbicides cinmethylin and 5-benzyloxymethyl-1, 2-isoxazolines: putative inhibitors of plant tyrosine aminotransferase. Pest Manage Sci 2012;68(3):482−92.

[89] Tresch S, Niggeweg R, Grossmann K. The herbicide flamprop-M-methyl has a new antimicrotubule mechanism of action. Pest Manage Sci 2008;64(11):1195−203.

[90] Diers JA, Bowling JJ, Duke SO, Wahyuono S, Kelly M, Hamann MT. Zebra mussel antifouling activity of the marine natural product aaptamine and analogs. Mar Biotechnol 2006;8(4):366−72.

[91] Meepagala KM, Sturtz G, Wedge DE, Schrader KK, Duke SO. Phytotoxic and antifungal compounds from two Apiaceae species, *Lomatium californicum* and *Ligusticum hultenii,* rich sources of Z-ligustilide and apiol, respectively. J Chem Ecol 2005;31 (7):1567–78.
[92] Białk-Bielińska A, Stolte S, Arning J, Uebers U, Böschen A, Stepnowski P, et al. Ecotoxicity evaluation of selected sulfonamides. Chemosphere 2011;85(6):928–33.
[93] Kołodziejska M, Maszkowska J, Białk-Bielińska A, Steudte S, Kumirska J, Stepnowski P, et al. Aquatic toxicity of four veterinary drugs commonly applied in fish farming and animal husbandry. Chemosphere 2013;92(9):1253–9.
[94] Maszkowska J, Stolte S, Kumirska J, Łukaszewicz P, Mioduszewska K, Puckowski A, et al. Beta-blockers in the environment: Part II. Ecotoxicity study. Sci Total Environ 2014;493:1122–6.
[95] Wagil M, Białk-Bielińska A, Puckowski A, Wychodnik K, Maszkowska J, Mulkiewicz E, et al. Toxicity of anthelmintic drugs (fenbendazole and flubendazole) to aquatic organisms. Environ Sci Pollut Res 2015;22(4):2566–73.
[96] Jäger T, Scherr C, Simon M, Heusser P, Baumgartner S. Development of a test system for homeopathic preparations using impaired duckweed (*Lemna gibba* L.). J Altern Complement Med 2011;17(4):315–23.
[97] Scherr C, Simon M, Spranger J, Baumgartner S. Duckweed (*Lemna gibba* L.) as a test organism for homeopathic potencies. J Altern Complement Med 2007;13(9):931–7.
[98] Naumann B, Eberius M, Appenroth K-J. Growth rate based dose–response relationships and EC-values of ten heavy metals using the duckweed growth inhibition test (ISO 20079) with *Lemna minor* L. clone St. J Plant Physiol 2007;164(12):1656–64.
[99] Khellaf N, Zerdaoui M. Growth response of the duckweed *Lemna minor* to heavy metal pollution. J Environ Health Sci Eng 2009;6(3):161–6.
[100] Gatidou G, Stasinakis AS, Iatrou EI. Assessing single and joint toxicity of three phenylurea herbicides using *Lemna minor* and *Vibrio fischeri* bioassays. Chemosphere 2015;119:S69–74.
[101] Binaeian E, Safekordi AA, Attar H, Saber R, Chaichi MJ, Kolagar AH. Comparative toxicity study of two different synthesized silver nanoparticles on the bacteria *Vibrio fischeri*. Afr J Biotechnol 2012;11(29):7554.
[102] Okamoto A, Yamamuro M, Tatarazako N. Acute toxicity of 50 metals to *Daphnia magna*. J Appl Toxicol 2015;35(7):824–30.
[103] Kousar S, Javed M. Heavy metals toxicity and bioaccumulation patterns in the body organs of four fresh water fish species. Pak Vet J 2014;34(2):161–4.
[104] Tahedl H, Häder D-P. Fast examination of water quality using the automatic biotest ECOTOX based on the movement behavior of a freshwater flagellate. Water Res 1999;33:426–32.
[105] Pavlić Ž, Stjepanović B, Ljubešić N, Puntarić D, Čulig J, editors. Electron microscope studies on Duckweed in toxicity tests with herbicides and metals. In: 2nd Croatin congress on microscopy with international participation; 2006.
[106] Rahman MA, Hogan B, Duncan E, Doyle C, Krassoi R, Rahman MM, et al. Toxicity of arsenic species to three freshwater organisms and biotransformation of inorganic arsenic by freshwater phytoplankton (*Chlorella* sp. CE-35). Ecotoxicol Environ Saf 2014;106:126–35.
[107] Munkittrick KR, Power EA, Sergy GA. The relative sensitivity of Microtox©, daphnid, rainbow trout, and fathead minnow acute lethality tests. Environ Toxicol Water Qual 1991;6:35–62.

[108] Vutukuru SS, Prabhath NA, Raghavender M, Yerramilli A. Effect of arsenic and chromium on the serum amino-transferases activity in Indian major carp, *Labeo rohita*. Int J Environ Res Public Health 2007;4(3):224–7.

[109] Clarke S, Barrick C, Samoiloff M. A bioassessment battery for use in an industrial setting: a new management approach. Environ Toxicol 1990;5(2):153–66.

[110] Ameer F, Javed M, Hayat S, Abdullah S. Growth responses of *Catla catla* and *Labeo rohita* under mixed exposure of dietary and water-borne heavy metals viz. Cu, Cd and Zn. J Anim Plant Sci 2013;23(5):1297–304.

[111] Khellaf N, Zerdaoui M, Faure O, Leclerc JC. Tolerance to heavy metals in the duckweed, *Lemna minor*. Environ Int 2008;34:1022–6.

[112] Teodorovic I, Planojevic I, Knezevic P, Radak S, Nemet I. Sensitivity of bacterial vs. acute *Daphnia magna* toxicity tests to metals. Central Eur J Biol 2009;4 (4):482–92.

[113] Meng Q, Li X, Feng Q, Cao Z, editors. The acute and chronic toxicity of five heavy metals on the *Daphnia magna*. In: Bioinformatics and Biomedical Engineering, ICBBE the 2nd International Conference; 2008: IEEE.

[114] Batool U, Javed M. Synergistic effects of metals (cobalt, chromium and lead) in binary and tertiary mixture forms on *Catla catla, Cirrhina mrigala* and *Labeo rohita*. Pak J Zool 2015;47(3):617.

[115] Cowgill U, Milazzo D, Landenberger B. The sensitivity of *Lemna gibba* G-3 and four clones of *Lemna minor* to eight common chemicals using a 7-day test. Res J Water Pollut Control Fed 1991;991–8.

[116] Backhaus T, Froehner K, Altenburger R, Grimme LH. Toxicity testing with *Vibrio fischeri*: a comparison between the long term (24 h) and the short term (30 min) bioassay. Chemosphere 1997;35:2925–38.

[117] Bošnir J, Puntarić D, Cvetković Ž, Pollak L, Barušić L, Klarić I, et al. Effects of magnesium, chromium, iron and zinc from food supplements on selected aquatic organisms. Coll Antropol 2013;37(3):965–71.

[118] Blaise C, Forghani R, Legault R, Guzzo J, Dubow MS. A bacterial toxicity assay performed with microplates, microluminetry and Microtox reagent. BioTechniques 1994;16:932–7.

[119] Abdullah S, Javed M, Javid A. Studies on acute toxicity of metals to the fish (*Labeo rohita*). Int J Agric Biol 2007;9(2):233–7.

[120] Dirilgen N. Mercury and lead: assessing the toxic effects on growth and metal accumulation by *Lemna minor*. Ecotoxicol Environ Saf 2011;74(1):48–54.

[121] Sigma Aldrich. Material safety data sheet for mercury. In: Region S-ACPS-A, editor. 2011.

[122] Kaiser KLE, Palabrica VS. *Photobacterium phosphoreum* toxicity data index. Water Pollut Res J Can 1991;26(3):361–431.

[123] Singh BO, Manjeet K. Determination of LC_{50} of lead nitrate for a fish, *Labeo rohita* (Hamilton Buchanan). Int Res J Biol Sci 2015;4(8):23–6.

[124] Manusadžianas L, Balkelyt L, Sadauskas K, Blinova I, Pollumaa L, Kahru A. Ecotoxicological study of Lithuanian and Estonian wastewaters: selection of the biotests, and correspondence between toxicity and chemical-based indices. Aquat Toxicol 2003;63(1):27–41.

[125] Maxam G, Rila J-P, Dott W, Eisentraeger A. Use of bioassays for assessment of water-extractable ecotoxic potential of soils. Ecotoxicol Environ Saf 2000;45(3):240–6.

[126] Isidori M, Lavorgna M, Nardelli A, Pascarella L, Parrella A. Toxic and genotoxic evaluation of six antibiotics on non-target organisms. Sci Total Environ 2005;346 (1):87–98.

[127] Castillo GC, Vila IC, Neild E. Ecotoxicity assessment of metals and wastewater using multitrophic assays. Environ Toxicol 2000;15(5):370–5.

[128] Kungolos A, Hadjispyrou S, Petala M, Tsiridis V, Samaras P, Sakellaropoulos G. Toxic properties of metals and organotin compounds and their interactions on *Daphnia magna* and *Vibrio fischeri*. Water Air Soil Pollut 2004;4(4):101–10.

[129] Choi K, Meier PG. Toxicity evaluation of metal plating wastewater employing the Microtox® assay: a comparison with cladocerans and fish. Environ Toxicol 2001;16 (2):136–41.

[130] Hernando MD, Fernández-Alba AR, Tauler R, Barceló D. Toxicity assays applied to wastewater treatment. Talanta 2005;65(2):358–66.

[131] Johnson I, Hutchings M, Benstead R, Thain J, Whitehouse P. Bioassay selection, experimental design and quality control/assurance for use in effluent assessment and control. Ecotoxicology 2004;13(5):437–47.

[132] Codina J, Perez-Garcia A, Romero P, De Vicente A. A comparison of microbial bioassays for the detection of metal toxicity. Arch Env Contamin Toxicol 1993;25(2):250–4.

[133] Zounková R, Klimešová Z, Nepejchalová L, Hilscherová K, Bláha L. Complex evaluation of ecotoxicity and genotoxicity of antimicrobials oxytetracycline and flumequine used in aquaculture. Environ Toxicol Chem 2011;30(5):1184–9.

[134] Yamane AN, Okada M, Sudo R. The growth inhibition of planktonic algae due to surfactants used in washing agents. Water Res 1984;18(9):1101–5.

[135] Blanck H, Wallin G, Wängberg S-Å. Species-dependent variation in algal sensitivity to chemical compounds. Ecotoxicol Environ Saf 1984;8(4):339–51.

[136] Lewis M. Chronic toxicities of surfactants and detergent builders to algae: a review and risk assessment. Ecotoxicol Environ Saf 1990;20(2):123–40.

[137] Warne MSJ, Schifko AD. Toxicity of laundry detergent components to a freshwater cladoceran and their contribution to detergent toxicity. Ecotoxicol Environ Saf 1999;44:196–206.

[138] Dowden BF, Bennett HJ. Toxicity of selected chemicals to certain animals. J Water Pollut Control Fed 1965;1308–16.

[139] Pettersson A, Adamsson M, Dave G. Toxicity and detoxification of Swedish detergents and softener products. Chemosphere 2000;41:1611–20.

[140] Herkovits J, Pérez-Coll CS, Herkovits FD. Evaluation of nickel-zinc interactions by means of bioassays with amphibian embryos. Ecotoxicol Environ Saf 2000;45:266–73.

[141] Danilov RA, Ekelund NGA. Effects of short-term and long-term aluminium stress on photosynthesis, respiration, and reproductive capacity in a unicellular green flagellate *(Euglena gracilis)*. Acta Hydrochim Hydrobiol 2002;30(4):190–6.

[142] Stallwitz E, Häder D-P. Motility and phototactic orientation of the flagellate *Euglena gracilis* impaired by heavy metal ions. J Photochem Photobiol B Biol 1993;18:67–74.

[143] Medina M, Barata C, Telfer T, Baird D. Age-and sex-related variation in sensitivity to the pyrethroid cypermethrin in the marine copepod *Acartia tonsa* Dana. Arch Env Contamin Toxicol 2002;42(1):17–22.

[144] Tsui MT, Chu L. Aquatic toxicity of glyphosate-based formulations: comparison between different organisms and the effects of environmental factors. Chemosphere 2003;52(7):1189–897.

[145] Folmar LC, Sanders H, Julin A. Toxicity of the herbicide glyphosate and several of its formulations to fish and aquatic invertebrates. Arch Env Contamin Toxicol 1979;8 (3):269–78.

[146] Bundy K, Mowat F. Speciation studies and toxicity assessment of complex heavy metal mixtures. Kansas State Univ., Manhattan, KS (United States), 1996; 1054–8564.

[147] Azizullah A, Richter P, Häder D-P. Sensitivity of various parameters in *Euglena gracilis* to short-term exposure to industrial wastewaters. J Appl Phycol 2012;24 (2):187–200.

[148] Aydin ME, Aydin S, Tongur S, Kara G, Kolb M, Bahadir M. Application of simple and low-cost toxicity tests for ecotoxicological assessment of industrial wastewaters. Environ Technol 2015;36(22):2825–34.

[149] Persoone G, van de Vel A. Cost-analysis of 5 current aquatic ecotoxicological tests. Commission of the European Communities, 1987. Contract No.: Report EUR 1134 EN, Brussels, Belgium.

[150] Persoone G, Baudo R, Cotman M, Blaise C, Thompson KC, Moreira-Santos M, et al. Review on the acute *Daphnia magna* toxicity test–Evaluation of the sensitivity and the precision of assays performed with organisms from laboratory cultures or hatched from dormant eggs. Knowl Manage Aquat Ecosyst 2009;393:01.

Ecotoxicological monitoring of wastewater

18

Donat-P. Häder
Friedrich-Alexander University, Erlangen-Nürnberg, Germany

18.1 Introduction

In the year 1900 1.6 billion people populated the Earth, by 2000 the number exceeded 5 billion and the latest UN predictions expect more than 9 billion for 2050 [1]. This explosive growth of our species has tremendous consequences for our planet including creating sustainable food sources [2], protection of the environment [3], and prevention of global diseases [4]. Freshwater is one of the major resources for human development and it is in critically short supply [5]. This is due to several facts: less than 1% of the global resources is accessible [6] and the rapidly growing population requires increasing amounts of freshwater for households, irrigation in agriculture, and industry [7]. Another factor is the increasing pollution by a multitude of toxic substances including heavy metals, persistent organic pollutants, pesticides, and fertilizers [8]. In addition, a growing number of pharmaceuticals and personal care products are released into the environment [9]. Many of these pollutants enter the terrestrial and aquatic environment due to uncontrolled release and accidental spills [10].

Especially in developing countries, industrial wastes, household effluents, and chemicals used in agriculture are dismissed into the air, soil, and water without previous treatment and control [11]. But even in developed countries toxic substances can enter the environment without treatment by accidental spills or voluntary pollution. However, most of the solid hazardous wastes are deposited in a controlled manner in deposits and are monitored to keep the toxicants under control [12]. Fluid wastes from households, industry, and mining are being collected and subjected to mechanical and biological cleansing [13]. The efficiency of the wastewater treatment needs to be closely controlled to warrant removal of toxic substances before the water reenters the aquatic environment, such as lakes or rivers [14]. This can be done by chemical analysis [15] or the employment of bioassays [16]. Due to the huge number of toxic pollutants in wastewater, chemical analysis is restricted to only a few classes of toxins and is both time consuming and expensive [17]. In addition, synergistic interactions between different toxic substances or the effects of environmental factors such as pH, UV, or temperature escape detection by chemical analysis. In contrast, bioassays cannot identify the toxic substances, but they can indicate a potential thread for the environment and human health when a hazardous level is reached [18].

Bioassays. DOI: http://dx.doi.org/10.1016/B978-0-12-811861-0.00018-8

18.2 Water treatment plants

Modern wastewater treatment plants are composed of a number of stages using mechanical (physical), biological, and chemical methods to remove contaminants from sewage and industrial effluents [19]. In many highly regulated countries, industrial wastewaters are pretreated at the factory to reduce pollutants which can often not be removed by conventional sewage treatment [20]. Due to the construction of the sewage system in cities and settlements the urban runoff may be diluted by rainwater which may contain soil particles, heavy metals, organic compounds, animal wastes as well as oil and grease and—in increasing concentrations—nanoparticles [21], when the rain runs over exposed surfaces [22]. During excessive rain flooding the extra water may be directly guided to natural reservoirs or rivers to protect the biological and chemical treatment systems from overload.

Construction of water treatment plants may vary due to local, state, or federal regulations, but usually sewage treatment consists of three stages. Before the wastewater enters the treatment plant, large objects such as tree parts, plastic and metal objects are removed by rake bars. The first step of a three-stage wastewater treatment plant comprises a mechanical cleaning in a storage basin to remove coarse and fine particles which may collect at the surface (oil, grease, and lighter solids) or sediment to the bottom (sediments and sand) [23]. The resulting sludge may be deposited or used for e.g., the production of concrete. The secondary treatment removes suspended and dissolved biological material such as human wastes, food wastes, soaps, and detergents by using aquatic microorganisms which break down the organic substances [24]. The bacteria and protozoa are often removed by separation before the water is discharged to the tertiary treatment. This final step often involves chemical or physical disinfection or microfiltration using e.g., sand filtration [25]. The resulting water output can be used for irrigation or discharged into streams, rivers, lagoons, or wetland [26]. If sufficiently clean it may also be used to recharge groundwater reservoirs. Some wastewater treatment plants may use additional steps such as nitrogen or phosphorous removal as well as biological nutrient removal. Disinfection is often done by UV radiation, ozone, chlorine, or sodium hypochlorite to remove bacteria, viruses, and other pathogens [27].

18.3 Measurement of toxicity in water treatment plants

Measurements of wastewater pollution using the ECOTOX system (cf. Chapter 10: Ecotox, this volume) with the bioindicator *Euglena gracilis* showed strong differences in the toxicity which depend on the dilution by rain water, composition of the wastewater of municipal and industrial effluents. Fig. 18.1 shows a typical example of results with water samples from the water treatment plant in Schwabach, Bavaria, Germany taken after the sedimentation stage. The ECOTOX system was programmed to make automatic dilutions of the sample. The first dilution is 1:1 with the cell suspension. The following dilutions contain 25%, 12.5%, 6.25% and 3.125% of the wastewater sample, respectively (Fig. 18.1). Motility shows the highest inhibition of up to about 30%, and even at the lowest concentration of the wastewater there was a remarkable inhibition. The percentage of upward swimming cells, the form factor, and swimming velocity were likewise inhibited.

In most cases the *r*-value for gravitaxis in *E. gracilis*, which is a statistical measure for the precision of orientation, shows the highest inhibition by wastewater and many pollutants [28,29]. Fig. 18.2 shows the effect of increasing concentrations of

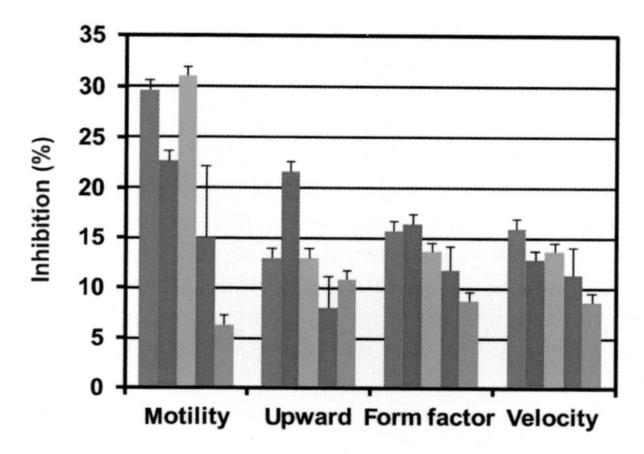

Figure 18.1 Inhibition of motility (percent motile cells), upward swimming (percent of cells swimming within ± 90 degrees from the vertical upward direction), form factor (deviation from a circular cell form), and swimming velocity by wastewater from the sedimentary stage of the water treatment plant in Schwabach, Bavaria, Germany at decreasing wastewater concentrations: 50% (*dark blue*), 25% (*red*), 12.5% (*green*), 6.25% (*purple*) and 3.125% (*light blue*). Bars indicate mean values ($n = 3$) with error bars.

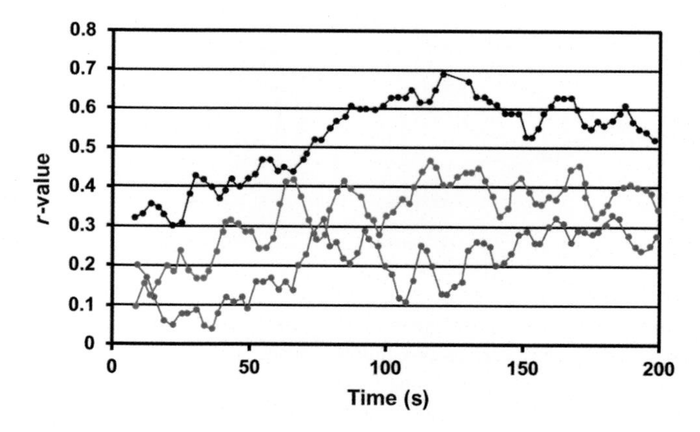

Figure 18.2 Kinetics of the precision of gravitactic orientation over time in a control (*black symbols and line*), in the presence of 50% (*red symbols and line*) and 25% (*green symbols and line*) of wastewater from the intermediary basin of the water treatment plant in Nürnberg, Bavaria, Germany.

wastewater from the intermediary basin of the water treatment plant in Nürnberg, Bavaria, Germany over time. Immediately after being transferred to the observation chamber of ECOTOX the gravitactic orientation as indicated by the *r*-value is rather low due to the turbulences during pumping. Over time the precision of orientation increases and reaches a stable value after about 100 s. Mixing the cell culture with 50% wastewater decreases the precision of orientation by about 65% and even a concentration of 25% of wastewater added to the cell culture significantly reduces the *r*-value.

When we look at the complete sequence of cleaning steps from the crude wastewater input to the release of the purified water into the environment we can follow the progress and effectiveness by taking samples at each step and analyze them using ECOTOX (Fig. 18.3). Samples of the crude wastewater inflow into the wastewater treatment plant in Herzogenaurach, Bavaria, Germany—with contributions of both industrial and domestic sewage—inhibits most parameters measured with ECOTOX by about 40% or more when the *Euglena* culture is diluted 1:1 with the sample (Fig. 18.3A). Further dilutions up to 1:16 decrease the inhibition by more than half, but even this low concentration of wastewater has a significant inhibition on motility, compactness, upward swimming, and the precision of gravitactic orientation (*r*-value). After the physical treatment (sedimentation of coarse and fine

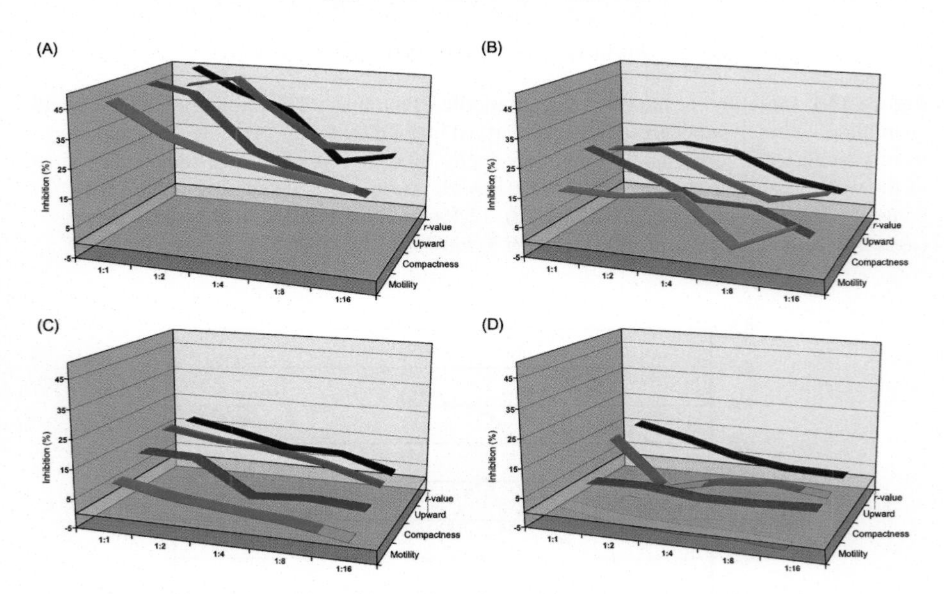

Figure 18.3 Inhibition (in percent of the controls) of motility, compactness, upward swimming, and gravitactic orientation (*r*-value) by crude wastewater (A), after physical treatment (B), biological treatment (C), and by the purified output water after chemical treatment (D) from the wastewater treatment plant in Herzogenaurach, Bavaria, Germany. Negative inhibition (augmentation of an endpoint) is shown in lighter colors.
Source: Based on data from Ahmed H. Biomonitoring of aquatic ecosystems [PhD thesis]. Erlangen-Nürnberg: Friedrich-Alexander-Universität; 2010.

particles and removal of floating substances) the measurements show an inhibition of about half of that of the crude wastewater (Fig. 18.3B), but also after this step the highest dilution results in an inhibition of all four endpoints. Samples taken after the next step (biological treatment) show a much lower inhibition of between 10% and 19% for the 1:1 dilution and the highest dilution causes inhibition by only a few percent (Fig. 18.3C). Motility is even slightly (but not significantly) higher than in the controls. Samples of the purified output after chemical treatment from the wastewater treatment plant result in inhibition values between 0.3% and 15% at the highest concentration which decrease to 1.5% (Fig. 18.3D). Motility is even enhanced by almost all dilutions with the exception of a dilution of 1:1.

18.4 Monitoring of industrial wastewaters

As indicated above, industrial effluents may require additional treatment before they are transferred to the municipal wastewater treatment plants [31]. Depending on the pollutant substance, nitrification [32], electrochemical oxidation [33], coagulation-flocculation [34], and membrane filtration [35], e.g., are being used.

Wastewater samples from a company producing catalytic converters for cars were obtained from the Bavarian office for the Environment (LFU, Munich, Germany) and tested with the ECOTOX instrument using motility, upward swimming, velocity, and compactness as endpoints at different concentrations. From the results EC_{50} curves have been calculated (Fig. 18.4). These three-way catalytic converters combine carbon monoxide with oxygen and unburned hydrocarbons to produce carbon dioxide and water. In addition, they reduce nitrogen oxides [36]. The support for the catalyst usually consists of a honeycomb ceramic monolith which is covered by a washcoat made from aluminum oxide, titanium dioxide, or silicon dioxide [37], which in turn is the basis of the catalyst consisting of a mixture of precious metals [38]. Platinum is the most active catalyst and widely used. From this list the ingredient of the wastewater can be deduced, mainly containing heavy metals.

For many years industrialized and developed countries have been separating and recycling household and industry garbage [39]. In Germany glass is recycled in two ways. One is the reusage of cleaned glass which is common practice for beer bottles and yoghurt containers. The other option is the collection of glass in containers located in the municipalities in which glass is separated according to its color (*white, brown, green*). In Germany about 85% of the glass is recycled [40]. This is cleaned, sorted, broken, and molten to form new containers. Paper and cardboard are collected in central containers or in individual containers in each household. In Germany 22.6 million tons of paper and cardboard were produced in 2011. For this purpose 15.3 million tons of scrap paper and cardboard have been used. In the United States about 65% of paper was recycled in 2012 [41]. After sorting the recycled paper it is deinked and liquefied to produce new paper [42]. Organic material is either recycled or collected to be converted to compost in centralized

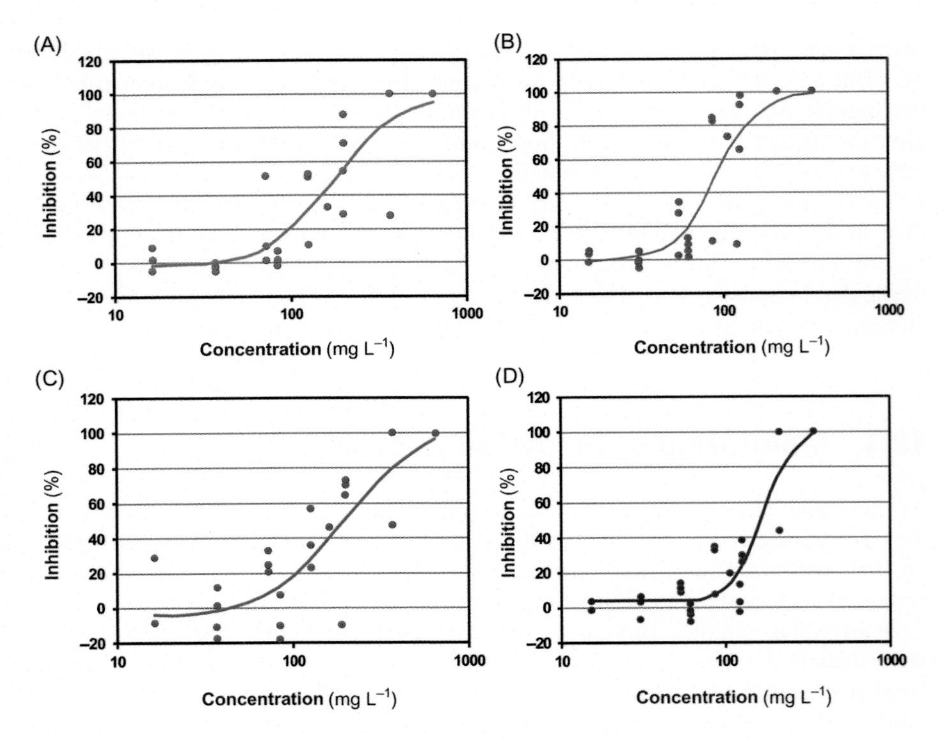

Figure 18.4 Inhibition (in percent of the controls) of four endpoints by different concentrations of wastewater from a car catalytic converter producing plant using *Euglena gracilis*. The calculated EC_{50} values were for (A) motility 113.3 mg L^{-1}; (B) upward swimming 89.7 mg L^{-1}; (C) velocity 124.5 mg L^{-1}; and (D) compactness 166.9 mg L^{-1}, respectively.
Source: Based on data from Ahmed H. Biomonitoring of aquatic ecosystems [PhD thesis]. Erlangen-Nürnberg: Friedrich-Alexander-Universität; 2010.

facilities [43]. Metal is collected in containers or by central agencies in the case of larger appliances. After disassembly and cleaning and sorting the reclaimed metal is recycled. Most of these recycling activities result in little pollution (for pollution from paper pulp industry see Chapter 21: Environmental monitoring using bioassays, this volume). In contrast, recycling of reclaimed plastic may result in considerable pollution problems. In Germany plastic and compound wrappings are collected in containers or plastic bags. This material is sorted mainly by hand and clean objects are recycled [44].

Studies show that microplastics are already widespread in both inland and marine aquatic ecosystems. However, most of the material is incinerated [45] which may result in considerable air pollution but results in the usage of the material in energy production. Unfortunately, a large fraction of the plastic wastes ends up in the oceans where it is broken into micro- and nanoparticles by sand, waves, and solar UV radiation [46]. The remaining waste material is usually collected and stored in garbage dumps. Care has to be taken that pollutants can neither reach the groundwater nor gaseous products reach the air [47].

Samples from a special wastewater treatment plant for chemical and solid wastes have been obtained from the Bavarian office for the Environment (Landesamt für Umwelt, LFU, Munich, Germany) and have been analyzed using Ecotox (Fig. 18.5). EC_{50} values for inhibition of gravitactic orientation, motility, upward swimming, and velocity were found in the range of 67–189 mg L^{-1} with gravitactic orientation being impaired with the lowest concentration and upward swimming with the highest.

Another example is the inhibition of several endpoints by wastewater from a textile producing plant (Fig. 18.6). Motility was inhibited by 50% at a concentration of 200 mg L^{-1}, which corresponds to a dilution of 5 times of the wastewater sample. Wastewater samples of an industrial services plant impair motility, compactness, upward swimming, and gravitactic orientation (*r*-value) (Fig. 18.7).

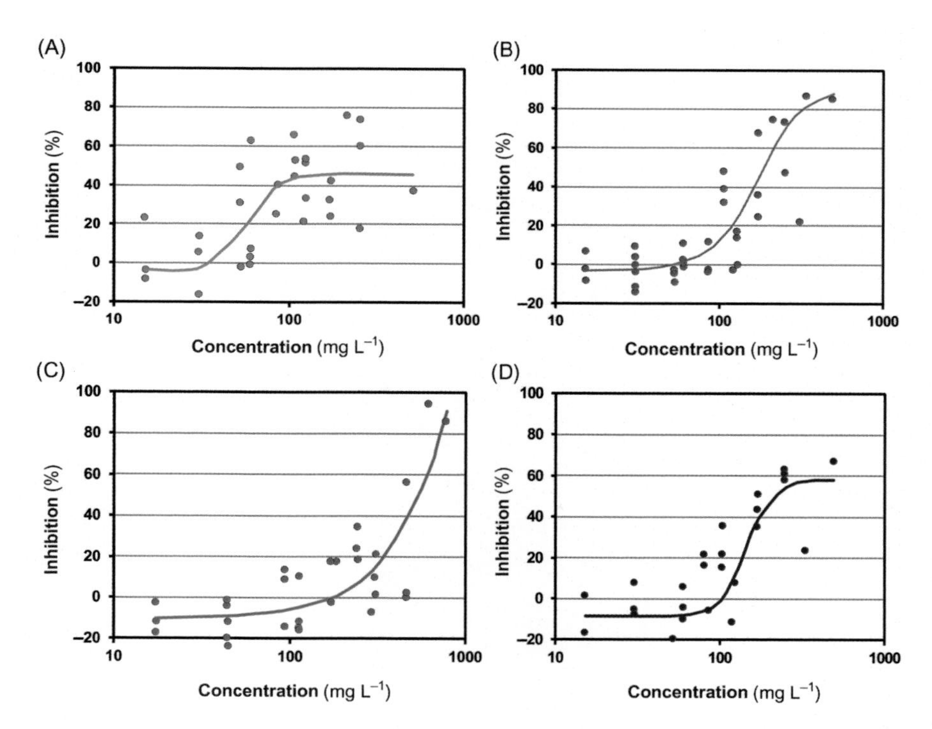

Figure 18.5 Inhibition (in percent of the controls) of four endpoints by different concentrations of wastewater from a special waste treatment plant using *Euglena gracilis*. The calculated EC_{50} values were for (A) gravitactic orientation (*r*-value) 57.2 mg L^{-1}; (B) motility 174.3 mg L^{-1}; (C) upward swimming 189 mg L^{-1}; and (D) swimming velocity 145.5 mg L^{-1}, respectively.
Source: Based on data from Ahmed H. Biomonitoring of aquatic ecosystems [PhD thesis]. Erlangen-Nürnberg: Friedrich-Alexander-Universität; 2010.

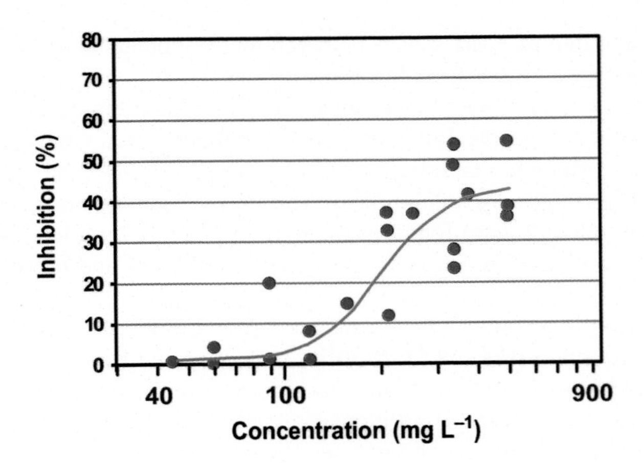

Figure 18.6 Inhibition (in percent of the controls) of the motility by wastewater samples from a textile producing industry at increasing concentrations. The calculated EC_{50} value is 200 mg L^{-1}.
Source: Based on data from Ahmed H. Biomonitoring of aquatic ecosystems [PhD thesis]. Erlangen-Nürnberg: Friedrich-Alexander-Universität; 2010.

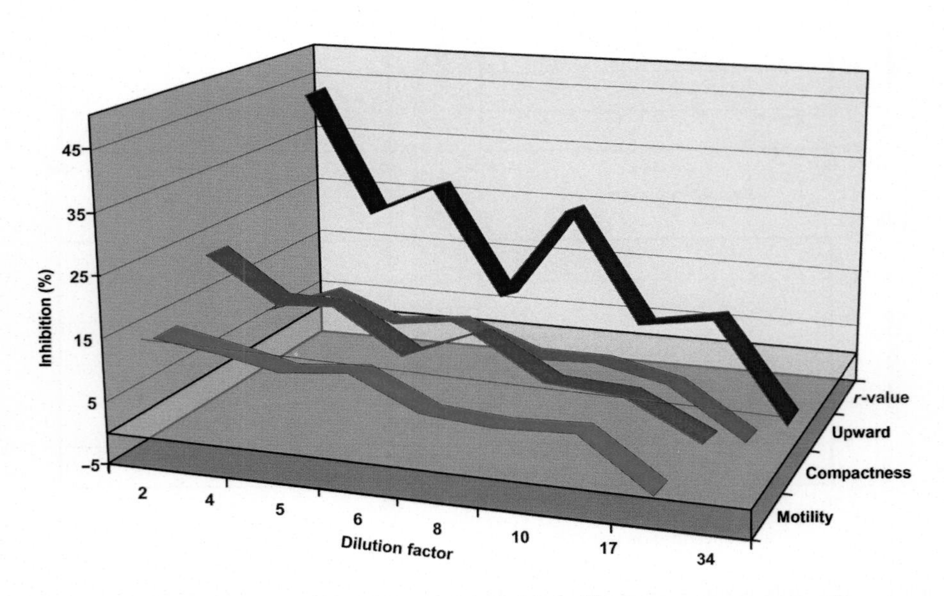

Figure 18.7 Inhibition (in percent of the controls) of motility, compactness, upward swimming, and gravitactic orientation (*r*-value) by wastewater from an industrial services plant. Negative inhibition (augmentation of an endpoint) is shown in lighter colors.
Source: Based on data from Ahmed H. Biomonitoring of aquatic ecosystems [PhD thesis]. Erlangen-Nürnberg: Friedrich-Alexander-Universität; 2010.

Most endpoints are only very little affected by the wastewater with the exception or gravitactic orientation. However, this endpoint shows a large variability. This example indicates that certain endpoints are more or less affected than others by specific pollutants. Wastewater samples from a technical paper production also result in little inhibition (Fig. 18.8). The same is true for wastewater samples taken from a car recycling plant (Fig. 18.9). Also here the gravitactic orientation is affected most. This may be due to the high content of heavy metals in the mixture of pollutants. Heavy metals have been found to affect especially gravitaxis [48–50].

During dying of textiles only 70%–90% of the dyes react with the fibers. In order to warrant high quality, used dyes are not recycled but leave the process as wastewater [51]. The Remazol family of dyes can induce allergies and respiratory problems [52]. Wastewater from a textile dying company had to be further diluted before analysis since even a concentration of 12.5% inhibited all four endpoints (motility, upward swimming, compactness, and velocity) in the ECOTOX measurements (Fig. 18.10). One option to clean the Remazol wastewater is by electrochemical treatment [53]. This results in a 90% reduction in the color absorption. However, even this treated wastewater resulted in an almost 70% inhibition of the motility and a 40% inhibition of velocity in *E. gracilis*.

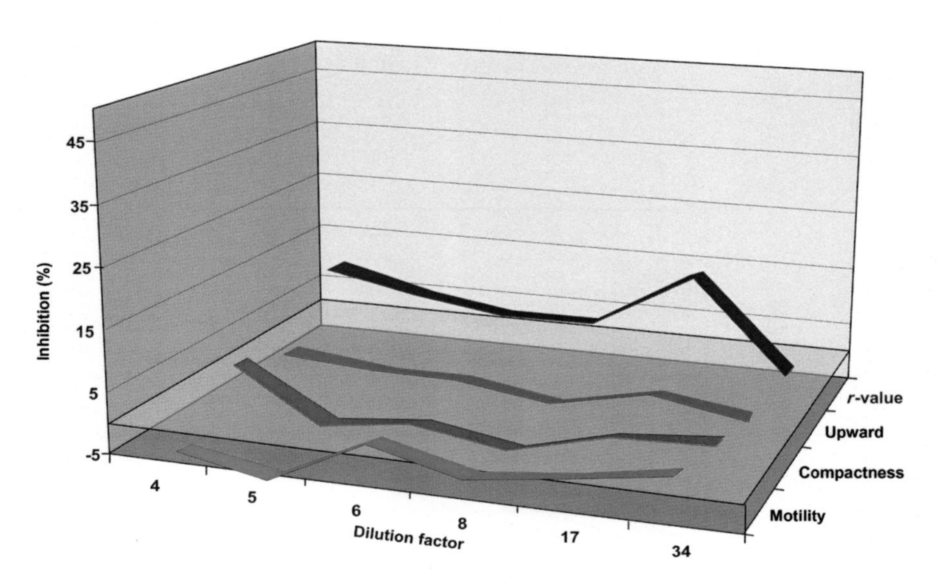

Figure 18.8 Inhibition (in percent of the controls) of motility, alignment, upward swimming, and gravitactic orientation (*r*-value) by wastewater from a technical paper plant.
Source: Based on data from Ahmed H. Biomonitoring of aquatic ecosystems [PhD thesis]. Erlangen-Nürnberg: Friedrich-Alexander-Universität; 2010.

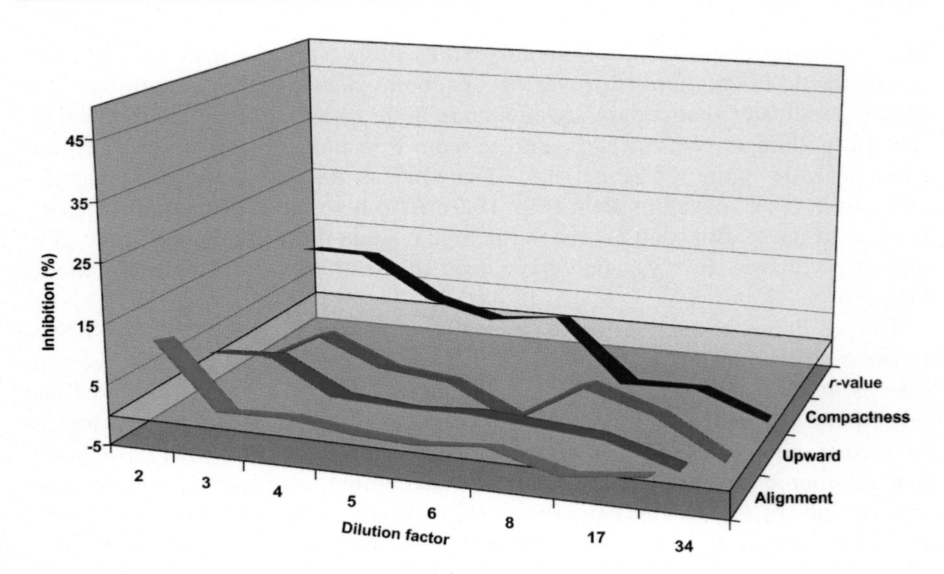

Figure 18.9 Inhibition (in percent of the controls) of alignment, upward swimming, compactness, and gravitactic orientation (r-value) by wastewater from a car recycling plant. *Source*: Based on data from Ahmed H. Biomonitoring of aquatic ecosystems [PhD thesis]. Erlangen-Nürnberg: Friedrich-Alexander-Universität; 2010.

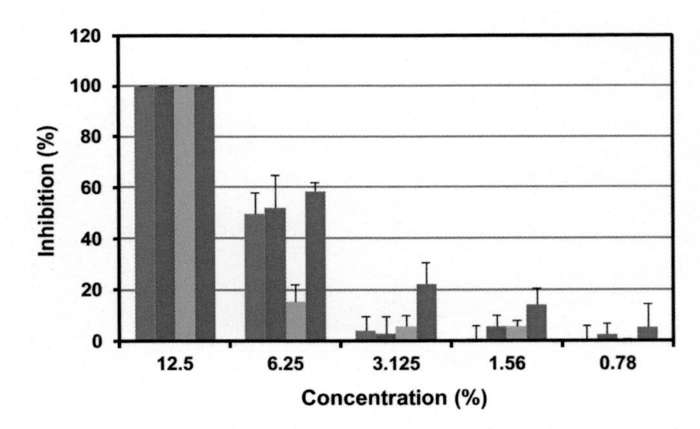

Figure 18.10 Inhibition (in percent of the controls) of motility (*blue bars*), upward swimming (*red bars*), compactness (*green bars*), and swimming velocity (*purple bars*) by a wastewater from a textile dying company containing Remazol. Values are means of 5 independent measurements and standard error. Unpublished data by C. Streb, Erlangen.

18.5 Comparison of bioassays

Wastewater samples have been provided by the Laus Ltd. company (Kirrweiler, Germany) for comparison. The samples had anonymous numbers. Fig. 18.11 shows the inhibition of the gravitactic orientation of *E. gracilis* measured with the ECOTOX system. Concentrations of 50% and 25% resulted in a strong inhibition of the *r*-value, while a concentration of 12.5% even caused an augmentation of the precision of orientation. Comparative measurements by the Laus company using the *Daphnia* test according to DIN 38412 [54] showed a G value of 1 which indicates no inhibition was determined by an undiluted sample. Table 18.1 shows the results of several tests carried out by Laus Ltd. using a *Daphnia magna* test according to DIN 28412 part 30, a bioluminescence test using *Vibrio fischeri* according to ISO 11348 [55] (cf. Chapter 12: Bioluminescence systems in environmental biosensors, this volume), and an algal growth test (using *Desmodesmus subspicatus* var. Chodat according to DIN 28412 L 33) [56], and compares them with those obtained with the ECOTOX instrument using the endpoints motility, upward swimming compactness, and swimming velocity.

Several technical oils have been found to impair various endpoints measured with the ECOTOX instrument. Methyl ester from rape is an additive for diesel fuel. Fig. 18.12 shows the inhibition of gravitactic orientation of *E. gracilis*. Plantohyd 40N is an oil used in hydraulics based on rape methyl ester. J20C is a conventional mineral oil derived from fossil oil [57], and Econa E46 contains synthetic ethers and is obtained from plant oils. The universal gear oil for tractors Bio-Hy-Gard is based on rape oil and contains natural esters from other plant oils. It has been distinguished as excellent by the Bavarian ministry of agriculture. However, in Ecotox the *r*-value was inhibited even at concentrations of 1.375 mg L^{-1}. Likewise, the polydisperse silicon oil Baysilone strongly affected the orientation parameters and

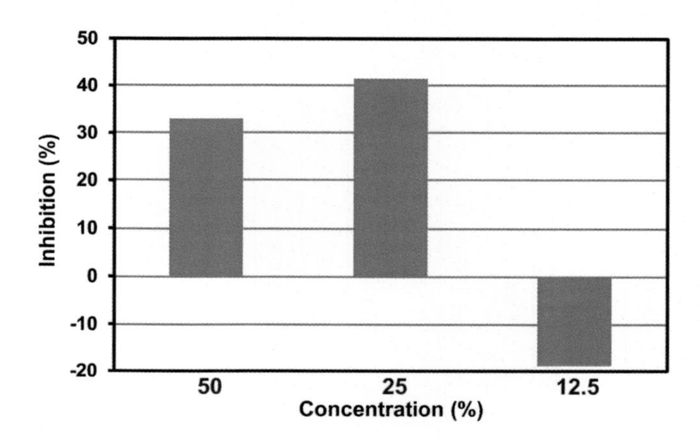

Figure 18.11 Inhibition of gravitactic orientation (*r*-value) by a wastewater sample (04030201R) provided by the company Laus Ltd. (Kirrweiler, Germany). Unpublished data by C. Streb, Erlangen.

Table 18.1 Comparison of G-values obtained by several bioassays (Laus Ltd.) with those obtained with the ECOTOX instrument using four different endpoints. Ecotox: unpublished data by C. Streb, Erlangen

	Test by Laus Ltd.		G-value measured with Ecotox			
Sample	**Test**	**G value**	**Motility**	**Upward**	**Compactness**	**Velocity**
04031703W	*Daphnia*	1	2	2	2	2
	Bioluminescence	≤2				
04031801J	Bioluminescence	4	2	2	2	n.a.
04030201R	*Daphnia*	1	n.a.	n.a.	2	n.a.
04030401J	Bioluminescence	≤2	2	2	2	n.a.
04031901R	Bioluminescence	≤2	2	2	2	n.a.
04031702W	Algae	1	2	8	2	2
	Daphnia	1				
04031001W	Algae	1	2	2	2	2
	Daphnia	1				
	Bioluminescence	≤2				
04031704W	*Daphnia*	1	2	2	2	2
	Bioluminescence	≤2				

G-values of 1 indicate that no result can be obtained with the automated biotests. A G-value of 2 indicates a response at a concentration of 50%, a G-value of 3 a response at 33% etc.

augmented the gravitactic orientation even at the lowest tested concentration (1.375 mg L^{-1}) by 40%. Also the chain saw oil Plantotac N, based on rape oil, inhibited the *r*-value by 50% at all measured concentrations down to 1.357 mg L^{-1}.

The results indicated above have been compared with the cress germination test [58]. For this purpose 100 μL of the test oil was spread on filter paper in a Petri dish. After addition of 1 mL water 50 garden cress seeds (*Lepidium sativum* L.) were distributed on the substratum. For each oil sample five independent tests were prepared, and after 24, 48, and 72 h the germinated seeds were counted (Fig. 18.13).

Lemna is one of the smallest aquatic angiosperms (Monocotyledonae). It is established as a bioindicator in a test which has been standardized and established in many countries [59]. It has been used to monitor the toxicity of 2,4-dichlorophenol, pharmaceuticals, heavy metals, herbicides, and many other pollutants [60–63].

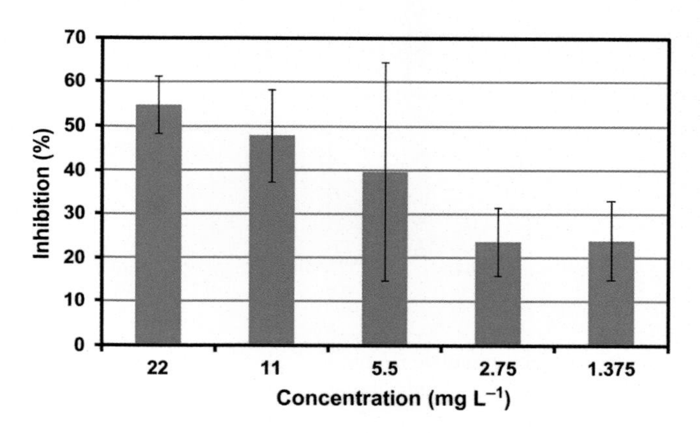

Figure 18.12 Inhibition (in percent of the controls) of the gravitactic orientation (*r*-value) by methyl ester from rape. Values are means of 3 independent measurements with standard error. Unpublished data by C. Streb, Erlangen.

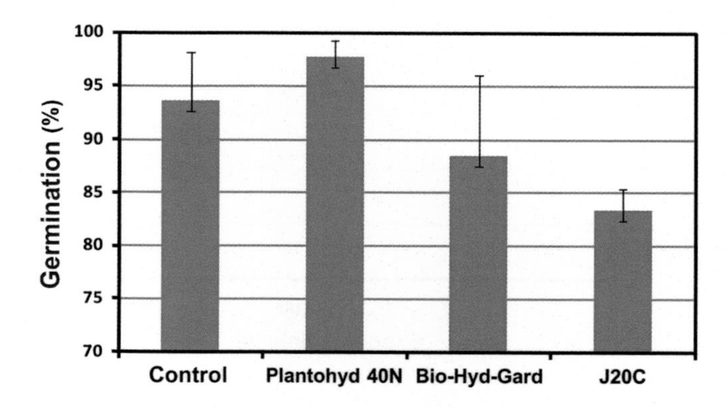

Figure 18.13 Effects of several commercial oils on the germination (in percent) of garden cress (*Lepidium sativum* L.) seeds determined 24 h after seeding. Unpublished data by C. Streb, Erlangen.

Several endpoints have been defined for tests with *Lemna* such as number of roots, root length, frond growth and pigmentation [64,65]. Exposing *Lemna minor* in water with copper granulate affected the number of roots and the root length after 9 days of growth (Fig. 18.14). In comparison, movement parameters of *E. gracilis*

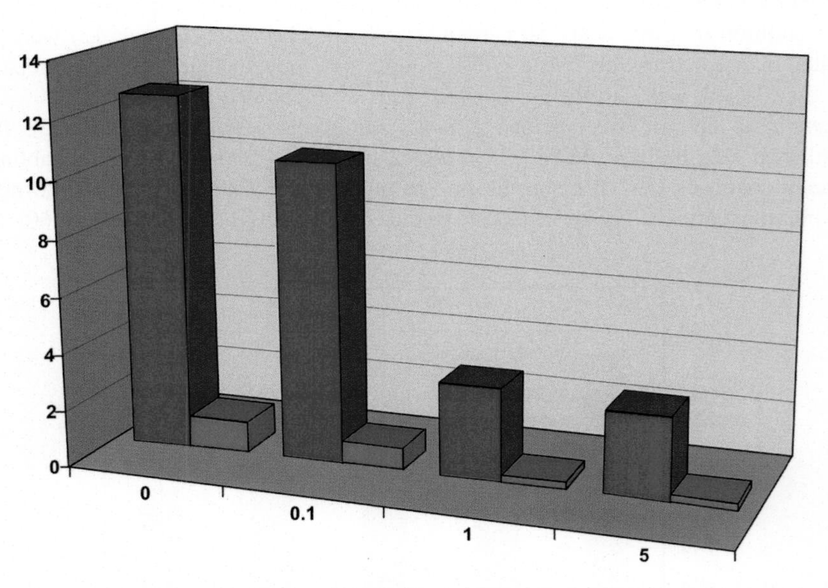

Figure 18.14 Effect of copper granulate (in mg L^{-1}, abscissa) on the root length (*blue* columns [mm]) and number of roots (*red* columns) of *Lemna minor* after 9 days of growth.

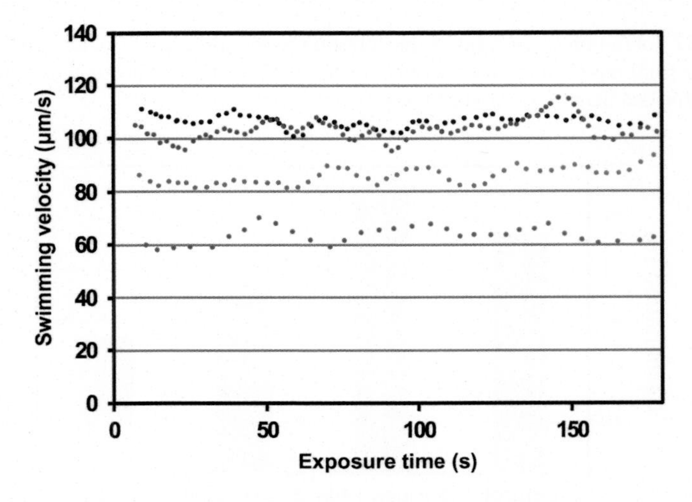

Figure 18.15 Swimming velocity of *Euglena gracilis* when exposed to copper granulate. Control (*black symbols*); 100 mg L^{-1} (*red symbols*); 500 mg L^{-1} (*blue symbols*); and 1000 mg L^{-1} (*green symbols*). Unpublished data by C. Streb, Erlangen.

were impaired at fairly higher concentrations than those impairing root growth and number in *Lemna*. This may be due to a different type of medium for *Lemna* and *Euglena*, which may leach more copper from the granulate in the case of *Lemna*. This is confirmed by the fact that copper salts have been found to strongly affect motility and orientation in *Euglena* [66] (Fig. 18.15).

18.6 Conclusions and outlook

Municipal and industrial wastewaters can have a high toxicity. While in developing countries these are often discharged into natural ecosystems without previous treatment, most developed nations employ water treatment plants using physical, biological, and chemical methods to clean the wastewater. The efficiency of the treatment can be determined using bioassays after each step and compared with the toxicity of the input wastewater. Industrial wastewaters usually have a considerably higher toxicity and need a pretreatment before they are sent to a municipal wastewater treatment plant. Because of their sensitivity, fast response time, and ease of use, these bioassays should be routinely employed to monitor wastewater at each step of the treatment in order to provide an early warning. Comparisons between several types of bioassays demonstrated in most cases good comparison and high sensitivity.

References

[1] Roberts L. 9 Billion? Science 2011;333(6042):540–3.
[2] Searchinger T, Hanson C, Ranganathan J, Lipinski B, Waite R, Winterbottom R, et al. Creating a sustainable food future. A menu of solutions to sustainably feed more than 9 billion people by 2050. In: World resources report 2013–14: interim findings. 2014.
[3] Sutherland WJ, Aveling R, Brooks TM, Clout M, Dicks LV, Fellman L, et al. A horizon scan of global conservation issues for 2014. Trends Ecol Evol 2014;29(1):15–22.
[4] Koff WC, Burton DR, Johnson PR, Walker BD, King CR, Nabel GJ, et al. Accelerating next-generation vaccine development for global disease prevention. Science 2013;340 (6136):1232910.
[5] Motoshita M, Ono Y, Pfister S, Boulay A-M, Berger M, Nansai K, et al. Consistent characterisation factors at midpoint and endpoint relevant to agricultural water scarcity arising from freshwater consumption. Int J Life Cycle Assess 2014;1–12.
[6] Müller Schmied H, Adam L, Döll P, Eisner S, Flörke M, Güntner A, et al., editors. Modelling global freshwater resources using WaterGAP 2.2-model overview, selected results and applications. In: EGU General Assembly Conference Abstracts; 2014.
[7] Bijl DL, Bogaart PW, Kram T, de Vries BJ, van Vuuren DP. Long-term water demand for electricity, industry and households. Environ Sci Pol 2016;55:75–86.
[8] Voulvoulis N, Georges K. Industrial and agricultural sources and pathways of aquatic pollution. In: Impact of Water Pollution on Human Health and Environmental Sustainability. 2015. p. 29.

[9] Liu J-L, Wong M-H. Pharmaceuticals and personal care products (PPCPs): a review on environmental contamination in China. Environ Int 2013;59:208–24.
[10] Tornero V, Hanke G. Chemical contaminants entering the marine environment from sea-based sources: A review with a focus on European seas. Mar Pollut Bull 2016;112 (1):17–38.
[11] Sikder MT, Kihara Y, Yasuda M, Mihara Y, Tanaka S, Odgerel D, et al. River water pollution in developed and developing countries: judge and assessment of physicochemical characteristics and selected dissolved metal concentration. Clean (Weinh) 2013;41(1):60–8.
[12] Pusch R. Waste disposal in rock. Oxford, Boston, MA: Newnes; 2013.
[13] Williams PT. Waste treatment and disposal. Hoboken, NJ: John Wiley & Sons; 2013.
[14] Gupta VK, Ali I, Saleh TA, Nayak A, Agarwal S. Chemical treatment technologies for waste-water recycling—an overview. RSC Adv 2012;2(16):6380–8.
[15] Jiang J-Q, Zhou Z, Sharma V. Occurrence, transportation, monitoring and treatment of emerging micro-pollutants in waste water—a review from global views. Microchem J 2013;110:292–300.
[16] Lionetto M, Caricato R, Calisi A, Erroi E, Giordano M, Schettino T. Ecotoxicological bioassays for treated wastewater monitoring and assessment, 2014.
[17] Gałuszka A, Migaszewski Z, Namieśnik J. The 12 principles of green analytical chemistry and the SIGNIFICANCE mnemonic of green analytical practices. TrAC Trends Anal Chem 2013;50:78–84.
[18] Waters M, Nesnow S, Huisingh JL, Sandhu SS, Claxton L. Short-term bioassays in the analysis of complex environmental mixtures II. Berlin: Springer Science & Business Media; 2012.
[19] Vesilind P. Wastewater treatment plant design. London, UK: IWA Publishing; 2003.
[20] Tran AT, Mondal P, Lin J, Meesschaert B, Pinoy L, Van der Bruggen B. Simultaneous regeneration of inorganic acid and base from a metal washing step wastewater by bipolar membrane electrodialysis after pretreatment by crystallization in a fluidized pellet reactor. J Membr Sci 2015;473:118–27.
[21] Limbach LK, Bereiter R, Müller E, Krebs R, Gälli R, Stark WJ. Removal of oxide nanoparticles in a model wastewater treatment plant: influence of agglomeration and surfactants on clearing efficiency. Environ Sci Technol 2008;42(15):5828–33.
[22] Rhee CH, Martyn PC, Kremer J. Removal of oil and grease in oil processing wastewater. Los Angeles, CA: Sanitation District of Los Angeles County, USA; 1989.
[23] Roccaro P., Franco A., Contrafatto L., Vagliasindi F., editors. Use of sludge from water and wastewater treatment plants in the production of concrete: an effective end-of-waste alternative. In: Proceedings of the 14th International Conference on Environmental Science and Technology Rhodes, Greece; 2015.
[24] Ramalho R. Introduction to wastewater treatment processes. San Diego, CA: Elsevier; 2012.
[25] Juang Y, Nurhayati E, Huang C, Pan JR, Huang S. A hybrid electrochemical advanced oxidation/microfiltration system using BDD/Ti anode for acid yellow 36 dye wastewater treatment. Sep Purif Technol 2013;120:289–95.
[26] Matos C, Pereira S, Amorim E, Bentes I, Briga-Sá A. Wastewater and greywater reuse on irrigation in centralized and decentralized systems—An integrated approach on water quality, energy consumption and CO_2 emissions. Sci Total Environ 2014;493:463–71.
[27] Lee O-M, Kim HY, Park W, Kim T-H, Yu S. A comparative study of disinfection efficiency and regrowth control of microorganism in secondary wastewater effluent using UV, ozone, and ionizing irradiation process. J Hazard Mater 2015;295:201–8.

[28] Streb C, Böcker R, Häder D-P. *Euglena gracilis*: an indicator organism for aquatic toxicity of xenobiotics. Curr Topics Toxicol 2006;3:17–24.
[29] Häder D-P, Jamil M, Richter P, Azizullah A. Fast bioassessment of wastewater and surface water quality using freshwater flagellate *Euglena gracilis*—a case study from Pakistan. J Appl Phycol 2014;26(1):421–31.
[30] Ahmed H. Biomonitoring of aquatic ecosystems [PhD thesis]. Erlangen-Nürnberg: Friedrich-Alexander-Universität; 2010.
[31] Rosenwinkel K-H, Verstraete W, Wagner M, Kipp S, Manig N. Industrial wastewater treatment. In: Activated sludge-100 years and counting. 2014. p. 343-367). IWA Publishing London.
[32] Daverey A, Su S-H, Huang Y-T, Chen S-S, Sung S, Lin J-G. Partial nitrification and anammox process: a method for high strength optoelectronic industrial wastewater treatment. Water Res 2013;47(9):2929–37.
[33] García-Morales M, Roa-Morales G, Barrera-Díaz C, Bilyeu B, Rodrigo M. Synergy of electrochemical oxidation using boron-doped diamond (BDD) electrodes and ozone (O_3) in industrial wastewater treatment. Electrochem Commun 2013;27:34–7.
[34] Amuda O, Amoo I. Coagulation/flocculation process and sludge conditioning in beverage industrial wastewater treatment. J Hazard Mater 2007;141(3):778–83.
[35] Lin H, Gao W, Meng F, Liao B-Q, Leung K-T, Zhao L, et al. Membrane bioreactors for industrial wastewater treatment: a critical review. Crit Rev Environ Sci Technol 2012;42(7):677–740.
[36] Sharma S, Goyal P, Maheshwari S, Chandra A., editors. A technical review of automobile catalytic converter: current status and perspective. In: International conference on energy technology, power engineering & environmental sustainability, strategic technologies of complex environment issues–a sustainable approach Jawaharlal Nehru University, New Delhi, India; 2014.
[37] Wang R. Fabrication and testing of low-temperature catalytically active washcoat materials for next-generation vehicle catalytic converters, 2014.
[38] de Aberasturi DJ, Pinedo R, de Larramendi IR, de Larramendi JR, Rojo T. Recovery by hydrometallurgical extraction of the platinum-group metals from car catalytic converters. Miner Eng 2011;24(6):505–13.
[39] McDougall FR, White PR, Franke M, Hindle P. Integrated solid waste management: a life cycle inventory. Hoboken, NJ: John Wiley & Sons; 2008.
[40] Lüdemann C. Ökologisches Handeln und Schwellenwerte: Ergebnisse einer Studie zum Recycling-Verhalten. ZUMA Nachrichten, 1995;19(37):63–75.
[41] Wikipedia. Papierrecycling. Available from: https://de.wikipedia.org/wiki/Papierrecycling.
[42] Brothers KJ, Krantz PJ, McClannahan LE. Office paper recycling: a function of container proximity. J Appl Behav Anal 1994;27(1):153–60.
[43] Polprasert C. Organic waste recycling: technology and management. London: IWA Publishing; 2007.
[44] Schallenberg G. Plastic recycling process and process for producing plastic materials. Google Patents; 2001.
[45] Huang SJ. Polymer waste management–biodegradation, incineration, and recycling. J Macromol Sci Pure Appl Chem 1995;32(4):593–7.
[46] Perkins S. Environment: oceans yields huge haul of plastic:'Garbage patches' more common and deeper than thought. Sci News 2010;177(7) 8.
[47] Kjeldsen P, Barlaz MA, Rooker AP, Baun A, Ledin A, Christensen TH. Present and long-term composition of MSW landfill leachate: a review. Crit Rev Environ Sci Technol 2002;32(4):297–336.

[48] Fuma S, Takeda H, Miyamoto K, Yanagisawa K, Inoue Y, Ishii N, et al. Ecological evaluation of gadolinium toxicity compared with other heavy metals using an aquatic microcosm. Bull Environ Contam Toxicol 2001;66:231–8.

[49] Stallwitz E, Häder D-P. Effects of heavy metals on motility and gravitactic orientation of the flagellate, *Euglena gracilis*. Eur J Protistol 1994;30:18–24.

[50] Ahmed H, Häder D-P. Rapid ecotoxicological bioassay of nickel and cadmium using motility and photosynthetic parameters of *Euglena gracilis*. Environ Exp Bot 2010;69:68–75.

[51] Yang Z, Wang F, Zhang C, Zeng G, Tan X, Yu Z, et al. Utilization of LDH-based materials as potential adsorbents and photocatalysts for the decontamination of dyes wastewater: a review. RSC Adv 2016;6(83):79415–36.

[52] Isaksson M. Contact urticaria syndrome from reactive dyes in textiles. In: Contact Urticaria Syndrome. 2014. p. 219.

[53] Rocha JHB, Solano AMS, Fernandes NS, da Silva DR, Peralta-Hernandez JM, Martínez-Huitle CA. Electrochemical degradation of remazol red BR and novacron blue CD dyes using diamond electrode. Electrocatalysis 2012;3(1):1–12.

[54] Witt U, Einig T, Yamamoto M, Kleeberg I, Deckwer W-D, Müller R-J. Biodegradation of aliphatic–aromatic copolyesters: evaluation of the final biodegradability and ecotoxicological impact of degradation intermediates. Chemosphere 2001;44(2):289–99.

[55] Peters C, Ahlf W. Validieren, Harmonisieren und Implementieren eines minimalen biologischen Testsets zur Bewertung mariner Wasser- und Sedimentproben. Umweltbundesamt, Forschungsbericht 2003;299(25):261.

[56] Freystein K, Salisch M, Reisser W. Algal biofilms on tree bark to monitor airborne pollutants. Biologia 2008;63(6):866–72.

[57] Lemke M, Fernández-Trujillo R, Löhmannsröbenc H-G. In-situ LIF analysis of biological and petroleum-based hydraulic oils on soil. Sensors 2005;5(1):61–9.

[58] Araújo ASF, Monteiro RTR. Plant bioassays to assess toxicity of textile sludge compost. Sci Agric 2005;62(3):286–90.

[59] Radić S, Stipaničev D, Cvjetko P, Mikelić IL, Rajčić MM, Širac S, et al. Ecotoxicological assessment of industrial effluent using duckweed (*Lemna minor* L.) as a test organism. Ecotoxicology 2010;19(1):216–22.

[60] Cleuvers M. Mixture toxicity of the anti-inflammatory drugs diclofenac, ibuprofen, naproxen, and acetylsalicylic acid. Ecotoxicol Environ Saf 2004;59(3):309–15.

[61] Wang W. Literature review on higher plants for toxicity testing. Water Air Soil Pollut 1991;59(3-4):381–400.

[62] Nitschke L, Wilk A, Schüssler W, Metzner G, Lind G. Biodegradation in laboratory activated sludge plants and aquatic toxicity of herbicides. Chemosphere 1999;39(13):2313–23.

[63] Naumann B, Eberius M, Appenroth K-J. Growth rate based dose–response relationships and EC-values of ten heavy metals using the duckweed growth inhibition test (ISO 20079) with *Lemna minor* L. clone St. J Plant Physiol 2007;164(12):1656–64.

[64] Arts GH, Belgers JDM, Hoekzema CH, Thissen JT. Sensitivity of submersed freshwater macrophytes and endpoints in laboratory toxicity tests. Environ Pollut 2008;153(1):199–206.

[65] Cedergreen N, Abbaspoor M, Sørensen H, Streibig JC. Is mixture toxicity measured on a biomarker indicative of what happens on a population level? A study with *Lemna minor*. Ecotoxicol Environ Saf 2007;67(3):323–32.

[66] Ahmed H, Häder D-P. A fast algal bioassay for assessment of copper toxicity in water using *Euglena gracilis*. J Appl Phycol 2010;22(6):785–92.

Marine toxicology: assays and perspectives for developing countries

19

Therezinha M. Novais Oliveira[1] *and Cleiton Vaz*[2]
[1]University of Joinville Region—UNIVILLE, Joinville, SC, Brazil, [2]Universidade do Estado de Santa Catarina—UDESC, Pinhalzinho, SP, Brazil

19.1 Introduction

Increasing human activities have been changing marine and coastal environments with high loads of pollutants. Examples of these toxicants are oil and fuels [1], nanomaterials [2–4], pesticides [5,6], domestic sewage [7,8], pharmaceuticals [9,10], and industrial wastewater [11]. This pollution could cause impacts on the aquatic ecosystems, unbalancing the environment and reducing aquaculture production.

The coastal regions are highly disturbed by human activities in the continental areas. The impacts are caused by the discharge of diffuse sources of organic and inorganic compounds, nutrients, habitat changes, acidification, and temperature changes, caused by climate change and anthropogenic activities [12]. This condition alerted scientists to the need to develop strategies for a proper managing of these regions. The toxicity bioassays are commonly used tools for detecting impacts in different environments that could help monitoring and to provide data for improving the ecosystem conditions.

In this case, for developing countries it is necessary to establish low-cost and effective methods to detect pollution in marine and estuarine environments. This chapter intents to show simple and cost-effective methods to be used in developing countries for monitoring marine and estuarine water.

19.2 *Artemia* sp. toxicity tests

Artemia sp. are divided into six bisexual species known in a large number of parthenogenic populations. They are characterized by high adaptability to a wide salinity range of 5–250 g L^{-1} as well as to temperatures from 6 to 35°C. They have a short lifecycle, high fecundity, and can be obtained from cysts hatching. Furthermore, they are adaptable for many nutrients and they are non-selective filterers. Thus, *Artemias* are ideal for toxicological studies, providing easy execution, effectiveness, and a high cost benefit ratio, considering the routines in laboratories [13].

Bioassays. DOI: http://dx.doi.org/10.1016/B978-0-12-811861-0.00019-X

Artemia spp. can be used for short- [14] and long-term toxicity tests [15] in order to detect potential hazardous chemical in aqueous samples [16–18], potent pharmaceuticals [19–23], plant extracts [2,24–26], residues of chemicals [27,28], arid environments [29], and recently to assess toxicity properties of nanomaterials [30–38]. *Artemia* spp. represent a suitable and easy to use euryhaline organisms that can adapt to salinities from 10 ppt [18] to hypersaline environments [34].

The acute toxicity test with brine shrimp (*Artemia* sp.) is designed to expose a known number of instar II and III of *Artemia* larvae during 24–48 h to the target chemical substance in an aqueous saline sample. After this time, the number of dead organisms is quantified to determine the lethal toxic effect [39,40].

Artemias could be obtained from hatching commercial cysts in the laboratory. The conditions for hatching are shown in Table 19.1. For hatching and separating the *Artemias* only simple equipment is necessary, as are shown in Fig. 19.1. Erlenmeyers covered with parafilm are recommended for hatching with gentle aeration during 48 h. The water used for this task must be of the same salinity that will be used to perform the bioassay. After this time the suspension of cysts and hatched *Artemias* are transferred to a separation funnel. A light bulb is used to attract the organisms to the bottom of the equipment, because brine shrimp is positive phototactic.

For testing, 10 *Artemias* nauplii are transferred into 100 mL flasks in triplicates. These flasks must be kept in a dark place for 48 h without aeration. Feeding *Artemias* is avoided. It is recommended to use 5–10 different concentrations of the target substance to perform the test and calculate the LC_{50} (48 h). Negative control is made using the same prepared water for hatching. Positive control is made using sodium dodecyl sulfate (SDS). Sensitivity tests are recommended weekly when the test is started to make a control chart with the LC_{50} (48 h) for SDS with at least 20 values (individual values and moving range charts are suggested). The results in the control chart must be within $\pm 2\sigma$. With these results, the concentration of positive control could be defined for the bioassays.

Table 19.1 Hatching conditions to obtain *Artemia* sp. nauplii

Item	Condition
Dilution water	Deionized water reconstituted from commercial marine salt (Instant Ocean, for example)
Hatching time	48 h
Water for hatching	400 mL g^{-1} of cysts
Aeration	Soft
Light	Not relevant
Temperature	$24 \pm 1°C$
Salinity	10 to 40 (depends on the test conditions)
Flask type	Covered Erlenmeyer
Separation	Separating funnel + light

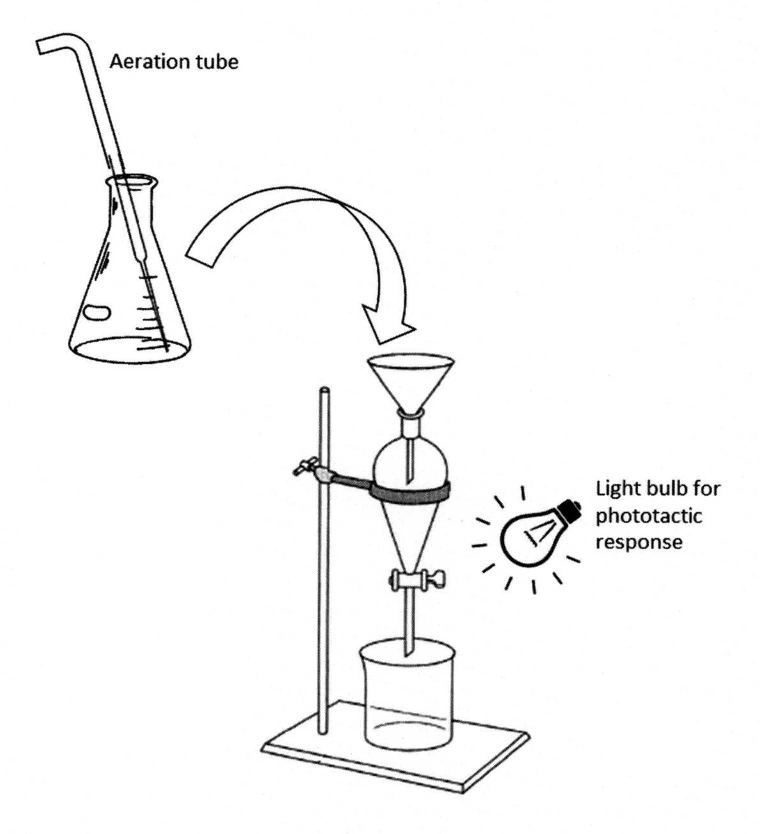

Figure 19.1 System for hatching and separation of instar II and III of *Artemia* larvae.

Water samples from Babitonga Bay, on the north coast of Santa Catarina, Brazil (Fig. 19.2), were studied to detect sites that show acute toxicity in this important estuary.

The study was made to help in identifying healthy areas for aquaculture development. Fourteen sampling sites were chosen to enable a wide coverage of the bay (Fig. 19.3). To collect samples, sanitized 5-L polyethylene bottles were used. All samples were obtained from 50 cm below the surface and the bottles were cooled with ice in an isolated box for transportation to the laboratory.

To perform the acute toxicity tests using *Artemia* nauplii, five different concentrations of Babitonga Bay waters were made (10%, 25%, 50%, 75%, and 100%) plus a negative control. For each concentration, three flasks were made with 100 mL of target dilution and 10 Artemias, after 48 h of hatching time, were added into them. The experimental conditions for the samples and control were as follows: salinity 34 ± 1 and temperature 24 ± 1°C. The negative control was prepared from deionized water with the addition of Red Sea salt salt to obtain a salinity value similar to the samples. After 48 h the living Artemias were quantified again to check

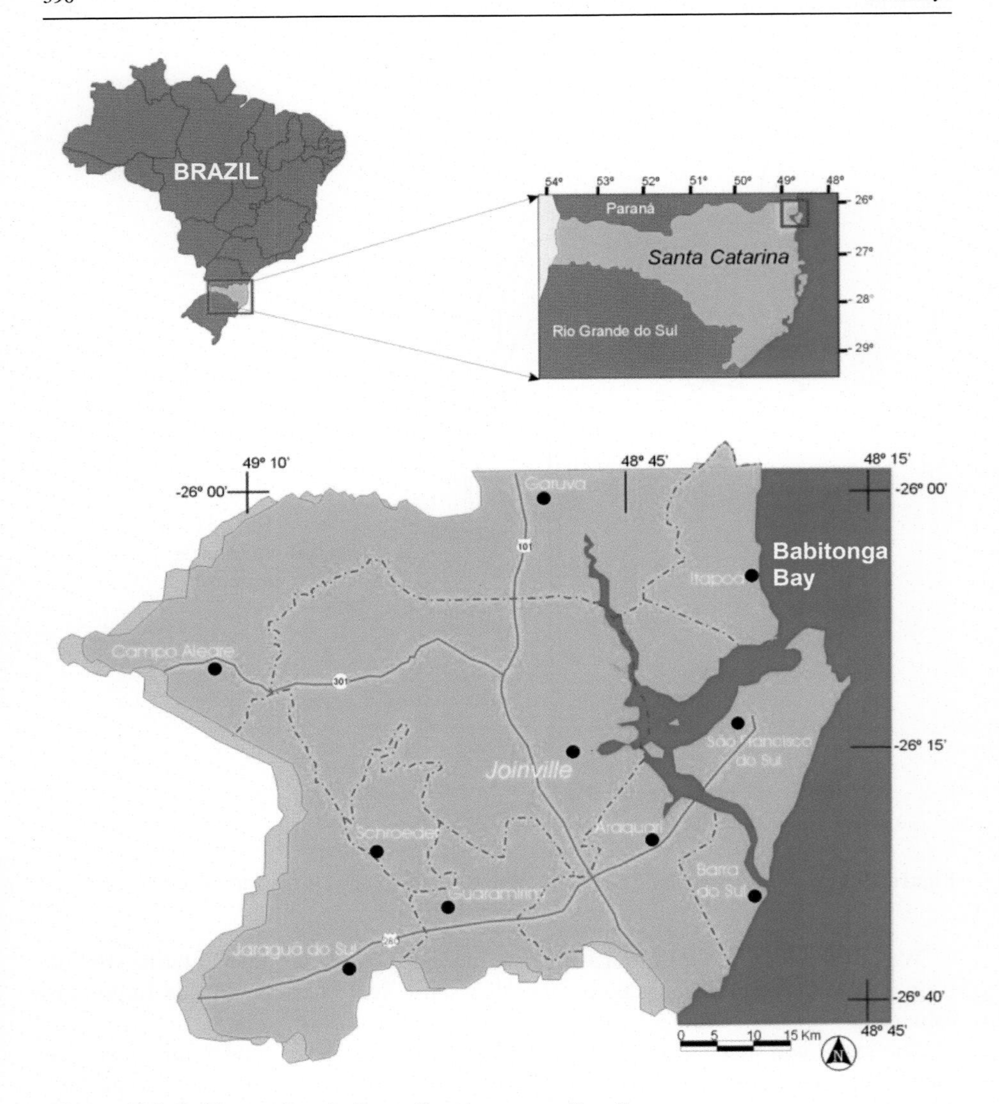

Figure 19.2 Babitonga Bay in Santa Catarina state—Brazil.

the toxicity of the samples. The data analysis was performed with EPA Probit Analysis software. The results indicated a LC_{50} (48 h) of 70.2% (E), 57.7% (D), 41.5% (B), 36.9% (A), and 16.7% (C). The other nine sampling sites did not show toxicity to the organism.

These results suggest a relationship between the anthropogenic activities and the use and occupation of land in the region, due to the presence of a harbor and the discharge of sewage into several rivers that flow to the Babitonga Bay. *Artemia* sp. could be considered as an effective indicator of toxicity, revealing that Babitonga Bay is a threatened system.

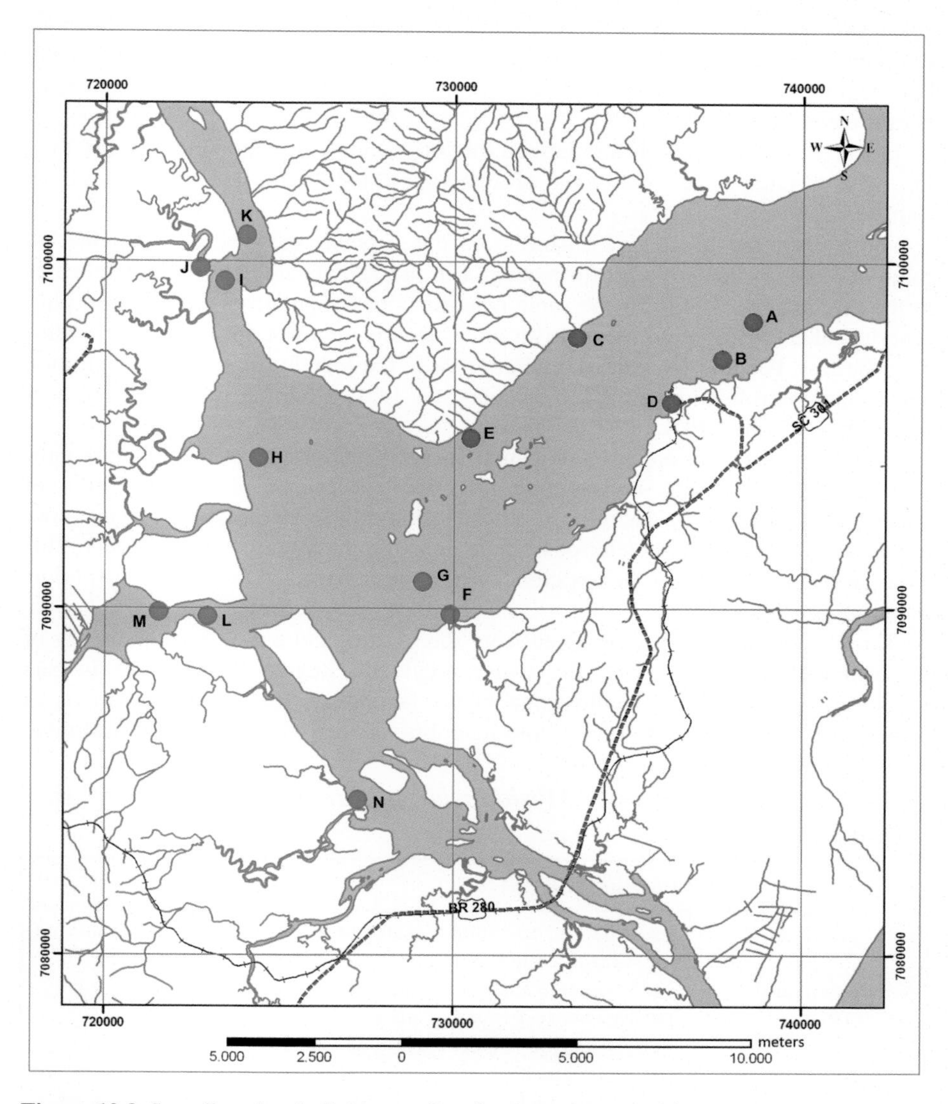

Figure 19.3 Sampling sites in Babitonga Bay for *Artemia* sp. toxicity test.

19.3 *Mysidopsis juniae* toxicity tests

The mysids (*Crustacea: Mysidae*) are small shrimp-like crustaceans. As specific characteristics they have the presence of a marsupium in which to keep the juveniles [41,42].

These organisms are used for toxicity testing because of their short life cycle of about 60 days, their high sensitivity to a wide range of toxicants, and their easy cultivation and handling [43].

Mysids are omnivorous [44] and can eat small particles stuck on their bodies or catch planktonic organisms. The geographical distribution is not entirely known, but they are found in the three biggest oceans at all latitudes and in different temperature conditions [42].

In Brazil, some standards are defined for performing cultivation and toxicity testing with mysids. The conditions of cultivation [41,45,46] are summarized in Table 19.2.

For testing, the procedure is similar to the *Artemia* sp. toxicity test. One- to three-day old mysids are used for testing. Ten individuals are exposed in triplicates to the target substance under the same conditions of temperature, salinity, and photoperiod of the cultivation. A positive and a negative control are necessary to validate the test. Feeding is avoided, and after 96 h, the lethality in each concentration and in the controls is quantified to calculate the LC_{50} (96 h). A scheme of the testing is shown in Fig. 19.4.

One of the uses of copper oxide nanoparticles (CuONPs) is to manufacture antifouling paints of boats, representing an important source of contamination of aquatic ecosystems. The decomposition of these dyes can release soluble ionic copper or as NPs, inducing deleterious effects in different aquatic trophic levels. This study aimed at evaluating the toxicity of nanocopper particles to *Mysidopsis juniae*, a marine organism.

The test organisms were obtained from the existing culture at the Laboratory of Environmental Toxicology at the Universidade da Região de Joinville in São Francisco do Sul city. For cultivation in the laboratory, the recommendations of Table 19.2 were followed. The culture medium was deionized water, reconstituted

Table 19.2 Summary of *Mysidopsis juniae* cultivation in the laboratory

Item	Conditions
Dilution water	Sea water or reconstituted water with commercial marine salt with pH from 7.8 to 8.3
Volume/organism	1 reproductive adult per 100 mL or 50 mL per juvenile
Water change	Totally change 1 time per week
Flask	Glass or plastic—inert to the target substance
Proportion	1:4 (male:female)
Feeding	*Artemia* sp. nauplii with 72 h, enriched with fish and codfish oil (0.1 mL/100 mL of water)
Salinity	33 ± 1
Light	Diffuse
Photoperiod	12 h dark/12 h light
Aeration	Constant and soft
Temperature	$24 \pm 1°C$
Daily controls	Temperature, dissolved oxygen, pH, salinity
Sensitivity control	Monthly. A control chart must be made from dodecyl sodium sulfate as reference substance. LC_{50}(96 h) must be in $\pm 2\sigma$.

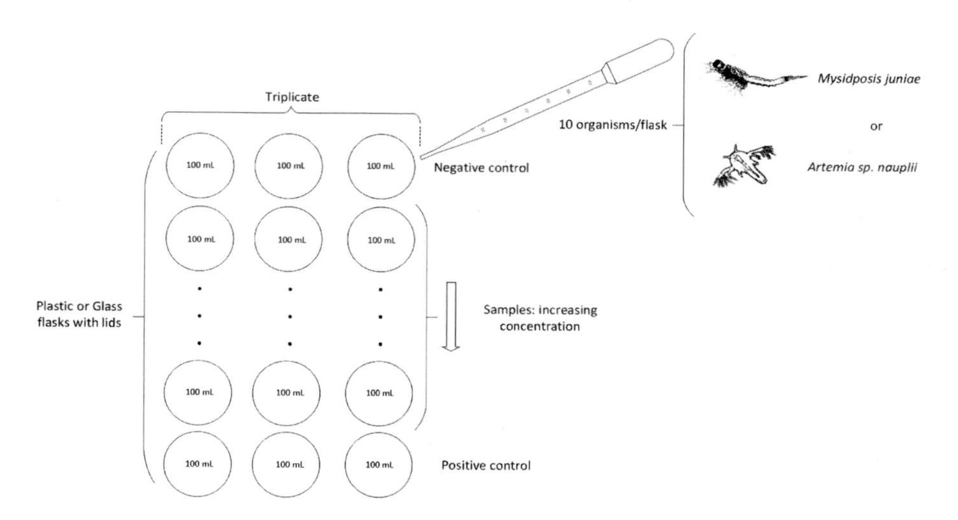

Figure 19.4 Toxic effects of copper oxide nanoparticles on *Mysidopsis juniae*.

with sea salt (Red Sea salt). The cultures were developed in a ratio of one male to four females in 2-L tanks with a total of 20 mysids (population density of 10 adults per liter of water in quantities of 16 females and 4 males). The water was renewed entirely once a week. The photoperiod was 12 h dark/12 h light, the water temperature was maintained at $24 \pm 1°C$, with soft and constant aeration. Salinity was maintained at 33 ± 1. The mysids were fed with *Artemia* sp. that was enriched with fish oil and liver oil from codfish, in amounts accordant with the age of mysids.

For the acute toxicity test, the concentrations of CuONPs used were 0 (control), 0.12, 0.24, 0.6, 1.2, and 2.4 mg L^{-1}, made with deionized water, reconstituted with sea salt. A total of 10 mysids aged between 3 and 5 days old were used in each dilution and control, in triplicate for 96 h. The organisms were fed every 24 h with *Artemia* sp., enriched with fish oil (omega 3), and codfish liver oil. After the test period the number of deaths was quantified. Data analysis was performed with statistical software Minitab.

The acute test showed LC_{50} (96 h) of 9.24 mg L^{-1}, using the probit method for estimation. It is possible to observe in Fig. 19.5 the distribution of the probability percentage of mortality in relation to exposure concentration.

Gao et al. [47] found values of LC_{50} for *Ceriodaphnia dubia* from 0.00214 to 0.048 mg L^{-1} and suggested that higher values of dissolved organic carbon cause lower toxicity, probably related to complexation of the metal (Cu) with dissolved organic compounds. It is suggested that this type of reaction may explain the lower toxicity of the CuONPs in mysids, because the presence of high concentration of chloride in the sample, facilitating the complexation of the metal.

Another possibility for the lower toxicity to *M. juniae* is the aggregation of NPs that occurs rapidly in the presence of saltwater, increasing the particle diameter from 40–500 nm to 1000 nm in 2 days [48], making them bio-unavailable, because

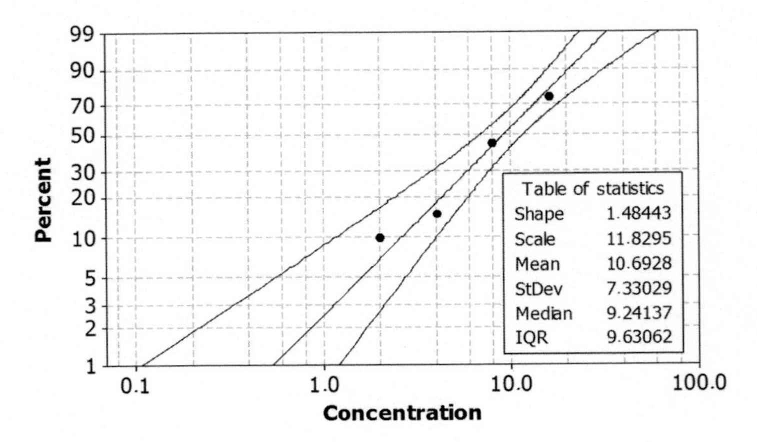

Figure 19.5 Representation of the LC_{50} (96 h) using the software Minitab 14.

then the NPs tend to aggregate [49] and the behavior of the block formed could be distinct from the nanoscale [50].

Then, considering the high number of ships and offshore extraction in oil platforms that may be sources of environmental contamination, when coated with paint containing CuONPs, *M. juniae* could be considered as an alternative to evaluate the toxicity of this kind of compound in marine environments because its sensitivity to the target substance.

19.4 *Perna perna* toxicity test

Perna perna (Linnaeus, 1758) is a mussel found in many places because it is used as source of food and income for coastal populations. The embryos of bivalve organisms are sensitive to a wide variety of compounds and *P. perna* is an organism that is easy to collect in the environment, widely distributed, has a known biology, and has a fast response to pollutants [51]. The toxicity testing is based on world standards [52] and other procedures [53,54].

A number of thirty mussels are recommended in developing one toxicity test and they must be selected according to the state of the shells, choosing only those that are not broken or deformed. All organisms must be washed in running water using a brush and spatula to remove accrustations.

To perform the tests it is suggested to use reconstituted water with commercial sea salt. The release of gametes is made by the thermal cycle method. The mussels are submerged in water baths at different temperatures, one with 10°C and another with 30°C. The organisms are divided into two groups and organized in Petri dishes that are submerged in the reconstituted sea water baths. The first group is placed in the 10°C bath and the second group is placed in the 30°C bath. Each 30 min, the thermal cycle is reached, and the groups are changed to the other temperature.

The released gametes are divided according to the sex of the individuals into beakers with 50 mL of reconstituted sea water. Female gametes are washed with reconstituted sea water and filtered in a 75 μm sieve. The fertilization is made by adding 1 mL of the sperm suspension until more than 90% of the ovules are fertilized. The fertilization is observed by the beginning of cell division using a microscope after the addition of each mL of male gametes into the water with female gametes.

After fertilization of the ovules, the density of eggs must be estimated by counting 1 mL of the suspension diluted in 99 mL of reconstituted sea water using a Neubauer chamber. The obtained mean is used to calculate the egg density. After these calculations, about 40–50 embryos are added in each well of a cell culture plate in triplicates, starting the incubation period.

The target substance must be divided between at least 5 different concentrations plus the positive and negative control. During the test, cultivation cell plates must be kept at 25 ± 1°C during 48 h, or until more than 70% of the embryos of the control are in the D-larvae stage. This condition is achieved by fixing embryos from the negative control. For this step it is necessary that the negative control has at least 8 replicates, to check the stage of the larvae. If the organisms are not in the D-larvae stage, the test must be extended for more time, checking the situation each hour, until the D-larvae stage is achieved.

The tests are completed using 0.5 mL of formalin (0.4% v/v) to steady the embryos for posterior counting. The samples can be counted using an inverted microscope with 40 times amplification. The 100 first found organisms must be analyzed. The normal conditions are the organisms in the D-larvae stage with symmetric closed valves with visceral mass present inside the shell. Any other condition is considered abnormal and caused by the target substance. The scheme for assembling the test is shown in Fig. 19.6.

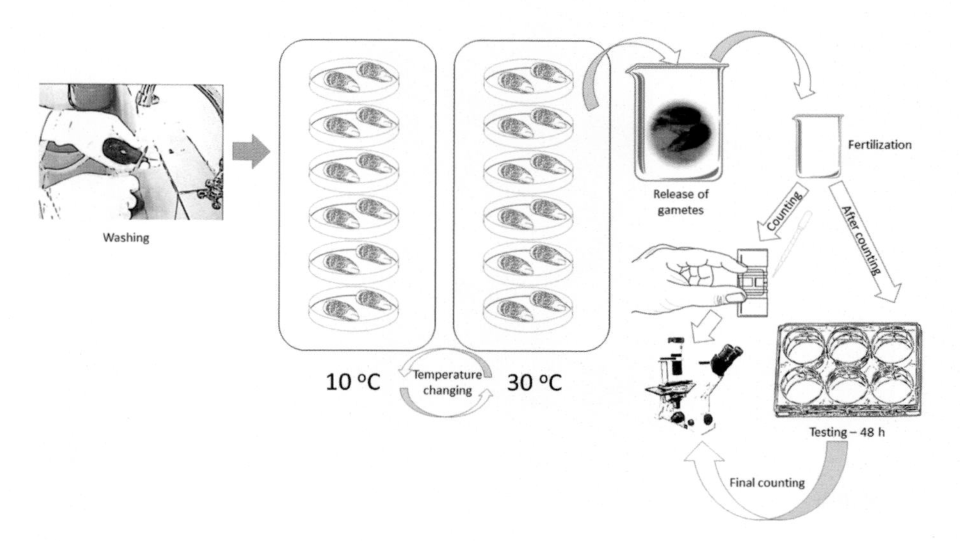

Figure 19.6 Toxicity of Babitonga Bay water samples for *Perna perna* mussel.

Babitonga Bay has six municipalities with a wide variety of economic activities like industries, agriculture, and a harbor with more than 700,000 people living in its surroundings. This situation could impact the water quality of this estuarine environment through the disposal of many pollutants.

Toxicity tests could help to detect the water quality of different places and are useful for planning and managing the natural resources.

P. perna (Linnaeus, 1758) larvae are used as a bioindicator in embryotoxicity tests and were chosen to investigate the estuarine waters of Babitonga Bay, close to aquaculture sites. Three different places were selected for the study. These sites are shown in Fig. 19.7.

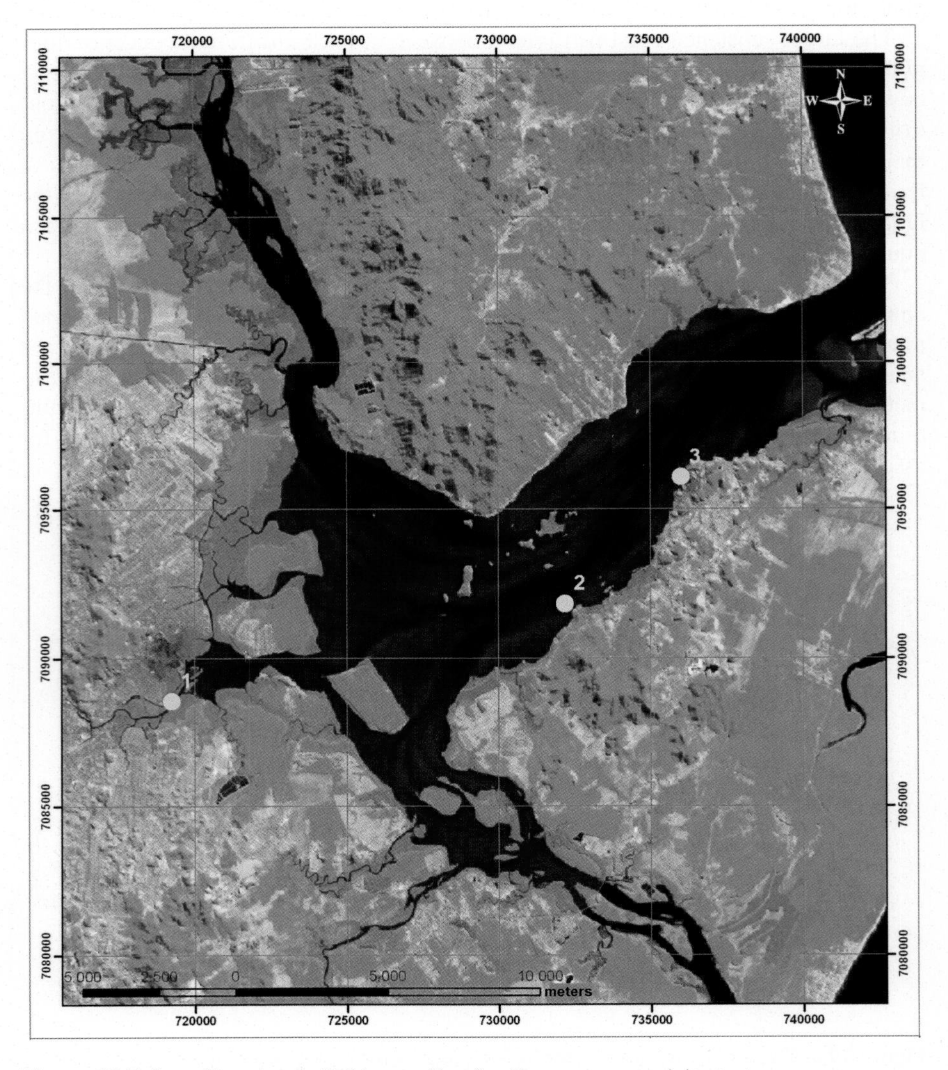

Figure 19.7 Sampling sites in Babitonga Bay for *Perna perna* toxicity test.

The samples were collected at 30 cm depth in 1-L bottles that were placed into an isolated box with ice until transport to the laboratory. After identification, samples were frozen to −20°C. The water samples were defrosted one day before the testing.

A number of 100 organisms without deformities in the shells were selected to promote the spawn. All organisms were washed with a brush in running water. The accrustations were removed with a spatula. Reconstituted water was made using deionized water and Red Sea salt salt to avoid contamination of the samples. This water was filtered with 0.2 μm membrane filters to remove salt crystals. Afterwards, the water was used to spawn the gametes. The gametes were obtained by the thermal cycle technique, exposing the organisms in alternate baths with 10°C and 30°C each 30 min. When the mussels released the gametes, the organisms were separated by sex and added into flasks with 50 mL of water.

The gametes were collected with a Pasteur pipette and mixed into a beaker until more than 90% of fertilization was observed under the microscope. SDS was used as positive control.

The three sites of Babitonga Bay were tested with 5 different concentrations (10%, 25%, 50%, 75% and 100%) (Fig. 19.8). It was not possible to calculate the LC_{50} (48 h), because less than 50% of the embryos showed deformities.

Site P1 is the most polluted, because of the discharge of Cachoeira river that receives industrial and sewage from Joinville city. The P2 site is close to the São Francisco do Sul harbor, a place disturbed by anthropogenic activities. The last site (P3) is known as Laranjeiras, a place less impacted by human activities. Analyzing

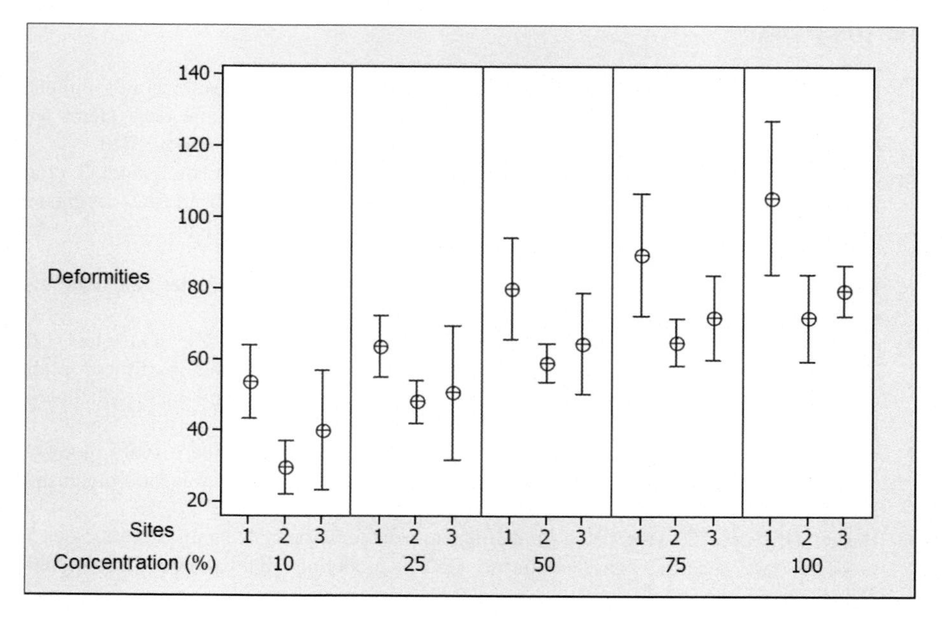

Figure 19.8 Number of organisms with deformities by sites.

the data with a Student t-test, it was possible to detect that the three places show significant differences between them.

The embriotocixity test was useful to detect the toxicity of the natural environment and could be useful to develop an assessment of the entire Babitonga Bay. It is suggested to investigate more places and repeat the tests in different weather conditions in different seasons to check the impact of natural changes.

19.5 Conclusions

Marine environments are impacted by human activities in the whole world and the coastal areas receive the highest load of pollutants because they are close to the continents that discharge sewage and industrial wastewater into the water resources.

The use of methodologies to detect the pollution is necessary for developing strategies that could help to improve the quality of marine environments. In developing countries, the cost of this monitoring is relevant and the use of low cost and effective assays is considered.

In this chapter three simple methods are shown to detect pollution using different invertebrate marine organisms that could be useful since they cover estuarine to hypersaline marine environments. These tests use protocols that are cited in this chapter to facilitate the implementation in the laboratory that could help to evaluate the health of marine and estuarine environments.

References

[1] Barrosa V, Oliveiraa T, Vaza C, Zuppi G, editors. Lead and stable isotopes in sediments of Babitonga Bay: an oil spill case. In: Isotopes in Hydrology, Marine Ecosystems and Climate Change Studies Vol I Proceedings of an International Symposium; 2013.

[2] Iyapparaj P, Revathi P, Ramasubburayan R, Prakash S, Palavesam A, Immanuel G, et al. Antifouling and toxic properties of the bioactive metabolites from the seagrasses *Syringodium isoetifolium* and *Cymodocea serrulata*. Ecotoxicol Environ Saf 2014;103:54–60.

[3] Hadrup N, Lam HR. Oral toxicity of silver ions, silver nanoparticles and colloidal silver–a review. Regulat Toxicol Pharmacol 2014;68(1):1–7.

[4] Lalau CM, de Almeida Mohedano R, Schmidt ÉC, Bouzon ZL, Ouriques LC, dos Santos RW, et al. Toxicological effects of copper oxide nanoparticles on the growth rate, photosynthetic pigment content, and cell morphology of the duckweed *Landoltia punctata*. Protoplasma 2015;252(1):221–9.

[5] Nowell LH, Norman JE, Moran PW, Martin JD, Stone WW. Pesticide toxicity index—a tool for assessing potential toxicity of pesticide mixtures to freshwater aquatic organisms. Sci Total Environ 2014;476:144–57.

[6] Basant N, Gupta S, Singh KP. Predicting aquatic toxicities of chemical pesticides in multiple test species using nonlinear QSTR modeling approaches. Chemosphere 2015;139:246–55.

[7] Ma XY, Wang XC, Ngo HH, Guo W, Wu MN, Wang N. Reverse osmosis pretreatment method for toxicity assessment of domestic wastewater using *Vibrio qinghaiensis* sp.-Q67. Ecotoxicol Environ Saf 2013;97:248–54.
[8] Chen T-H, Chou S-M, Tang C-H, Chen C-Y, Meng P-J, Ko F-C, et al. Endocrine disrupting effects of domestic wastewater on reproduction, sexual behavior, and gene expression in the brackish medaka *Oryzias melastigma*. Chemosphere 2016;150:566–75.
[9] Quinn B, Gagné F, Blaise C. Evaluation of the acute, chronic and teratogenic effects of a mixture of eleven pharmaceuticals on the cnidarian, *Hydra attenuata*. Sci Total Environ 2009;407(3):1072–9.
[10] Hu H, Jiang C, Ma H, Ding L, Geng J, Xu K, et al. Removal characteristics of DON in pharmaceutical wastewater and its influence on the N-nitrosodimethylamine formation potential and acute toxicity of DOM. Water Res 2017;109:114–21.
[11] Bilińska L, Gmurek M, Ledakowicz S. Comparison between industrial and simulated textile wastewater treatment by AOPs–Biodegradability, toxicity and cost assessment. Chem Eng J 2016;306:550–9.
[12] Halpern BS, Walbridge S, Selkoe KA, Kappel CV, Fiorenza Micheli F, Caterina D'Agrosa C, et al. A global map of human impact on marine ecosystems. Science 2008;319:948–52.
[13] Nunes BS, Carvalho FD, Guilhermino LM, Van Stappen G. Use of the genus *Artemia* in ecotoxicity testing. Environ Pollut 2006;144(2):453–62.
[14] Vaz C, Oliveira TMND, Böhm RF, Spitzner EC, Simm M, Barros VG. Use of *Artemia salina* to identify sites with risk of contamination in the waters of Babitonga Bay. Elsevier Ireland Ltd; 2010 [2 Dec 2014]. Available from: http://linkinghub.elsevier.com/retrieve/pii/S0378427410005527.
[15] Manfra L, Savorelli F, Pisapia M, Magaletti E, Cicero AM. Long-term lethal toxicity test with the crustacean *Artemia franciscana*. J Vis Exp 2012;(62):3790.
[16] Manfra L, Tornambè A, Savorelli F, Rotini A, Canepa S, Mannozzi M, et al. Ecotoxicity of diethylene glycol and risk assessment for marine environment. J Hazard Mater 2015;284:130–5.
[17] Wall JM, Wood SA, Orlovich DA, Rhodes LL, Summerfield TC. Characterisation of freshwater and marine cyanobacteria in the Hokianga region, Northland, New Zealand. N Z J Mar Freshw Res, 48. 2014. p. 37–41.
[18] Kokkali V, van Delft W. Overview of commercially available bioassays for assessing chemical toxicity in aqueous samples. TrAC Trends Anal Chem 2014;61:133–55.
[19] Zou Y-F, Ho GTT, Malterud KE, Le NHT, Inngjerdingen KT, Barsett H, et al. Enzyme inhibition, antioxidant and immunomodulatory activities, and brine shrimp toxicity of extracts from the root bark, stem bark and leaves of *Terminalia macroptera*. J Ethnopharmacol 2014;155(2):1219–26.
[20] Kpoviessi BGK, Kpoviessi SD, Ladekan EY, Gbaguidi F, Frédérich M, Moudachirou M, et al. In vitro antitrypanosomal and antiplasmodial activities of crude extracts and essential oils of *Ocimum gratissimum* Linn from Benin and influence of vegetative stage. J Ethnopharmacol 2014;155(3):1417–23.
[21] Oloyede GK, Willie IE, Adeeko OO. Synthesis of mannich bases: 2-(3-Phenylaminopropionyloxy)-benzoic acid and 3-phenylamino-1-(2, 4, 6-trimethoxy-phenyl)-propan-1-one, their toxicity, ionization constant, antimicrobial and antioxidant activities. Food Chem 2014;165:515–21.
[22] Patel MN, Patel CR, Joshi HN, Thakor KP. DNA interaction and cytotoxic activities of square planar platinum (II) complexes with N, S-donor ligands. Spectrochim Acta A Mol Biomol Spectrosc 2014;127:261–7.

[23] Patel MN, Patel CR, Joshi HN, Vekariya PA. Square planar platinum (II) complexes with N, S-donor ligands: synthesis, characterisation, DNA interaction and cytotoxic activity. Appl Biochem Biotechnol 2014;172(4):1846–58.

[24] Carro RT, Isla MI, Ríos JL, Giner RM, Alberto MR. Anti-inflammatory properties of hydroalcoholic extracts of Argentine Puna plants. Food Res Int 2015;67:230–7.

[25] Cock IE, van Vuuren S. The potential of selected South African plants with anti-*Klebsiella* activity for the treatment and prevention of ankylosing spondylitis. Inflammopharmacology 2015;23(1):21–35.

[26] Martins MC, Silva MC, Silva LR, Lima VL, Pereira EC, Falcao EP, et al. Usnic acid potassium salt: an alternative for the control of *Biomphalaria glabrata* (Say, 1818. PLoS One 2014;9(11):e111102.

[27] Marković M, Jović M, Stanković D, Kovačević V, Roglić G, Gojgić-Cvijović G, et al. Application of non-thermal plasma reactor and Fenton reaction for degradation of ibuprofen. Sci Total Environ 2015;505:1148–55.

[28] Jović MS, Dojčinović BP, Kovačević VV, Obradović BM, Kuraica MM, Gašić UM, et al. Effect of different catalysts on mesotrione degradation in water falling film DBD reactor. Chem Eng J 2014;248:63–70.

[29] Metcalf J, Banack S, Richer R, Cox PA. Neurotoxic amino acids and their isomers in desert environments. J Arid Environ 2015;112:140–4.

[30] Pretti C, Oliva M, Di Pietro R, Monni G, Cevasco G, Chiellini F, et al. Ecotoxicity of pristine graphene to marine organisms. Ecotoxicol Environ Saf 2014;101:138–45.

[31] Priyaragini S, Veena S, Swetha D, Karthik L, Kumar G, Rao KVB. Evaluating the effectiveness of marine actinobacterial extract and its mediated titanium dioxide nanoparticles in the degradation of azo dyes. J Environ Sci 2014;26(4):775–82. Available from: doi:10.1016/S1001-0742(13)60470-2.

[32] Becaro AA, Jonsson CM, Puti FC, Siqueira MC, Mattoso LH, Correa DS, et al. Toxicity of PVA-stabilized silver nanoparticles to algae and microcrustaceans. Environ Nanotechnol Monit Manage 2015;3:22–9.

[33] Baumerte A, Sakale G, Zavickis J, Putna I, Balode M, Mrzel A, et al. Comparison of effects on crustaceans: carbon nanoparticles and molybdenum compounds nanowires. J Phys Conf Ser 2013;429(1):012041.

[34] Arulvasu C, Jennifer SM, Prabhu D, Chandhirasekar D. Toxicity effect of silver nanoparticles in brine shrimp *Artemia*. Scientific World J 2014;2014:10.

[35] Clemente Z, Castro V, Jonsson C, Fraceto L. Minimal levels of ultraviolet light enhance the toxicity of TiO_2 nanoparticles to two representative organisms of aquatic systems. J Nanopart Res 2014;16(8):2559.

[36] Ates M, Daniels J, Arslan Z, Farah IO, Rivera HF. Comparative evaluation of impact of Zn and ZnO nanoparticles on brine shrimp (*Artemia salina*) larvae: effects of particle size and solubility on toxicity. Environ Sci Process Impacts 2013;15(1):225–33.

[37] Rodd AL, Creighton MA, Vaslet CA, Rangel-Mendez JR, Hurt RH, Kane AB. Effects of surface-engineered nanoparticle-based dispersants for marine oil spills on the model organism *Artemia franciscana*. Environ Sci Technol 2014;48(11):6419–27.

[38] Ates M, Demir V, Arslan Z, Daniels J, Farah IO, Bogatu C. Evaluation of alpha and gamma aluminum oxide nanoparticle accumulation, toxicity, and depuration in *Artemia salina* larvae. Environ Toxicol 2015;30(1):109–18.

[39] Libralato G. The case of *Artemia* spp. in nanoecotoxicology. Mar Environ Res 2014;101:38–43.

[40] Artoxkit M. Artemia toxicity screening test for estuarine and marine waters, standard operational procedure. Deinze, Belgium: Creasel; 2014.

[41] Badaró-Pedroso C, Reynier MV, Prósperi VAA. Testes de toxicidade aguda com misidáceos – Ênfase nas espécies Mysidopsis juniae e Mysidium gracile (Crustacea: Mysidacea). Ecotoxicologia Marinha – Aplicações no Brasil, São Paulo, Artes Gráficas 2002;123–39.
[42] Mauchline J. The predation of mysids by fish of the Rockall Trough, northeastern Atlantic Ocean. In: Morgan MD, editor. Ecology of Mysidacea, Dordrecht. Netherlands: Springer; 1982. p. 85–99.
[43] Nimmo DR, Hammaker TL. Mysids in toxicity testing – a review. Hydrobiologia 1982;93(1):171–8.
[44] Ortega-Salas AA, Núñez-Pastén A, Camacho M, Humberto A. Fecundity of the crustacean *Mysidopsis californica* (Mysida, Mysidae) under semi-controlled conditions. Rev Biol Trop 2008;56(2):535–9.
[45] ABNT, Toxicidade aguda – Método de ensaio com misidáceos (Crustacea), Brasil: Associação Brasileira de Normas Técnicas, 2011.
[46] CETESB. Água do mar teste de toxicidade aguda com Mysidopsis juniae SILVA, 1979 (crustacea mysidacea). São Paulo: Companhia ambiental do estado de São Paulo; 1992.
[47] Gao J, Youn S, Hovsepyan A, Llaneza VL, Wang Y, Bitton G, et al. Dispersion and toxicity of selected manufactured nanomaterials in natural river water samples: effects of water chemical composition. Environ Sci Technol 2009;43(9):3322–8.
[48] Buffet P-E, Tankoua OF, Pan J-F, Berhanu D, Herrenknecht C, Poirier L, et al. Behavioural and biochemical responses of two marine invertebrates Scrobicularia plana and Hediste diversicolor to copper oxide nanoparticles. Chemosphere 2011; 84(1):166–74.
[49] Vaz C, Melegari SP, Costa CHD, Popovic R, Kleine T, Böhm RFS, et al. The effects of copper oxide nanoparticles on *Mysidopsis juniae*. Toxicol Lett 2011;205:S290.
[50] Peralta-Videa JR, Zhao L, Lopez-Moreno ML, de la Rosa G, Hong J, Gardea-Torresdey JL. Nanomaterials and the environment: a review for the biennium 2008–2010. J Hazard Mater 2011;186(1):1–15.
[51] Carvalho C, Cavalcante M, Gomes M, Faria V, Rezende C. Distribuição de metais pesados em mexilhões (*Perna perna*, L.) da Ilha de Santana, Macaé, SE, Brasil. Ecotoxicol Environ Res 2001;4(1):1–5.
[52] ASTM. Standard Guide for Conducting Static Acute Toxicity Tests Starting with Embryos of Four Species of Saltwater Bivalve Molluscs. American Society for Testing and Materials; 2004. p. E724–98.
[53] Simm M. Avaliação da qualidade da água em amostras provenientes da baía da Babitonga – SC, através de ensaios de embriotoxicidade e de exposição prolongada ao ar, utilizando mexilhão da espécie Perna perna (Linnaeus, 1758) na fase larval e adulta [Mestrado]. Joinville: Universidade da Região de Joinville; 2009.
[54] Zaroni L, Abessa D, Lotufo G, Sousa E, Pinto Y. Toxicity testing with embryos of marine mussels: protocol standardization for *Perna perna* (Linnaeus, 1758). Bull Environ Contam Toxicol 2005;74(4):793–800.

Applications for the real environment

20

Peter-Diedrich Hansen
Technical University of Berlin, Berlin, Germany

20.1 From the bioassay in the laboratory to the field situation

20.1.1 Bridging the lab-field gap

From bioasssays in the laboratory to field bioassays and field monitoring an increasing environmental realism is signified [1,2], but under decreasing experimental control [3]. There are several scaling-up steps required in the transferal of bioassay techniques to the real environment. One interesting approach is the in situ bioassays using "transplanted" organisms [4]. Transplants are animals collected from natural populations or cultures, in clean environments, which are placed for exposure in areas of concern with elevated concentrations of chemicals of concern to characterize exposure and/or effects. The advantage of transplants is a realistic environmental exposure over a well-defined exposure period and a close integration of exposure and effects. This design allows a mapping of chemical gradients, for example plume mapping. Another advantage of these field bioassays is a long-term steady-state concerning the chemical equilibrium and clear defined effect endpoints like estrogenic [1], cholinergic [3,5], or immunicity [3]. There is a quite long history of techniques and development by caging mussels with the classical endpoints byssal thread production and bioaccumulation [6]. For the byssal thread endpoint the prototype cages are individual compartments. The byssal threads are counted on individual mussels. In so-called microcosm cages juvenile and adult mussels are used for the endpoint measurements: bioaccumulation and growth. The juvenile mussels are for growth and the adults for bioaccumulation, and the exposure time is about 90 days. The mussels for bioassays in the field have many advantages; they integrate bioavailable chemicals and bioconcentrate chemicals at the same time, and the bioaccumulation is a link between the environment and the organisms. There were many biological monitoring exercises called "mussel watch" [7,8] guided by the marine conventions of Oslo and Paris (OSPARCOM) and supported by the European Commission DG XII D-1, funded by the EC Environment and Climate Programme and ICES (International Council for the Exploration of the Sea). After progress in field testing a so-called site-specific bioassay with mussels

Bioassays. DOI: http://dx.doi.org/10.1016/B978-0-12-811861-0.00020-6

became of interest because the whole animal as well as the tissues of the mussels integrate biological processes, giving environmentally significant responses and providing the link to population effects. As a result, guidelines after OECD were drafted [6]: bioaccumulation testing with the common mussel, *Mytilus edulis*. It became clear during these standardization procedures that the bioaccumulation kinetics depends as well on the stressor parameters of mussels. The measurements of the phagocytosis [9] generate an "indicator value" to guarantee the quality of the test organisms [6] collected from musselbeds. These quality parameters to check unstressed wildlife organisms for exposure experiments became of relevance with the site-specific bioassay system [10,11] testing onsite effluents.

20.1.2 The on site "WaBoLu Aquatox" monitoring system

The "WaBoLu Aquatox" in situ monitoring system [12,13] has been designed to recognize the sublethal effects on the aquatic ecosystem resulting from discharges of wastewater which contains critical and hazardous substances (i.e., EDC—Endocrine Disrupting Compounds), in particular discharges from industrial sources like effluents from pulp mills [14,15] but also communal wastewater. Fish were exposed in four circular flow-through tanks of the WaBoLu Aquatox filled with relevant concentrations of effluent and river water for one to six months. After 6 months in the exposed male rainbow trout (*Oncorhynchus mykiss*), the vitellogenin expression was measured [13] against a control tank stocked with male fish as well (Fig. 20.1).

The WaBoLu Aquatox is a mobile onsite monitoring system and it permits a realistic performance of in situ bioassays in which organisms from phytoplankton, zooplankton, fish, fish-eggs, and shellfish as well as macrophytes can be exposed,

Figure 20.1 WaBoLu Aquatox—onsite monitoring system.

under flow conditions and corresponding to the prevailing situation in relevant dilution ratios of wastewater and river water. The compact set-up covers a floor area of 6.5 m^2 and is easily transported by truck to be used at discharge sites.

20.2 Real time bioassays for online monitoring in the environment

("Bioassays online" linked to the environment at river sites, buoys, research platforms, and research vessels.)

With a strong vision of a cleaner aquatic ecosystem in rivers and lakes the EU Member States and political decision makers are aiming for a good ecological status as well as a good physical-chemical status of the freshwater and marine aquatic environment. In Europe there is a demand by the Water Frame Work Directive (WFD) to achieve a better understanding of priority pollutants and other physiological stressors entering the EU waters [16], claiming monitoring tools like bioassays, screening devices, and real time bioassays relevant for risk assessment and management [16,17], protecting the aquatic communities, protecting drinking water [18], and serving because of safety issues, risk of freshwater fish [17], and seafood for human consumption. The implementation of bioassays for environmental monitoring on river sites provides relevant data for control and management strategies as well as precautionary measures. Only with real time bioassays and in situ bioassays like the WaBoLu Aquatox [10,12,13] close to the aquatic environment should many of the nonregulated contaminants be detected in the biota and sediment [16], including endocrine disruptors, pharmaceutical products, metal species, biotoxins, microplastics, accumulated nanoparticles from personal care products in fish food, off-flavors in fish by, e.g., cyanobacteria blooms, and pollution from parasites and bacteria [18]. Early warning signals from real time bioassays and other sensing systems like whole-cell biosensors [19] close to the real environment would not only tell us the initial levels of damage, but these signals will also provide us with answers to develop control strategies and precautionary measures with respect to the WFD and the marine conventions OSPARCOM, the convention from Oslo and Paris (1998) which protects and conserves the North-East Atlantic and its resources, and HELCOM, the Helsinki convention (1992) on the protection of the marine environment of the Baltic Sea—including inland waters as well as the waters of the sea itself and the seabed.

The real time bioassays [20,21] are bioassays integrated in a powerful online instrument with a real time detection of a wide range of toxicants and also including their effects (Fig. 20.2).

These so-called Toximeters [22–24] (Algae Toximeter II, Daphnia Toximeter II, ToxProtect64) allow long-term monitoring in the real environment and are designed as an early warning system to detect the entire range of dissolved emerging toxic compounds such as insecticides, pesticides, algicides, herbicides, hydrocarbons, chlorinated hydrocarbons, Polychlorinated biphenyls (PCBs), neurotoxins,

Figure 20.2 Real time bioassay: Algae bbe—Toximeter.

biotoxins, heavy metals, and warfare agents. The toximeters are helpful to detect cyanobacteria and their spills with upcoming off-flavors of fish.

The Algae Toximeter [23] for "on-line monitoring" use green algae, and the end-point measurements are the inhibition of photosynthesis (μg Chl *a*/l) by spectral fluorometry. The cultivation is controlled through active chlorophyll measurements. The *Daphnia* Toximeter [24] allows a continuous visual analysis of the *Daphnia* swimming behavior (speed observation, altitude, turns, and circling movements) by the passing of the light barrier. The swimming activity of fish in the "Fish Biomonitor" (ToxProtect64) [22] is defined by the number of light barriers passing per minute, and fish. The fish are kept under physiologically correct conditions by feeding every 4 h. A similar automatic intake protection system (early warning) with eight free-swimming fish (*O. mykiss*) exposed as an "organismic-sensor" was applied after the SANDOZ disaster in 1986 at the River Rhine [25]. The automatic real time WRC MK III-Fish Monitor [26] was tested at the international monitoring station in Kleve at the river Rhine close to the Netherlands. Over 1 ½ years more than 10 disasters and alarms were activated by disasters of increasing toxicities [25,27,28] and detected by real time bioassay-monitoring-systems. The monitor is based on the well established principle of changes in fish ventilation frequency.

The fish produce small oscillating voltages in the water in which they swim and these are picked up by electrodes in the fish tanks. Every minute eight signals are coming from eight fish. These signals are digitized for data transfer over a 20s period [26].

A whole suite of in-house built instrumental real time bioassay concepts are assembled in the so-called 4H-FerryBox, an autonomous flow-through measuring system for online measurements on vessels, applications on ferries, and/or Cargo-liners equipped with GPS-guided sampling, monitoring, and data transmission. The FerryBox [29,30] is especially developed for continuous deployment on ships routing on sea, on buoys, on research platforms, and monitoring stations on river sites. The concept allows the integration of various bioassays and biosensors [19,31]. Due to the integrated automatic cleaning and antifouling for all of the automatic working bioassays and biosensors, maintenance is kept to a minimum. In connection with a communication module remote control and tele-maintenance there are online programs, series, and position-dependent measurements possible [32].

20.3 Application of the Early-Life-Stage Test (ELST) for in situ testing on, a research vessel

Fish embryos (*Limanda limanda*) were experimentally exposed to ultraviolet-B (UV-B) radiation in a solar radiation simulator and in exposure experiments on board RV Walther Herwig III. The experimental design [33] tried to simulate present and future conditions with increased UV-B exposure due to Northern Hemisphere ozone loss. Employing mainly two scenarios, a reduction to 270 (S1) and to 180 (S2) Dobson units (DU) in single or repetitive exposure of 2, 4, or 6 h. The following endpoints of the exposed fish embryos were studied: mortality [34], malformations [35], the changes in buoyancy of embryos measured as changes in osmolarity of the perivitelline fluid [33], permeability, and genotoxicity [36–38] (DNA unwinding and thymine dimers). Results show a time- and dosage-dependent influence of UV-B radiation on the mortality and the sublethal parameter, i.e., genotoxic effects (DNA damages). The impact of low dosages of UV-B indicates an important ecological threat concerning ecosystem integrity status at the population level [39]. Monitoring the effects of environmental genotoxic substances beside the UV-B exposure has gained increasing importance in recent times. Assessment of environmental samples for the presence of genotoxins (mutagens and carcinogens) has a high priority in protecting inland and coastal waters. Many tools have been meanwhile standardized and harmonized [40,41] but the missing tools are real time assays in genotoxicity. In the field of bio-analytical approaches [42] promising online formats are under development and in progress (Fig. 20.3).

Figure 20.3 In situ bioassays with fish embryos on board of the RV Walther Herwig III. *Source*: Copyright © Thünen/Daniel Stepputtis.

20.4 Applications of bioassays in an extreme environment

The phagocytosis bioassay [9] with the Blue Mussel (*M. edulis*) was modified for an onsite experiment (TRIPLE LUX-B) under microgravity at the International Space Station (ISS). The bioassay in the extreme environment in orbit on the ISS [43] was performed with cryo-conservated hemocytes of *M. edulis*. The Blue Mussels were collected from mussel beds near the Alfred-Wegener-Institute, Wadden Sea Station in List (Island of Sylt). For isolating the hemocytes, the hemolymph (mussel plasma) was taken via a syringe from the posterior adductor muscle of the animals. The procedure is not harmful to the animals and they were put back to their natural environment into the mussel bed. The collected hemolymph of approx. 330 animals per preparation was pooled and kept on ice. The hemocytes were concentrated by centrifugation and frozen at −80°C in their own hemolymph with an addition of 10% polyvenylpyrrholidone (PVP). Usually five batches were made from one hemocyte preparation. Each batch contains the same volume and cell concentration of the pooled hemocytes of approx. 66 animals which were needed for the preparation of the hemocytes for one space experiment. Mussel hemocytes behave in many ways like human macrophages. Therefore this experiment helps to understand the immunotoxic effects after long-mission spaceflights by astronauts. From the phylogenetic point these cells are older and less specialized. From our own observations it seems that mussels are able to sense gravity but nearly nothing is known about the cellular mechanism of gravity sensing. The results obtained at this time indicate that a hemocyte key function (reactive oxygen species (ROS) production) is effected by alterations in gravity [43,44]. It was unknown so far whether the same effect on the whole

Figure 20.4 Real time bioassay and AEC (Advanced Experimental Containment).

animal is able to cause further indirect changes in the ability of hemocytes to produce ROS [44]. As a robot system, and well protected in the flight hardware, the so-called Advanced Experimental Containment (AEC) applied phagocytosis bioassay has contributed to the question of whether this ability is limited to the whole organism or whether isolated hemocytes can sense changes in gravity on their own. For this purpose whole mussels were exposed to hypergravity (1.8 g) on a multisample incubator centrifuge (MuSIC) or to simulated microgravity in a submersed clinostat [44]. In the clinostat experiments new investigations concerning the induced nongravitational effects by shear forces [45] helps to avoid misinterpretation of the results. Therefore it is so important to use a suite of GBF-Facilities to validate the gravitational effects. Hypergravity during parabolic flights led to a decrease in ROS production in isolated hemocytes, whereas the centrifugation of whole mussels on a Short Arm Human Centrifuge (SAHC) did not influence the ROS response as all [44]. This study is a good example how ground-based facility experiments can be used to prepare for an ISS experiment [43], in this case the TRIPLE LUX-B experiment. During the preparations for the bioassay experiments for the in-orbit exposure a mini-bioreactor and whole-cell biosensor with immobilized hemocytes were developed [31] (Fig. 20.4).

20.5 Bioassays and mesocosms: artificial streams and ponds

The transfer of bioassays to the field situation is very complex. There are many up-scaling tools available for the ecotoxicological evaluation of chemicals in the field and effects monitoring by bioassays in large-scale ecosystems: mesocosm studies for example in the Experimental Lake Area (ELA) [46,47], outdoor artificial streams [48] and ponds [49–51], as well as in microcosms and/or fish studies in enclosures. To work with artificial streams there are many advantages concerning the transfer and coming closer to the real environment but there are also disadvantages because of the increasing complexity [48,49], costs, and of course manpower [47–51]. For simulation of the environmental conditions in surface waters like rivers, lakes, or ponds, on a technical scale, it will be necessary to arrive at a compromise for example between the natural depth of major rivers which may be 2–3 m or even more and a minimum depth to keep the influence of artificial streams, their bottom and walls, within reasonable limits. The depth proposed by Sanders and Falco [52], namely 0.60 m, is exactly the same as the one that has been operated in Berlin-Marienfelde for more than 10 years as an artificial stream facility with eight artificial streams each 500 m long with a water depth of 0.60 m. Very important for a relevant "scaling-up bioassay" for field experiments in artificial streams and/or ponds, the relationship between the surface and volume of the water column [52] are important. In the 1990s many research facilities were using artificial streams for chemical testing but at a much smaller size [53,54]. The smaller streams are easier to handle but the relationship between surface and volume is difficult to compare and the scaling of lower flow velocity in comparison with the real environment is even more complex. To keep the water together with the biota (i.e., plankton) carefully moving without mechanical stress (shear forces) in the 500-m streams there were screw pumps installed.

The pumping station allowed any conceivable mixing to be made and controlled. It was possible, for example, to have water circulate within the individual streams or let the water and biota pass in a defined sequence. The total running length for one passage was 4 km (see Fig. 20.5(A)). The maximum flow velocity was about 0.25 m s^{-1}. More than 10 years experience with these in total eight streams showed clearly that the high flow velocities were not a problem but the velocities are not relevant for the simulation of real rivers. To get more relevant data close to the environment the very slow flow velocities are of great importance but difficult to calibrate and to keep running in the long-term (Fig. 20.6).

In the vicinity of the artificial streams there are eight ponds with a surface of 100 m^2 and a water depth of about 1 m. They may be passed by water and biota in sequence as well as in parallel (in groups of two or four), with variable times of passage (Figs. 20.5--20.7).

The idea was to establish in these streams and ponds various stable aquatic ecosystems for test-batteries of bioassays [55]. These bioassays should be selected for the evaluation of the degradation potential, and accumulation and effects of new chemicals [56,57] especially of high production volumes ($>$ 1000 t per year, level 2) in aquatic ecosystems according to the Federal Chemical Act (ChemG). The ChemG in the FRG came into power in 1982 [57]; later it was replaced by the REACH Regulation [58]. REACH stands for Registration, Evaluation, Authorisation of Chemicals, the regulation under REACH was adopted in 2007 (Fig. 20.6).

Figure 20.5 Artificial streams and ponds at the Test Site for Special Problems of Environmental Hygiene in Berlin Marienfelde (Institute for Water, Soil, and Air Hygiene of the Federal Health Office): the starting and developing of biocoenosis substrate in the streams and ponds under standardized and controlled conditions for ecotoxicological testing of chemicals with bioassays under field situations. (A) Artificial streams (1980–94) made out of concrete with eight streams in parallel with a total of 4 km in length. The streams are 50 cm wide and the depth of the filled water in the streams is 50 cm. (B) Artificial stream with fish-compartments and sediment (feces and pellets) trapping between the individual compartments under realistic water flow. Rainbow trout were stocked in equal numbers simulating a progressive increase in stocking density in terms of water quality and stress indicator values. (C) the new artificial streams (1995–present day) made out of GfK-Polymers. (D) ponds divided in four segments for chemical testing. Simulation of the transitional phase of the water matrix, biota, and riverine vegetation.

The first projects with the artificial streams in 1985 was on phosphorus load and eutrophication [59] of the Berlin's waterways and the so-called "Wirkstudie" [60] (Quality Objectives for the inorganic nitrogen and phosphorus compounds (nutrients) in running waters). Both studies [59,60] confirmed that the biocoenosis of the planktonic organisms in the artificial streams and ponds should be a tool for ecotoxicological testing (ChemG) for the real environment in the sense of a "field bioassay testing" [55,57,58]. Another up-scaled experiment with the artificial streams and ponds followed because of the high volume chemical nitrilo acetic acid (NTA). The artificial streams were filled to certain extents with sediment of different origins and analytical characterizations. In the streams the biodegradation of H_3NTA was measured and calculated ($k = 0.95\ d^{-1}$) after the Monod-Kinetik [61] as well as the metal remobilization potential and the growth of phyto- and zooplankton.

Figure 20.6 Experimental setup to simulate the input of seven fish farms at a river site: accumulation of the suspended solids measured in the outlets from each of the seven system compartments. Sequential feeding of individual simulated farms and the resulting water quality, input of pellets for medical treatment, and health status and stress indicators of the exposed fish.

The results provided clear findings and recommendations on the use of NTA to replace phosphorus by NTA in washing powders (Fig. 20.7).

A third example of applications using the artificial streams was a classical approach for such streams to evaluate the water quality in relation to simulated stocking density and its sublethal effects on rainbow trout [62]. Rainbow trout were stocked in seven equally large compartments of a narrow stream. Since stocking density was identical in each compartment, density-related stress had been excluded while water quality decreased from the first towards the last compartment, each receiving the effluents of the previous one, thereby simulating a higher stocking density through reduced water quality. The daily fluctuation of the water quality and the concentrations of suspended solids were followed several times during the experimental period. Growth, condition factors, and health status (changes in SSI—Spleen Somatic Index) were observed as the responses to deteriorated water quality (Fig. 20.8).

Many studies were done in the artificial streams to develop in situ and real time bioassays as well as whole-cell biosensors like the EuCyanobacterial-Electrode [31,63] for effects related online Monitoring [55].

Figure 20.7 (A) and (B) The newly built indoor and outdoor artificial streams and ponds facility (FSA) at the Federal Protection Agency of Germany in Berlin-Marienfelde made out of GfK-polymer-segments connected with ponds.

Figure 20.8 The new enclosures at the Lake Stechlin (Lake Laboratory of the IGB in New Globsow).

In the year 1995 the artificial streams were replaced by a new facility on the scale of a mesocosm: indoor and outdoor artificial streams and ponds made out of GfK-polymer segments with a length of 300 m, a width of 100 cm, and a depth of 40 cm. In total there are 16 artificial streams with a total length of 1.6 km and 16 ponds [64].

The new artificial streams and ponds are relevant for the registration of new chemicals of high volume production and the artificial stream facilities also contribute to the WFD to confirm the "good physical-chemical status" in rivers according to the WFD by the environmental Quality Norms [17].

References

[1] Gagné F, Blaise C, Salazar M, Salazar S, Hansen P. Evaluation of estrogenic effects of municipal effluents to the freshwater mussel *Elliptio complanata*. Comp Biochem Physiol C Toxicol Pharmacol 2001;128(2):213–25.

[2] Blaise C, Gagne F, Salazar M, Salazar S, Trottier S, Hansen P. Experimentally-induced feminisation of freshwater mussels after long-term exposure to a municipal effluent. Fresen Environ Bull 2003;12(8):865–70.

[3] Yaqin K, Lay BW, Riani E, Masud ZA, Hansen P-D. Hot spot biomonitoring of marine pollution effects using cholinergic and immunity biomarkers of tropical green mussel (*Perna viridis*) of the Indonesian waters. J Toxicol Environ Health Sci 2011;3 (14):356–66.

[4] Salazar MH, Salazar SM. In-situ bioassays using transplanted mussels: I. Estimating chemical exposure and bioeffects with bioaccumulation and growth. In: Hughes JS, Biddinger GR, Mones E, editors. Environmental Toxicology and Risk Assessment - Third Volume. Philadelphia: American Society for Testing and Materials; 1995. p. 216–41.

[5] Yaqin K. The use of cholinergic biomarker, cholinesterase activity of blue mussel *Mytilus edulis* to detect the effects of organophosphorous pesticides. Afr J Biochem Res 2010;4(12):265–72.

[6] Ernst W, Weigelt S, Rosenthal H, Hansen P-D. Testing bioconcentration of organic chemicals with the common mussel (*Mytilus edulis*). In: Nagel R, Loskill R, editors. Bioaccumulation in aquatic systems. Weinheim: VCH Verlagsgesellschaft; 1991. p. 99–131.

[7] Goldberg ED. The international mussel watch. Washington, DC: National Academy of Sciences; 1980.

[8] Goldberg ED, Bowen VT, Farrington JW, Harvey G, Martin JH, Parker PL, et al. The mussel watch. Environ Conserv 1978;5(02):101–25.

[9] Hansen P-D, Bock R, Brauer F. Investigations of phagocytosis concerning the immunological defence mechanism of *Mytilus edulis* using a sublethal luminescent bacterial assay (*Photobacterium phosphoreum*). Comp Biochem Physiol C Comparat Pharmacol 1991;100(1-2):129–32.

[10] Salazar S, Davidson B, Salazar M, Stang P, Meyers-Schulte K, editors. Effects of TBT on marine organisms: field assessment of a new site-specific bioassay system. In: OCEANS'87; 1987: IEEE.

[11] Salazar MH, Salazar SM. Using caged bivalves to characterize exposure and effects associated with pulp and paper mill effluents. Water Sci Technol 1997;35(2-3):213–20.

[12] Hansen P-D. Das "WaBoLu-Aquatox" zur integralen Erfassung von Schadstoffen im Wasser. The "Wabolu-Aquatox" for integral monitoring of water pollutants. Vom Wasser 1987;67:221–35.

[13] Hansen P-D. Unerwünschte wirkungen. In: Höll K, editor. Wasser, nutzung im kreislauf, hygiene, analyse und bewertung. Berlin, New York: Walter de Gruyter; 2010. p. 602–15.

[14] Hansen P-D. Wirkungsbezogene Biotestverfahren. Ökologische Texte. In: Aurand K, Irmer H, editors. Zellstoffabwasser und Umwelt Schriftenreihe des Verein für Wasser-, Boden- und Lufthygiene eV. 56. Stuttgart: Gustav Fischer Verlag; 1983. p. 203–11.

[15] Hansen P-D, Dizer H, Hock B, Marx A, Sherry J, McMaster M, et al. Vitellogenin–a biomarker for endocrine disruptors. TrAC Trends Anal Chem 1998;17(7):448–51.

[16] Huschek G, Hansen P-D. Ecotoxicological classification of the Berlin river system using bioassays in respect to the European Water Framework Directive. Environ Monit Assess 2006;121(1):15–31.

[17] Hansen P-D. Risk assessment of emerging contaminants in aquatic systems. TrAC Trends Anal Chem 2007;26(11):1095–9.

[18] Hansen P-D. Biosensors for environmental and human health. In: Kim YJ, Platt U, editors. Advanced Environmental Monitoring. 4. Heidelberg, New York: Springer Publisher; 2007. p. 297–311.

[19] Hansen P-D. Biosensors and ecotoxicology. Eng Life Sci 2008;8(1):1–7.

[20] Gunatilaka A, Diehl P. A brief review of chemical and biological continuous monitoring of rivers in Europe and Asia. In: Gunatilaka A, Gosebatt B, editors. Biomonitors and biomarkers as indicators of environmental change. Volume 2, New York: Plenum Press; 2001. p. 9–28

[21] Reed RE, Burkholder JM, Allen EH. Recent developments in online monitoring technology: algae from source to treatment. In: Allen EH, editor. Manual of water supply practices – M57. Denver: American Water Works Association; 2010. p. 3–24.

[22] Moldaenke C, Baganz D, Staaks G. Monitoring of toxins in drinking water by the ToxProtec64 fish monitor. In: TECHNAU report D.3.4.12, European Commission Sixth Framework Programme, sustainable development, global change and ecosystems thematic priority area (contract number 018320), Brussels/Kiel; 2010, p. 1–8.

[23] Penders E, Wagenvoort, de Hoogh C, Frijns N, Kamps R. Standardization, quality assurance and data-evaluation of on-line biological monitoring systems: bbe-Algae Toximeter II. In: Kiwa NV, editor. BTO report 2005.036, Nieuwegein; 2006, p. 1–41.

[24] Lechelt M, Blohm W, Kirschneit B, Pfeiffer M, Gresens E, Liley J, et al. Monitoring of surface water by ultrasensitive *Daphnia* toximeter. Environ Toxicol 2000;15 (5):390–400.

[25] Irmer U. Continuous biotest for water monitoring of the river rhine – summary, recommendations, description of test methods, uba Texte 58/94 (ISS 0722-186X), Berlin, p. 12–14, Annex XXI–XXII.

[26] Evans GP, Johnson D, Withell C. Development of the WRc, Mk III fish monitor: description of the system and its response to some commonly encountered pollutants. Medmenham; 1986, WRc Publication, TR 233.

[27] Löbbel H-J, Stein P. Messung der Kiemendeckelbewegung mit dem WRc-Fischmonitor im on-line Betrieb am Niederrhein. In: Steinhäuser KG, Hansen P-D, editors. Biologische Testverfahren. Schriftenreihe des Vereins für Wasser-, Boden- und Lufthygiene. Stuttgart/New York: Gustav Fischer Verlag; 1992. p. 323–32.

[28] Stein P, Hansen P-D, Löbbel J. Erprobung des WRc–Fischmonitors zur Störfallüberwachung am Rhein. In: Pluta H-J, Knie J, Leschber R, editors. Biomonitore in der Gewässerüberwachung. Stuttgart, Jena, New York: Gustav Fischer Verlag, 1994. p. 75–85.

[29] Petersen W, Petschatnikov M, Schroeder F, Colijn F. FerryBox systems for monitoring coastal waters. Elsev Oceanogr Ser 2003;69:325–33.

[30] Schroeder F, Knauth H-D, Pfeiffer K, Nohren I, Duwe K, Jennerjahn T, et al., editors. Water quality monitoring of the Brantas Estuary, Indonesia. In: Oceans'04 MTTS/IEEE Techno-Ocean'04; 2004: IEEE.

[31] Hansen P-D. Whole-cell biosensors and bioasays. In: Palchetti I, Hansen P-D, Barceló D, editors. Past, present and future challenges of biosensors and bioanalytical tools in analytical chemistry: A Tribute to Professor Mascini, Comprehensive analytical chemistry (CAC). Oxford, Amsterdam: Elsevier; 2017, Volume 77, in print.

[32] Franke H-D, Buchholz F, Wiltshire KH. Ecological long-term research at Helgoland (German Bight, North Sea): retrospect and prospect—an introduction. Helgoland Mar Res 2004;58(4):223.

[33] Dethlefsen V, von Westernhagen H, Tüg H, Hansen PD, Dizer H. Influence of solar ultraviolet-B on pelagic fish embryos: osmolality, mortality and viable hatch. Helgoland Mar Res 2001;55(1):45–55.

[34] von Westernhagen H. Erbrütung der Eier von Dorsch (*Gadus morhua*), Flunder (*Pleuronectes flesus*) und Scholle (*Pleuronectes platessa*) unter kombinierten Temperatur- und Salzgehaltsbedingungen. Helgoland Mar Res 1970;21(1):21–102.

[35] von Westernhagen H. Sublethel effects of pollutants on fish eggs and larvae. Fish Physiol 1988;XIA(4):253–346.
[36] Browman HI, Vetter RD, Rodriguez CA, Cullen JJ, Davis RF, Lynn E, et al. Ultraviolet (280-400 nm)-induced DNA damage in the eggs and larvae of *Calanus finmarchicus* G. (Copepoda) and Atlantic cod (*Gadus morhua*). Photochem Photobiol 2003;77:397–404.
[37] Herbert A, Hansen P-D. Genotoxicity in fish eggs/embryos. In: Wells KLaCB PG, editor. Microscale aquatic toxicology - Advances, techniques and practice. Boca Raton, FL: CRC Lewis Publishers; 1998. p. 491–505.
[38] Kanter PM, Schwartz HS. A hydroxylapatite batch assay for quantitation of cellular DNA damage. Anal Biochem 1979;97(1):77–84.
[39] Hansen P, Wittekindt E, Sherry J, Unruh E, Dizer H, Tüg H, et al. Genotoxicity in the environment (eco-genotoxicity). Biosensors for environmental monitoring of aquatic systems. Berlin: Springer; 2009. p. 203–26.
[40] ISO 21427-2. Water quality – evaluation of genotoxicity by measurement of the induction of micronuclei - Part 2: mixed population method using the cell line V79. Berlin: Beuth Verlag; 2006. p. 1–18.
[41] al-Sathi K, Metcalf DD. Fish micronuclei for assign geneotoxicity in water. Mutat Res 1995;343(2-3):121–35.
[42] Bilitewski U, Brenner-Weiss G, Hansen P-D, Hock B, Meulenberg E, Müller G, et al. Bioresponse-linked instrumental analysis. TRAC Trends Anal Chem 2000;19 (7):428–33.
[43] Hansen P-D. TRIPLELUX-B: immunotoxicity and genotoxicity testing for in-flight experiments under microgravity at the ISS. In: Schmaling U, editor. Raumfahrt Concret, Neubrandenburg 2017;96(1):14–9.
[44] Unruh E, Brungs S, Langer S, Bornemann G, Frett T, Hansen P-D. Comprehensive study of the influence of altered gravity on the oxidative burst of mussel (*Mytilus edulis*) hemocytes. Microgravity Sci Technol 2016;28(3):275–85.
[45] Hauslage J, Cevik V, Hemmersbach R., *Pyrocystis noctiluca* an excellent bioassay for shear forces induced in ground-based microgravity simulators (clinostst and random positioning machine). Npj Nat Microgravity. 2017, 3, 12.
[46] Johnson WE, Vallentyne JR. Rationale, background, and development of experimental lake studies in northwestern Ontario. J Fish Board Can 1971;28(2):123–8.
[47] Schindler D. A hypothesis to explain differences and similarities among lakes in the Experimental Lakes Area, northwestern Ontario. J Fish Board Can 1971;28 (2):295–301.
[48] Clark J, Rodgers J, Dickson K, Cairns J. Using artificial streams to evaluate perturbation effects on aufwuchs structure and function. J Am Water Res Assoc 1980;16 (1):100–4.
[49] Crossland N, Wolff C. Outdoor ponds: their construction, management and use in experimental ecotoxicology. In: Hutzinger, editor. Environmental chemistry, vol. 2, Part D. Heidelberg: Springer-Publishers; 1988. p. 1–53.
[50] Crossland N, Wolff C. Fate and biological effects of pentachlorophenol in outdoor ponds. Environ Toxicol Chem 1985;4(1):73–86.
[51] Neugebaur K, Zieris F-J, Huber W. Ecological effects of atrazine on two outdoor artificial freshwater ecosystems. Z Wasser Abwasser-Forschung 1990;23(1):11–17.
[52] Sanders WM, Falco IW. Ecosystem simulation for water pollution research. In: Proceedings of the Sixth International Conference on Water Pollution Research, 1972.

[53] Zischke JA, Arthur JW, Hermanutz RO, Hedtke SF, Helgen JC. Effects of pentachlorophenol on invertebrates and fish in outdoor experimental channels. Aquat Toxicol 1985;7(1-2):37–58.

[54] Sturm A, Wogram J, Hansen PD, Liess M. Potential use of cholinesterase in monitoring low levels of organophosphates in small streams: Natural variability in three-spined stickleback (*Gasterosteus aculeatus*) and relation to pollution. Environ Toxicol Chem 1999;18(2):194–200.

[55] Hansen P-D. Assessment of ecosystem health: development of tools and approaches. In: Rapport DJ, Gaudet CL, Calow P, editors. Evaluating and monitoring the health of large-scale ecosystems. Berlin, Heidelberg: Springer; 1995. p. 195–217.

[56] Rudolph P. Erkenntnisgrenzen biologischer Testverfahren zur Abbildung ökologischer Wirklichkeiten. In: Steinhäuser K, Hansen P-D, editors. Biologische Testverfahren. 89. Stuttgart, Jena, New York: Gustav Fischer Verlag; 1992. p. 25–34. and 867–869.

[57] Rudolph P, Boje R. Ökotoxikologie – Grundlagen für die ökotoxikologische Bewertung von Umweltchemikalien nach dem Chemikaliengesetz. In: Vogl J, Heigl A, Schäfer K, editors. Handbuch des Umweltschutzes: ecomed Verlagsgesellschaft mbH, Landsberg, München; 1986. p. 105.

[58] REACH-Regulation (EC) No. 1907/2006 of the European Parliament and of the council concerning the registration, evaluation, authorisation and restriction of chemicals (REACH). Official Journal of the European Union; 2006, L396, p. 1–852.

[59] Hansen P-D. Phosphorus load and eutrophication of Berlin Waterways simulated in artificial streams. In: Lester J, editor. Management strategies for phosphorus in the environment. London: Selper Ltd; 1985. p. 151–9.

[60] Hamm A. Studie über Wirkungen und Qualitätsziele von Nährstoffen in Fließgewässern. Sankt Augustin: Arbeitskreis "Wirkstudie" im Hauptausschuß "Phosphate und Gewässer" in der Fachgruppe Wasserchemie in der Gesellschaft Deutscher Chemiker, 1991.

[61] Hansen P-D, Stehfest H. Ermittlung von NTA-Abbaukonstanten für Untersuchungen zur Umweltverträglichkeit von Nitrilotriessigsäure. Sankt Augustin: Bernhardt; 1984.

[62] Rosenthal H, Hansen P-D, Peters G, Hoffmann R. Water quality in relation to simulated stocking density and its effect on rainbow trout. In: International Council for the Exploration of the Sea (ICES), 1984.

[63] Hansen P-D, Usedom AV. New biosensors for environmental analysis. In: Scheller FW, Schubert F, Fedrowitz J, editors. Frontiers in biosensorics. Basel, Boston, Berlin: Birkhäuser Verlag; 1997. p. 109–20.

[64] Meinecke S., Feibicke M. Fließ- und Stillgewässer-Simulationsanlage (FSA), 2006.

Environmental monitoring using bioassays

21

Nils G.A. Ekelund[1] and Donat-P. Häder[2]
[1]Malmö University, Malmö, Sweden, [2]Friedrich-Alexander University, Erlangen-Nürnberg, Germany

21.1 Introduction

The degree and rapid deterioration of water quality in aquatic ecosystems has been in the focus for many decades since the growth of industrial activity and use of many environmental toxicants. The problems with public health caused by water pollution were already recognized in the early 1800s and at this time the first legislation was made as a consequence to water pollution [1]. Even during the 21st century it is one of the greatest challenges to achieve more sustainable water management, which is among many countries in Europe recognized by the Water Framework Directive (WFD). Until the 1990s in Sweden, untreated wastewater from the pulp and paper industries caused extensive pollutions in coastal aquatic ecosystems [2]. Today the situation has improved in developed countries indicating that less untreated wastewater is discharged into aquatic environments. In developing countries the amount of effluent or untreated wastewater into water sources is still very high. The large quantities of wastewater contain toxic metals, organic and inorganic pollutants, which all could be harmful for organisms in aquatic environments [3]. Even if acute lethality caused by pollutants and chemical compounds in aquatic ecosystems has become rare there is still a significant risk for negative effects on ecosystems even if the levels of the chemicals are low [4]. The low levels and high complexity of chemical compounds are nearly impossible to monitor and regulate which makes it even more important to develop methods to evaluate contamination to aquatic ecosystems [4,5].

One of the components that play a decisive role in the pollution of aquatic ecosystems is the ongoing urbanization. It is the change of land use and agricultural activities which increases the amount of fertilizers, pesticides, herbicides, and chemical compounds of different origin that contaminate water bodies. The use of herbicides in agriculture and forestry has increased during the last decades [6,7]. Due to the water solubility of herbicides the breakdown products are often found in aquatic ecosystems where they have the possibility to affect nontarget algal species [7]. In Portugal Primextra Gold TZ is one of the best-selling herbicides and for the synthesis of this pesticide copper is used [8]. The two main active ingredients in Primextra Gold TZ are terbuthylazine (TBA) and S-metolachlor where TBA inhibits photosynthesis in photosystem II (PS II) and S-metolachlor blocks cell division and cell enlargement [8]. The effects and damage on species and

Bioassays. DOI: http://dx.doi.org/10.1016/B978-0-12-811861-0.00021-8

ecosystems from pollutants like Primextra Gold TZ do not only show the complexity of the compound itself with different active ingredients and heavy metals but are also related to the use of herbicides in agricultural areas near aquatic systems. Impacts from pollution on aquatic systems differ depending on whether the outlet into the aquatic systems is from point sources or from nonpoint sources [9]. It is more straightforward and easier to use ecotoxicological test assessments of point sources than of nonpoint sources where the complexity is higher due to multiple stressor effects [10].

A group of compounds which are a part of all pollutants are heavy metals. The pollution of heavy metals is strongly related to industrial development and these compounds cannot be easily detoxified, which results in their persistence in the environment [11]. Even if most of the heavy metals are toxic they also act as essential elements in the living system and for cellular metabolism and growth [12]. Living organisms require trace amounts of some metals, although increased levels of these compounds may be highly toxic [13,14]. The group of heavy metals contains copper (Cu), zinc (Zn), lead (Pb), chrome (Cr), nickel (Ni), cobalt (Co), arsenic (As), and mercury (Hg). Some of these elements have also been found in the wood ash, which is derived from different biofuel sources in the forest industry [15]. Metal ions may also bind to soil particles but additionally to components of herbicides like glyphosate [16].

The contamination of aquatic ecosystems with heavy metals affects the organisms within this ecosystem in the way that heavy metals tend to be taken up by the organisms. Microalgae, which are the primary producers and the basis of the food chain in aquatic systems, are one of the first groups to be contaminated by the pollution. The effects of excess heavy metals on microalgae are suppression of photosynthesis, cell growth, chlorophyll synthesis, and motility [17]. Since microalgae are the first step in the food web, the contamination affects other organisms on a higher level in the food chain. A change in microalgae populations or in their chemical composition due to contamination may have a harmful impact on higher trophic levels in the environment [8]. As microalgae are very sensitive to toxic elements they could be used as an early warning system when monitoring pollution. The advantage of using microalgae is that there is no need to consider ethical constraints associated with higher organisms such as vertebrates when testing toxicity of different compounds [18]. The important role of microalgae is not only of interest for environmental monitoring but also within the fields of waste treatment, biodiesel production, and the use of biomass to obtain commercial products.

21.2 Bioassay tests

During the last years the need for monitoring aquatic ecosystems in order to evaluate and manage the risk of environmental damage has led to the development of a number of different bioassay tests. These toxicity tests use a range of microorganisms, algae, invertebrates, and fish [19–21]. To detect biological effects often small scaled biological systems on a molecular level, cellular (in vitro) or whole organism (in vitro) levels are being used [4]. The main aim is to generate protocols for toxicity, standards and limits imposed by regulators which are fast and economical. Bioassay methods are therefore based on the response of living organisms in order to analyze and evaluate potential effects of toxicants. In order to improve the ecological relevance of these tests it is of great importance to establish a scientific base when extrapolating from bioassay tests in the laboratory to field situations [22].

When using microalgae in bioassays the dominant parameters for study are physiological responses. Single-celled algae are very sensitive to toxic compounds and this makes it possible to identify the presence of a pollutant above a potentially harmful threshold, which is important when determining the suitability of the bioassays. In addition to the sensitivity of microalgae to toxicity, the cells respond more quickly than larger organisms, have a short generation time, and are easy to handle. The cells are small and only separated from the surrounding toxic medium by membranes or a cell wall, which enables rapid uptake of toxic compounds by their cells [23]. The high diversity of microalgae makes toxic effects of aquatic pollutants very heterogeneous due to the complexity and interactions between species, physiological responses, and environmental factors.

Physiological parameters such as photosynthesis, quantum yield, chlorophyll content, growth, motility, and fluorescence respond differently to pollutants and therefore it is of high value to set the criteria of the toxicity for the bioassays. The criteria include the standardization of culture medium and conditions and experimental parameters [24]. Factors which could affect the toxicity of a compound are physico–chemical properties such as pH, temperature, and light (irradiance and exposure time) [25]. Other biomarkers which have been proven to be excellent when studying environmental stressors are biomolecular and biochemical measures [8,9].

There is an urgent need for the development of early warning systems that can monitor, detect, and evaluate effects of toxicants and are rapid, reliable, and sensitive to a wide range of compounds. However, it must be pointed out that there are some problems concerning the evaluation of laboratory studies in relation to the effects of stressors on the health of aquatic ecosystems [9]. One important issue is to assess the ecological significance of laboratory early-warning indicators and to establish cause and effects relationships between specific stressors and the types and levels of environmental pollution and damage [9]. Fast responses could be observed either within hours or even during short-term incubation within minutes. Ahmed and Häder [26] defined end points for short tests after 0–24 h, and for long-term tests after 1–5 days. In a study by Pettersson and Ekelund [6] the short-term tests were 3 min and the long-term tests were 7 days when studying the motility in *Euglena gracilis*. Short-term and small-scaled bioassays are generally favored over long-term, large-scale bioassays [4].

A fast method is the noninvasive measurement of the in vivo fluorescence which facilitates monitoring of activity changes affecting PS II in microalgae. Fluorescence is one of three processes where the energy from light absorption by chlorophyll molecules is re-emitted as red fluorescence [7]. The two other processes are photochemistry and dissipation of the excess energy as heat. The degree of fluorescence indicates the quantum yield of both photochemistry and heat emission from living cells. Brayner et al. [27] used biosensors when studying the effects of herbicides on three different green microalgae; the best results and most accurate responses were obtained with *Dictyosphaerium chlorelloides*.

One of the main advantages of quantifying fluorescence from single cells is that it does not require any preparation of the algae. Toxicity effects of the tested chemicals, even at low concentrations, can be obtained after 20 min of incubation [28]. The rapid techniques of pulse amplitude modulation (PAM) fluorometry [29] has a great advantage because it provides direct information on the photosynthetic activity of the algae, in comparison to measurements of cell numbers which requires several hours of incubation [7,26,30]. Other tests with microalgae and photosynthesis which are defined as short-term tests use 6 h to monitor and evaluate the toxic effects of metals [31].

In a study by Ahmed and Häder [17] PAM fluorescence was used for measurements of both photosynthetic yield and electron transport rate (ETR) when evaluating the effects of nickel and cadmium of photosynthetic parameters of *E. gracilis*. The duration of exposure to heavy metals at different concentrations varied from direct (20 min) exposure to long-term exposure of 5 days. The long-term monitoring using days was initiated because of the fact that heavy metals are nondegradable pollutants with an ability of bioaccumulation. Similar photosynthetic parameters were used in a study by Ekelund and Aronsson [32] where *E. gracilis* was shown to be more resilient to high pH than the green alga Chlamydomonas *reinhardtii* after 7 days at different concentrations of non-pH adjusted wood-ash solutions.

Another rapid method based on photosynthetic activity which is used in environmental monitoring is the use of biosensors that reflect the real physiological impact of active compounds [33]. A biosensor can be defined as an integrated bioreceptor-physicochemical transducer device which consists of different elements: a bioreceptor or biological recognition element, which interacts with the pollutant molecules, a physicochemical transducer, which converts the biological response into a detectable physicochemical signal, and a microelectronic processor of this signal, which amplifies it and converts it into a numeric record [13]. Promising investigations using biosensors for photosynthetic activity and PS II could be attractive and sensitive enough when studying the effects of chemical pollutants [33]. The effects of pollutants or chemical compounds using biosensors will modify the signal from the photosynthetic activity [27].

To identify the toxicity of a substance in different organisms requires the determination of a predicted no-effects concentration (PNEC) [34,35]. The PNEC for a substance is regulated by the European Commission [5]. The PNCE data are obtained in short-term tests using common endpoints. Common endpoints for the assessment of toxicity for algal growth tests and photosynthetic activity in microalgae are the proportion of the dead algae or the photosynthetic activity after a defined exposure time to the toxic compound. This could either be described as an EC_{50} value, which is the concentration of the test substance that results in a 50% inhibition [36], or by a the NOEC (No Observed Effect Concentration) value [37], which is the highest tested concentration at which no significant inhibition occurs [6,22,38]. EC_{50} and NOEC values are important values for the application of environmental quality standards.

Besides the common bioassay parameters of photosynthesis and growth of microalgae, swimming behavior of phytoplankton organisms is a reliable parameter for ecotoxicological tests when studying the effects of water quality. Swimming behavior has been observed in a broad context from ecological phenomena, phototactic orientation, physical stratification, avoidance, and sex [18]. Feng et al. [18] argue that the swimming behavior of microalgae can be used as a sensor for environmental signals and a link between biochemical processes and consequences on the ecosystem level. One method which has been used to identify toxicity effects on the motility of microalgae is the determination of toxicity by using different types of chips. A chip can be used to perform many parallel tests simultaneously [18].

A method based on image analysis that operates in real time and tracks the numbers of motile cells is ECOTOX [39–41]. The effects of wastewater, heavy metals, herbicides, and other pollutants are evaluated with the ECOTOX system by using the parameters motility and orientation of *E. gracilis* [3,6,42].

21.3 ECOTOX

Using the ECOTOX method allows for calculating the number of motile cells, the percentage of cells moving upwards (gravitaxis), the mean velocity, the compactness (form factor, which is the ratio of the circumference to the area normalized to a circle) of the cells, and the precision of orientation (*r*-value) (cf. Chapter 10: Ecotox, this volume). The system operates in real time and tracks of up to 1500 cells in parallel. The cells of *E. gracilis* are pumped into an observation chamber. This chamber could hold up to 10^6 cells mL^{-1} and the test sample of toxic compounds can be automatically diluted with increasing dilutions of 1:1, 1:3, 1:7, 1:15, and 1:31 [6].

An important factor before starting a measurement is the filling time of the cuvette. This time must be long enough that fresh cells of *E. gracilis* are transferred from the reservoir where the cells are located before the beginning of the measurement. The tracking time of the analysis shows variations but is often within in the range of 3 to 5 min. Depending on the cells being used in the ECOTOX system there is a need to set minimum and maximum areas of the cells. The setup of areas makes it possible to exclude debris or other species which may be contained within in the sample cuvette. The setup of a minimum speed of the organisms is of importance because it allows distinguishing between dead or nonmotile cells and active motile cells. In a study by Pettersson and Ekelund [6] cells were considered motile when they moved at 7 $\mu m\ s^{-1}$ or faster but in a study by Azizullah et al. [3] the minimal velocity was chosen as 15 $\mu m\ s^{-1}$. Within the system there is also a possibility to set different incubation times in darkness before measurements start.

21.4 Short-term vs long-term tests

ECOTOX measurements of the effects of herbicides using *E. gracilis* in short-term tests of 0, 30, or 60 s showed strong effects on the velocity and upward swimming at concentrations higher than 1.25 g L^{-1} [6] (Fig. 21.1). The results showed different effects dependent on what type of herbicide was used. Avans had a stronger effect than Roundup. Avans showed lower NOEC values than Roundup for the different parameters but after 7 days (long-term) the NOEC values were similar except for upward swimming.

However, the effects on motility were less pronounced; this is probably due to the fact that cells of *E. gracilis* were still motile but that the orientation behavior was affected. In a study by Azizullah et al. [3] the gravitactic orientation of *E. gracilis* showed similar negative effects when exposed to wastewater. In comparison to cell shape and motility the gravitactic orientation was significantly impaired. The variation in sensitivity to pollutions of the different parameters in *E. gracilis* indicates that gravitactic orientation and cell shape are dependent of each other. Several studies show discrepancies between the effects of pollution on motility, cell shape, speed, and gravitactic orientation, which could be the result of changes in density between the cell body and the surrounding medium. Gravitactic orientation shows stronger negative effects when exposed to pollution which alter the density thus impairing the capacity to orient in the water column. In contrast to motility, velocity of *E. gracilis* showed a strong inhibition after short tests of only 30 and 60 s incubation with herbicides (Fig. 21.2) [6]. The effects were most pronounced with Avans and less with Roundup.

When testing the effects of industrial wastewater on *E. gracilis* the most sensitive parameters were gravitactic orientation and cell compactness [43]. The content of industrial wastewater is very complex depending on the large amount of different pollutants and toxic compounds. This complexity of wastewater makes it difficult to relate an impairment of motility parameters to a specific compound. Heavy metals which can occur in wastewater have different effects on motility parameters. After 24 h exposure to different heavy metals the motility and cell shape of *E. gracilis* showed no significant negative effects at the different concentrations (Table 21.1) [44].

Probably the concentrations were too low and the exposure time too short to induce significant effects of the different heavy metals. However, in a study with the heavy metals nickel (Ni) and cadmium (Cd) the effects were stronger after 24 h in comparison to tests immediately after exposure [17]. The EC_{50} values for upward swimming showed large differences, and the value after direct exposure was 292 mg L^{-1} and after 24 h incubation was 12.7 mg L^{-1}. The results from this study showed that longer incubation times did not show any significant difference for the different parameters, indicating that shorter incubation times are recommended for toxicological studies. However, it was concluded that acute toxicity is useful for rapid assessment but not enough to produce pronounced inhibition for all parameters [41].

Like wastewater, wood ash solutions contain toxic compounds [15]. Many different heavy metals are found in wood ash and the most common compound is cadmium (Cd). The potential release of these compounds into forest ecosystems

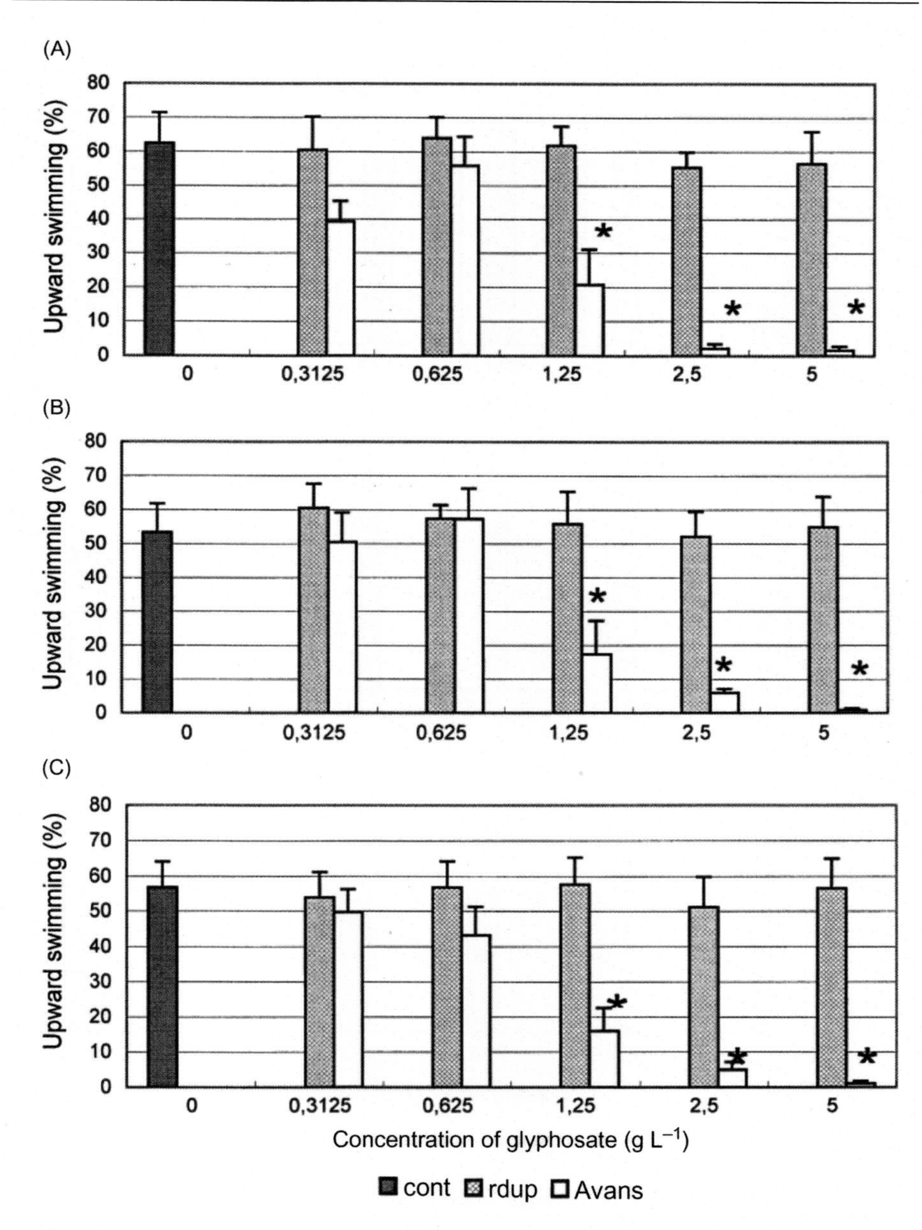

Figure 21.1 Upward swimming (%) of *Euglena gracilis* at different concentrations (g L^{-1}) of glyphosate with the herbicides Roundup (rdup) and Avans after different incubation times of 60 (A), 30 (B), and 0 s (C). Control (cont) tests are made without the addition of glyphosate. Columns indicate mean values with standard deviation, and asterisks indicate statistical significance between treatment and control.

Source: From Pettersson M, Ekelund NG. Effects of the herbicides Roundup and Avans on Euglena gracilis. Arch Env Contamin Toxicol 2006;50(2):175–181 with permission.

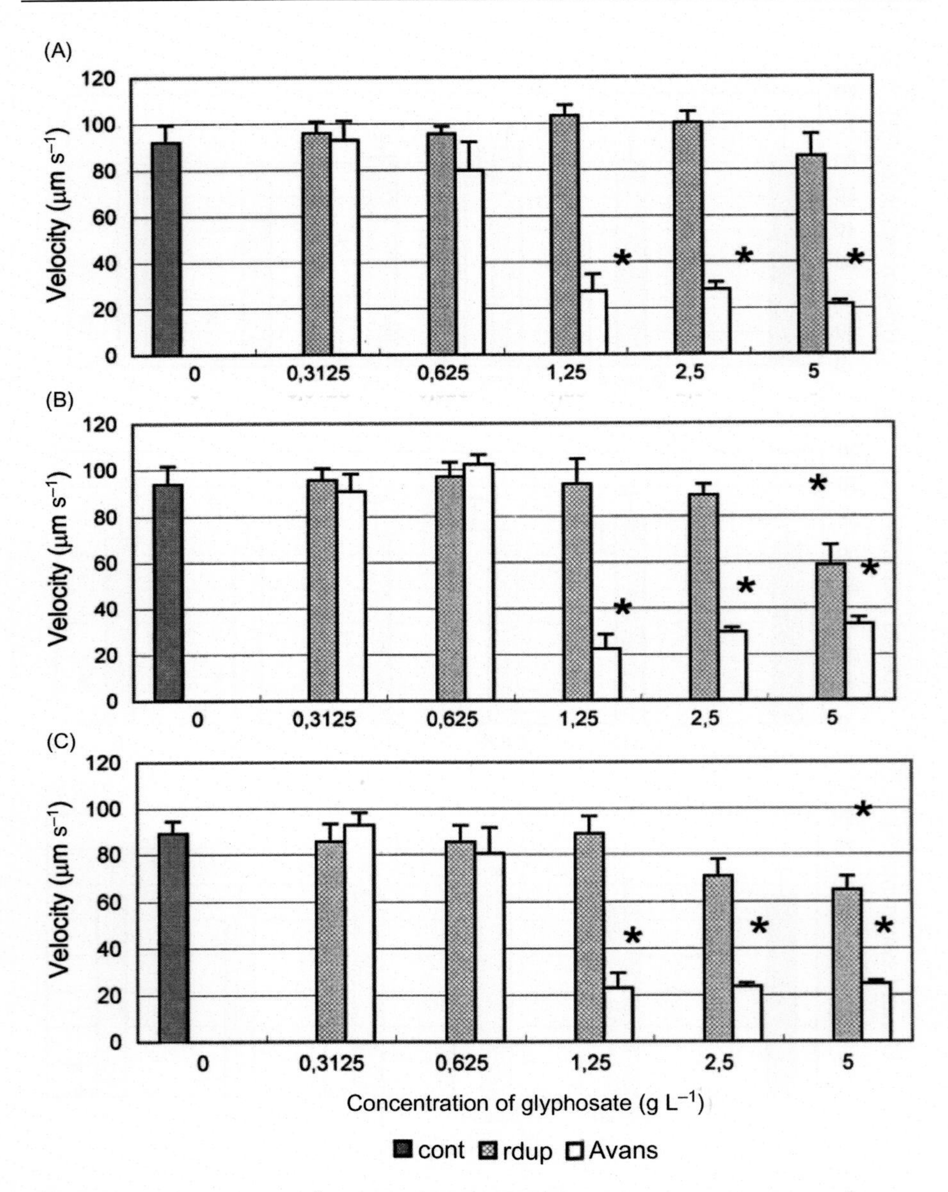

Figure 21.2 Velocity (μm s^{-1}) of *Euglena gracilis* at different concentrations (g L^{-1}) of glyphosate with the herbicides Roundup (rdup) and Avans. Short-term tests with different incubation times of 60 (A), 30 (B), and 0 (C). Control tests are made without the addition of glyphosate. Columns indicate mean values with standard deviation, and asterisks indicate statistical significance between treatment and control.

Source: From Pettersson M, Ekelund NG. Effects of the herbicides Roundup and Avans on Euglena gracilis. Arch Env Contamin Toxicol 2006;50(2):175–181 with permission.

Table 21.1 Effects of different concentrations of heavy metals and pentachlorophenol (PCP) on the cell shape (circularity) of *Euglena gracilis* after 24 h exposure.

	Heavy metal				
Concentration (mg L^{-1})	Copper	Nickel	Lead	Zinc	PCP
0	4.5 ± 0.8	4.5 ± 0.8	4.5 ± 0.8	4.5 ± 0.8	4.5 ± 0.8
0.1	3.7 ± 0.6	3.9 ± 0.7	3.9 ± 1.1	3.0 ± 0.5	4.4 ± 0.7
0.5	3.4 ± 0.5	4.2 ± 0.6	3.8 ± 0.9	3.1 ± 0.6	3.6 ± 0.8
1.0	2.7 ± 0.6	4.2 ± 0.6	3.6 ± 1.0	3.3 ± 0.6	3.7 ± 1.1
1.5	3.7 ± 0.8	4.3 ± 1.1	2.9 ± 0.7	3.7 ± 0.7	
2.0	3.4 ± 0.8	4.3 ± 0.7	3.4 ± 0.7	4.0 ± 0.8	
5.0					4.3 ± 1.2
10.0					1.7 ± 0.2

Mean values and standard deviation are shown.

could have negative effects on microorganisms. In contrast to the content of heavy metals, wood ash is also enriched with nutrients, which may have positive effects for the forest ecosystems. The most pronounced effects from wood ash application are an increased pH and an elevated content of cations. In order to study the effects of wood ash *E. gracilis* was used in both short- and long-term tests.

The tests were performed with ECOTOX and the time intervals for short-term tests were 0, 24, and 48 h. In the experiments the effects of pH on cells of *E. gracilis* were tested which ranged from 7.0 to 11.2. The effects of wood ash on motility parameters, when no adjustment of pH was made, showed strong inhibition at higher concentrations (above 10 g L^{-1}, Fig. 21.3).

Especially upwards swimming and velocity after 48 h showed significant inhibition. Under natural conditions *E. gracilis* is limited to habitats with low acidity and has growth rates between 0.33 to 1.12 at pH values ranging from 2.0–6.0 [45]. Above pH 6 the growth rate shows an inhibition which looks similar to the effects of high pH values on the motility parameters of *E. gracilis*. Similar results were shown in a study by Danilov and Ekelund [46]; cells of *E. gracilis* below pH 4 and above pH 8 did not survive.

In the tests with wood ash and no adjustments of pH it is difficult to distinguish between effects from either high pH or from toxic compounds in the wood ash. However, when the pH of the wood ash solution was adjusted to pH 7 the results showed a stimulation of the motility of *E. gracilis* (Fig. 21.4) [15]. Higher motility of *E. gracilis* in the presence of wood ash at an adjusted pH is probably an effect of the availability of nutrients in wood ash such as calcium (Ca), potassium (K), and magnesium (Mg) [15].

The observed differences in the motility parameters and growth rate are a combination of the addition of nutrients, organic compounds, and pH changes [45]. A long-term (7 days) growth test using *E. gracilis* with both adjusted pH 7 and no

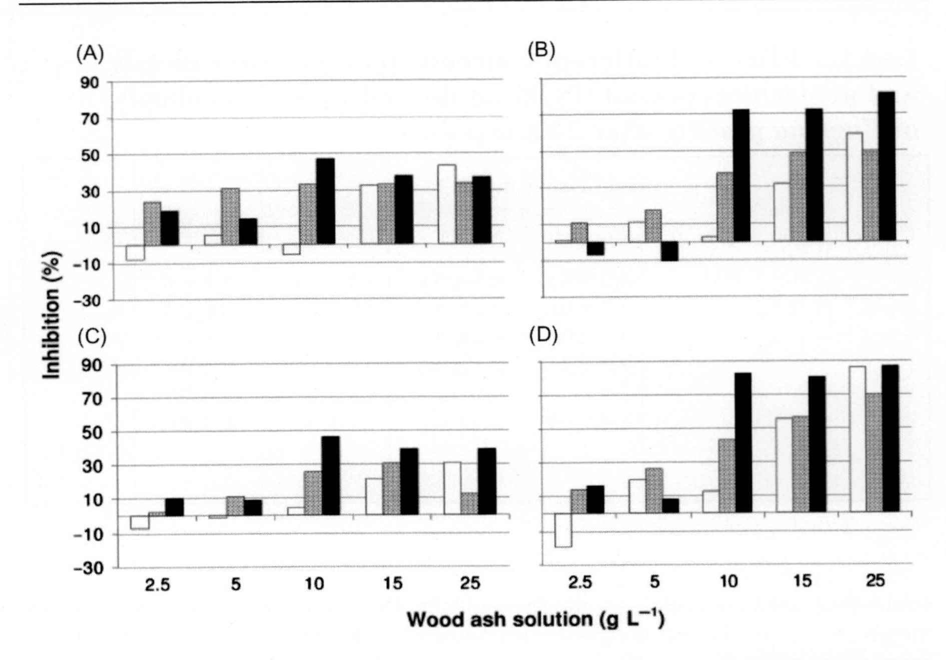

Figure 21.3 Inhibition of motility parameters in *Euglena gracilis* after 0 h (white bars), 24 h (gray bars), and 48 h (black bars) exposure to wood ash solution (2.5, 5, 10, 15, 25 g L^{-1}); pH was not adjusted. Graphs show motility (A), upwards swimming (B), compactness (C), and velocity (D). Inhibition values are shown to those of the controls of each test.

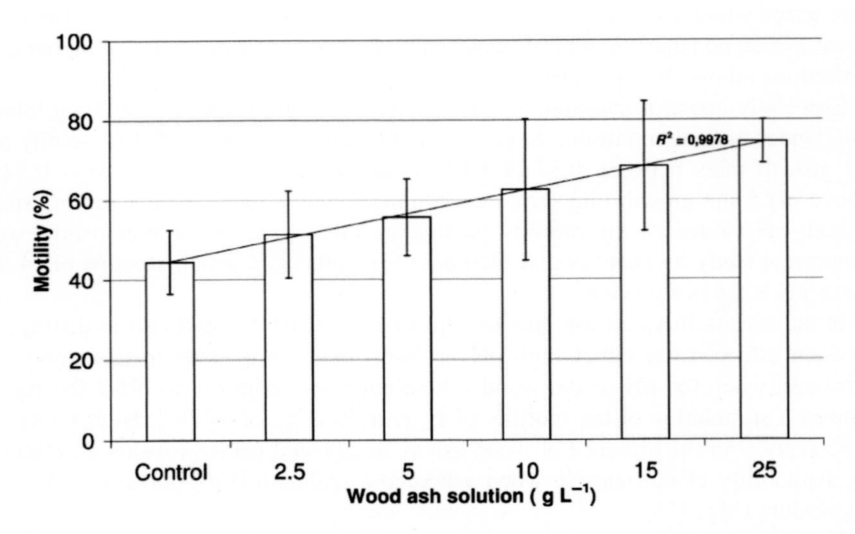

Figure 21.4 Mean values and standard deviation of *Euglena gracilis* motility tests in wood ash solution with pH adjusted. Controls were prepared with distilled water.

adjustment of pH showed only a strong inhibition at the highest concentration of wood ash (15 and 25 g L^{-1}) and a pH value of around 11 [15]. This indicates that *E. gracilis* has a high sensitivity at high pH values.

Copper (Cu) is an essential microelement and has also been shown to have toxic effects at higher concentrations on different parameters in *E. gracilis* [26,46], The immediate response to copper was rapid when studying motility parameters with ECOTOX [26]. Already after 3 min copper showed a disturbance of the different motility parameters. The sensitivity of the different parameters was highest for motility followed by *r*-value, velocity, and upward swimming. The same order of sensitivity was found after 24 h but the EC_{50} values changed for motility and upward swimming in comparison to the short-term tests. The EC_{50} values for the short-term tests in this study were in the concentration range of from 19 to 50 mg L^{-1} which is much higher than the values found for copper in industrial wastewater in Pakistan where copper concentration varied from 0.01 to 0.18 mg L^{-1} [43]. In contrast, the results from the effects of wastewater in Pakistan on the different motility parameters of *E. gracilis* showed strong inhibition on cell compactness and gravitactic orientation. These negative effects of wastewater were not only due to copper but also to a large mixture of heavy metals including Cd, Fe, Ni, Cr, Zn, and Pb [43].

21.5 ECOTOX measurements in Egyptian lakes

In the Northern part of Egypt near the Mediterranean coast several lakes and lagoons are located. Alexandria is the second largest city in Egypt and has about 4.1 million inhabitants (http://en.wikipedia.org/wiki/Alexandria). The lakes are polluted by industrial and domestic waste from the city and the surrounding areas. These wastes undergo only primary treatment, which removes suspended solid particles by filtration. Solubilized pollutants and toxic substances are not removed from the wastewater. Samples were taken from the wastewater treatment plant west of Alexandria before and after treatment and analyzed using the ECOTOX system.

Located to the Southeast of Alexandria is a large brackish lake (Lake Mariout), which has an area of ca. 250 km^2. This lake is the source of fish (mainly *Tilapia*) for the local market in Alexandria. This economically important fish is heavily affected by pollution since about two thirds of the city sewage water is drained into the lake. Only half of this wastewater undergoes primary treatment before it is discarded into the lake. Over the years the input of almost untreated domestic waste and industrial effluents has resulted in elevated concentrations of heavy metals such as Cd, Pb, and Cu, which accumulate in the water and sediments of lake Mariout [47].

The wastewater samples were collected in sterile 50-mL tubes and transferred rapidly to the laboratory where they were frozen until subsequent analysis. Prior to the ECOTOX measurements the samples were thawed at room temperature and subsequently centrifuged at $1000 \times g$ for 5 min. When necessary, the samples were diluted with deionized water before key chemical parameters were determined such as pH, nitrate (NO_3^-), nitrite (NO_2^-), total and carbonate hardness using colorimetric tests (Merck, Germany). Dissolved oxygen was measured (DO-5509, Lutron electronics enterprise Co., LTD Taiwan).

Analysis of the input wastewater to the treatment plant using the ECOTOX instrument shows a significant pollution from mainly domestic effluents. As also found in other pollution analyses [48], gravitactic orientation (*r*-value) shows the highest inhibition of up to 30% at a 1:1 dilution of the sample (Fig. 21.5A).

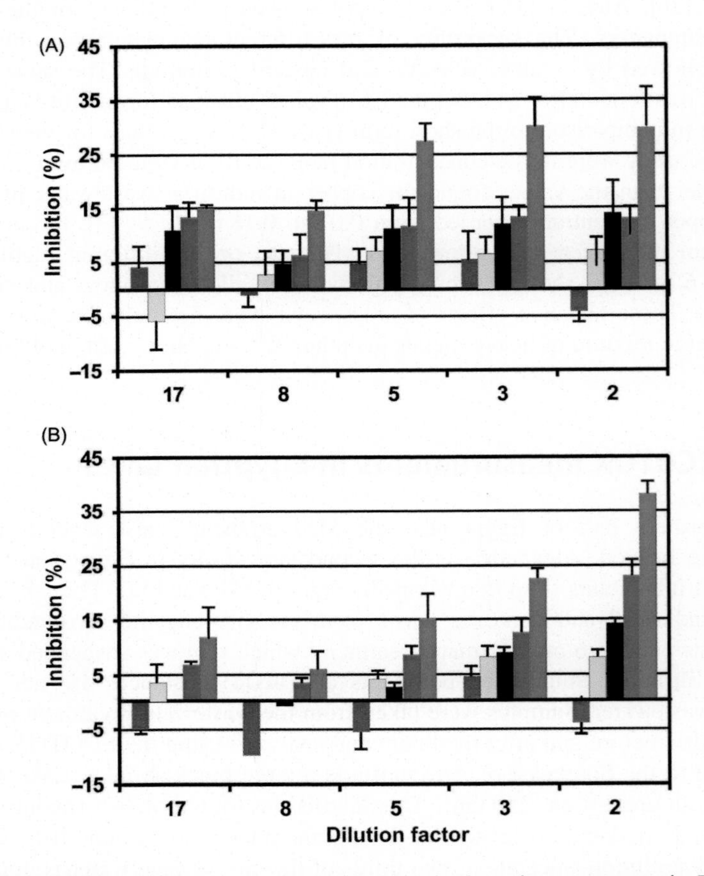

Figure 21.5 Inhibition (ordinate, %) of movement and orientation parameters in *Euglena gracilis* by samples from a wastewater treatment plant in Alexandria at decreasing dilution comparing input (A) and output wastewater (B): compactness (*red*), motility (*yellow*), alignment (*black*), upward swimming (*green*), *r*-value (*blue*). Negative values indicate an augmentation in the parameter. Each data point represents the mean of three independent measurements with standard deviation.
Source: redrawn after Ahmed H. Biomonitoring of aquatic ecosystems [PhD thesis]. Erlangen-Nürnberg: Friedrich-Alexander-Universität; 2010 [49].

The other parameters (compactness, motility, alignment and upward swimming) are less affected and some are not significant. The comparison with the samples from the output from the wastewater plant after mechanical filtration of solid particles reveals that the toxicity is not reduced by the treatment (Fig. 21.5B). In fact, most of the inhibition values are even higher and the *r*-value shows an inhibition of 37% at the lowest dilution of 1:1. Among the lakes in the Nile Delta, Lake Mariout is the most polluted one [47]. It can be subdivided into four basins—the main basin, the fish farm basin, the Southwest basin, and the Southeast basin [50]. The main basin, located to the Northwest, contains the most polluted water. In addition to domestic wastewater, it receives industrial effluents and agricultural runoff. This combined pollution increases the sulfur concentration in the water and reduces the dissolved oxygen concentration. In contrast, the Southeast basin contains the least polluted water which enters from the El-Umoum drain and the Noubaria canal [50–52].

Even though the domestic and industrial wastewaters from Alexandria are being strongly diluted in the lake, water samples taken from the Southeast basin of Lake Mariout resulted in a significant inhibition of the tested endpoints. E.g., the *r*-value was inhibited by 11%–24% at the lowest and highest concentration, respectively, and also alignment and upward swimming were significantly impaired by increasing sample concentrations (Fig. 21.6A).

Samples from the main basin even resulted in a higher inhibition with the *r*-value being reduced by 15%–57.5% at the highest and lowest dilution, respectively (Fig. 21.6B). Also the motility, upward swimming, compactness, and alignment were strongly inhibited with increasing concentration of the polluted water.

For comparison, a pond with the name Rakta situated east of Alexandria was analyzed. This pond is strongly polluted by untreated effluents from nearby industries. At the highest dilution factors (17 to 5) no significant inhibition could be detected in all movement parameters (Fig. 21.7). At higher concentrations (lower dilution factors 2 and 3) motility was not significantly affected or even enhanced, while compactness, alignment, and upward swimming showed an inhibition of about 11%. As in other tests of aquatic ecosystems, the *r*-value showed the highest inhibition of up to 29% by water samples of pond Rakta.

In addition to toxic pollutants such as heavy metals, organic and inorganic toxins, the physicochemical parameters of the water samples affect the inhibition of the movement, orientation and form parameters of the test organism *E. gracilis* in the ECOTOX measurements. These are summarized in Table 21.2. Especially the pH and total and carbonate hardness influence the water quality and affect the toxicity and thereby modify the biological risk [53]. In the analyzed samples pH and nitrite were in the normal range for freshwater lakes and drinking water. The samples from the input and output of the wastewater treatment plant as well as pond Rakta and the main basin of Lake Mariout had decreased dissolved oxygen (D.O.) values (4.1–4.6 $mg\,L^{-1}$) as compared with the recommended values ($\geq$ 5 $mg\,L^{-1}$). In addition, the analyzed samples had moderate to increased water hardness.

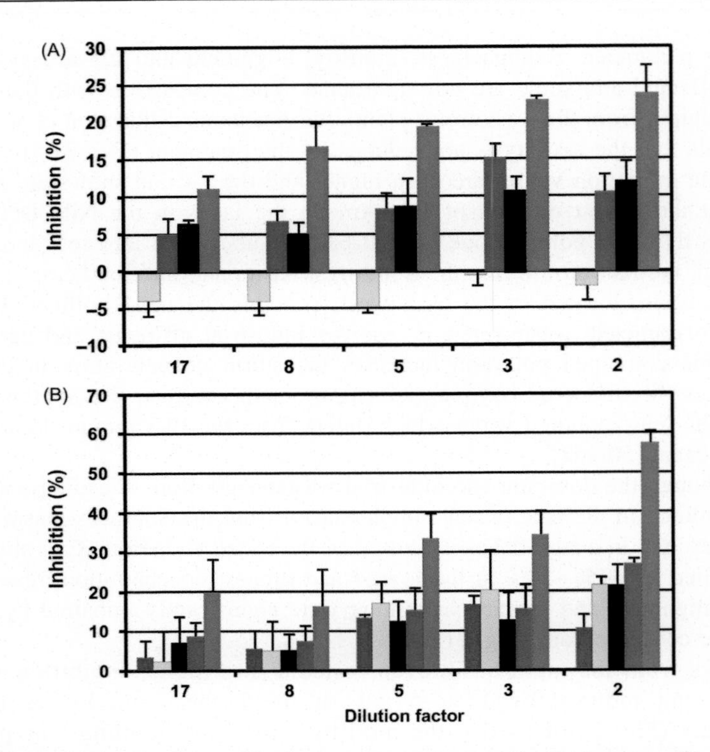

Figure 21.6 Inhibition (ordinate, %) of movement parameters in *Euglena gracilis* by samples from the Southeast basin (A) and the main basin (B) of Lake Mariout at decreasing dilution: compactness (*red*), motility (*yellow*), alignment (*black*), upward swimming (*green*), *r*-value (*blue*). Negative values indicate an augmentation in the parameter. Each data point represents the mean of three independent measurements with standard deviation.
Source: redrawn after Ahmed H. Biomonitoring of aquatic ecosystems [PhD thesis]. Erlangen-Nürnberg: Friedrich-Alexander-Universität; 2010.

21.6 Conclusions

Aquatic ecosystems are still affected by the contamination of industrial and municipal wastes. The pathways for pollutants and wastes to enter aquatic ecosystems are often via nonpoint sources, runoff from land or direct discharges from industries. Toxic compounds associated with these wastes are heavy metals, organic and inorganic compounds, pesticides and herbicides, which all may have negative impacts on aquatic organisms and function of aquatic ecosystems. Therefore it is of importance to establish monitoring systems which have the possibility and sensitivity to get an early warning signal of environmental damage. For testing the water quality the OECD guidelines promote microalgae and cyanobacteria as bioindicators [54].

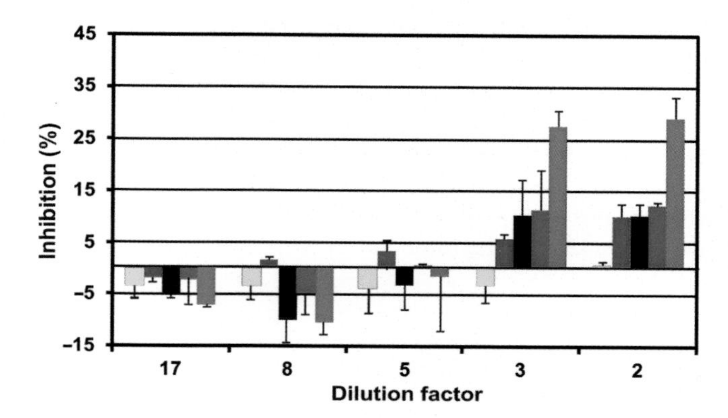

Figure 21.7 Inhibition (ordinate, %) of the movement and orientation parameters of *Euglena gracilis* by samples from the Rakta pond at decreasing dilution: compactness (*red*), motility (*yellow*), alignment (*black*), upward swimming (*green*), *r*-value (*blue*). Negative values indicate an augmentation in the parameter. Each data point represents the mean of three independent measurements with standard deviation.
Source: redrawn after Ahmed H. Biomonitoring of aquatic ecosystems [PhD thesis]. Erlangen-Nürnberg: Friedrich-Alexander-Universität; 2010.

Table 21.2 Chemical parameters of wastewater samples from different sources near Alexandria, Egypt.

Source of water samples	pH	CH ($mg\ L^{-1}$)	TH ($mg\ L^{-1}$ Ca)	NO_3- ($mg\ L^{-1}$)	NO_2- ($mg\ L^{-1}$)	D.O. ($mg\ L^{-1}$)
Lake Mariout						
Main basin	8	160	143	0	0	4.6
South east basin	8	160	143	10	0.5	6.6
Rakta Pond	8	160	143	10	0	4.5
Wastewater treatment plant in Alexandria						
Input	8	160	143	0	0.5	4.1
Output	8	160	143	0	0.5	4.1

TH, total hardness ($mg\ L^{-1}$ Ca); CH, carbonate hardness $CaCO_3$ ($mg\ L^{-1}$); D.O., dissolved oxygen ($mg\ L^{-1}$).

In the OECD tests monitoring growth over a period of normally 72 h is recommended. As a complement to toxicological growth studies swimming behavior of motile microorganisms could be used. The present study shows that by using the ECOTOX method there is a possibility to calculate the number of motile cells, gravitaxis, the mean velocity, the compactness, and the precision of orientation of the cells. In most of the studies with the ECOTOX method the freshwater flagellate *E. gracilis* is used. This flagellate is very appropriate for this method due to its well-developed motility and precise gravitactic and phototactic orientation. Especially gravitactic orientation seems to be the most sensitive parameter when testing the effects of wastewater and heavy metals. The results indicate that ECOTOX, with its possibility of performing measurements within short periods (min), could be very useful in the future as an early warning biomonitoring system.

References

[1] Rossi SH. The toxic impacts of wastes from source to ultimate fate. In: Tapp JF, Hunt SM, Wharfe JR, editors. Toxic impacts of wastes on the aquatic environment. Cambridge: Royal Society of Chemistry; 1996. p. 1–8.

[2] Danilov RA, Ekelund NGA. Influence of wastewater from the paper industry and UV-B radiation on the photosynthetic efficiency of *Euglena gracilis*. J Appl Phycol 1999;11:157–63.

[3] Azizullah A, Jamil M, Richter P, Häder D-P. Fast bioassessment of wastewater and surface water quality using freshwater flagellate *Euglena gracilis*—a case study from Pakistan. J Appl Phycol 2014;26(1):421–31.

[4] Brack W, Ait-Aissa S, Burgess RM, Busch W, Creusot N, Di Paolo C, et al. Effect-directed analysis supporting monitoring of aquatic environments—An in-depth overview. Sci Total Environ 2016;544:1073–118.

[5] European Commission. Directive 2013/39/EU of the European Parliament and the Council of 12 August 2013 amending Directive 2000/60/EC and 2008/105/EC as regards priority substances in the field of water policy. Brussels: Official Journal of the European Union, 2013.

[6] Pettersson M, Ekelund NG. Effects of the herbicides Roundup and Avans on *Euglena gracilis*. Arch Env Contamin Toxicol 2006;50(2):175–81.

[7] Fai PB, Grant A, Reid B. Chlorophyll *a* fluorescence as a biomarker for rapid toxicity assessment. Environ Toxicol Chem 2007;26(7):1520–31.

[8] Filimonova V, Gonçalves F, Marques JC, De Troch M, Gonçalves AM. Biochemical and toxicological effects of organic (herbicide Primextra® Gold TZ) and inorganic (copper) compounds on zooplankton and phytoplankton species. Aquat Toxicol 2016;177:33–43.

[9] Adams SM. Biomarker/bioindicator response profiles of organisms can help differentiate between sources of anthropogenic stressors in aquatic ecosystems. Biomarkers 2001;6(1):33–44.

[10] Adams S, Greeley M. Ecotoxicological indicators of water quality: using multi-response indicators to assess the health of aquatic ecosystems. Water Air Soil Pollut 2000;123(1-4):103–15.

[11] Khan S, Cao Q, Zheng Y, Huang Y, Zhu Y. Health risks of heavy metals in contaminated soils and food crops irrigated with wastewater in Beijing, China. Environ Pollut 2008;152(3):686–92.

[12] Steffens J. The heavy metal-binding peptides of plants. Annu Rev Plant Biol 1990;41 (1):553–75.

[13] Gutiérrez JC, Amaro F, Martín-González A. Heavy metal whole-cell biosensors using eukaryotic microorganisms: an updated critical review. Front Microbiol 2015;6:48.

[14] Khan Z, Rehman A, Hussain SZ. Resistance and uptake of cadmium by yeast, *Pichia hampshirensis* 4Aer, isolated from industrial effluent and its potential use in decontamination of wastewater. Chemosphere 2016;159:32–43.

[15] Aronsson AK, Ekelund NG. Effects on motile factors and cell growth of *Euglena gracilis* after exposure to wood ash solution; assessment of toxicity, nutrient availability and pH-dependency. Water Air Soil Pollut 2005;162(1-4):353–68.

[16] Busse MD, Ratcliff AW, Shestak CJ, Powers RF. Glyphosate toxicity and the effects of long-term vegetation control on soil microbial communities. Soil Biol Biochem 2001;33(12):1777–89.

[17] Ahmed H, Häder D-P. Rapid ecotoxicological bioassay of nickel and cadmium using motility and photosynthetic parameters of *Euglena gracilis*. Environ Exp Bot 2010;69:68–75.

[18] Feng C-Y, Wei J-F, Li Y-J, Yang Y-S, Wang Y-H, Lu L, et al. An on-chip pollutant toxicity determination based on marine microalgal swimming inhibition. Analyst 2016;141(5):1761–71.

[19] Ahlf W, Calmano W, Erhard J, Förstner U, editors. Comparison of five bioassay techniques for assessing sediment-bound. In: Environmental Bioassay Techniques and their Application: Proceedings of the 1st International Conference held in Lancaster, England, 11–14 July 1988; 2013: Springer Science & Business Media.

[20] Chavan M, Thacker N, Tarar J. Toxicity evaluation of pesticide industry wastewater through fish bioassay.. IRA-Int J Appl Sci 2016;3:3.

[21] Udovic M, Drobne D, Lestan D. An in vivo invertebrate bioassay of Pb, Zn and Cd stabilization in contaminated soil. Chemosphere 2013;92(9):1105–10.

[22] Cartwright NG, Lewis S. The role of environmental quality standards in controlling chemical contaminants in the environment. In: Tapp JF, Hunt SM, Wharfe JR, editors. Toxic impacts of wastes on the aquatic environment. Cambridge: Royal Society of Chemistry; 1996. p. 149–56.

[23] Li M, Gao X, Wu B, Qian X, Giesy JP, Cui Y. Microalga *Euglena* as a bioindicator for testing genotoxic potentials of organic pollutants in Taihu Lake, China. Ecotoxicology 2014;23(4):633–40.

[24] Stratton GW, Giles J. Importance of bioassay volume in toxicity tests using algae and aquatic invertebrates. Bull Environ Contam Toxicol 1990;44(3):420–7.

[25] Singh S, Singh N, Kumar V, Datta S, Wani AB, Singh D, et al. Toxicity, monitoring and biodegradation of the fungicide carbendazim. Environ Chem Lett 2016;14 (2):317–29.

[26] Ahmed H, Häder D-P. A fast algal bioassay for assessment of copper toxicity in water using *Euglena gracilis*. J Appl Phycol 2010;22(6):785–92.

[27] Brayner R, Couté A, Livage J, Perrette C, Sicard C. Micro-algal biosensors. Anal Bioanal Chem 2011;401(2):581–97.

[28] Leunert F, Grossart H-P, Gerhardt V, Eckert W. Toxicant induced changes on delayed fluorescence decay kinetics of cyanobacteria and green algae: a rapid and sensitive biotest. PLoS One. 2013;8(4):e63127.

[29] Herlory O, Bonzom J-M, Gilbin R. Sensitivity evaluation of the green alga *Chlamydomonas reinhardtii* to uranium by pulse amplitude modulated (PAM) fluorometry. Aquat Toxicol 2013;140:288–94.

[30] Sjollema SB, van Beusekom SA, van der Geest HG, Booij P, de Zwart D, Vethaak AD, et al. Laboratory algal bioassays using PAM fluorometry: effects of test conditions on

the determination of herbicide and field sample toxicity. Environ Toxicol Chem 2014;33(5):1017–22.
[31] Machado MD, Soares EV. Use of a fluorescence-based approach to assess short-term responses of the alga *Pseudokirchneriella subcapitata* to metal stress. J Appl Phycol 2015;27(2):805–13.
[32] Ekelund NGA, Aronsson KA. Assessing *Euglena gracilis* motility using the automatic biotest ECOTOX application to evaluate water toxicity (cadmium). Vatten 2004;60: 77–83.
[33] Giardi MT, Koblızek M, Masojıdek J. Photosystem II-based biosensors for the detection of pollutants. Biosens Bioelectron 2001;16(9):1027–33.
[34] Girling AE, Pascoe D, Janssen CR, Peither A, Wenzel A, Shäfer H, et al. Development of methods for evaluating toxicity to freshwater ecosystems. Ecotoxicol Environ Saf 2000;45.
[35] Araújo C, Souza-Santos L. Use of the microalgae *Thalassiosira weissflogii* to assess water toxicity in the Suape industrial-port complex of Pernambuco, Brazil. Ecotoxicol Environm Saf 2013;89:212–21.
[36] Sebaugh J. Guidelines for accurate EC50/IC50 estimation. Pharm Stat 2011;10 (2):128–34.
[37] Delignette-Muller ML, Forfait C, Billoir E, Charles S. A new perspective on the Dunnett procedure: filling the gap between NOEC/LOEC and ECx concepts. Environ Toxicol Chem 2011;30(12):2888–91.
[38] Haglund K. The use of algae in aquatic toxicity assessment. In: Round FE, Chapman DJ, editors. Progress in phycological research. Bristol: Lubrecht and Cramer Ltd; 1995. p. 181–212.
[39] Tahedl H, Häder D-P. Fast examination of water quality using the automatic biotest ECOTOX based on the movement behavior of a freshwater flagellate. Water Res 1999;33:426–32.
[40] Tahedl H, Häder D-P. The use of image analysis in ecotoxicology. In: Häder D-P, editor. Image Analysis: Methods and Applications. Boca Raton, FL: CRC Press; 2001. p. 447–58.
[41] Tahedl H, Häder D-P. Automated biomonitoring using real time movement analysis of *Euglena gracilis*. Ecotoxicol Environ Saf 2001;48(2):161–9.
[42] Azizullah A, Richter P, Häder D-P. Effects of long-term exposure to industrial wastewater on photosynthetic performance of *Euglena gracilis* measured through chlorophyll fluorescence. J Appl Phycol 2014;27:303–10.
[43] Azizullah A, Richter P, Häder D-P. Sensitivity of various parameters in *Euglena gracilis* to short-term exposure to industrial wastewaters. J Appl Phycol 2012;24:187–200.
[44] Danilov RA, Ekelund NGA. Responses of photosynthetic efficiency, cell shape and motility in *Euglena gracilis* (Euglenophyceae) to short-term exposure to heavy metals and pentachlorophenol. Water Air Soil Pollut 2001;132:61–73.
[45] Olaveson MM, Nalewajko C. Effects of acidity on the growth rate of two *Euglena* species. Hydrobiologia 2000;433:39–56.
[46] Danilov R, Ekelund N. Effects of pH on the growth rate, motility and photosynthesis in *Euglena gracilis*. Folia Microbiol 2001;46(6):549–54.
[47] Saad MAH. Impact of diffuse pollution on the socio-economic development opportunities in the costal Nile delta lakes. In: Diffuse Pollution Conference; Dublin 2003. p. 81–86.
[48] Azizullah A, Richter P, Jamil M, Häder D-P. Chronic toxicity of a laundry detergent to the freshwater flagellate *Euglena gracilis*. Ecotoxicology 2012;21(7):1957–64.

[49] Ahmed H. Biomonitoring of aquatic ecosystems [PhD thesis]. Erlangen-Nürnberg: Friedrich-Alexander-Universität; 2010.
[50] Matta CA, Kheirallah A-MM, Abdelmeguid NE, Abdel-Moneim AM. Effects of water pollution in lake Mariut on gonadal free amino acid compositions in *Oreochromis niloticus* fish. Pak J Biol Sci 2007;10:1257–63.
[51] Adham K, Khairalla A, Abu-Shabana M, Abdel-Maguid N, Moneim AA. Environmental stress in lake Maryut and physiological response of *Tilapia zilli* Gerv. J Environ Sci Health Part A 1997;32(9-10):2585–98.
[52] Adham KG, Hamed SS, Ibrahim HM, Saleh RA. Impaired functions in Nile Tilapia, *Oreochromis niloticus* (Linnaeus, 1757), from polluted waters. Acta Hydrochim Hydrobiol 2001;29:278–88.
[53] Schuler LJ, Hoang TC, Rand GM. Aquatic risk assessment of copper in freshwater and saltwater ecosystems of South Florida. Ecotoxicology 2008;17(7):642–59.
[54] OECD Guidelines for the Testing of Chemicals 201: Freshwater Alga and Cyanobacteria. Growth Inhibition Test, OECD Guidelines for the Testing of Chemicals.

Index

Note: Page numbers followed by "*f*" and "*t*" refer to figures and tables, respectively.

D

E

F

G

J

K

Q

R